T0156058

NON-COMMUTATIVE ANALYSIS

NON-COMMUTATIVE ANALYSIS

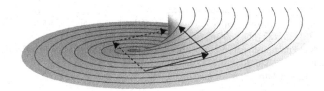

Palle Jorgensen
The University of Iowa, USA

Feng Tian
Hampton University, USA

World Scientific

NEW JERSEY · LONDON · SINGAPORE · BEIJING · SHANGHAI · HONG KONG · TAIPEI · CHENNAI · TOKYO

Published by

World Scientific Publishing Co. Pte. Ltd.

5 Toh Tuck Link, Singapore 596224

USA office: 27 Warren Street, Suite 401-402, Hackensack, NJ 07601

UK office: 57 Shelton Street, Covent Garden, London WC2H 9HE

Library of Congress Cataloging-in-Publication Data
Names: Jørgensen, Palle E. T., 1947– | Tian, Feng (Mathematics professor)
Title: Non-commutative analysis / by Palle Jorgensen (The University of Iowa, USA),
    Feng Tian (Hampton University, USA).
Description: New Jersey : World Scientific, 2017. |
    Includes bibliographical references and index.
Identifiers: LCCN 2016047613| ISBN 9789813202115 (hardcover : alk. paper) |
    ISBN 9789813202122 (pbk : alk. paper)
Subjects: LCSH: Functional analysis. | Hilbert space. | Mathematical physics. | Stochastic processes.
Classification: LCC QA320 .J638 2017 | DDC 515/.7--dc23
LC record available at https://lccn.loc.gov/2016047613

British Library Cataloguing-in-Publication Data
A catalogue record for this book is available from the British Library.

Printed in Singapore

Dedicated to the memory of William B. Arveson (see [MJD$^+$15])
(22 November 1934 – 15 November 2011).

"What goes around has come around, and today quantum information theory (QIT) has led us back into a finite-dimensional context. Completely positive maps on matrix algebras are the objects that are dual to quantum channels; in fact, the study of quantum channels reduces to the study of completely positive maps of matrix algebras that preserve the unit. This is an area that is still undergoing vigorous development in efforts to understand entanglement, entropy, and channel-capacity in QIT."

— William B. Arveson (written around 2009.)[1,2]

---

[1]For a summary of the terms used in the above quote (QIT, entanglement, and related), see Appendix B.

[2]The cover image illustrates a geometric aspect of non-commutativity. It is from Figure 10.3 (pg. 371) which deals with certain implications for Riemann surfaces of non-commutativity for unbounded operators.

Foreword

by Wayne Polyzou, Professor of Physics, University of Iowa

> If one had to synthesize the novelty of Physics of the XX century with a single magic word, one possibility would be "Noncommutativity".
> — Sergio Doplicher and Roberto Longo. From [CCG$^+$04]

Progress in science, engineering and mathematics comes fast and it often requires a significant effort to keep up with the advances in other fields that impact applications. Functional analysis (especially operators in Hilbert space, unitary representations of Lie groups, and spectral theory) is one discipline that impacts my physics research. Bringing my students up to speed with the subject facilitates their ability to efficiently perform research, however the typical curriculum in functional analysis courses is not directed to practitioners whose primary objective is applications. This is also reflected in the many excellent available texts on the subject, which primarily focus on the mathematics, and are directed at students aspiring to a career in mathematics.

I have been fortunate to have Palle Jorgensen as a colleague. He participates in a weekly joint mathematical physics seminar that is attended by faculty and students from both departments. It provides a forum to address questions related to the role of mathematics in physics research. Professor Jorgensen has a healthy appreciation of applications of functional analysis; in these seminars he has been at the center of discussions on a diverse range of applications involving wavelets, reflection positivity, path integrals, entanglement, financial mathematics, and algebraic field theory.

A number of the mathematically inclined students in my department have benefited from taking the functional analysis course taught by Professor Jorgensen. These students are motivated to enroll in his class because the course material includes a significant discussion of applications of functional analysis to subjects that interest them.

This book is based on the course that Professor Jorgensen teaches on functional analysis. It fills in a gap that is not addressed by the many excellent available texts on functional analysis, by using applications to motivate basic results in functional analysis. The way that it uses applications makes the material more accessible to students; particularly for students who will eventually find careers in related disciplines. The book also points to additional reference material for students who are motivated to learn more about a specific topic.

W. Polyzou

Preface

There are already many books in Functional Analysis, so why another?

The main reason is that we feel there is a need: in the teaching at the beginning graduate level; more flexibility, more options for students and instructors in pursuing new directions. And aiming for a book which will help students with primary interests elsewhere to acquire a facility with tools of a functional analytic flavor, say in spectral theory for operators in Hilbert space, in commutative and non-commutative harmonic analysis, in PDE, in numerical analysis, in stochastic processes, or in physics.

The book   Over the decades, Functional Analysis, and the theory of operators in Hilbert space, have been enriched and inspired on account of demands from neighboring fields, within mathematics, *harmonic analysis* (wavelets and signal processing), *numerical analysis* (finite element methods, discretization), *PDEs* (diffusion equations, scattering theory), *representation theory*; *iterated function systems* (fractals, Julia sets, chaotic dynamical systems), *ergodic theory, operator algebras*, and many more. And neighboring areas, *probability/statistics* (for example stochastic processes, Itō and Malliavin calculus), *physics* (representation of Lie groups, quantum field theory), and *spectral theory* for Schrödinger operators.

The book is based on a course sequence (two-semesters 313-314) taught, over the years at the University of Iowa, by the first-named author. The students in the course made up a mix: some advanced undergraduates, but most of them, first or second year graduate students (from math, as well as some from physics and stats.)

We have subsequently expanded the course notes taken by the second-named author: we completed several topics from the original notes, and we added a number of others, so that the book is now self-contained, and it covers a unified theme; and yet it stresses a multitude of applications. And it offers flexibility for users.

A glance at the table of contents makes it clear that our aim, and choice of topics, is different from that of more traditional Functional Analysis courses. This is deliberate. For example, in our choice of topics, we placed emphasis on the use of Hilbert space techniques which are then again used in our treatment of central themes of applied functional analysis.

We have also strived for a more accessible book, and yet aimed squarely at applications; — we have been serious about motivation: Rather than beginning with the four big theorems in Functional Analysis, our point of departure is an initial choice of topics from applications. And we have aimed for flexibility of use; acknowledging that students and instructors will invariably have a host of diverse goals in teaching beginning analysis courses. And students come to the course with a varied background. Indeed, over the years we found that students have come to the Functional Analysis sequence from other and different areas of math, and even from other departments; and so we have presented the material in a way that minimizes the need for prerequisites. We also found that well motivated students are easily able to fill in what is needed from measure theory, or from a facility with the four big theorems of Functional Analysis. And we found that the approach "learn-by-using" has a comparative advantage.

Our present approach is dictated by a need in the teaching of analysis in beginning grad-courses; especially in the US. In the US in the past decade, there has been quite a dramatic shift in the approach to graduate courses. And the existing literature no longer meets the new demands well. At least not in the area of our book. Our aim is to draw connections between a number of current themes from non-commutative analysis which previously have not been unified in the book literature, at the beginning level.

In the Introduction below, our aim is to initially go into depth with only a selection of central core themes, and then subsequently, we show how these core ideas have diverse, perhaps surprising, and seemingly disparate implications, and applications.

In the front matter we explain how a number of the applications are best sketched first in outline early on (Brownian motion etc, see Section 1.2.4), and then developed more fully later in the book (Chapter 7 in the case of Brownian motion) when the reader has been guided and prepared for them via early chapters on core themes. As we do this, we supply forward references in the book, and back-references too. We also help readers with a literature guide to the topics which are not treated in depth.

This approach, we have found, meets a definite demand in the teaching of functional analysis in standard graduate courses, at least in the US. Students in most grad programs in the US need to know these applied topics, but there isn't time for an in-depth treatment within the allotted time. Nonetheless our students who come from neighboring areas need to have some understanding of these related branches of applied analysis.

So it is by design that these spin-off topics inside the book, e.g., Brownian motion, large matrices, and Kadison-Singer, are treated only in outline. — We have presented these outlines with the purpose of stressing connections to the core themes early in the book.

The applied topics such as (i) Brownian motion (Chapter 7), (ii) large matrices (Section 7.5), (iii) Kadison-Singer (Chapter 9), (iv) signal processing (Section 6.6), (v) wavelets and fractals (Chapter 6), (vi) quantum physics (Section 2.5.2 and Chapter 8), in their own right could easily justify six separate books, but there simply isn't room in most US grad programs for them. Nonetheless, students must have some level of familiarity with topics (i)-(vi) and related. And so, yes, to serve this purpose, we do deviate from what is perhaps the expected norm in some circles, and in some math programs, but there is a good reason for it.

An Apology The central themes in our book are as follows: (i) *Operators in Hilbert Space* with emphasis on *unbounded* operators and non-commutativity, (ii) *Multivariable Spectral Theory*, (iii) *Non-commutative Analysis*, (iv) *Probability* with emphasis on Gaussian processes, and (v) *Unitary Representations*. But more importantly, it is our goal to stress the mutual interconnection between these five themes, or in fact central areas of modern analysis. And these are interrelationships which in the literature, at least up to now, have usually not been thought of as especially related. But here we stress, and elaborate in detail, on how a number of key theorems in anyone of these five areas crucially impact advances in the others. Nonetheless, for readers expecting a rehash of the standard list of topics from books in Functional Analysis of past generations, therefore perhaps an apology is in order.

The number of topics making up Functional Analysis is vast; and when applications are added, the size and diversity are daunting. A glance at the many books out there (see the partial list in Appendix A) will give readers an idea of the vast scope. It is by necessity that we have made choices; and that readers will in all likelihood have favorite topics not covered here. And there are probably surprises too; — things we cover here that are not typically included in standard Functional Analysis books. We apologize to readers who had expected a different table of contents. But we hope our choices are justified in our discussion in Part I below.

Our glaring omissions among the big classical areas of Functional Analysis include more technical aspects of the theory of Banach spaces. Even in our consideration of L^p spaces we have favored $p = 1, 2$, or ∞. Although we have included some fundamentals from Banach space theory, in this, we made a selection of only a few topics which are of direct relevance to the concrete applications that we do include. As for our bias in the choice of L^p spaces, we can excuse this in part by the familiar availability of interpolation theorems (the interpolation refers to values of p), starting with the *Riesz-Thorin theorem* and related; see e.g., [Kru07, HMU80]. Moreover, there is a host of books out there dealing with the exciting and deep areas of Banach space theory, both new and classical; and we refer readers to [JLS96, Joh88] for a sample.

Emphasis: In our applications, such as to physics, and statistics, we have concentrated on those analysis tools that are directly needed for the goal at hand. To fit the material into a single volume, we have been forced to omit a number of classical areas of functional analysis, and to concentrate on those that serve the applications we have selected. And in particular we have omitted a number of proofs, or reduced our discussion of some proofs to a few hints, or to exercises. We feel this is justified as there are many great books out there (see Appendix A) which contain complete proofs of the big theorems in functional analysis; for example, the books by W. Rudin, or by P. Lax.

Note on Presentation and Exercises In presenting our results, we have aimed for a reader-friendly account: We found it helpful to include *worked examples* in order to illustrate abstract ideas. Main theorems hold in various degrees of generality, but when appropriate, we have not chosen to present details in their highest level of generality. Rather, we typically give the result *in a setting where the idea is more transparent*, and easier to grasp. But we do include comments about the more general versions; sketching them in rough outline. The more general versions will typically be easier for readers to follow, and to appreciate, after ideas have been fleshed out in simpler contexts. We have made a second choice in order to make it easier for students to grasp both ideas and the technical details: We have included a lot of worked examples. At the end of each of these examples, we then outline how details (from the example in question) serve to illustrate one or more features in the general theorems elsewhere in the book. Finally, we have made generous use of both *tables* and *figures*. These are listed with page-references at the end of the book.

We shall be using some terminology from neighboring areas. And in order to help readers absorb that, we have included in Appendix B a summary, with cited references, of some key notions from *infinite dimensional analysis*[3], *probability theory, quantum theory, signal processing, stochastic processes, unitary representations*, and from *wavelet theory*.

Our selection of Exercises varies in level of difficulty, and they vary in purpose as well. Some are really easy, aimed mainly for the benefit of beginners; getting used to a definition, or a concept. Others are more traditional homework Exercises that can be assigned in a standard course. And yet others are quite demanding.

We believe the material covered here is accessible to beginning graduate students. In our choice of topics and presentation, we have aimed for a user friendly book which hopefully will also be of interest to both pure and applied mathematicians, as well as to students and scientists, in anyone of a number of neighboring areas. In our presentation, we have stressed applications, and also interconnections between disparate areas.

The book is subdivided into five parts, I through V. For each Part, the cover page contains title, and images (dingbats, ornament), intended to convey ideas from inside the book. And for each Part, the ornaments are from figures used in the book. Readers wondering about their captions can look them up in the List of Figures (Appendix G), and also in the discussion inside the text.

In addition to the appendices at the end of the book, we have included appendices for some individual chapters. They are chapters 2, 3, 5, and 8. The appendix matter at the end of Chapter 2 deals with Hahn-Banach extension, and related issues. In the appendix at the end of Chapter 3, we outline statement and proof sketch for Stone's theorem characterizing all strongly con-

---

[3]A summary of some highpoints from infinite-dimensional analysis is contained in Appendix B. See also Sections 1.2.3 (pg. 8), 2.1 (pg. 25), 2.4 (pg. 39), and Chapter 7 (pg. 249).

tinuous unitary one-parameter groups, and related. The appendix to Chapter 5 has full proofs, and it deals with an important representation of the canonical commutation relation (CCR)-algebra. An analysis of this representation yields an important part of Malliavin's infinite-dimensional calculus. Finally, the appendix to Chapter 8 covers the Stone-von Neumann Uniqueness Theorem.

Reader's guide to References In the Reference list, and in citations, we have included both books and research papers. For the various themes, we have aimed at citing both original sources, as well as timely papers; but we also cite brand new research. As for the latter, i.e., the cited papers in the References dealing with recent research (relating to the present topics), we mention a few, followed by citations:

- Spectral theory: [AH13, CJK$^+$12, HdSS12, Hel13, JP13b, JPT12, JPT14].

- The theory of frames, including Kadison-Singer: [Wea03, Cas13, Cas14, CFMT11, KOPT13, MSS15, SWZ11]. (The paper [MSS15] is the solution to K-S.)

- Stochastic processes and applications: [SS09, AJ12, AJS14, Jør14].

- Analysis of infinite networks: [RAKK05, BKS13, AJSV13, JP13a, JT15b].

- Representations of groups and algebras: [CM06, CM13, DHL09, JÓ00, JP12, JPS05, Boc08, DJ08, HJL06].

- Quantization and quantum information: [OH13, ARR13, CJK$^+$12, CM07, Fan10, Maa10, OR07].

Analysis of Continuous Systems vs Discrete (Networks and Graphs) A new theme here, going beyond traditional books in the subject, is applications of functional and harmonic analysis to "*large networks*," so to discrete problems. (See Figure 1.) More precisely, we study infinite network models. Such models can often be represented as follows: By a pair of sets, V (*vertices*), and E, (*edges*). In addition, one specifies a positive function c defined on the edge set E. (In electrical network models, c represents *conductance*, see Definition 11.1.)

There are then two associated operators Δ and P (Definition 11.2), each depending on the triple (V, E, c). Both operators represent actions (i.e., operations) on appropriate spaces of functions, more precisely functions defined on the infinite vertex set V. For the networks of interest to us, the vertex set V will be infinite, reflecting statistical and stochastic properties; and it will have additional geometric and ergodic theoretic properties. We are therefore faced with a variety of choices of infinite-dimensional function spaces. Many questions are of spectral theoretic flavor, and as a result, the useful choices of function spaces will be Hilbert spaces.

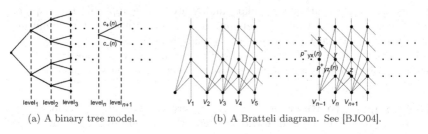

(a) A binary tree model.   (b) A Bratteli diagram. See [BJO04].

Figure 1: Examples of infinite weighted network; see Chapter 11.

But even restricting to Hilbert spaces, there are at least three natural (and useful) candidates (see Definition 11.6, and Lemma 11.4):

(i) the plain l^2 sequence space, so an l^2-space of functions on V,

(ii) a suitably weighted l^2-space, and finally,

(iii) an energy Hilbert space $\mathscr{H}_E$. (The latter is an abstraction of more classical notions of Dirichlet spaces, see Definition 11.4.) Which one of the three to use depends on the particular operator considered, and also on the questions asked.

In *infinite network models*, both the Laplacian Δ (Definition 11.2), and the Markov operator P (see Definition 11.7 and Lemma 11.8), will have $\infty \times \infty$ matrix representations (see eq (11.3)). Each of these infinite by infinite matrices will have the following property: it will have non-zero entries localized only in finite bands containing the infinite matrix-diagonal (i.e., they are *infinite banded matrices*.) See Section 2.5 and Figure 2.16. Thus, the standard algebraic matrix operations will be well defined.

Functional analytic and spectral theoretic tools now enter as follows: In passing to appropriate Hilbert spaces, we arrive at various classes of Hilbert space-operators. In the present setting, the operators in question will be Hermitian, some unbounded, and some bounded. The *Laplacian* Δ will typically be an unbounded operator, albeit semibounded. When Δ is realized in the energy Hilbert space $\mathscr{H}_E$, we must introduce boundary value considerations in order to get selfadjoint extensions. By contrast, for *the Markov operator* P, there is a weighted l^2-space such that P is a bounded, selfadjoint operator[4]. Moreover, its spectrum is then contained in the finite interval $[-1, 1]$. In all of the operator realizations by selfadjoint operators, Δ or P, the corresponding spectra may be continuous, or may have a mix of spectral types, *continuous* (singular or Lebesgue), and *discrete*.

For the operator theory, and graph Laplacians, for infinite network models, we refer to Chapter 11.

---

[4]The reader will find the precise definition of selfadjoint operators in Section 2.4 (pg. 39); see also Sections 2.7 (pg. 83), 3.1 (pg. 90), and Chapters 4 (pg. 119), 10 (pg. 349).

Acknowledgments

The first named author thanks his students in the Functional Analysis sequence 313-314. Also he thanks postdocs, other colleagues, collaborators over the years, and his mentors; — the combined list includes the following: D. Alpay, W. Arveson, L. Baggett, S. Bezuglyi, T. Branson, O. Bratteli, P. Casazza, I. Cho, D. Dutkay, B. Fuglede, S. Gelfand, T. Hida, R. Kadison, W. Klink, K. Kornelson, D. Larson, G. Mackey, P. Muhly, E. Nathanson, K.-H. Neeb, E. Nelson, R. Niedzialomski, G. Ólafsson, B. Ørsted, J. Packer, E. Pearse, R.S. Phillips, W. Polyzou, R.T. Powers, C. Radin, D. Robinson, W. Rudin, S. Sakai, R. Schrader, I.E. Segal, K. Shuman, Myung-Sin Song, M. H. Stone, R. Werner.

We are especially grateful to Professor Wayne Polyzou (of The University of Iowa Physics Department) for reading our manuscript, giving us corrections, and making suggestions. And for graciously agreeing to write a Foreword.

The authors are extremely grateful to their colleagues for suggestions while the book was being written and tested in the class room. They are especially grateful to their editor Rochelle Kronzek, and to five anonymous reviewers. The constructive criticism and very helpful suggestions we received are much appreciated. We have made every effort in a last round of revisions, following all these suggestions. Nonetheless, any flaws that may remain are solely the responsibility of the authors.

Contents

Foreword   vii

Preface   ix

Acknowledgments   xv

Abbreviations, Notation, Some Core Theorems   xxiii

I   Introduction and Motivation   1

1   Subjects and User's Guide   3
1.1   Motivation 3
1.2   Key Themes in the Book: A Bird's-eye Preview 5
    1.2.1   Operators in Hilbert Space 5
    1.2.2   Multivariable Spectral Theory 7
    1.2.3   Noncommutative Analysis 8
    1.2.4   Probability 9
    1.2.5   Other Neighboring Areas 12
    1.2.6   Unitary Representations 13
1.3   Note on Cited Books and Papers 13
1.4   Reader Guide 15
1.5   A Word About the Exercises 16
1.6   List of Applications 17
1.7   Groups and Physics 18

II   Topics from Functional Analysis and Operators in Hilbert Space: A Selection   21

2   Elementary Facts   23
2.1   A Sample of Topics 25
2.2   Duality 27
    2.2.1   Duality and Measures 33
    2.2.2   Other Spaces in Duality 36

2.3   Transfinite Induction (Zorn and All That)   37
2.4   Basics of Hilbert Space Theory   39
       2.4.1   Positive Definite Functions   42
       2.4.2   Orthonormal Bases   46
       2.4.3   Bounded Operators in Hilbert Space   50
       2.4.4   The Gram-Schmidt Process and Applications   55
2.5   Dirac's Notation   63
       2.5.1   Three Norm-Completions   67
       2.5.2   Connection to Quantum Mechanics   70
       2.5.3   Probabilistic Interpretation of Parseval in Hilbert Space .   75
2.6   The Lattice Structure of Projections   77
2.7   Multiplication Operators   83
2.A   Hahn-Banach Theorems   85
2.B   Banach-Limit   86

3 **Unbounded Operators in Hilbert Space**   **89**
3.1   Domain, Graph, and Adjoints   90
3.2   Characteristic Matrix   97
       3.2.1   Commutants 101
3.3   Unbounded Operators Between Different
       Hilbert Spaces 102
       3.3.1   An application to the Malliavin derivative 109
3.4   Normal Operators 111
3.5   Polar Decomposition 113
3.A   Stone's Theorem 114

4 **Spectral Theory**   **119**
4.1   An Overview 120
4.2   Multiplication Operator Version 125
       4.2.1   Transformation of Measures 128
       4.2.2   Direct Integral Representation 131
       4.2.3   Proof of Theorem 4.1 continued: 132
4.3   Projection-Valued Measure (PVM) 136
4.4   Convert M_φ to a PVM (projection-valued measure) 140
4.5   The Spectral Theorem for Compact Operators 144
       4.5.1   Preliminaries 144
       4.5.2   Integral operators 150

III **Applications**   **153**

5 **GNS and Representations**   **155**
5.1   Definitions and Facts: An Overview 158
5.2   The GNS Construction 163
5.3   States, Dual and Pre-dual 168
5.4   New Hilbert Spaces From "old" 174

        5.4.1   GNS 174
        5.4.2   Direct sum $\bigoplus_\alpha \mathscr{H}_\alpha$ 175
        5.4.3   Hilbert-Schmidt operators (continuing the discussion in 1) 175
        5.4.4   Tensor-Product $\mathscr{H}_1 \otimes \mathscr{H}_2$ 176
        5.4.5   Contractive Inclusion 176
        5.4.6   Inflation (Dilation) 176
        5.4.7   Quantum Information 177
        5.4.8   Reflection Positivity (or renormalization) $(\mathscr{H}_+/\mathscr{N})^\sim$ . . . 179
    5.5 A Second Duality Principle: A Metric on the Set of Probability
         Measures 184
    5.6 Abelian C^*-algebras 187
    5.7 States and Representations 189
        5.7.1   Normal States 196
        5.7.2   A Dictionary of operator theory and quantum mechanics 197
    5.8 Krein-Milman, Choquet, Decomposition of States 198
        5.8.1   Noncommutative Radon-Nikodym Derivative 202
        5.8.2   Examples of Disintegration 202
    5.9 Examples of C^*-algebras 203
    5.10 Examples of Representations 211
    5.11 Beginning of Multiplicity Theory 214
    5.A The Fock-state, and Representation of CCR, Realized as Malli-
         avin Calculus 220

6 Completely Positive Maps   223
    6.1 Motivation 224
    6.2 CP v.s. GNS 226
    6.3 Stinespring's Theorem 228
    6.4 Applications 233
    6.5 Factorization 239
    6.6 Endomorphisms, Representations of $\mathscr{O}_N$, and Numerical Range . 241

7 Brownian Motion   249
    7.1 Introduction, Applications, and Context for Path-space Analysis 250
    7.2 The Path Space 259
    7.3 Decomposition of Brownian Motion 266
    7.4 The Spectral Theorem, and Karhunen-Loève Decomposition . . . 270
    7.5 Large Matrices Revisited 271

8 Lie Groups, and their Unitary Representations   273
    8.1 Motivation 276
    8.2 Unitary One-Parameter Groups 280
    8.3 Group - Algebra - Representations 281
        8.3.1   Example $- ax + b$ group 286
    8.4 Induced Representations 288
        8.4.1   Integral operators and induced representations 297
    8.5 Example - Heisenberg group 302

8.5.1 $ax + b$ group 305
8.6 Co-adjoint Orbits 306
8.6.1 Review of some Lie theory 306
8.7 Gårding Space 310
8.8 Decomposition of Representations 315
8.9 Summary of Induced Representations, the Example of d/dx . . . 318
8.9.1 Equivalence and imprimitivity for induced representations 319
8.10 Connections to Nelson's Spectral Theory 322
8.11 Multiplicity Revisited 326
8.A The Stone-von Neumann Uniqueness Theorem 329

9 The Kadison-Singer Problem   333
9.1 Statement of the Problem 334
9.2 The Dixmier Trace 340
9.3 Frames in Hilbert Space 341

IV Extension of Operators   347

10 Selfadjoint Extensions   349
10.1 Extensions of Hermitian Operators 350
10.2 Cayley Transform 362
10.3 Boundary Triple 364
10.4 The Friedrichs Extension 371
10.5 Rigged Hilbert Space 376

11 Unbounded Graph-Laplacians   385
11.1 Basic Setting 387
11.1.1 Infinite Path Space 389
11.2 The Energy Hilbert Spaces $\mathscr{H}_E$ 389
11.3 The Graph-Laplacian 392
11.4 The Friedrichs Extension of Δ, the Graph Laplacian 394
11.5 A 1D Example 395

12 Reproducing Kernel Hilbert Space   403
12.1 Fundamentals 403
12.2 Application to Optimization 407
12.2.1 Application: Least square-optimization 409
12.3 A Digression: Stochastic Processes 413
12.4 Two Extension Problems 415
12.5 The Reproducing Kernel Hilbert Space $\mathscr{H}_F$ 416
12.6 Type I v.s. Type II Extensions 425
12.7 The Case of $e^{-|x|}$, $|x| < 1$ 426
12.7.1 The Selfadjoint Extensions $A_\theta \supset -iD_F$ 427
12.7.2 The Spectra of the s.a. Extensions $A_\theta \supset -iD_F$ 431

V   Appendix   439

A   An Overview of Functional Analysis Books (Cast of Characters)   441

B   Terminology from Neighboring Areas   451
    Classical Wiener measure/space
    Hilbert's sixth problem
    Infinite-dimensional analysis
    Monte Carlo (MC) simulation
    Multiresolution analysis (MRA)
    Quantum field theory (QFT)
    Quantum Information (QI)
    Quantum mechanics(QM)
    Quantum probability (QP)
    Signal processing (SP)
    Stochastic processes (SP)
    Uncertainty quantification (UQ)
    Unitary representations (UR)
    Wavelets

C   Often Cited   459

D   Prizes and Fame   475

Quotes: Index of Credits   477

E   List of Exercises   479

F   Definitions of Frequently Occurring Terms   485

G   List of Figures   489

List of Tables   493

Bibliography   495

Index   525

Abbreviations, Notation, Some Core Theorems

$\aleph_0$	aleph-sub 0, cardinality of $\mathbb{N}$ (pg. 48, 160, 338)
$\Im\{z\}$	the imaginary part of $z \in \mathbb{C}$ (pg. 173, 357)
$\Re\{z\}$	the real part of $z \in \mathbb{C}$
$\mathscr{O}_N$	the Cuntz-algebra, i.e., generators $\{s_i\}_{i=1}^N$ and relations, $s_i^* s_j = \delta_{i,j}\mathbb{1}$, and $\sum_1^N s_i s_i^* = \mathbb{1}$. (pg. 158, 209, 214, 241, 243)
BMO	bounded mean oscillation (pg. 29, 173, 174)
CAR	canonical anti-commutation relations (pg. 213)
CCR	canonical commutation relations (pg. 26, 50, 72, 212, 296)
conv	convex hull (pg. 143, 198)
CP	completely positive map (pg. 201, 224, 226, 228, 236, 238)
ext	set of extreme-points (pg. 159, 198, 199)
i.i.d	independent identically distributed (system of random variables) (pg. 263, 268, 271, 413)
IFS	iterated function system (pg. 105, 185)
ind	induced representation (pg. 299, 301, 303, 319)
irrep	irreducible representation (pg. 13)
KS	Kadison-Singer (pg. 18, 333, 344)
Nbh_x	neighborhood of a point x (pg. 198)
NR_T	numerical range of a given operator T (pg. 65, 143, 206, 207, 247)
ODE	ordinary differential equation (pg. 94, 447)

ONB   orthonormal basis (in Hilbert space) (pg. 5, 32, 46, 50, 69, 151, 170, 197, 343, 436)

PDE   partial differential equation; examples: the heat equation, diffusion equation, the wave equation, the Laplace equation, the Schrödinger equation. (pg. 5, 15, 27, 39, 90, 447)

PDO   partial differential operator (pg. 459)

Proj   projection (pg. 77, 80, 82, 115, 136, 143, 280)

PVM   projection valued measure (the condition $P(A \cap B) = P(A)P(B)$ is part of the definition) (pg. 25, 70, 77, 115, 124, 136, 140, 281, 331, 343, 380, 426)

$\mathrm{Rep}(\mathfrak{A}, \mathscr{H})$ representations of an algebra $\mathfrak{A}$ acting on a Hilbert space $\mathscr{H}$ (pg. 164, 189, 216, 236, 236, 278)

$\mathrm{Rep}(G, \mathscr{H})$ representations of a group G acting on a Hilbert space $\mathscr{H}$ (pg. 92, 273, 278, 312, 319, 323)

RKHS   reproducing kernel Hilbert space (pg. 5, 403, 405, 409, 417, 426)

SDE   stochastic differential equation (pg. 255, 265, 266)

span   linear span (pg. 41, 48, 55, 64, 67, 204, 231, 393, 394, 394)

supp   support of a function, a measure, or a distribution (pg. 122, 133, 418, 421)

Operators in Hilbert space

$A = A^*$ .... selfadjoint (pg. 97, 127, 136, 379)

$A \subset A^*$ .... symmetric (also called Hermitian) (pg. 74, 351)

$A \subset -A^*$ .... skew-symmetric (pg. 422)

$AA^* = A^*A$ .... normal (pg. 23, 51, 100)

$UU^* = U^*U = I$ .... unitary (pg. 67)

I, or $I_{\mathscr{H}}$ .... identity operator in a given Hilbert space $\mathscr{H}$, i.e., $I(v) = v, \forall v \in \mathscr{H}$. (pg. 68, 158, 390)

$P = P^* = P^2$ .... projection (pg. 80, 107)

$\mathscr{G}(A)$ .... graph of operator (pg. 90, 353)

$\langle \cdot, \cdot \rangle$ .... inner product of a given Hilbert space $\mathscr{H}$, i.e., $\langle v, w \rangle$ for $v, w \in \mathscr{H}$; linear in the second variable. (pg. 40, 149)

$|v\rangle\langle w|$ .... Dirac ket-bra vectors for rank-one operator (pg. 63, 67)

$\wedge$ .... lattice operation "minimum" applied to projections (pg. 79)

$\vee$ .... lattice operation "maximum" applied to projections (pg. 79)

$\cap$ .... set-theoretic intersection (pg. 79)

$\cup$ .... set-theoretic union (pg. 79)

ess sup .... essential supremum (pg. 30)

l^p .... sequence space, l^p-summable (29, 168)

$L^p(\mu)$ .... L^p-integrable functions on a μ measure space (pg. 29)

span .... all linear combination of a specified subset (pg. 41, 48, 55, 64, 67, 204, 231, 393, 394, 394)

$\overline{\text{span}}$ .... closure of span (pg. 43, 47, 55, 79, 133, 164, 214, 228)

E^* .... the dual of a given normed space E (E^* is a Banach space) (pg. 28, 32, 173)

E^{**} .... double-dual (pg. 168, 315)

$(..)'$ .... commutant of a set of operators (pg. 191, 214, 330)

$(..)''$ .... double-commutant (pg. 192, 214)

$\widehat{}$ .... Fourier transform, or Gelfand transform (pg. 122, 188, 211, 212, 323)

χ_E .... indicator function of a set E (pg. 61, 142, 122, 283)

δ .... Dirac delta "function" (pg. 160, 183, 188, 418)

$*$ .... convolution (pg. 122, 251, 278)

sp, or spec .... spectrum (pg. 143, 145, 430, 435)

res .... resolvent set (pg. 97)

$\mathcal{B}(X)$ .... Borel sets, i.e., the sigma-algebra generated by the open sets in a topological space X (pg. 34, 41, 45, 280)

σ-algeb$\{\cdots\}$ .... the sigma-algebra generated by a given set of random variables (pg. 34, 128, 259, 281)

$\mathscr{B}(\mathscr{H})$ .... all bounded linear operators $\mathscr{H} \longrightarrow \mathscr{H}$ (pg. 51, 136, 161, 189)

$\mathscr{F}R\left(\mathscr{H}\right)$............ all finite-rank operators in $\mathscr{H}$ (pg. 64, 67, 69)

tr (trace)............ the trace functional (pg. 67, 72, 171, 307)

$\mathscr{T}_1\left(\mathscr{H}\right)$ .... all trace class operators in $\mathscr{B}\left(\mathscr{H}\right)$ (pg. 169, 196)

Proj $\left(\mathscr{H}\right)$............ the lattice of all orthogonal projections P in a fixed Hilbert space, i.e., $P = P^2 = P^*$. (pg. 77, 80, 82, 115, 136, 143, 280)

dom (A) or $\mathscr{D}\left(A\right)$.... the domain of some linear operator A (pg. 84, 91, 97, 359)

ran (A) or $R\left(A\right)$ .... the range of A (pg. 90, 96, 379)

Ker (A) or $\mathscr{K}\left(A\right)$ ... the kernel of A (pg. 90, 145)

M_φ .... the operator of multiplication by some function acting in some $L^2\left(\mu\right)$, or a multiplier in some RKHS. (pg. 83, 127, 219, 406)

$T_\varphi = P_+ M_\varphi P_+$ ...... Toeplitz-operator with symbol φ (pg. 208, 209)

$\mu \circ T^{-1}$ .... transformation of measure, $\left(\mu \circ T^{-1}\right)\left(\triangle\right) = \mu\left(T^{-1}\left(\triangle\right)\right)$, $\triangle \in$ sigma-algebra, $T^{-1}\left(\triangle\right) = \{x : Tx \in \triangle\}$. (pg. 34, 128)

$\int^{\oplus}$ .... direct integral decomposition (pg. 131, 203, 316)

$\oplus$ .... orthogonal sum (pg. 132, 175, 204, 411)

$\otimes$ .... tensor product (pg. 176, 231, 239)

$\perp$ .... orthogonal (pg. 47, 96, 164)

P, Q .... notation used for pairs of projections, but also for the momentum and position operators from quantum mechanics (pg. 26, 73, 77, 296)

$\mathbb{P}$.................... some probability measure (pg. 34, 251)

$\mathbb{E} = \mathbb{E}_{\mathbb{P}}$............ expectation $\mathbb{E}_{\mathbb{P}}\left(X\right) = \int_\Omega X d\mathbb{P}$ (pg. 45, 260, 263, 270)

G .... Lie group (pg. 92, 273, 288, 319)

$\mathfrak{g}$.................... Lie algebra (pg. 277, 291, 296, 312)

$\mathfrak{g} \xrightarrow{\exp} G$............ exponential mapping (pg. 291, 291, 310)

$\mathcal{U}$ .... representation of some Lie group G (pg. 91, 162, 279, 303)

$d\mathcal{U}$ .... representation of the Lie algebra $\mathfrak{g}$ corresponding to G; the derived representation. (pg. 310, 315)

C^*-algebra .... an algebra $\mathfrak{A}$ with involution $\mathfrak{A} \ni a \to a^* \in \mathfrak{A}$, $a^{**} = a$, $(ab)^* = b^*a^*$, $a, b \in \mathfrak{A}$; and norm $\|\cdot\|$, such that $(\mathfrak{A}, \|\cdot\|)$ is complete; and $\|ab\| \le \|a\| \, \|b\|$, $a, b \in \mathfrak{A}$, holds, as well as $\|a^*a\| = \|a\|^2$, $a \in \mathfrak{A}$. (pg. 158, 164, 187, 191)

W^*-algebra .... (also called *von Neumann algebra*, or a *ring of operators*) A W^*-algebra is a C^*-algebra $\mathfrak{A}$ which has the following additional property: There is a Banach space $(X_*, \|\cdot\|_*)$ such that $\mathfrak{A}$, with its C^*-norm (i.e., $\|a^*a\| = \|a\|^2$, $a \in \mathfrak{A}$), is the dual, $\mathfrak{A} = (X_*)^*$. When such a Banach space X_* exists, it is called a *pre-dual*. (This characterization of W^*-algebra is due to Sakai [Sak98].) (pg. 101, 114, 211)

Operations on subspaces of Hilbert spaces $\mathcal{H}$

$\mathcal{T} \subset \mathcal{H}$ .... some subspace in $\mathcal{H}$ (pg. 52, 103, 164, 208)

$\mathcal{T}^\perp$ .... ortho-complement (pg. 103, 236)

$$\mathcal{T}^\perp = \{h \in \mathcal{H} \ : \ \langle h, s \rangle = 0, \ \forall s \in \mathcal{T}\} = \mathcal{H} \ominus \mathcal{T}$$

$\mathcal{T}^{\perp\perp} = \overline{span}\,\mathcal{T}$ (pg. 101, 351, 353)

Normal or not! It depends:

- An *operator* T (bounded or not) is normal iff (Def.) (pg. 23, 51, 100)

$$T^*T = TT^*$$

- A *state* s on a $*$-algebra $\mathfrak{A}$ is normal if it allows a representation $(\mathcal{H}, \rho)$ where $\mathcal{H}$ is a Hilbert space, and ρ is a positive trace-class operator in $\mathcal{H}$ such that $trace\,(\rho) = 1$, and

$$s\,(A) = trace\,(\rho A)\,, \ \forall A \in \mathfrak{A}. \quad \text{(pg. 172)}$$

- A *random variable* X on a probability space $(\Omega, \mathcal{F}, \mathbb{P})$ is said to be normal iff (Def.) its distribution is normal, i.e., $\exists m \in \mathbb{R}$, $\sigma > 0$ such that

$$\mathbb{P}\,(\{\omega \in \Omega \mid a \le X\,(\omega) \le b\}) = \int_a^b \frac{1}{\sigma\sqrt{2\pi}} e^{-\frac{1}{2}\left(\frac{x-m}{\sigma}\right)^2} dx. \text{ (pg. 34, 252)}$$

Graph models (pg. xiv, 387, 396)

V .... set of vertices, $x \in V$

E .... set of edges, $(x, y) \in E$, $x \neq y$, $x \sim y$

c_{xy} .... conductance function defined for $(x, y) \in E$

$p_{xy} = \dfrac{c_{xy}}{\sum_{z \sim x} c_{xz}}$ ..... transition probability

Δ .... graph Laplacian, $(\Delta f)(x) = \sum_{y \sim x} c_{xy}(f(x) - f(y))$

P .... graph Markov operator, $(Pf)(x) = \sum_{y \sim x} p_{xy} f(y)$

SOME CORE THEOREMS

- Existence of ONBs (Theorem 2.6, pg. 47)

- The Hahn-Banach theorems (Theorems 2.10-2.14, pg. 86)

- The Polar Decomposition/Factorization (Theorem 3.9, pg. 106)

- Stone's theorem (Theorem 3.17, pg. 115),

- The Spectral Representation Theorem (Theorem 4.1, pg. 127)

- The Spectral theorems (Theorem 4.3, pg. 140; Theorem 4.6, pg. 145)

- The GNS construction (Theorem 5.1, pg. 158; Theorem 5.3, pg. 165)

- The Gelfand-Naimark Theorem (Theorem 5.4, pg. 167)

- The Banach-Alaoglu theorem (Theorem 5.6, pg. 169)

- The Krein-Milman theorem (Theorem 5.12, pg. 198)

- Choquet's theorem (Theorem 5.13, pg. 199)

- Stinespring's theorem (Theorem 6.1, pg. 228)

- The Imprimitivity theorem (Theorem 8.3, pg. 302)

- von Neumann's double commutant theorem (Theorem 5.16, pg. 216)

- The Pontryagin duality theorem (Theorem 8.6, pg. 315)

- The Stone-von Neumann uniqueness theorem (Theorem 8.8, pg. 330)

- The Friedrichs extension theorem (Theorem 10.11, pg. 372)

Part I

Introduction and Motivation

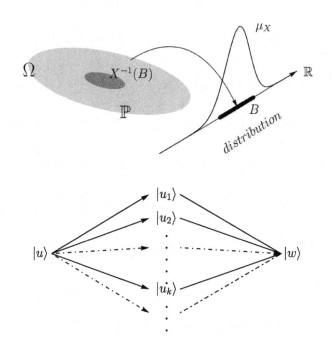

"I saw that noncommutation was really the dominant characteristic of Heisenberg's new theory. ... So I was led to concentrate
on the idea of noncommutation, and on how to modify ordinary dynamics (which people had been using until then) in order to include
it."
— P. A. M. Dirac

Below we outline the main areas covered inside the book. We offer some tips
for the reader, and conclude with a list of applications.

Chapter 1

Subjects and User's Guide

1.1 Motivation

More traditional books on Functional Analysis, and operators in Hilbert space, tend to postpone applications till after all the big theorems from the theory have been covered. The purpose of the present book is to give students a tour of a selection of applications. We aim to do this by first offering a crash course in theory topics tailor made for the purpose (part II). In order to stress the interplay between theory and applications (part III) we have emphasized the traffic in both directions. We believe that the multitude of new applications makes Functional Analysis both a powerful, versatile, and timeless tool in mathematics.

The notion of Hilbert space[1] has become an indispensable part of modern analysis. As a tool, it is exceptionally successful in numerous areas of both pure and applied mathematics. A number of central themes in the present book have found particularly elegant and powerful formulations in the setting of Hilbert space. For the reader to get a bird's-eye view of some of them, we need only refer to the Table of Contents above, and to the more detailed overview-discussion below. The setting of Hilbert space offers geometric and metric tools, especially well-suited for problems involving an infinite number of dimensions; — so clearly of relevance to applications since function spaces in analysis, in statistics, and in physics, are typically infinite-dimensional.

The early pioneers in the area date from the beginning of 20th century, David Hilbert, Erhard Schmidt, Frigyes Riesz, Max Born, John von Neumann, Marshall Stone, Mark Krein, and A. N. Kolmogorov[2]. And early motivation came from quantum theory (W. Heisenberg and E. Schrödinger), and from thermodynamics; but many more came later, right up to the current date; e.g., mathematical statistics, representation theory, optimization, partial differential equations,

---

[1]The axioms for Hilbert space are in Definitions 2.11 and 2.12 inside the chapter, see Section 2.4.

[2]The reader will find biographical sketches of Hilbert, Riesz, von Neumann and Stone in Appendix C.

Fourier analysis, the study of holomorphic functions, harmonic analysis (both commutative and non-commutative), signal processing, transformation groups, and ergodic theory. It was von Neumann who coined the term Hilbert space for the axiom system which underlies these diverse applications. By "linear operators" we refer to linear transformations between Hilbert spaces. Part of the success of the approach is that it allows one, with relative ease, to carry over basic techniques from finite dimensions to infinite-dimensional Hilbert space. A case in point is the Spectral Theorem for selfadjoint operators (now extended to infinite dimensions), and to multiplicity theory for unitary representations of compact groups, now extending to "most" non-compact groups.

A glance at existing books in Functional Analysis and related areas (see list of reviews in the Appendix A) shows that related books so far already display a rich variety, even if they may have the same title "Functional Analysis" or "Functional Analysis with a subtitle, or a narrowing of the focus."

Still *the aims, and the contents* of these other books go in a different directions than ours. One thing they have in common is an emphasis on the Four Big Theorems in Functional Analysis, The *Hahn-Banach Theorem*, The *Open Mapping Theorem*, The *Uniform Boundedness Principle*, The *Closed Range Theorem*, and duality principles.

By contrast, we do as follows; rather we select a list of topics and applications that acquire a degree of elegance when presented in a functional analytic setting. There are several reasons for this different approach, the main ones are as follows:

(i) The subject is ever changing on account of demands from neighboring fields;

(ii) Students come to our graduate functional analysis course with a diversity of backgrounds, and a "one-size fits all" approach is not practical;

(iii) Well-motivated students can pick up on their own what is needed from the Four Big Theorems;

(iv) Concentrating on the Four Big Theorems leaves too little time for a variety of neighboring areas, both within mathematics, and in neighboring sciences;

(v) Also the more traditional approach, beginning with the Four Big Theorems is already in many existing books (see the Appendix A).

In Appendix B, we give a bird's-eye view of some current areas of applied mathematics with direct connections to our present functional analytic topics.

A glance at the *Table of Contents* will reflect our aim: beginning with tools from Hilbert space in Chapters 2 & 3, but motivated by quantum physics; a preview of the Spectral Theorem in Chapter 4; some basic tools from the theory of operator algebras in Chapter 5, with an emphasis on the Gelfand[3]-Naimark-Segal (GNS) construction; and stressing the many links between properties of

---

[3]Appendix C includes a short biographical sketch of I. M. Gelfand. See also [GJ60, GS60, GG59].

states, and the parallel properties of the representations, and the operator algebras they generate.

In Chapter 6, and motivated by physics and harmonic analysis, we discuss dilation theory. This is the general phenomenon (pioneered by Stinespring and Arveson) of studying problems in an initial Hilbert space by passing to an enlarged Hilbert space.

Chapter 7 (Brownian motion), while different from the others, still fits perfectly, and inviting application of the tools already discussed in the book. The applications we cover in Chapter 8 are primarily to representations of groups and algebras. Chapter 9 is an application of theorems from Chapters 4-5 to the problem named after Kadison and Singer, now abbreviated the KS-problem. It is 50 years old, motivated by Dirac's formulation of quantum physics (observables, states, and measurements); and it was solved only a year ago (as of present).

The last three chapters are, 10: s*elfadjoint extensions*, 11: *graph-Laplacian*, and 12: *reproducing kernel Hilbert spaces* (RKHSs), and they are somewhat more technical, but they are critical for a host of the questions inside the book. Some readers may be familiar with this material already. If not, a quick reading of Chapters 3, 10, and 11 may be useful. Similarly, in Appendix A, to help students orient themselves, we give a bird's-eye view of, in the order of 20 books out there, all of which cover an approach to Functional Analysis, and its many applications.

1.2 Key Themes in the Book: A Bird's-eye Preview

While each of our central themes has found book presentations, the particular interconnections and applications that are the focus of the present book, have not previously been explored in a textbook form. To the extent they are in the literature at all, it will be in the form of research papers.

When making selection of topics, we aimed for timeliness. Even if some themes are disparate, we have made connections, and it is our hope that tools from one area will enlighten and advance the other. To help readers make connections back to fundamental themes, we have opened each chapter with a selection of Quotations. We hope they serve to place key ideas in a historic context.

1.2.1 Operators in Hilbert Space

The notion of a *Hilbert space*, is one of the most successful axiomatic constructions in modern analysis. It was John von Neumann who coined the term Hilbert space. While, historically, the concept originated with problems from partial differential equations (PDE), potential theory, quantum physics, and ergodic theory, it has since found a host of other applications involving the part of functional analysis dealing with infinite-dimensional function spaces; such areas as: the study of unitary representations of groups (for example symmetry groups from physics), complex function theory (Hardy spaces of holomorphic functions), applications to probability, to stochastic processes, to signal processing, to thermodynamics (heat transfer, ergodic theory.) One reason the von Neumann-Hilbert axioms have proved especially successful is their versatility in dealing with optimization problems arising in the study of infinite-dimensional function spaces. This is so despite the fact that the Hilbert space axioms themselves are formulated in the abstract, independently of the particular context where they are applied. Specifically, the axioms entail a given vector space $\mathscr{H}$, equipped with an inner product (part of the axiom system), which in turn induces a norm. The last axiom is that $\mathscr{H}$ must be complete with respect to this norm.

With this one then proceeds to devise a host of coordinate systems, orthonormal bases (ONB). A noteworthy family of ONBs of more recent vintage are wavelet bases.

Among more recent areas of application, we mention *machine learning*; a sub-area of artificial intelligence. In its current version, machine learning models are formulated in the setting of *reproducing kernel Hilbert space* (RKHS); see Chapter 12 below, and [SZ07]. Indeed, in modern machine learning theory, the RKHSs play a critical role in the construction of *optimization algorithms*. A second use of RKHSs is in the solution of maximum-likelihood problems from probability theory.

As for the study of *linear transformations* (operators)[4], our present dual emphasis will be unbounded operators, and non-commutativity. Specifically, we study systems of densely defined linear operators. A key motivation for this emphasis is again quantum mechanics: Indeed quantum mechanical observables (momentum, position, energy, etc) correspond to non-commuting selfadjoint unbounded operators in Hilbert space.

The first two Hilbert spaces most students encounter are l^2 and $L^2(\mathbb{R})$:

l^2: sequences $(x_n)_{n=1}^{\infty}$ such that

$$\|x\|_{l^2}^2 = \sum_{n=1}^{\infty} |x_n|^2 < \infty.$$

$L^2(\mathbb{R})$: measurable functions on $\mathbb{R}$ such that

$$\|f\|_{L^2}^2 = \int_{\mathbb{R}} |f(x)|^2 \, dx < \infty.$$

These two examples serve to illustrate the axiom system for Hilbert space which we shall study in Section 2.4 below.

---

[4]The reader will find the precise definition of linear operators in Definition 2.5 (pg. 33), see also Sections 2.4 (pg. 39), 3.1 (pg. 90).

The norms for l^2 and for $L^2(\mathbb{R})$ come from associated inner products $\langle \cdot, \cdot \rangle$, for example, $\langle x, y \rangle_{l^2} = \sum_{n=1}^{\infty} \bar{x}_n y_n$, $\forall x, y \in l^2$. The system of vectors $\delta_1, \delta_2, \cdots$ in l^2, given by

$$\delta_k(n) = \delta_{k,n} = \begin{cases} 1 & \text{if } n = k \\ 0 & \text{if } n \neq k \end{cases}$$

satisfies $\langle \delta_k, \delta_l \rangle_{l^2} = \delta_{k,l}$ (the orthonormality property); and, for all $x = (x_n)_1^{\infty} \in l^2$, we have

$$\lim_{N \to \infty} \left\| x - \sum_{k=1}^{N} x_k \delta_k \right\|_{l^2}^2 = \lim_{N \to \infty} \sum_{k=N+1}^{\infty} |x_k|^2 = 0;$$

the second property (called "*total*") that orthonormal bases (ONBs) have.

One naturally wonders "what are analogous ONBs for the second mentioned Hilbert space $L^2(\mathbb{R})$?" We shall turn to this question in Chapter 2 below. There are two classes: (i) *special functions*, of which the best known are the Hermite functions (Section 2.4.3); and (ii) *wavelet-bases* (Sections 2.4 and 6.6).

The chapters of special relevance to these topics are: Chapters 2, 3, 9, 10, and 12. Sections of special relevance include 2.6, 2.7, 3.4, 4.4, 10.1, 10.2, and 10.4.

1.2.2 Multivariable Spectral Theory

In this setting we are dealing with more than one operator at a time. A host of applications naturally present themselves, again with the same applications as mentioned in Section 1.2.1 above. From the late nineteen thirties we have the study of selfadjoint algebras in the works of *Murray-von Neumann* and of Gelfand-Naimark [MvN43, GN94]. Later, this was followed up with systematic studies of non-selfadjoint algebras; e.g., the work of *Kadison-Singer*. Other studies of multivariate operator theory emphasize analogues of analyticity, both in the commutative as well as in the non-commutative settings. It has had remarkable successes, including applications in other areas of mathematics such as complex and algebraic geometry, and non-commutative geometry. In the multivariable case, some researchers consider either n-tuples of operators, or representations of algebras with generators and relations; while others have adopted the language of Hilbert modules; for example, modules over algebras of holomorphic functions, polynomials or entire functions depending on the given number n of (commuting) complex variables.

Historically, the first important multivariable problem in operator theory was perhaps the relations of *Heisenberg* for a pair of linear operators P and Q with common dense domain $\mathscr{D}$ in a fixed Hilbert space $\mathscr{H}$. The relations require that

$$PQf - QPf = -i\,f \tag{1.1}$$

holds for all $f \in \mathscr{D}$.

By now (1.1) is well understood, but there are many subtle points; all of which make important connections to what we call multivariable spectral theory; for example (1.1) does not have solutions for bounded operators in $\mathscr{H}$.

We shall also consider a variety of multivariable systems of bounded operators; — in this case, it is usually in the setting of non-normal operators (so in particular non-selfadjoint), for example for:

(i) finite sets of bounded commuting operators in a fixed Hilbert space $\mathcal{H}$;

(ii) algebras $\mathfrak{A}$ of operators on $\mathcal{H}$ such that the pair $(\mathfrak{A}, \mathcal{H})$ forms a module; and

(iii) finite systems of isometries in some Hilbert space $\mathcal{H}$; and

(iv) sets of isometries subject to the added condition that the ranges forms a system of orthogonal subspaces of $\mathcal{H}$ with sum equal to $\mathcal{H}$.

The relations on sets of isometries described in (iv) are called the Cuntz relations (Section 5.1) and [Cun77, BJ02, BJO04]. The Cuntz relations correspond to representations of a C^*-algebra, called the Cuntz-algebra. It has many applications, some of which will be studied, see e.g., Section 5.9. A good reference for (i) is [Arv98].

The chapters of special relevance to these topics are: Chapters 6, 8, and 10. Sections of special relevance include 6.2, 6.3, 8.4, 8.5, 8.8, and 10.4.

1.2.3   Noncommutative Analysis

The above multivariable settings are part of a wider theme: *noncommutative analysis*, a field which extends (classical commutative) Fourier analysis. This began with the study of locally compact groups from physics, and their unitary representations. The case of compact groups encompasses the Peter-Weyl theorem from the 1920s, but needs from number theory (mathematics), and from relativistic quantum physics, have dictated extensions to non-compact (and noncommutative) groups, typically Lie groups.

A more recent area of noncommutative analysis is the study of *free probability*, which we shall only touch on tangentially inside the book. It is an exciting and new, rapidly growing, research direction; with new advances in theory as well as in applications. Fortunately, there are already nice and accessible book treatments, see e.g., [Spe11] and the sources cited there. In free probability, we study systems of *non-commutative random variables*. As stochastic processes, they are not Gaussian. Rather the notion of free independence dictates the semicircle-law (not the Gaussian distribution). The rigorous study of free probability entails such operator algebraic notions as free products. We emphasize that the important new notion of free independence is dictated by non-commutativity, and that it generalizes the more familiar notion of independence which was used previously in probability. The subject was initiated by Dan Voiculescu in the 1980ties. Its applications up to now include: random matrix theory, representations of symmetric groups, large deviations of stochastic processes, and quantum information theory.

We use the term "noncommutative analysis" more broadly than the related one, "noncommutative geometry." The latter owes much to the pioneering work of Alain Connes[5], see e.g., [Con07]. In broad outline, it covers the role von Neumann algebra theory plays in noncommutative considerations in geometry and in quantum physics (the Standard Model); in noncommutative metric theory and spaces, noncommutativity in topology, spectral triples, differential geometry, cyclic cohomology, cyclic homology, K-theory, and M-theory. In more detail, noncommutative geometry (NCG) is concerned with a geometric approach to the construction of spaces that are locally presented via noncommutative algebras of operators. This is the framework of, what in physics, is referred to as "local quantum field theory."[6] The prime applications of NCG are to particle physics where A. Connes has developed a noncommutative standard model. Some of the other successes of NCG include extensions of known topological invariants to formal duals of noncommutative operator algebras. Via a Connes-Chern character map, this has led to the discovery of a new homology theory of noncommutative operator algebras; and to a new non-commutative theory of characteristic classes; and to generalizations of the classical index theorems.

The Standard Model of particle physics deals with the electromagnetic, weak, and strong nuclear interactions, and with classifications of all the known subatomic particles, the "theory of almost everything." It received a boost in the mid-1970s after an experimental confirmation of the existence of quarks; and later of the tau neutrino, and the Higgs boson (2013).

Helpful references here are [Arv76, BD91, BR79, DJ08, Gli61, JM84, Jor94, Jor11, Pow75, Tak79].

The chapters of special relevance to these topics are: Chapters 5, 6, and 8. Sections of special relevance include 5.7, 5.8, 6.3, 6.6, 8.4, 8.8, and 8.11.

1.2.4 Probability

Probability theory originated with the need for quantification of uncertainty, as it arises for example in quantum physics, and in financial markets. In the 1930s, Kolmogorov formulated the precise mathematical axioms of *probability space* Ω, sample points, events as specified subsets, in a prescribed sigma-algebra $\mathcal{F}$ of subsets of Ω, and a probability measure $\mathbb{P}$, defined on $\mathcal{F}$[7]. (See Figure 1.1 for an illustration.)

Our present focus will be a subclass of *stochastic processes*, the Gaussian processes, especially those which are derived from stochastic integration defined relative to Brownian motion.

Brownian motion is the simplest of the continuous-time stochastic (meaning probabilistic) processes. It is a limit of simpler stochastic processes going by the name random walks; a fact which reflects the universality of the normal distribution, the Gaussians.

---

[5]Appendix C includes a short biographical sketch of A. Connes. See also [Con07].

[6]A summary of some highpoints from quantum information and quantum field theory is contained in Appendix B. See also Sections 1.6 (pg. 17), 2.5 (pg. 63), 4.4 (pg. 140), 5.4 (pg. 174), 8.2 (pg. 280); and Chapters 5 (pg. 155), 6 (pg. 223), 9 (pg. 333).

[7]The reader will find the precise definition of linear operators in Definition 2.7 (pg. 34); see also Sections 7.1 (pg. 250), 7.2 (pg. 259), 12.3 (pg. 413).

It is not an accident that we have focused on problems from quantum physics and from probability. With some over simplification, it is fair to say that Hilbert's 6th problem asked for a mathematical rigorous treatment of these two areas. In 1900, when Hilbert formulated his 23 problems, these two areas did not yet have mathematically rigorous foundations.

The topic from probability that shall concern us the most is that of *Brownian motion*. In a nutshell, a Brownian motion may be thought this way: There is a probability measure $\mathbb{P}$ on a sigma-algebra of subsets of the continuous functions ω on $\mathbb{R}$ such that

$$B_t(\omega) = \omega(t), \quad t \in \mathbb{R}, \ \omega \in C(\mathbb{R}) \tag{1.2}$$

satisfy a number of axioms of which we mention here only that for each $t \in \mathbb{R}$, B_t has a Gaussian distribution relative to $\mathbb{P}$ such that

$$\int_{C(\mathbb{R})} |B_t(\omega) - B_s(\omega)|^2 \, d\mathbb{P}(\omega) = |t - s|, \quad s, t \in \mathbb{R}, \tag{1.3}$$

and

$$\int_{C(\mathbb{R})} B_t(\omega) \, d\mathbb{P}(\omega) = 0, \quad \forall t \in \mathbb{R}.$$

Notation. We note that these formulas are often written alternatively as

$$\mathbb{E}(B_t) = 0, \text{ and}$$

$$\mathbb{E}(|B_t - B_s|^2) = |t - s|,$$

in the probability literature.

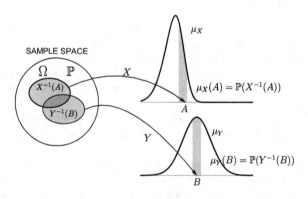

Figure 1.1: Two random variables X, Y, and their distributions.

A Technical Point. Our summary presentation of Brownian motion via formulas (1.2)-(1.3) above makes things look very easy. It is an over-simplification; and, for the moment, a number of technical points are omitted. Our present overview is only part of the story. Full details will be included in Chapter 7 below. Brownian motion is in fact characterized by a set of axioms. The main thing we swept under the rug in the present quick summary, and our use of formulas (1.2)-(1.3), is the measure $\mathbb{P}$ on path space. The existence of this measure, and its properties, is in fact subtle. The measure $\mathbb{P}$ is called the *Wiener measure*, path-space measure, or white noise measure (see the cited references below, and our discussion of the literature at the end of Chapter 7.) In our present very brief summary, we have not gone into a discussion of $\mathbb{P}$, and of the choice of probability space Ω, the *sample space of paths*. Using for Ω all the continuous functions is a good choice, but it requires a justification, which we have chosen to postpone. In technical terms, if for Ω we take instead all functions, it can be shown that the subset of the continuous functions will have outer measure 1 with respect to $\mathbb{P}$. In fact, it is known that even smaller subsets (e.g., spaces of Lipschitz functions with Lipschitz constant $< 1/2$) also have outer measure 1 with respect to $\mathbb{P}$. By contrast, $\mathbb{P}$ assigns outer measure 0 to the subset of Ω consisting of all differentiable functions. The latter fact is a theorem due to N. Wiener and A. Kolmogorov, but it is supported intuitively by simulations. Note from a glance at the figures in Chapter 7, the "wiggly" appearance of our Monte Carlo simulation[8] of Brownian motion, and of processes derived from Brownian motion; see especially Figures 7.3 and 7.9.

Chapter 7 serves to highlight the use of spectral theoretic tools (the Spectral Theorem[9], and spectral analysis) to stochastic processes, and especially to the theory of Gaussian processes. Via Itō's calculus, of course, the Brownian motion is a "building block" going into the solution of general stochastic differential equations (SDE). And more generally, a stochastic process is an indexed family of random variables; for example, indexed by time, or some other deterministic index variable. (In the latter case, the random process is called a "random field.")

The first tool in the analysis of stochastic process is a so called *covariance kernel*. In the case of a Gaussian process, or Gaussian field, the Karhunen-Loève expansion (Section 7.3) serves as a powerful tool for separating variables into deterministic vs stochastic. The proof of the Karhunen-Loève theorem (Section 7.4), in its many variants, in turn is an early application of Hilbert space geometry, the Spectral Theorem, and its analysis refinements (see Chapters 3 and 4.)

Helpful references here are [AJ12, AJS14, GJ60, Itô04, Itô06, Nel67, Par05, Sla03].

The chapters of special relevance to these topics are: Chapters 7, and 12. Sections of special relevance include 7.3, 12.3-12.6, and Figures 7.3, 7.4, 7.6, 12.2, and 12.3.

---

[8]A summary of some highpoints from Wiener measure and Monte Carlo simulation is contained in Appendix B. See also Sections 2.4, 7.2, and Chapters 7 (pg. 249), 8 (pg. 273).

[9]The reader will find the precise formulation in Theorem 4.1 (pg. 127); see also Sections 5.3 (pg. 168), 5.7 (pg. 189), 7.4 (pg. 270), and Chapters 4 (pg. 119), 8 (pg. 273).

1.2.5   Other Neighboring Areas

Other areas making contact to our main themes inside the book include harmonic analysis, non-commutative harmonic analysis; the latter including the theory of unitary representations. In more detail:

Harmonic analysis. In rough outline, the study of harmonic analysis deals with representation of functions, or of signals, as the superposition of basic functions, "building-blocks," usually of a "simpler" form; for example represented as "waves," or as wavelets. This includes the study of Fourier series, Fourier transforms, and orthogonal polynomials.

Over two centuries, harmonic analysis has flourished, and it now includes applications to for example signal processing, quantum mechanics, tidal analysis, and to neuroscience. The term "harmonics" is from the Greek word, "harmonikos," meaning "skilled in music." In physics, eigenvalue problems entail the study of waves, and frequency distributions. In the simplest case, we have frequencies arising as integer multiples of one another, arithmetic progressions. This is the case of frequencies of the harmonics of music notes. The classical Fourier transform on $\mathbb{R}^n$, and on other Abelian and non-Abelian groups, is concerned with Fourier transforms in the setting of tempered distributions (notation of L. Schwartz). Among other areas we mention here: (i) uncertainty principles, (ii) criteria for convergence of Fourier series, and (iii) analysis and synthesis.

Non-commutative harmonic analysis. In noncommutative harmonic analysis, results from Fourier analysis are extended to a variety of generally non-Abelian topological groups. In the non-commutative setting one deals with extensions of the classical theory from the commutative case to that of non-commutative but locally compact groups G. The reason for "locally compact" is that locally compact groups have Haar measure. Caution: There are separate Haar measures with reference to left-"translation-invariance" and right-invariance. Each is unique up to scalar multiples. If the two Haar measures (left and right) agree, we say that G is unimodular. This includes the case of compact groups, and semisimple Lie groups. In the compact case, a cornerstone is the Peter-Weyl theorem. It accounts for all the irreducible representations and their multiplicities. The relevant examples in the non-compact case include many Lie groups (e.g., semisimple groups, and certain classes of solvable Lie groups), where a detailed theory is available. Lie groups and algebraic groups over p-adic fields are used in physics, in number theory, and in the study of automorphic representations.

Sample theorems. If the von Neumann group algebra of G is of type I, then $L^2(G)$, as a unitary representation of G, is a direct integral of irreducible representations. We say that "the unitary dual" of G is the set of isomorphism classes of irreducible representations of G. For general non-type I groups, including countable discrete groups G, the von Neumann group algebra cannot be written in terms of irreducible representations.

1.2.6 Unitary Representations

An early motivation (see also Section 1.2.3 above) is work of J. von Neumann and I. E. Segal. They showed that, if G is a locally compact unimodular group such that the associated von Neumann group algebra is of type I, then the regular representation of G, acting on the Hilbert space $L^2(G)$ relative to Haar measure, as a unitary representation[10], is a direct integral of irreducible unitary representations ("irreps" for short.) This leads to a notion of a unitary dual for G, defined as the set of equivalence classes (under unitary equivalence) of such representations, the "irreps."

But for general locally compact groups, including countable discrete groups, the von Neumann group algebra typically is not of type I and the regular translation-representation of G cannot be expressed in terms of building blocks of "irreps."

The applications of our present results to unitary representations will include those discussed above, so in particular, applications to quantum physics, and to probability, especially to Gaussian stochastic processes.

The chapters of special relevance to these topics are: Chapters 3, 5, and especially 8. Sections of special relevance include 3.1, 3.2, 5.2, 5.4, 5.6, 8.2, 8.7, and 8.8.

1.3 Note on Cited Books and Papers

For readers looking for references on the foundations, our suggestions are as follows: Operators in Hilbert space: [Arv72, Arv76]. Quantum mechanics[11]: [GG02, OR07, Wei03], and [CP82, PK88, Pol02]. Non-commutative functional analysis and algebras of operators: [BR79, BR81a, BJKR84]. Unitary representations of groups: [Mac52, Mac85, Mac92].

In our use of citations we adopted the following dual approach. Inside the chapters, as the material is developed, we include citations to key sources that we rely on; — but this is done sparingly so as not to interrupt the narrative too much.

To remedy sparse citations inside chapters, and, in order to help the reader orient herself in the literature, each of the 11 chapters concludes with a little bibliographical section, summarizing papers and books of special relevance to the topic inside the text. Thus there is a separate list of citations which concludes each chapter. Readers who do not find a particular citation inside the chapter itself will likely be able to locate it from the *end-of-chapter-list*.

Some cited books, organized by topic. (While a number of books cover multiple topics, we have still subdivided them, mainly by the respective author's emphasis.)

---

[10]The reader will find the precise definition of unitary representations in Section 5.1 (pg. 158); see also Sections 5.7 (pg. 189), 5.10 (pg. 211), and Chapter 8 (pg. 273).

[11]A summary of some highpoints from quantum mechanics is contained in Appendix B. See also Sections 2.5.2 (pg. 70), 3.1 (pg. 90), 4.1 (pg. 120), and Chapter 9 (pg. 333).

Algebras of operators. [Arv76, BR79, Dix81, Dou80, KR97a, KR97b, Sak98, Tak79].

Convex analysis/optimization. [AH13, And79a, GBD94, JS07, KR57].

Functional analysis; linear spaces. [Ban93, Con90, Die75, GS77, JL01, KF75, Lax02, Mac09, Rud91, Shi96, SS11, Trè06b, Yos95].

Generalized functions and Schwartz distributions. [DM72, GJ60, GS77, Sch64b].

Geometric measure theory. [BHS05, BKS13, GG59, Hut81, Kan58, Rüs07, SZ09].

Harmonic analysis, complex analysis, Hilbert space, spectral theory, perturbation theory. [Akh65, dB68, DM72, Fri80, Hel13, Kat95, LP89, Mac92, Nel69, Rud90].

Infinite-dimensional analysis. [Akh65, CW14, Gro64, Gro70, Jor11, Nel69].

Lie groups and symmetric spaces. [GG59, Hal15, HS68].

Mathematical physics. [BR81b, CM06, Dir47, Emc84, GJ87, Hal13, Wig76].

Non-commutative geometry/analysis. [Con07, Tay86].

Normed linear spaces. [AG93, BN00, Con90, DS88a, JL01, Joh88, RSN90, Yos95].

Numerical analysis. [AH09, GvN51, Pre89].

Operators in Hilbert space, including unbounded. [AG93, DS88c, DS88a, DS88b, Hal82, Sch12, Sto90].

Partial differential operators. [Trè06a].

Real and complex analysis. [Phe01, Rud87].

Representation theory. [JM84, Jor88, Mac85].

Reproducing kernels and applications. [AD86, Alp01, CZ07].

Stochastic processes, Brownian motion, and related. [CW14, Hid80, HØUZ10, Itô04, Loè63, Nel67, Par05].

Wavelets and applications. [BJ02, Jor06].

1.4 Reader Guide

Below we explain chapter by chapter how the six areas in Table 1.1 are covered. Also see Figure 1.2.

Ch 1: Areas A, E, F.

Ch 2: Areas A, C, F.

Ch 3: Areas B, C, E, F.

Ch 4: Areas A, B, E, F.

Ch 5: Areas A, E, F.

Ch 6: Area E.

Ch 7: Areas D, F.

Ch 8: Areas E, F.

Ch 9: Areas B, C, D.

Ch 10: Areas A, F.

Ch 11: Areas A, B, C, D, E.

	Subject	Example
A	analysis	$f(x) - f(0) = \int_0^x f'(y)\, dy$
B	dynamical systems	functions on fractals, Cantor sets, etc.
C	PDE	Sobolev spaces
D	numerical analysis	discretization
E	measures / probability theory	probability space $(\Omega, \mathcal{F}, \mathbb{P})$
F	quantum theory	Hilbert spaces of quantum states

Table 1.1: Examples of Linear Spaces: Banach spaces, Banach algebras, Hilbert spaces $\mathscr{H}$, linear operators act in $\mathscr{H}$.

In more detail, the six areas in Table 1.1 may be fleshed out as follows:

Examples of subjects within *area A* include measure theory, transforms, construction of bases, Fourier series, Fourier transforms, wavelets, and wavelet transforms, as well as a host of operations in analysis.

Subjects from *area B* include solutions to ordinary differential equations (ODEs), and the output of iteration schemes, such as the Newton iteration algorithm. Also included are ergodic theory; and the study of fractals, including harmonic analysis on fractals.

Area C encompasses the study of the three types of linear PDEs, elliptic, parabolic and hyperbolic. Sample questions: weak solutions, *a priori* estimates, diffusion equations, and scattering theory.

Area D encompasses discretization, algorithms (Newton etc.), estimation of error terms, approximation (for example wavelet approximation, and the associated algorithms.)

Area E encompasses probability theory, stochastic processes (including Brownian motion), and path-space integration.

Finally, *area F* includes the theory of unbounded operators in Hilbert space, the three versions of the Spectral Theorem, as well as representations of Lie groups, and of algebras generated by the commutation relations coming from physics.

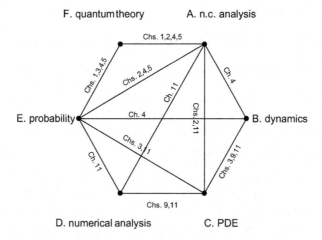

Figure 1.2: The lines illustrate connections between the six themes, A through F; different chapters making separate connections between these related and intertwined themes. See also Table 1.1, and the discussion before Table 1.1.

1.5   A Word About the Exercises

Following the dictum, "learn mathematics by doing it" in the spirit of [Hal82], we have presented some themes in the book in the form of exercises, and readers are encouraged to absorb the gist of each exercise even if completing it in detail is postponed.

The topics in the last three chapters are more specialized, and exercises seem less natural there. The purpose of the exercises is to improve and facilitate the use of the book in courses; — to help students and instructors. There is a total of 174 exercises. To help with classroom use, we have listed them in the back, numbering chapter-by-chapter. Each exercise is given a name identification. Here is a sample: Exercise 2.18 (Lax-Milgram), 2.24 (The Haar wavelet), 3.1 (the resolvent identity), 4.14 (Powers-Størmer), 5.17 (time-reflection), 5.28 (extreme measures), 8.16 (multiplicity), 8.21 (a formula from Peter-Weyl), and so on; ... 12.3 (Szegö-kernel).

The degree of difficulty of the exercises varies from one to the next, some are relatively easy; for example, serving to give the reader a chance to practice definitions or new concepts; and some are quite challenging. But of the exercises all interact naturally with the topics developed in the various chapters. This is why we have integrated them into the development of the topics, chapter for chapter. And this is also why some chapters have many exercises, such as Chapter 2 with a total of 40 exercises; — Chapter 4 has 16 exercises; and Chapter 5 has 46 exercises in all. In all of the chapters, we have mixed and interspaced the placement of exercises with the central themes: some exercises supplement examples, and some theorems, within each chapter.

There are two lists after the Appendices, a *List of Exercises*, and a *List of Figures*. The second should help readers with cross-references; and the first with use of the exercises in course-assignments.

The Appendices themselves serve to aid readers navigate the book-literature. Appendix A includes *telegraphic reviews*, and Appendix C is a collection of *biographical sketches* of the pioneers in the subject.

1.6 List of Applications

Part of the discussion below will make use of terminology from neighboring areas, such as uncertainty quantification, and more generally from physics, engineering, and statistics. For readers who might be encountering this for the first time, we have a terminology section in the back. It is a section in the Appendix B, called "Terminology from neighboring areas." The Appendix also includes other lists: a list of telegraphic reviews of related books; biographical sketches; diagrams illustrating interconnections between disparate areas; a list of figures, and a list of all the exercises inside the text. Each exercise is given a descriptive name.

Each of the chapters is illustrated with examples and applications. A recurrent theme is the important notion of *positive definite functions*, and their realization in Hilbert space. Applications to *Wiener measure* and path-space are included in Chapters 2, 7, and 12.

More applications, starting in Chapter 2 are: (i) a variance formula for the Haar-wavelet basis in $L^2(0,1)$; (ii) a formula for perturbations of diagonal operators; and (iii) the $\infty \times \infty$ matrix representation of *Heisenberg's commutation relations* in the case of the canonical pair, momentum and position operators from quantum mechanics.

The case of *Heisenberg's commutation relations* motivates the need for a systematic study of *unbounded operators* in Hilbert space[12]. This is started in Chapter 3, and resumed then systematically in Chapters 4 (the *Spectral Theorem*), and 10 (the theory of *extensions of symmetric operators with dense domain; — von Neumann indices, and All That*). Both our study of selfadjoint operators and normal operators, and their spectral theory, throughout the book is motivated by the axioms from quantum theory: *observables, states, measurements,* and *the uncertainty principle.* Our systematic treatment of *projection valued measures,* and *quantum states,* in Chapter 4 is a case in point.

This goes for our theme in Chapter 5 as well, the *Gelfand-Naimark-Segal* (GNS) representation. *Quantum states* must be realized in Hilbert space, but what is the relevant Hilbert space, when a *quantum observable* is prepared in a state? To answer this we must realize the observables as selfadjoint operators affiliated with a suitable C^*-algebra, say $\mathfrak{A}$, or von Neumann algebra. States on these algebras then become positive linear functionals. The GNS construction is a device for constructing representation in Hilbert space for every state, defined as a *positive linear functional* on $\mathfrak{A}$. In this construction, the pure states are matched up with irreducible representations.

In Section 5.4, our application is to the subject of *"reflection-positivity"* from quantum physics. This notion came up first in a renormalization question in physics: "How to realize observables in relativistic quantum field theory (RQFT)?"

The material in Chapter 6 has applications to signal processing; — to the construction of sub-band filters, and filter banks. These applications are discussed in Chapter 6; — included as one of the applications of a certain family of representations of the Cuntz relations. Other applications of these representations include wavelet filters.

1.7   Groups and Physics

"The importance of group theory was emphasized very recently when some physicists using group theory predicted the existence of a particle that had never been observed before, and described the properties it should have. Later experiments proved that this particle really exists and has those properties."
— Irving Adler

Recall that in relativistic quantum field theory (RQFT), the symmetry group is the Poincaré group, but its physical representations are often illusive. Starting

---

[12]The reader will find the precise definition of unbounded operators in Section 3.1 (pg. 90), see also Chapters 4 (pg. 119), 10 (pg. 349), 11 (pg. 385).

with papers by Osterwalder-Schrader[13] in the 1970s (see e.g., [OS75, GJ87, JÓ00]), it was suggested to instead begin with representations of the Euclidian group, and then to get to the Poincaré group through the back door, via an analytic continuation (a c-dual group construction), and a *renormalization.* This leads to a systematic study of renormalizations for the Hilbert space of quantum states. The "c-dual" here refers to an *analytic continuation* which links the two groups. This in turn is accomplished with the use of a certain reflection, and a corresponding change in the inner product. In a simplified summary, the construction is as follows:

Starting with the inner product in the initial Hilbert, say $\mathcal{H}$, and a unitary representation admitting a reflection $\mathcal{J}$, we then pass to a certain invariant subspace of $\mathcal{H}$, and use $\mathcal{J}$ in the definition of *the new inner product.* The result is a physical energy operator (dual of time) with the correct positive spectrum for the relativistic problem, hence "*reflection-positivity.*" The invariant subspace refers to invariance only in a positive time direction. All of this is presented in Section 5.4, and illustrated with an example.

Chapter 6 deals with the same theme; only there the states are *operator valued.* From the theory of Stinespring and Arveson we know that there is then a different positivity notion, *complete positivity* (CP).

Among the Hilbert spaces we encounter are the L^2-spaces of random variables on a probability space $(\Omega, \mathcal{F}, \mathbb{P})^{14}$. The case of Brownian motion is studied in Chapter 7, and again in Chapter 12.

In Chapter 8, we introduce families of *unitary representations* of groups, and ∗-representations of algebras; each one motivated by an application from physics, or from signal processing. We are stressing examples as opposed to general theory.

Chapter 9 is devoted to the *Kadison-Singer* problem (KS). It is a problem from operator algebras[15], but originating with Dirac's presentation of quantum mechanics. By choosing a suitable orthonormal basis (ONB) we may take for Hilbert space the sequence $l^2(\mathbb{N})$ space, square-summable sequences. Dirac was interested in the algebra $\mathcal{B}(l^2(\mathbb{N}))$ of all bounded operators in $l^2(\mathbb{N})$. With the $\infty \times \infty$ matrix representation for elements in $\mathcal{B}(l^2(\mathbb{N}))$, we can talk about the *maximal Abelian subalgebra* $\mathcal{D}$ of all diagonal operators in $\mathcal{B}(l^2(\mathbb{N}))$. Note $\mathcal{D}$ is just a copy of $l^\infty(\mathbb{N})$. The Dirac-KS question is this: "Does every pure state on $\mathcal{D}$ have a *unique* pure-state extension to $\mathcal{B}(l^2(\mathbb{N}))$?"

The problem was solved in the affirmative; just a year ago (see [MSS15]). We sketch the framework for the KS problem. However, the details of the solution are far beyond the scope of our book.

The application in Chapter 11 is to potential theory and random walk models

---

[13]Appendix C includes a short biographical sketch of R. Schrader.

[14]A summary of some highpoints from quantum probability is contained in Appendix B. See also Sections 2.2 (pg. 27), 2.5.2 (pg. 70), 4.1 (pg. 120), 12.3 (pg. 413); and Chapters 2 (pg. 23), 5 (pg. 155), 7 (pg. 249), 12 (pg. 403).

[15]The reader will find the precise definition of C^*-algebras in Definition 5.1 (pg. 158), and von Neumann algebras in Theorem 5.16 (pg. 216); see also Section 3.2 (pg. 97), and Chapters 5 (pg. 155), 9 (pg. 333).

(Markov processes) for *infinite networks*; mathematically infinite graphs $G = (V, E)$, V the specified set of vertices, and E the edges. Our emphasis is electrical networks, and the functions include energy, conductance, resistance, voltage, and current. The main operator here is the so called *graph Laplacian.*

The new applications in Chapter 12 include *scattering theory, learning theory* (as it is used in machine learning and in pattern recognition.)

Part II

Topics from Functional Analysis and Operators in Hilbert Space: A Selection

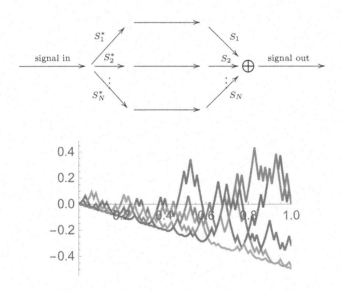

Chapter 2

Elementary Facts

"... the [quantum mechanical] observables are operators on a Hilbert space. The algebra of operators on a Hilbert space is noncommutative. It is this noncommutativity of operators on a Hilbert space that provides a precise formulation of [Heisenberg's] uncertainty principle: There are operator solutions to equations like $pq - qp = 1$. This equation has no commutative counterpart. In fact, it has no solution in operators p, q acting on a finite dimensional space. So if you're interested in the dynamics of quantum theory, you must work with operators rather than functions and, more precisely, operators on infinite dimensional spaces."

— William B. Arveson (1934-2011. The quote is from 2009.)

"I received an early copy of Heisenberg's first work a little before publication and I studied it for a while and within a week or two I saw that the noncommutation was really the dominant characteristic of Heisenberg's new theory. It was really more important than Heisenberg's idea of building up the theory in terms of quantities closely connected with experimental results. So I was led to concentrate on the idea of noncommutation and to see how the ordinary dynamics which people had been using until then should be modified to include it."

— P. A. M. Dirac

Problems worthy
 of attack
prove their worth
 by hitting back.
— Piet Hein

Outline Below we outline some *basic concepts, ideas,* and *examples* which will be studied inside the book. While they represent only a sample, and we favor the setting of Hilbert space, the details below still tie in nicely with diverse tools and techniques not directly related to Hilbert space.

The discussion below concentrates on topics connected to Hilbert space, but we will also have occasion to use some other basic facts from functional analysis; e.g., duality and Hahn-Banach (see Appendix 2.A.) For related results, see Appendix 2.B below.

From linear algebra we know precisely what square matrices M can be diagonalized; the *normal matrices,* i.e., $M^*M = MM^*$. More precisely, a matrix is normal if and only if it is conjugate to a diagonal matrix. More general square matrices do not diagonalize, but they admit a Jordan form.

In the infinite dimensional case, — while infinite matrices are useful, the axiomatic setting of *Hilbert space* and *linear operators* has proved more successful than an infinite matrix formulation; and, following von Neumann and Stone, we will make precise the notion of normal operators. Because of applications, the case of *unbounded operators* is essential. In separate chapters, we will prepare the ground for this.

The *Spectral Theorem* (see [Sto90, Yos95, Nel69, RS75, DS88b]) states that a linear operator T (in Hilbert space) is *normal,* i.e., $T^*T = TT^*$, if and only if it is unitarily equivalent to a *multiplication operator* in some L^2-space, i.e., multiplication by a measurable function, and the function may be unbounded. The implied Hilbert space L^2 is with respect to some measure space, which of course will depend on the normal operator T, given at the outset. Hence the classification of *normal operators* is equivalent to the classification of *measure spaces;* — a technically quite subtle problem.

There is a second (and equivalent) version of the Spectral Theorem, one based on *projection valued measures* (PVMs), and we will present this as well. It is a powerful tool in the theory of unitary representations of locally compact groups (Chapter 8), and in a host of areas of pure and applied mathematics.

It is natural to ask whether there is an analogue of the finite dimensional Jordan form; i.e., extending from finite to the infinite dimensional case. The short answer is "no," although there are partial results. They are beyond the scope of this book.

In our first two chapters below we prepare the ground for the statement and proof of the *Spectral Theorem,* but we hasten to add that there are several versions. In the bounded case, for *compact selfadjoint operators* (Section 4.5), the analogue to the spectral theorem from linear algebra is closest, i.e., eigenvalues and eigenvectors. Going beyond this will entail an understanding of *continuous spectrum* (Section 4.4), and of *multiplicity theory* in the measure theoretic category (Section 5.11).

With a few exceptions, we will assume that all of the Hilbert spaces considered are separable; i.e., that their orthonormal bases (ONBs) are countable. The exceptions to this will include the L^2-space of the Bohr completion of the reals $\mathbb{R}$. (See, e.g., Exercise 5.13.)

2.1 A Sample of Topics

"Too many people write papers that are very abstract and at the end they may give some examples. It should be the other way around. You should start with understanding the interesting examples and build up to explain what the general phenomena are."
— Sir Michael Atiyah

Classical functional analysis is roughly divided into branches, each with a long list of subbranches:

- Study of function spaces (Banach space, Hilbert space).

- Semigroups of operators.
  The theory of semigroups of operators is a central theme in functional analysis. It plays a role both in pure and applied mathematics; with the applications including quantum physics (especially scattering theory), partial differential equations (PDE), Markov processes, diffusion, and more generally the study of stochastic processes[1], Itō-calculus, and Malliavin calculus. It centers around a theorem named after Hille, Yosida, and Phillips[2]. They proved a fundamental theorem giving a precise correspondence between: (i) a given strongly continuous semigroup of operators, and (ii) its associated infinitesimal generator. It is a necessary and sufficient condition for a given linear operator with dense domain to be the infinitesimal generator for a semigroup; thus characterizing the particular unbounded operators which are generators of strongly continuous one-parameter semigroups of bounded operators acting on Banach space. Of special note is a recent application to Malliavin calculus. The latter is an extension of the classical calculus of variations (for deterministic functions) to random variables and stochastic processes. In this path-space setting, it makes precise the computation of "derivatives" of random processes; and it allows for integration by parts in such an infinite dimensional setting. Applications include mathematical finance and stochastic filtering.

- Applications in physics, statistics, and engineering.

Within pure mathematics, it is manifested in the list below:

- Representation theory of groups and algebras, among a long list of diverse topics.

We will consider three classes of algebraic objects of direct functional analytic relevance: (i) generators and relations; (ii) algebras, and (iii) groups.

---

[1] A summary of some highpoints from stochastic processes is contained in Appendix B. See also Sections 2.2.1 (pg. 33), 12.3 (pg. 413); and Chapters 7 (pg. 249), 8 (pg. 273).

[2] Appendix C includes a short biographical sketch of R. S. Phillips. See also [LP89, Phi94].

In the case of (i), we illustrate the ideas with the *canonical commutation relation*

$$PQ - QP = -i\,I, \quad i = \sqrt{-1}. \tag{2.1}$$

The objective is to build a Hilbert space such that the symbols P and Q are represented by unbounded essentially selfadjoint operators (see [RS75, Nel69, vN32a, DS88b]), each defined on a common dense domain in some Hilbert space, and with the operators satisfying (2.1) on this domain. (See technical points inside the present book, and in the cited references.)

In class (ii), we consider both C^*-*algebras* and *von Neumann algebras* (also called W^*-algebras); and in case (iii), our focus is on *unitary representations* of the group G under consideration. The group may be Abelian or non-Abelian, continuous or discrete, locally compact or not. Our present focus will be the case when G is a *Lie group*. In this case, we will study its representations with the use of the corresponding *Lie algebra*[3].

. . . .

- C^*-algebras, von Neumann algebras

We will be considering C^*-, and W^*-algebras axiomatically. In doing this we use the theorem by S. Sakai to the effect that the W^*-algebras consist of the subset of the C^*-algebras that are the dual of a Banach space. If the W^*-algebra is given, the Banach space is called the pre-dual. Representations will be studied with the use of states, and we stress the theorem of Gelfand, Naimark, and Segal (GNS) linking states with cyclic representations.

- wavelets theory

A wavelet is a special basis for a suitable L^2-space which is given by generators and relations, plus self-similarity. Our approach to wavelets will be a mix of functional analysis and harmonic analysis, and we will stress a correspondence between a family of representations of a particular C^*-algebra, called the Cuntz-algebra, on one side and wavelets on the other.

- harmonic analysis

Our approach to harmonic analysis will be general, — encompassing anyone of a set of direct sum (or integral) decompositions. Further our presentation will rely on representations.

- analytic number theory

Our notions from analytic number theory will be those that connect to groups, and representations; such as the study of automorphic forms, and of properties of generalized zeta-functions; see e.g., [CM07, CM06, OPS88, LPS88].

---

[3]The reader will find the precise definition of Lie groups and Lie algebras in Section 8.1 (pg. 276), see also Section 8.6.1 (pg. 306), and Chapter 12 (pg. 403).

Our brief bird's-eye view of the topics above is only preliminary, only hints; and most questions will be addressed in more detail inside the book.

As for references, the literature on the commutation relations (2.1) is extensive, and we refer to [Sza04, Nel59a, Fug82, Pou73].

Some of the questions regarding the *commutation relations* involve the subtle difference between (2.1) itself vs its group version, — often referred to as the *Weyl relations*, or the integrated form. As for the other themes mentioned above, operator algebras, mathematical physics, wavelets and harmonic analysis, the reader will find selected references to these themes at the end of this chapter.

A glance at the table of contents makes it clear that we venture into a few topics at the cross roads of mathematics and physics; and a disclaimer is in order. In the 1930s, David Hilbert encouraged the pioneers in quantum physics to axiomatize the theory that was taking shape then with the initial papers by Heisenberg, Schrödinger, Dirac. Others like J. von Neumann joined into this program. These endeavors were partly in response to *Hilbert's Sixth Problem*[4] [Wig76];

"Give a mathematical treatment of the axioms of physics"

in his famous list of 23 problems announced in 1900 [Hil02]. At this time, quantum physics barely existed. Max Planck's hypothesis on discreteness of atomic energy-measurements is usually dated a little after the turn of the Century.

Quantum mechanics is a first quantized quantum theory that supersedes classical mechanics at the atomic and subatomic levels. It is a fundamental branch of physics that provides the underlying mathematical framework for many fields of physics and chemistry. The term "quantum mechanics" is sometimes used in a more general sense, to mean quantum physics.

With hindsight, we know that there are considerable limitations to the use of axioms in physics. While a number of important questions in quantum physics have mathematical formulations, others depend on physical intuition. Hence in our discussion of questions at the cross-roads of mathematics and physics, we will resort to hand-waiving.

> "For those who are not shocked when they first come across quantum theory cannot possibly have understood it."
>
> Niels Bohr, — quoted in W. Heisenberg, Physics and Beyond (1971).

2.2 Duality

Duality theory lies at the foundations of a host of powerful tools in both pure and applied mathematics. While they have their roots in applications to such areas as quantum physics, partial differential equations (PDE), probability theory, mathematical statistics, and to optimization, most textbooks take the abstract

---

[4]A summary of some highpoints from Hilbert's Sixth Problem is contained in Appendix B. See also Sections 1.6 (pg. 17), 2.5.2 (pg. 70), 7.2 (pg. 259).

approach, and the connections to applications tend to get lost. Here we aim to turn the table, and stress the many ways problems from outside functional analysis, in fact make direct connection to elegant duality principles. To illustrate the point, we have organized some of the interconnections in tables, see especially Table 2.1, and Figures 2.1, 2.3, 2.4, and 2.5. Table 2.1 summarizes some uses of duality principles in Schwartz' distribution theory[5], also called "the theory of generalized functions" (terminology from the Gelfand school.)

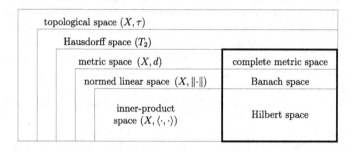

Figure 2.1: A few spaces in analysis, and their inter-relations; ranging from general spaces to classes of linear spaces.

The "functional" in the name "Functional Analysis" derives from the abstract notion of a *linear functional*: Let E be a vector space over a field $\mathbb{F}$ (we shall take $\mathbb{F} = \mathbb{R}$, or $\mathbb{C}$ below.)

Definition 2.1. A function $\varphi : E \longrightarrow \mathbb{R}$ (or $\mathbb{C}$) is said to be a *linear functional*, if we require

$$\varphi\left(u + \lambda v\right) = \varphi\left(u\right) + \lambda\varphi\left(v\right), \ \forall \lambda \in \mathbb{R}, \ \forall u, v \in E.$$

If E comes with a topology (for example from a norm, or from a system of seminorms), we will consider continuous linear functionals. Occasionally, continuity will be implicit.

Definition 2.2. The set of all continuous linear functionals is denoted E^*, and it is called the *dual space*. (In many examples there is a natural identification of E^* as illustrated in Table 2.1.)

Definition 2.3. If E is a normed vector space, and if it is complete in the given norm, we say that E is a *Banach space*.

Lemma 2.1. *Let E be a normed space with dual E^*. For $\varphi \in E^*$, set*

$$\|\varphi\|_* := \sup_{\|x\|=1} |\varphi\left(x\right)|.$$

Then $(E^*, \|\cdot\|_*)$ is a Banach space.

---

[5]Appendix C includes a short biographical sketch of L. Schwartz. See also [Sch95, Sch58, Sch57, Sch97].

Proof. An exercise.   □

Given a Banach space E (Definition 2.3), there are typically three steps involved in the discovery of an explicit form for the *dual Banach space* E^* (Definition 2.2). Table 2.1 illustrates this in two examples, but there are many more to follow; — for example, the case when $E =$ the Hardy space $\mathbb{H}_1$, or $E =$ the trace-class operators on a Hilbert space.

Moreover, the same idea based on a duality-pairing applies *mutatis mutandis*, to other topological vector spaces as well, for example, to those from Schwartz' theory of distributions.

The three steps are as follows:

Finding "the" dual:

Step 1. Given E, then first come up with a second Banach space F as a candidate for the dual Banach space E^*. (Note that E^* is so far, *a priori*, only an abstraction.)

Step 2. Set up a bilinear and non-degenerate pairing, say p, between the two Banach spaces E and F, and check that $p(\cdot, \cdot)$ is continuous on $E \times F$. Rescale such that

$$|p(x, y)| \leq \|x\|_E \|y\|_F, \ \forall x \in E, \ y \in F.$$

This way, via p, we design a linear and isometric embedding of F into E^*.

Step 3. Verify that the embedding from step 2 is "onto" E^*. If "yes" we say that F "is" the dual Banach space. (Example, the dual of $\mathbb{H}_1$ is BMO [Fef71].)

Examples of Banach spaces include: (i) l^p, $1 \leq p \leq \infty$, and (ii) $L^p(\mu)$ where μ is a positive measure on some given measure space; details below.

Example 2.1. l^p: all p-summable sequences.

A sequence $x = (x_k)_{k \in \mathbb{N}}$ is in l^p iff $\sum_{k \in \mathbb{N}} |x_k|^p < \infty$, and then

$$\|x\|_p := \left(\sum_{k \in \mathbb{N}} |x_k|^p \right)^{\frac{1}{p}}.$$

Example 2.2. L^p: all p-integrable functions with respect to a fixed measure μ.

Let $F : \mathbb{R} \to \mathbb{R}$ be monotone increasing, i.e., $x \leq y \implies F(x) \leq F(y)$; then there is a Borel measure μ on $\mathbb{R}$ (see [Rud87]) such that $\mu((x, y]) = F(y) - F(x)$; and $\int \varphi d\mu$ will be the limit of the Stieltjes sums:

$$\sum_i \varphi(x_i)(F(x_{i+1}) - F(x_i)), \text{ where } x_1 < x_2 < \cdots < x_n.$$

E	E^*	how?
l^p, $1 \le p < \infty$ with l^p-norm	l^q, $\frac{1}{p} + \frac{1}{q} = 1$	$x = (x_i) \in l^p$, $y = (y_i) \in l^q$, $\varphi_y(x) = \sum_i \overline{x_i} y_i$
$C(I)$, $I = [0,1]$ with max-norm	signed Borel measures μ on I, of bounded variation	$\varphi_\mu(f) = \int_0^1 f(x)\, d\mu(x)$, $\forall f \in C(I)$.
$C^\infty(\mathbb{R})$, system of seminorms	$\mathcal{E}'$ all Schwartz distributions D on $\mathbb{R}$ of compact support	$\varphi_D(f) = D$ applied to f, $f \in C^\infty(\mathbb{R})$.

Table 2.1: Examples of dual spaces.

We say that $\varphi \in L^p(\mu)$ iff $\int_{\mathbb{R}} |\varphi|^p\, d\mu$ is well defined and finite; then

$$\|\varphi\|_p = \left(\int_{\mathbb{R}} |\varphi(x)|^p\, d\mu(x) \right)^{\frac{1}{p}}.$$

By Stieltjes integral, $\int |\varphi|^p\, d\mu = \int |\varphi|^p\, dF$. Here, we give the definition of $L^p(\mu)$ in the case where $\mu = dF$, but it applies more generally.

For completeness of l^p and of $L^p(\mu)$, see [Rud87].

Remark 2.1. At the foundation of analysis of L^p-spaces (including l^p for the case of counting-measure) is Hölder's inequality; see e.g., [Rud87, ch 3]. Recall conjugate pairs $p, q \in [1, \infty)$, $\frac{1}{p} + \frac{1}{q} = 1$, or equivalently $p + q = pq$; see Figure 2.2.

We present *Hölder's inequality* without proof: Fix a measure space $(X, \mathcal{F}, \mu)$. If p, q are conjugate, $1 < p < \infty$, then for measurable functions f, g we have:

$$\left| \int_X fg\, d\mu \right| \le \left(\int_X |f|^p\, d\mu \right)^{\frac{1}{p}} \left(\int_X |g|^q\, d\mu \right)^{\frac{1}{q}}. \tag{2.2}$$

If $p = 1$, $q = \infty$, and we have:

$$\left| \int_X fg\, d\mu \right| \le \left(\int_X |f|\, d\mu \right) \text{ess sup}_{x \in X} |g(x)|; \tag{2.3}$$

where

$$\|g\|_\infty := \text{ess sup } |g|$$

denotes essential supremum, i.e., neglecting sets of μ-measure zero.

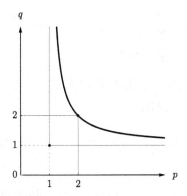

Figure 2.2: Dual exponents for the L^p spaces, $\frac{1}{p} + \frac{1}{q} = 1$.

Inside the book we shall make use of a variety of duality operations, e.g., if E is a normed space, E^* is the dual; E^* is a Banach space. In Hilbert space, $(\cdots)^\perp$ is the perpendicular subspace;

$A \longrightarrow A^*$ denotes the adjoint operator;

$\mathscr{T} \longrightarrow \mathscr{T}'$ denotes the commutant of a set $\mathscr{T}$ of operators; and

$G \longrightarrow \widehat{G}$ in the case when G is a locally compact Abelian group, and $\widehat{G}$ is the group of continuous characters on G. Table 8.2 in Section 8.11 offers an overview of five duality operations and their properties.

The following result is basic in the subject.

Theorem 2.1 (Hahn-Banach). *Let E be a Banach space, and let $x \in E\backslash\{0\}$, then there is a $\varphi \in E^*$ such that $\varphi(x) = \|x\|$, and $\|\varphi\|_{E^*} = 1$.*

Proof. The proof relies on a Zorn lemma argument. This has been postponed till the two appendices at the end of the present chapter. There we will also take up related extension theorems. $\qquad\square$

Below we resume our discussion of examples and applications.

Remark 2.2. In Examples 2.1 and 2.2 above, i.e., l^p and $L^p(\mu)$, it is possible to identify the needed elements in E^*. But the power of Theorem 2.1 is that it yields existence for all Banach spaces, i.e., when E is given only by the axioms from Definitions 2.1-2.2.

Definition 2.4. The *weak-* topology* on E^* is the weakest topology which makes all the linear functionals

$$E^* \ni l \longrightarrow l(x) \in \mathbb{C}$$

continuous, as x ranges over E.

Exercise 2.1 (An ergodic theorem). Let E be a Banach space, and let $A :$ $E \longrightarrow E$ be a contractive linear operator, i.e.,

$$\|A\| = \sup \{\|Ax\| \; ; \; x \in E, \|x\| = 1\} \leq 1$$

holds. Set $A^0 = I =$ the identity operator in E $(Ix = x, \forall x \in E.)$

1. Show that the following limit

$$\lim_{n \to \infty} \frac{1}{n+1} \sum_{k=0}^{n} A^k x = Px, \quad x \in E, \tag{2.4}$$

exists in the norm of E; and moreover that P satisfies $\|P\| \leq 1$, and $P^2 = P$. The limit in (2.4) is called a Cesaro limit, or "limit in mean."

2. Show the following: For vector $x \in E$, we have:

$$Ax = x \iff Px = x. \tag{2.5}$$

Exercise 2.2 (weak-∗ neighborhoods). Let E be a Banach space, and let E^* be the dual Banach space; see Lemma 2.1. We resume our study of the week ∗-topology on E^*; see Definition 2.4 above.

Show that the neighborhoods of 0 in E^* have a basis of open sets $\mathcal{N}$ indexed as follows:

Let $\varepsilon \in \mathbb{R}_+$, $n \in \mathbb{N}$, and $x_1, \ldots, x_n \in E$, and set

$$\mathcal{N}_{\varepsilon, x_1, \ldots, x_n} := \{l \in E^* \; : \; |l(x_i)| < \varepsilon, \; i = 1, \cdots, n\}. \tag{2.6}$$

Terminology. The subsets of E^* in Exercise 2.2 are often called *cylinder sets.* They form a basis for the weak-∗ topology. They also generate a sigma algebra of subsets of E^*, often called the *cylinder sigma algebra.* We will be using it in Sections 7.2 (pg. 259), 7.3 (pg. 266), and 12.3 (pg. 413) below.

Exercise 2.3 (weak-∗ vs norm). Let $1 < p \leq \infty$ be fixed. Set $l^p = l^p(\mathbb{N})$, and show that $\{x \in l^p \; : \; \|x\|_{l^p} \leq 1\}$ is weak-∗ compact, but not norm compact.

<u>Hint</u>: By weak-∗, we refer to $l^p = (l^q)^*$, $\frac{1}{p} + \frac{1}{q} = 1$.

Exercise 2.4 (Be careful with weak-∗ limits.). Settings as in the previous exercise, but now with $p = 2$. Let $\{e_k\}_{k \in \mathbb{N}}$ be the standard ONB in l^2, i.e.,

$$e_k(i) = \delta_{i,k}, \; \forall i, k \in \mathbb{N}. \tag{2.7}$$

Show that 0 in l^2 is a weak ∗-limit of the sequence $\{e_k\}_{k \in \mathbb{N}}$. Conclude that $\{x \in l^2 \; : \; \|x\|_2 = 1\}$ is *not* weak-∗ closed.

<u>Hint</u>: By Parseval, we have, for all $x \in l^2$,

$$\|x\|_2^2 = \sum_{k \in \mathbb{N}} |\langle e_k, x \rangle_2|^2,$$

so $\lim_{k \to \infty} \langle e_k, x \rangle_2 = 0$.

2.2.1 Duality and Measures

Definition 2.5. Let E_i, $i = 1, 2$, be Banach spaces, and let $T : E_1 \longrightarrow E_2$ be a linear mapping. We say that T is bounded (continuous) iff (Def.) $\exists C < \infty$, such that

$$\|Tx\|_2 \leq C \|x\|_1, \ \forall x \in E_1. \tag{2.8}$$

Definition 2.6. Define $T^* : E_2^* \longrightarrow E_1^*$ by

$$(T^*\varphi_2)(x) = \varphi_2(Tx), \ \forall x \in E_1, \ \forall \varphi_2 \in E_2^*. \tag{2.9}$$

We shall adopt the following equivalent notation:

$$\langle T^*\varphi_2, x \rangle = \langle \varphi_2, Tx \rangle, \ \forall x \in E_1, \ \forall \varphi_2 \in E_2^*. \tag{2.10}$$

(Here E^* denotes "dual Banach space.") It is immediate that (2.8) implies

$$\|T^*\varphi_2\|_* \leq C \|\varphi_2\|_*, \ \forall \varphi_2 \in E_2^*. \tag{2.11}$$

Application. Let Ω_k, $k = 1, 2$, be compact spaces, and let $\Psi : \Omega_2 \to \Omega_1$, be a continuous function. Set

$$Tf = f \circ \Psi, \ \forall f \in C(\Omega_1). \tag{2.12}$$

Recall the dual Banach spaces:

$$
\begin{aligned}
C(\Omega_k)^* \ &= \ \text{the respective signed measures on } \Omega_k & (2.13)\\
& \quad \text{of bounded variation, } k = 1, 2;\\
\|\mu\|_* \ &= \ |\mu|(\Omega) \, (= \text{variation of } \mu) & (2.14)\\
&= \ \sup \sum_i |\mu(E_i)|, & (2.15)
\end{aligned}
$$

where $\{E_i \mid E_i \in \mathcal{B}(\Omega)\}$ in (2.15) runs over all partitions of Ω.

Remark 2.3. The assertion in (2.13) is not trivial, but we shall not prove it here. It lies at the foundation of measure theory; beginning with *Riesz' theorem* for positive Borel measures; see e.g., [RSN90, Rud87]. While Riesz' theorem applies in the generality of locally compact Hausdorff space Ω, we shall limit the present discussion to the case where Ω is given, assumed compact. If then $\varphi : C(\Omega) \longrightarrow \mathbb{R}$ is a given linear functional s.t. $\varphi(f) \geq 0$ for all $f \geq 0$ in $C(\Omega)$, then *Riesz' theorem* states that there is a unique positive regular Borel measure μ on Ω s.t.,

$$\varphi(f) = \int_\Omega f(x) \, d\mu(x), \ \forall f \in C(\Omega). \tag{2.16}$$

Now $C(\Omega)$ is a complex Banach space with

$$\|f\| = \sup_{x \in \Omega} |f(x)| = \max_{x \in \Omega} |f(x)| \tag{2.17}$$

and so the natural question is if (2.16) has a generalization to all $\psi \in C(\Omega)^*$ where $(\cdot)^*$ refers to the dual Banach space with its norm

$$\|\psi\|_* = \sup \{|\psi(f)| \; ; \; f \in C(\Omega), \|f\| = 1\}. \tag{2.18}$$

Indeed, *the extended Riesz' theorem* (see (2.13)) states that, for every $\psi \in C(\Omega)^*$ there is a unique complex Borel measure ν such that the representation

$$\psi(f) = \int_\Omega f(x) \, d\nu(x), \; \forall f \in C(\Omega) \tag{2.19}$$

holds. But now the measure ν is $\mathbb{C}$-valued, and it is of bounded variation, i.e., $\|\nu\|_* < \infty$; see (2.14)-(2.15). Moreover, $\|\psi\|_* = \|\nu\|_*$. One further shows that there are four finite positive measures $\{\mu_i\}_{i=1}^4$ such that $\nu = \mu_1 - \mu_2 + i(\mu_3 - \mu_4)$.

Exercise 2.5 (Transformation of measures). Apply (2.10)-(2.11) to show that

$$(T^* \mu_2)(E) = \mu_2 \left(\Psi^{-1}(E)\right), \; \forall E \in \mathcal{B}(\Omega_1),$$

or stated equivalently

$$\int_{\Omega_1} f \, d(T^* \mu_2) = \int_{\Omega_2} (f \circ \Psi) \, d\mu_2, \; \forall f \in C(\Omega_1), \; \forall \mu_2 \in C(\Omega_2)^*.$$

See Figures 2.4 and 2.5 on the next page.

Remark 2.4. We shall make use of the following special case of *pull-back of measures*. It underlies the notion of "the distribution of a *random variable* (math lingo, a measurable function)" from statistics. See Figures 2.3 and 2.4. We shall make use of it below, both in the case of a single random variable, or an indexed family (called a *stochastic process*[6].)

Definition 2.7. Let $(\Omega, \mathcal{F}, \mathbb{P})$ be a *probability space*:

Ω: a set, called "the sample space".

$\mathcal{F}$: a *sigma-algebra* of subsets of Ω. Elements in $\mathcal{F}$ are called events.

$\mathbb{P}$: a *probability measure* defined on $\mathcal{F}$, so $\mathbb{P}$ is positive, sigma-additive, and $\mathbb{P}(\Omega) = 1$.

We say a function $X : \Omega \to \mathbb{R}$ is a *random variable* iff (Def.) the following implication holds:

$$B \in \mathcal{B}(\mathbb{R}) \implies X^{-1}(B) \in \mathcal{F}; \text{ see Fig 2.3.} \tag{2.20}$$

---

[6]A summary of some highpoints from stochastic processes is contained in Appendix B. See also Sections 2.2.1 (pg. 33), 12.3 (pg. 413); and Chapters 7 (pg. 249), 8 (pg. 273).

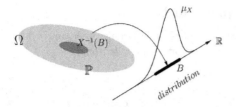

Figure 2.3: A measurement X; $X^{-1}(B) = \{\omega \in \Omega \, : \, X(\omega) \in B\}$. A random variable and its distribution.

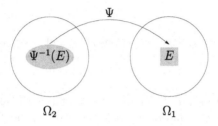

Figure 2.4: $\Psi^{-1}(E) = \{\omega \in \Omega_2 \, : \, \Psi(\omega) \in E\}$, pull-back.

$\longrightarrow$	compact spaces	$\Omega_2 \xrightarrow{\ \Psi\ } \Omega_1$
$\longleftarrow$	Banach spaces	$C(\Omega_2) \xleftarrow{\ T\ } C(\Omega_1)$
$\longrightarrow$	duals: measures	$\mathcal{M}(\Omega_2) \xrightarrow{\ T^*\ } \mathcal{M}(\Omega_1)$

Figure 2.5: Contra-variance (from point transformations, to transformation of functions, to transformation of measures.) See Exercise 2.5.

Definition 2.8. Let $(\Omega, \mathcal{F}, \mathbb{P})$ be a probability space, and let $X : \Omega \longrightarrow \mathbb{R}$ be a random variable on Ω. The *induced measure* μ_X is a positive Borel measure on $\mathbb{R}$, given by

$$\mu_X(B) = \mathbb{P}\left(X^{-1}(B)\right), \ \forall B \in \mathcal{B}(\mathbb{R}); \tag{2.21}$$

i.e., μ_X is the pull-back via X.

The measure μ_X is called the "distribution" of the random variable X, or the "law" of X. (Note that the word "distribution" also has a different meaning, that of generalized function in the sense of L. Schwartz; see below.)

If μ_X is Gaussian, see Section 2.7, we say that X is a *Gaussian random variable*. If μ_X is uniform, we say that X is uniformly distributed; and similarly for the other probability distributions on $\mathbb{R}$; see Table 7.1 in Section 3.1 below.

2.2.2   Other Spaces in Duality

Below we consider three spaces of functions on $\mathbb{R}$, and their duals. These are basics of the L. Schwartz' theory of *distributions*:

Definition 2.9. We set

$\mathcal{D} := C_c^\infty(\mathbb{R}) = $ all C^∞-functions on $\mathbb{R}$ having compact support;

$\mathcal{S} := \mathcal{S}(\mathbb{R}) = $ all C^∞-functions on $\mathbb{R}$ such that $x^k f^{(n)} \in L^2(\mathbb{R})$ for all $k, n \in \mathbb{N}$;

$\mathcal{E} := C^\infty(\mathbb{R}) = $ all C^∞-functions on $\mathbb{R}$ (without support restriction).

Each of the three spaces of test functions $\mathcal{D}$, $\mathcal{S}$, and $\mathcal{E}$ have countable families of *seminorms*, turning them into topological vector spaces (TVS). The two, $\mathcal{S}$ and $\mathcal{E}$ are Fréchet spaces, while $\mathcal{D}$ is an inductive limit of Fréchet spaces (abbreviated LF.)

For $\mathcal{S}$, the seminorms are the max of the absolute value of the above listed functions, so indexed by k and n. For the other two, $\mathcal{E}$, and $\mathcal{D}$, the seminorms are indexed by a number n of derivatives, and by compact intervals, say $[-k, k]$. For each n and k, we take $\max |f^{(n)}(x)|$ over $[-k, k]$. As TVSs, these three spaces in turn are the building blocks of Schwartz' theory of distributions, see [Sch57] and [Trè06a]. In each case, the dual space will be defined with reference to the respective topologies. See details below.

Clearly,

$$\mathcal{D} \hookrightarrow \mathcal{S} \hookrightarrow \mathcal{E}; \tag{2.22}$$

but all of the three spaces come with a natural system of seminorms turning them into topological vector spaces, and we have the continuous inclusions $\mathcal{D} \hookrightarrow \mathcal{S}$, and $\mathcal{S} \hookrightarrow \mathcal{E}$.

Hence for the duals, we have the following three distribution spaces

$$\mathcal{E}' \hookrightarrow \mathcal{S}' \hookrightarrow \mathcal{D}'; \text{ where} \tag{2.23}$$

$\mathcal{E}' = $ the space of all compactly supported distributions on $\mathbb{R}$;

$\mathcal{S}' = $ the space of all tempered distributions on $\mathbb{R}$; and

$\mathcal{D}' = $ all distributions on $\mathbb{R}$.

Note that the mappings indicated in the system (2.23) are obtained from dualizing the corresponding system of continuous inclusions for the Schwartz-test function spaces from (2.22). Moreover, in the system (2.22) each of the smaller topological vector spaces is dense in the bigger one. This implies that the corresponding dual system of mappings in the chain (2.23) will in fact be 1-1 (i.e., zero kernel). With this identification, we shall therefore interpret the maps in (2.23) as inclusions.

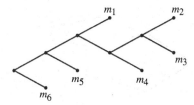

Figure 2.6: Finite Tree (natural order on the set of vertices). Examples of maximal elements: $m_1, m_2, \ldots$

Exercise 2.6 (Gelfand triple).

1. Using Table 2.1, show that $L^2(\mathbb{R})$ is contained in $\mathcal{S}'$ (= tempered distributions.)

2. Using self-duality of L^2, i.e., $(L^2)^* \simeq L^2$ (by Riesz), make precise the following double inclusions:

$$\mathcal{S} \hookrightarrow L^2 \hookrightarrow \mathcal{S}' \tag{2.24}$$

where each inclusion mapping in (2.24) is continuous with respect to the respective topologies; the Fréchet topology on $\mathcal{S}$, the norm topology on L^2, and the weak-$*$ (dual) topology on $\mathcal{S}'$. (The system (2.24) is an example of a *Gelfand triple*, see Section 10.5.)

2.3 Transfinite Induction (Zorn and All That)

Let $(X, \leq)$ be a *partially ordered* set. By partial ordering, we mean a binary relation "$\leq$" on the set X, such that (i) $x \leq x$; (ii) $x \leq y$ and $y \leq x$ implies $x = y$; and (iii) $x \leq y$ and $y \leq z$ implies $x \leq z$.

A subset $C \subset X$ is said to be a *chain*, or *totally ordered*, if $x, y \in C$ implies that either $x \leq y$ or $y \leq x$. *Zorn's lemma* says that if every chain has a majorant then there exists a maximal element in X.

Theorem 2.2 (Zorn). *Let $(X, \leq)$ be a partially ordered set. If every chain C in X has a majorant (upper bound), then there exists an element m in X so that $x \geq m$ implies $x = m$.*

An illuminating example of a partially ordered set is the binary tree model (Figs 2.6-2.7). Another example is when X is a family of subsets of a given set, partially ordered by inclusion.

Zorn's lemma lies at the foundation of set theory. It is in fact an axiom and is equivalent to the axiom of choice, and to Hausdorff's maximality principle.

Theorem 2.3 (Hausdorff Maximality Principle). *Let $(X, \leq)$ be a partially ordered set, then there exists a maximal totally ordered subset L in X.*

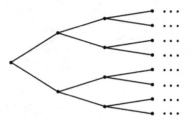

Figure 2.7: Infinite Tree (no maximal element!) All finite words in the alphabet $\{0, 1\}$ continued indefinitely.

The axiom of choice is equivalent to the following statement on infinite products, which itself is extensively used in functional analysis.

Theorem 2.4 (Axiom of choice). *Let A_α be a family of nonempty sets indexed by $\alpha \in I$. Then the infinite Cartesian product*

$$\Omega = \prod_{\alpha \in I} A_\alpha = \{\omega : I \to \cup_{\alpha \in I} A_\alpha \mid \omega(\alpha) \in A_\alpha\}$$

is nonempty.

The point of using the axiom of choice is that, if the index set is uncountable, there is no way to verify whether (x_α) is in Ω, or not. It is just impossible to check for each α that x_α is contained in A_α.

Remark 2.5. A more dramatic consequence of the axiom of choice is the mind-boggling *Banach-Tarski paradox*; see e.g., [MT13]. It states: For the solid ball B in 3-dimensional space, there exists a decomposition of B into a finite number of disjoint subsets, which can then in turn be put back together again, but in a different way which will yield two identical copies of the original ball B; — stated informally as: "A pea can be chopped up and reassembled into the Sun." The axiom of choice allows for the construction of nonmeasurable sets, i.e., sets that do not have a volume, and that for their construction would require performing an uncountably infinite number of choices.

In case the set is countable, we simply apply the down to earth standard induction. Note that the standard mathematical induction is equivalent to the Peano's axiom: *Every nonempty subset of the set of natural numbers has a unique smallest element.* The power of transfinite induction is that it applies to uncountable sets as well.

In applications, the key of using the transfinite induction is to cook up, in a clear way, a partially ordered set, so that the maximal elements turn out to be the objects to be constructed.

Examples include the *Hahn-Banach extension theorem, Krein-Milman's the*orem on compact convex sets, existence of orthonormal bases in Hilbert space,

Tychonoff's theorem on infinite Cartesian product of compact spaces (follow immediately from the axiom of choice.)

Theorem 2.5 (Tychonoff). *Let A_α be a family of compact sets indexed by $\alpha \in I$. Then the infinite Cartesian product $\prod_\alpha A_\alpha$ is compact with respect to the product topology.*

While Zorn's lemma is non-constructive, it still yields a number of powerful existence theorems throughout mathematics. Even if the conclusion at the initial application of a Zorn construction might not give answers that are easy to relate directly to geometric objects, it is often possible later to realize solutions obtained to something more concrete, for example maximal ideals, extreme points, generalized boundaries, maximal proper subgroups, certain linear functionals (or states) arising from extension (e.g., the Hahn-Banach theorem), pure states, irreducible representations, or maximal entropy measures. Even the proof of existence of Haar measure relies on a Zorn (transfinite induction) argument. And a variety of constructions of bases depends on Zorn: orthonormal bases (ONBs) in Hilbert space, and more generally Hamel bases in vector space. We turn to ONBs below, and we include other applications of transfinite induction in the appendices.

We will now apply transfinite induction (Zorn's lemma) to show that every infinite dimensional Hilbert space has an orthonormal basis (ONB).

2.4   Basics of Hilbert Space Theory

Key to functional analysis is the idea of *normed vector spaces*. The interesting ones are infinite dimensional. To use them effectively in the solution of problems, we must be able to take limits, hence the assumption of completeness. A complete normed linear space is called a *Banach space*. But for applications in physics, statistics, and in engineering it often happens that the norm comes from an *inner product*; — this is the case of *Hilbert space*. With an inner product, one is typically able to get much more precise results, than in the less structured case of Banach space. (Many Banach spaces are not Hilbert spaces.)

The more interesting Hilbert spaces typically arise in concrete applications as infinite dimensional spaces of function. And as such, they have proved indispensable tools in the study of partial differential equations (PDE), in quantum mechanics, in Fourier analysis, in signal processing, in representations of groups, and in ergodic theory. The term Hilbert space was originally coined by John von Neumann, who identified the axioms that now underlie these diverse applied areas. Examples include spaces of *square-integrable functions* (e.g., the L^2 random variables of a probability space), *Sobolev spaces*, Hilbert spaces of Schwartz distributions, and *Hardy spaces* of holomorphic functions; — to mention just a few.

One reason for their success is that geometric intuition from finite dimensions often carries over: e.g., the Pythagorean Theorem, the parallelogram law; and, for optimization problems, the important notion of "*orthogonal projection.*" And the idea (from linear algebra) of diagonalizing a normal matrix; — the *spectral theorem*.

Linear mappings (transformations) between Hilbert spaces are called *linear operators*, or simply "operators." They include partial differential operators (PDOs), and many others.

Definition 2.10. Let X be a vector space over $\mathbb{C}$.

A norm on X is a mapping $\|\cdot\| : X \to \mathbb{C}$ such that

- $\|cx\| = |c| \, \|x\|$, $c \in \mathbb{C}$, $x \in X$;

- $\|x\| \geq 0$; $\|x\| = 0$ implies $x = 0$, for all $x \in X$;

- $\|x + y\| \leq \|x\| + \|y\|$, for all $x, y \in X$.

Remark 2.6. Let $(X, \|\cdot\|)$ be a normed space. X is called a *Banach space* if it is complete with respect to the induced metric (see Definition 2.3)

$$d(x, y) := \|x - y\|, \ x, y \in X.$$

Definition 2.11. Let X be vector space over $\mathbb{C}$. An inner product on X is a function $\langle \cdot, \cdot \rangle : X \times X \to \mathbb{C}$ so that for all $x, y \in \mathscr{H}$, and $c \in \mathbb{C}$, we have

- $\langle x, \cdot \rangle : X \to \mathbb{C}$ is linear (linearity)

- $\langle x, y \rangle = \overline{\langle y, x \rangle}$ (conjugation)

- $\langle x, x \rangle \geq 0$; and $\langle x, x \rangle = 0$ implies $x = 0$ (positivity)

Remark 2.7. The abstract formulation of Hilbert space was invented by von Neumann in 1925. It fits precisely with the axioms of quantum mechanics (spectral lines, etc.) A few years before von Neumann's formulation, Max Born had translated Heisenberg's quantum mechanics into modern mathematics. In 1924, in a break-through paper, Heisenberg[7] had invented quantum mechanics, but he had not been precise about the mathematics. His use of "matrices" was highly intuitive. It was only in the subsequent years, with the axiomatic language of Hilbert space, that the group of physicists and mathematicians around Hilbert in Göttingen were able to give the theory the form it now has in modern textbooks. (See Appendix C for more detail.)

Lemma 2.2 (Cauchy-Schwarz). [8]*Let $(X, \langle \cdot, \cdot \rangle)$ be an inner product space, then*

$$|\langle x, y \rangle|^2 \leq \langle x, x \rangle \langle y, y \rangle, \ \forall x, y \in X. \tag{2.25}$$

---

[7]Appendix C includes a short biographical sketch of W. Heisenberg. See also [Hei69].

[8]Hermann Amandus Schwarz (1843 – 1921), German mathematician, contemporary of Weierstrass, and known for his work in complex analysis. He is the one in many theorems in books on analytic functions. We will often refer to (2.25) as simply "Schwarz". The abbreviation is useful because we use it a lot.

There are other two "Schwartz" (with a "t"):

Laurent Schwartz (1915 – 2002), French mathematician, Fields Medal in 1950 for his work of distribution theory.

Jack Schwartz (1930 – 2009), American mathematician, author of the famous book "Linear Operators", [DS88a, DS88b, DS88c].

Proof. By the positivity axiom in the definition of an inner product, we see that

$$\sum_{i,j=1}^{2} \overline{c_i} c_j \langle x_i, x_j \rangle = \left\langle \sum_{i=1}^{2} c_i x_i, \sum_{j=1}^{2} c_j x_j \right\rangle \geq 0, \ \forall c_1, c_2 \in \mathbb{C};$$

i.e., the matrix

$$\begin{bmatrix} \langle x_1, x_1 \rangle & \langle x_1, x_2 \rangle \\ \langle x_2, x_1 \rangle & \langle x_2, x_2 \rangle \end{bmatrix}$$

is positive definite. Hence the above matrix has nonnegative determinant, and (2.25) follows. □

Corollary 2.1. *Let* $(X, \langle \cdot, \cdot \rangle)$ *be an inner product space, then*

$$\|x\| := \sqrt{\langle x, x \rangle}, \ x \in X \tag{2.26}$$

defines a norm.

Proof. It suffices to check the triangle inequality (Definition 2.11). For all $x, y \in X$, we have (with the use of Lemma 2.2):

$$\begin{aligned}
\|x + y\|^2 &= \langle x + y, x + y \rangle \\
&= \|x\|^2 + \|y\|^2 + 2\Re \{\langle x, y \rangle\} \\
&\leq \|x\|^2 + \|y\|^2 + 2 \|x\| \|y\| \quad \text{(by (2.25))} \\
&= (\|x\| + \|y\|)^2
\end{aligned}$$

and the corollary follows. □

Definition 2.12. An inner product space $(X, \langle \cdot, \cdot \rangle)$ is called a *Hilbert space* if X is complete with respect to the metric

$$d(x, y) = \|x - y\|, \ x, y \in X;$$

where the RHS is given by (2.26).

Exercise 2.7 (Hilbert completion). Let $(X, \langle \cdot, \cdot \rangle)$ be an inner-product space (Definition 2.11), and let $\mathscr{H}$ be its *metric completion* with respect to the norm in (2.26). Show that $\langle \cdot, \cdot \rangle$ on $X \times X$ extends by limit to a sesquilinear form $\langle \cdot, \cdot \rangle^{\sim}$ on $\mathscr{H} \times \mathscr{H}$; and that $\mathscr{H}$ with $\langle \cdot, \cdot \rangle^{\sim}$ *is a Hilbert space.*

Exercise 2.8 (L^2 of a measure-space). Let $(M, \mathcal{B}, \mu)$ be as follows:

M : locally compact Hausdorff space;

$\mathcal{B}$: the Borel sigma-algebra, i.e., generated by the open subsets of M;

μ : a fixed positive measure defined on $\mathcal{B}$.

Let $\mathcal{F} := span\{\chi_E \mid E \in \mathcal{B}\}$, and on linear combinations, set

$$\left\|\sum_{\text{finite}} c_i \chi_{E_i}\right\|_{\mathscr{H}}^2 = \sum_i |c_i|^2 \mu(E_i) \tag{2.27}$$

where $E_i \in \mathcal{B}$, and $E_i \cap E_j = \emptyset$ $(i \neq j)$, are assumed.

Show that the Hilbert-completion of $\mathcal{F}$ with respect to to (2.27) agrees with the standard definitions [Rud87, Par05] of the $L^2(\mu)$-space.

2.4.1   Positive Definite Functions

An extremely useful method to build Hilbert spaces is the GNS construction. For details, see Chapter 5.

The idea is to start with a positive definite function defined on an arbitrary set X.

Definition 2.13. We say $\varphi : X \times X \to \mathbb{C}$ is *positive definite*, if for all $n \in \mathbb{N}$,

$$\sum_{i,j=1}^{n} \overline{c_i} c_j \varphi(x_i, x_j) \geq 0 \tag{2.28}$$

for all system of coefficients $c_1, \ldots, c_n \in \mathbb{C}$, and all $x_1, \ldots, x_n \in X$.

Remark 2.8. Given φ, set

$$H_0 := \left\{\sum_{\text{finite}} c_x \delta_x : x \in X, c_x \in \mathbb{C}\right\} = span_{\mathbb{C}}\{\delta_x : x \in X\},$$

and define a sesquilinear form on H_0 by

$$\left\langle \sum c_x \delta_x, \sum d_y \delta_y \right\rangle_{\varphi} := \sum \overline{c_x} d_y \varphi(x, y).$$

Note that

$$\left\|\sum c_x \delta_x\right\|_{\varphi}^2 := \left\langle \sum c_x \delta_x, \sum c_x \delta_x \right\rangle_{\varphi} = \sum_{x,y} \overline{c_x} c_y \varphi(x, y) \geq 0$$

by assumption. (All summations are finite.)

However, $\langle \cdot, \cdot \rangle_{\varphi}$ is in general not an inner product since the strict positivity axiom may not be satisfied. Hence one has to pass to a quotient space by letting

$$N = \left\{f \in H_0 \mid \langle f, f \rangle_{\varphi} = 0\right\},$$

and set $\mathscr{H} :=$ completion of the quotient space H_0/N with respect to $\|\cdot\|_{\varphi}$. (The fact that N is really a subspace follows from (2.25).) $\mathscr{H}$ is a Hilbert space.

Corollary 2.2. *Let X be a set, and let $\varphi : X \times X \to \mathbb{C}$ be a function. Then φ is positive definite if and only if there is a Hilbert space $\mathscr{H} = \mathscr{H}_\varphi$, and a function $\Phi : X \to \mathscr{H}$ such that*

$$\varphi(x, y) = \langle \Phi(x), \Phi(y) \rangle_{\mathscr{H}} \tag{2.29}$$

for all $(x, y) \in X \times X$, where $\langle \cdot, \cdot \rangle_{\mathscr{H}}$ denotes the inner product in $\mathscr{H}$.

Given a solution Φ satisfying (2.29), then we say that $\mathscr{H}$ is minimal if

$$\mathscr{H} = \overline{span}\{\Phi(x) : x \in X\}. \tag{2.30}$$

Given two minimal solutions, $\Phi_i : X \to \mathscr{H}_i$, $i = 1, 2$ (both satisfying (2.29)); then there is a unitary isomorphism $\mathcal{U} : \mathscr{H}_1 \to \mathscr{H}_2$ such that

$$\mathcal{U}\Phi_1(x) = \Phi_2(x)\mathcal{U}, \ \forall x \in X. \tag{2.31}$$

Proof. These conclusions follow from Remark 2.8, and the definitions. (The missing details are left as an exercise to the student.) □

Remark 2.9. It is possible to be more explicit about choice of the pair $(\Phi, \mathscr{H})$ in Corollary 2.2, where $\varphi : X \times X \to \mathbb{C}$ is a given positive definite function. We may in fact choose $\mathscr{H}$ to be $L^2(\Omega, \mathcal{F}, \mathbb{P})$ where $\mathbb{P} = \mathbb{P}_\varphi$ depends on φ, and $(\Omega, \mathcal{F}, \mathbb{P})$ is a probability space.

Example 2.3 (Wiener-measure). In Remark 2.9, take $X = [0, \infty) = \mathbb{R}_+ \cup \{0\}$, and set

$$\varphi(s, t) = s \wedge t = \min(s, t), \tag{2.32}$$

see Figure 2.8. In this case, we may then take $\Omega = C(\mathbb{R}) = $ all continuous functions on $\mathbb{R}$, and $\Phi_t(\omega) := \omega(t)$, $t \in [0, \infty)$, $\omega \in C(\mathbb{R})$.

Further, the sigma-algebra in $C(\mathbb{R})$, $\mathcal{F} := Cyl$, is generated by cylinder-sets, and $\mathbb{P}$ is the *Wiener-measure*; and Φ on $L^2(C(\mathbb{R}), Cyl, \mathbb{P})$ is the standard *Brownian motion*, i.e., $\Phi : [0, \infty) \to L^2(C(\mathbb{R}), \mathbb{P})$ is a Gaussian process with

$$\mathbb{E}_\mathbb{P}(\Phi(s)\Phi(t)) = \int_{C(\mathbb{R})} \Phi_s(\omega)\Phi_t(\omega)\, d\mathbb{P}(\omega) = s \wedge t.$$

The process $\{\Phi_t\}$ is called the Brownian motion; its properties include that each Φ_t is a Gaussian random variable. We refer to Chapter 7 for full details. Figure 2.9 shows a set of sample path of the *standard Brownian motion*.

If the positive definite function φ_1 in (2.32) is replaced with

$$\varphi_2(s, t) = s \wedge t - st = \varphi_1(s, t) - st, \tag{2.33}$$

we still get an associated Gaussian process $\Phi_2 : [0, 1] \longrightarrow L^2(\Omega, \mathbb{P})$, but instead with

$$\mathbb{E}_\mathbb{P}(\Phi_2(s)\Phi_2(t)) = \varphi_2(s, t) = s \wedge t - st; \tag{2.34}$$

it is called *pined Brownian motion*. See Exercise 4.11, and Figure 4.5.

Exercise 2.9 (Product of two positive definite functions). Let φ and ψ be positive definite functions $X \times X \longrightarrow \mathbb{C}$ (see (2.28)) and set

$$\xi(x,y) = \varphi(x,y)\psi(x,y), \ \forall(x,y) \in X \times X.$$

Show that $\xi = \varphi \cdot \psi$ is again positive definite.

Hint: Use Remark 2.8 and the fact that every positive $n \times n$ matrix B has the form $B = A^*A$. Fix n, and $x_1, \ldots, x_n \in X$. Apply this to $B_{ij} := \varphi(x_i, x_j)$.

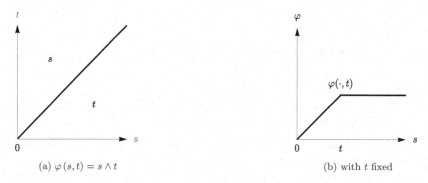

(a) $\varphi(s,t) = s \wedge t$ (b) with t fixed

Figure 2.8: Covariance function of Brownian motion.

Figure 2.9: A set of Brownian sample-paths generated by a Monte-Carlo computer simulation.

We resume the discussion of stochastic processes in Chapter 12 below.

Remark 2.10. We see from the proof of Lemma 2.2 that the Cauchy-Schwarz inequality holds for all positive definite functions.

Definition 2.14. Let $\mathscr{H}_i$, $i = 1, 2$, be two Hilbert spaces.

A linear operator $J : \mathscr{H}_1 \to \mathscr{H}_2$ is said to be an *isometry* iff (Def.)

$$\|Jx\|_{\mathscr{H}_2} = \|x\|_{\mathscr{H}_1}, \ \forall x \in \mathscr{H}_1.$$

Note that J is *not* assumed "onto."

Exercise 2.10 (An Itō-isometry). Let T be a locally compact Hausdorff space; and let μ be a positive Borel measure, i.e., consider the measure space $(T, \mathcal{B}(T), \mu)$, where $\mathcal{B}(T)$ is the Borel sigma-algebra.

1. Show that there is a measure space $(\Omega, \mathcal{F}, \mathbb{P}^{(\mu)})$ depending on μ; and a function (*Gaussian process*)

$$\Phi : \mathcal{B}(T) \longrightarrow L^2(\Omega, \mathbb{P}) \tag{2.35}$$

such that every Φ_A, for $A \in \mathcal{B}(T)$, is a Gaussian random variable, such that $\mathbb{E}(\Phi_A) = 0$, and

$$\mathbb{E}(\Phi_A \Phi_B) = \mu(A \cap B) \tag{2.36}$$

holds for all $A, B \in \mathcal{B}(T)$. The expectation $\mathbb{E}$ in (2.36) is with respect to $\mathbb{P} = \mathbb{P}^{(\mu)}$.

2. Show that there is an *isometry* $J = J_{(\text{Itō})}$ from $L^2(T, \mu)$ into $L^2(\Omega, \mathbb{P})$ such that

$$\mathbb{E}\left(\left|\int_T f(t)\,d\Phi_t\right|^2\right) = \int_T |f|^2\,d\mu \tag{2.37}$$

where the expression $\int_T f(t)\,d\Phi_t$ on the LHS in (2.37) is the L^2-limit of finite sums (simple functions):

$$\sum_i c_i \Phi_{A_i}, \tag{2.38}$$

$c_i \in \mathbb{R}$, finite indexing; and $A_i \in \mathcal{B}(T)$, $A_i \cap A_j = \emptyset$, $i \neq j$ (i.e., disjointness.)

Because of (2.37), we can set

$$Jf = \int_T f(t)\,d\Phi_t \in L^2(\Omega, \mathbb{P}^{(\mu)}). \tag{2.39}$$

3. Extend (2.37) to *adapted* measurable fields $F : T \longrightarrow L^2(\Omega, \mathcal{F}, \mathbb{P}^{(\mu)})$ as follows: Assume that, for all $A, B \in \mathcal{B}_T$ (the sigma-algebra on T) s.t. $A \cap B = \emptyset$, we have *independence* of

$$F\big|_A \text{ and } \sigma\text{-alge}\{\Phi_C \mid \forall C \subset B,\, C \in \mathcal{B}_T\}.$$

Then show that a limit, as in (2.38) exists and defines

$$\underbrace{\int_T F_t\,d\Phi_t}_{\text{Itō-integral}} \text{ in } L^2(\Omega, \mathcal{F}, \mathbb{P}^{(\mu)}),$$

and that

$$\mathbb{E}\left(\left|\int_T F_t\,d\Phi_t\right|^2\right) = \int_T \mathbb{E}\big(|F_t|^2\big)\,d\mu(t).$$

Hint:

1. This is an application of Corollary 2.2 (Section 2.4), applied to $X = \mathcal{B}(T)$. Note that

$$\mathcal{B}(T) \times \mathcal{B}(T) : (A, B) \longmapsto \mu(A \cap B) \qquad (2.40)$$

   is positive definite. Hence, the existence of the Gaussian process

$$(\Omega, \mathcal{F}, \mathbb{P}, \Phi)$$

   subject to the conditions in part 1 of the exercise, follows from Corollary 2.2.

2. Consider the simple functions in (2.38), and use (2.36). Then derive the following:

$$\mathbb{E}\left(\left|\sum_i c_i \Phi_{A_i}\right|^2\right) = \sum_i |c_i|^2 \mu(A_i) \qquad (2.41)$$

Hence, the *Itō-isometry*[9], defined initially only on simple functions, is isometric. Justify the extension by limits to all of $L^2(T, \mu)$. In your last step, taking the limit over partitions, make use of the conclusion from Exercise 2.8.

Remark 2.11. The construction in the exercise is an example of a *stochastic process*, and a *stochastic integral*. Both subjects are resumed in Chapters 7 and 12 below.

2.4.2   Orthonormal Bases

Definition 2.15. Let $\mathscr{H}$ be a Hilbert space. A family of vectors $\{u_\alpha\}$ in $\mathscr{H}$ is said to be an *orthonormal basis* if

1. $\langle u_\alpha, u_\beta \rangle_{\mathscr{H}} = \delta_{\alpha\beta}$ and

2. $\overline{span}\,\{u_\alpha\} = \mathscr{H}$. (Here "$\overline{span}$" means "closure of the linear span.")

   We are now ready to prove the existence of *orthonormal bases* for any Hilbert space. The key idea is to cook up a partially ordered set satisfying all the requirements for transfinite induction, so that each maximal element turns out to be an orthonormal basis (ONB). Notice that all we have at hand are the abstract axioms of a Hilbert space, and nothing else. Everything will be developed out of these axioms. A separate issue is *constructive* ONBs, for example, wavelets or orthogonal polynomials.

   There is a new theory which generalizes the notion of ONB; called *"frame"*, and it is discussed in Chapter 9 below, along with some more applications in Chapter 11.

---

[9]Appendix C includes a short biographical sketch of K. Itō. See also [Itō07, Itō04].

Theorem 2.6 (Existence of ONBs). *Every Hilbert space $\mathscr{H}$ has an orthonormal basis.*

To start out, we need the following lemma; its proof is left to the reader.

Lemma 2.3. *Let $\mathscr{H}$ be a Hilbert space and $S \subset \mathscr{H}$. Then the following are equivalent:*

1. *$x \perp S$ implies $x = 0$*

2. *$\overline{span}\{S\} = \mathscr{H}$.*

Proof. Now we prove Theorem 2.6. If $\mathscr{H} = \{0\}$ then the proof is done. Otherwise, let $u_1 \in \mathscr{H}$. If $\|u_1\| \neq 1$, it can be normalized by $u_1/\|u_1\|$. Hence we may assume $\|u_1\| = 1$. If $span\{u_1\} = \mathscr{H}$ the proof finishes again, otherwise there exists $u_2 \notin span\{u_1\}$. By Lemma 2.3, we may assume $\|u_2\| = 1$ and $u_1 \perp u_2$. It follows that there exists a collection S of orthonormal vectors in $\mathscr{H}$.

Let $\mathbb{P}(S)$ be the set of all orthonormal sets partially ordered by set inclusion. Let $C \subset \mathbb{P}(S)$ be any chain and let $M := \bigcup_{E \in C} E$. M is clearly a majorant of C.

In fact, M is in the partially ordered system. For if $x, y \in M$, there exist E_x and E_y in C so that $x \in E_x$ and $y \in E_y$. Since C is a chain, we may assume $E_x \leq E_y$. Hence $x, y \in E_y$, and so $x \perp y$. This shows that M is in the partially ordered system and a majorant.

By Zorn's lemma, there exists a maximal element $m \in \mathbb{P}(S)$. It remains to show that the closed span of m is $\mathscr{H}$. Suppose this is false, then by Lemma 2.3 there exists a vector $x \in \mathscr{H}$ so that $x \perp \overline{span}\{m\}$.

Since $m \cup \{x\} \geq m$, and m is assumed maximal, it follows that $x \in m$. This implies $x \perp x$. Therefore $x = 0$, by the positivity axiom of the definition of Hilbert space. $\square$

Corollary 2.3. *Let $\mathscr{H}$ be a Hilbert space, then $\mathscr{H}$ is isomorphic to the l^2 space of the index set of an ONB of $\mathscr{H}$. Specifically, given an ONB $\{u_\alpha\}_{\alpha \in J}$ in $\mathscr{H}$, where J is some index set, then*

$$v = \sum_{\alpha \in J} \langle u_\alpha, v \rangle \, u_\alpha, \text{ and} \tag{2.42}$$

$$\|v\|^2 = \sum_{\alpha \in J} |\langle u_\alpha, v \rangle|^2, \ \forall v \in \mathscr{H}. \tag{2.43}$$

Moreover,

$$\langle u, v \rangle = \sum_{\alpha \in J} \langle u, u_\alpha \rangle \langle u_\alpha, v \rangle, \ \forall u, v \in \mathscr{H}. \tag{2.44}$$

In Dirac's notation (see Section 2.5), (2.42)-(2.44) can be written in the following operator identity:

$$I_{\mathscr{H}} = \sum_{\alpha \in J} |u_\alpha \rangle \langle u_\alpha|. \tag{2.45}$$

(Note: eq. (2.43) is called the Parseval identity.)

Proof. Set $\mathscr{H}_0 := span\{u_\alpha\}$. Then, for all $v \in \mathscr{H}_0$, we have

$$v = \sum_{\text{finite}} \langle u_\alpha, v \rangle u_\alpha$$

and

$$\|v\|^2 = \sum_{\text{finite}} |\langle u_\alpha, v \rangle|^2 .$$

Thus, the map

$$\mathscr{H}_0 \ni v \longmapsto \widehat{v} := (\langle u_\alpha, v \rangle) \in C_c(J) \tag{2.46}$$

is an *isometric isomorphism*; where C_c denotes all the l^2-sequences indexed by J, vanishing outside some finite subset of J.

Since $\mathscr{H}_0$ is dense in $\mathscr{H}$, and $C_c(J)$ is dense in $l^2(J)$, it follows that (2.46) extends to a unitary operator from $\mathscr{H}$ onto $l^2(A)$, see Exercise 2.11. Thus, (2.42)-(2.43) hold.

Using the *polarization identity* (Lemma 4.10) in both $\mathscr{H}$ and $l^2(J)$, with $i = \sqrt{-1}$, we conclude that

$$
\begin{aligned}
\langle u, v \rangle_{\mathscr{H}} &= \frac{1}{4} \sum_{k=0}^{3} i^k \left\| v + i^k u \right\|_{\mathscr{H}}^2 \\
&\underset{(2.46)}{=} \frac{1}{4} \sum_{k=0}^{3} i^k \left\| \widehat{v} + i^k \widehat{u} \right\|_{l^2(J)}^2 \\
&= \langle \widehat{u}, \widehat{v} \rangle_{l^2(J)} \\
&= \sum_{\alpha \in J} \langle u, u_\alpha \rangle_{\mathscr{H}} \langle u_\alpha, v \rangle_{\mathscr{H}}, \ \forall u, v \in \mathscr{H},
\end{aligned}
$$

which is the assertion in (2.44).   □

Exercise 2.11 (Fischer). Fix an ONB $\{u_\alpha\}_{\alpha \in J}$ as in Corollary 2.3, and set

$$Tv = (\langle u_\alpha, v \rangle_{\mathscr{H}})_{\alpha \in J}.$$

Then show that $T : \mathscr{H} \longrightarrow l^2(J)$ is a unitary isomorphism of $\mathscr{H}$ onto $l^2(J)$.

Remark 2.12. The correspondence $\mathscr{H} \longleftrightarrow l^2$ (index set of an ONB) is *functorial*, and an *isomorphism*. Hence, there seems to be just one Hilbert space. But this is misleading, because numerous interesting realizations of an abstract Hilbert space come in when we make a choice of the ONB. The question as to which Hilbert space to use is equivalent to a good choice of an ONB; in $L^2(\mathbb{R})$, for example, a wavelet ONB.

Definition 2.16. A Hilbert space $\mathscr{H}$ is said to be *separable* iff (Def.) it has an ONB with cardinality of $\mathbb{N}$, (this cardinal is denoted $\aleph_0$).

Many theorems stated first in the separable case also carry over to non-separable; but in the more general cases, there are both surprises, and, in some cases, substantial technical (set-theoretic) complications.

As a result, we shall make the blanket assumption that our Hilbert spaces *are* separable, unless stated otherwise.

Exercise 2.12 (ONBs and cardinality). Let $\mathscr{H}$ be a Hilbert space, not necessarily assumed separable, and let $\{u_\alpha\}_{\alpha \in A}$ and $\{v_\beta\}_{\beta \in B}$ be two ONBs for $\mathscr{H}$.

Show that A and B have the same cardinality, i.e., that there is a set-theoretic bijection of A onto B.

Definition 2.17.

1. Let A be a set, and $p : A \to \mathbb{R}_+$ a function on A. We say that the sum $\sum_{\alpha \in A} p(\alpha)$ is well-defined and finite iff (Def.)

$$\sup_{F \subset A, \, F \text{ finite}} \sum_{\alpha \in F} p(\alpha) < \infty;$$

and we set $\sum_{\alpha \in A} p(\alpha)$ equal to this supremum.

2. Let A be a set. By $l^2(A)$ we mean the set of functions $f : A \to \mathbb{C}$, such that

$$\sum_{\alpha \in A} |f(\alpha)|^2 < \infty.$$

Exercise 2.13 ($l^2(A)$). Let A be a set (general; not necessarily countable) then show that $l^2(A)$ is a Hilbert space.

Hint: For $f, g \in l^2(A)$, introduce the inner product $\sum_{\alpha \in A} \overline{f(\alpha)} g(\alpha)$, by using Cauchy-Schwarz for every finite subset of A.

Exercise 2.14 (A functor from sets to Hilbert space). Let A and B be sets, and let $\psi : A \to B$ be a bijective function, then show that there is an induced unitary isomorphism of $l^2(A)$ onto $l^2(B)$.

Example 2.4 (Wavelets). [10] Suppose $\mathscr{H}$ is separable (i.e., having a countable ONB), for instance let $\mathscr{H} = L^2(\mathbb{R})$. Then

$$\mathscr{H} \cong l^2(\mathbb{N}) \cong l^2(\mathbb{N} \times \mathbb{N}),$$

and it follows that potentially we could choose a doubly indexed basis

$$\{\psi_{j,k} : j, k \in \mathbb{N}\}$$

for $L^2(\mathbb{R})$. It turns out that this is precisely the setting of wavelet bases! What's even better is that in the l^2-spaces, there are all kinds of diagonalized

---

[10] A summary of some highpoints from wavelets is contained in Appendix B. See also Sections 2.4.2 (pg. 46), 5.8 (pg. 198), 6.4 (pg. 233), and Chapter 6 (pg. 223).

operators, which correspond to selfadjoint (or normal) operators in L^2. Among these operators in L^2, we single out the following two:

$$\text{scaling:} f(x) \xrightarrow{U_j} 2^{j/2} f(2^j x) \tag{2.47}$$

$$\text{translation:} f(x) \xrightarrow{V_k} f(x - k) \tag{2.48}$$

for all $j, k \in \mathbb{Z}$. However, U_j and V_k are NOT diagonalized *simultaneously* though. See below for details!

Remark 2.13. The two unitary actions U_j and V_k, $j, k \in \mathbb{Z}$, in (2.47) and (2.48) satisfy the following important commutation relation:

$$V_k U_j = U_j V_{2^j k}; \tag{2.49}$$

or equivalent:

$$U_j^{-1} V_k U_j = V_{2^j k}. \tag{2.50}$$

Verify details!

Definition 2.18. We say a rational number is a *dyadic fraction* or *dyadic rational* if it has the form of $\frac{a}{2^b}$, where $a \in \mathbb{Z}$, and $b \in \mathbb{N}$.

In the language of groups, the pair in (2.49) & (2.50) forms a representation of a *semidirect product*; or, equivalently, of the discrete dyadic $ax + b$ group (see Section 8.5.1 for more details): The latter group consists of all 2×2 matrices

$$\begin{bmatrix} 2^j & \frac{k}{2^l} \\ 0 & 1 \end{bmatrix}; \quad j, k \in \mathbb{Z}, \, l \in \mathbb{N}.$$

This group is often referred to as one of the *Baumslag-Solitar groups*; see e.g., [Dud14, DJ08].

2.4.3   Bounded Operators in Hilbert Space

The theory of operators in Hilbert space was developed by J. von Neumann[11], M. H. Stone (among others) with view to quantum physics; and it is worth to stress the direct influence of the pioneers in quantum physics such as P. A. M. Dirac, Max Born, Werner Heisenberg, and Erwin Schrödinger. Our aim here is to stress these interconnections. While the division of the topic into the case of bounded operators vs unbounded operators in Hilbert space is convenient from the point of view of mathematics, in applications "most" operators are unbounded, i.e., they are defined and linear on suitable dense domains in the respective Hilbert spaces. There are some technical advantages of restricting the discussion to the bounded case, for example, we get normed algebras, and we avoid technical issues with domains. In both the bounded and case, and the unbounded theory, the theory is still parallel to the way we use transformations

---

[11]Appendix C includes a short biographical sketch of von Neumann. See also [vN31, vN32a].

and matrices in linear algebra. But while the vector spaces of linear algebra are finite dimensional, the interesting Hilbert spaces will be *infinite dimensional*. Nonetheless we are still able to make use of matrices, only now they will be $\infty \times \infty$ matrices, see e.g., Figures 2.14-2.18, and the discussion around this theme. Among the benefits of the $\infty \times \infty$ matrices are their uses in the analysis of orthogonal polynomials; see e.g., Table 2.2 below.

Definition 2.19. A bounded operator in a Hilbert space $\mathscr{H}$ is a linear mapping $T : \mathscr{H} \to \mathscr{H}$ such that

$$\|T\| := \sup \left\{ \|Tx\|_{\mathscr{H}} : \|x\|_{\mathscr{H}} \leq 1 \right\} < \infty.$$

We denote by $\mathscr{B}(\mathscr{H})$ the algebra of all bounded operators in $\mathscr{H}$.

Setting $(ST)(v) = S(T(v))$, $v \in \mathscr{H}$, $S, T \in \mathscr{B}(\mathscr{H})$, we have

$$\|ST\| \leq \|S\| \|T\|.$$

Lemma 2.4 (Riesz). *There is a bijection $\mathscr{H} \ni h \longmapsto l_h$ between $\mathscr{H}$ and the space of all bounded linear functionals on $\mathscr{H}$, where*

$$
\begin{aligned}
l_h(x) &:= \langle h, x \rangle, \ \forall x \in \mathscr{H}, \ and \\
\|l_h\| &:= \sup \left\{ |l(x)| : \|x\|_{\mathscr{H}} \leq 1 \right\} < \infty.
\end{aligned}
$$

Moreover, $\|l_h\| = \|h\|$.

Corollary 2.4. *For all $T \in \mathscr{B}(\mathscr{H})$, there exists a unique operator $T^* \in \mathscr{B}(\mathscr{H})$, called the adjoint of T, such that*

$$\langle x, Ty \rangle = \langle T^*x, y \rangle, \ \forall x, y \in \mathscr{H};$$

and $\|T^*\| = \|T\|$.

Proof. Let $T \in \mathscr{B}(\mathscr{H})$, then it follows from the Cauchy-Schwarz inequality that

$$|\langle x, Ty \rangle| \leq \|x\| \|Ty\| \leq \|T\| \|x\| \|y\|.$$

Hence the mapping $y \longmapsto \langle x, Ty \rangle$ is a bounded linear functional on $\mathscr{H}$. By Riesz's theorem, there exists a unique $h_x \in \mathscr{H}$, such that $\langle x, Ty \rangle = \langle h_x, y \rangle$, for all $y \in \mathscr{H}$. Set $T^*x := h_x$. One checks that T^* linear, bounded, and in fact $\|T^*\| = \|T\|$. □

Exercise 2.15 (The C^* property). Let $T \in \mathscr{B}(\mathscr{H})$, then prove

$$\|T^*T\| = \|T\|^2. \tag{2.51}$$

Definition 2.20. Let $T \in \mathscr{B}(\mathscr{H})$. Then,

- T is *normal* if $TT^* = T^*T$

- T is *selfadjoint* if $T = T^*$

- T is *unitary* if $T^*T = TT^* = I_{\mathscr{H}}$ (= the identity operator)

- T is a (selfadjoint) *projection* if $T = T^* = T^2$.

For $T \in \mathscr{B}(\mathscr{H})$, we may write

$$R = \frac{1}{2}(T + T^*)$$
$$S = \frac{1}{2i}(T - T^*)$$

where both R and S are selfadjoint, and

$$T = R + iS.$$

This is similar to the decomposition of a complex number into its real and imaginary parts. Notice also that T is normal if and only if R and S commute. (Prove this!) Thus the study of a family of commuting normal operators is equivalent to the study of a family of commuting selfadjoint operators.

Exercise 2.16 (The group of all unitary operators). Let $\mathscr{H}$ be a fixed Hilbert space, and denote by $G_{\mathscr{H}}$ the unitary operators in $\mathscr{H}$ (see Definition 2.20).

1. Show that $G_{\mathscr{H}}$ is a group, and that $T^{-1} = T^*$ for all $T \in G_{\mathscr{H}}$.

2. Let $\{u_\alpha\}_{\alpha \in J}$ be an ONB in $\mathscr{H}$, and let $v_\alpha := T(u_\alpha)$, $\alpha \in J$; then show that $\{v_\alpha\}_{\alpha \in J}$ is also an ONB.

3. Show that, for any pair of ONBs $\{u_\alpha\}_J$, $\{w_\alpha\}_J$ with the same index set J, there is then a unique $T \in G_{\mathscr{H}}$ such that $w_\alpha = T(u_\alpha)$, $\alpha \in J$. We say that $G_{\mathscr{H}}$ acts *transitively* on the set of all ONBs in $\mathscr{H}$.

Theorem 2.7. *Let $\mathscr{H}$ be a Hilbert space. There is a one-to-one correspondence between selfadjoint projections and closed subspaces of $\mathscr{H}$ (Figure 2.11),*

$$[Closed\ subspace\ \mathscr{M} \subset \mathscr{H}] \longleftrightarrow Projections.$$

Proof. Let P be a selfadjoint projection in $\mathscr{H}$, i.e., $P^2 = P = P^*$. Then

$$\mathscr{M} = P\mathscr{H} = \{x \in \mathscr{H} : Px = x\}$$

is a closed subspace in $\mathscr{H}$. Let $P^\perp := I - P$ be the complement of P, so that

$$P^\perp \mathscr{H} = \{x \in \mathscr{H} : P^\perp x = x\} = \{x \in \mathscr{H} : Px = 0\}.$$

Since $PP^\perp = P(1 - P) = P - P^2 = P - P = 0$, we have $P\mathscr{H} \perp P^\perp \mathscr{H}$.

Conversely, let $\mathscr{W} \subsetneq \mathscr{H}$ be a closed subspace. Note the following "parallelogram law" holds:

$$\|x + y\|^2 + \|x - y\|^2 = 2(\|x\|^2 + \|y\|^2), \ \forall x, y \in \mathscr{H}; \qquad (2.52)$$

see Figure 2.10 for an illustration.

Let $x \in \mathscr{H} \backslash \mathscr{W}$, and set

$$d := \inf_{w \in \mathscr{W}} \|x - w\|.$$

The key step in the proof is showing that the infimum is attained; see Figure 2.11.

By definition, there exists a sequence $\{w_n\}$ in $\mathscr{W}$ so that $\|w_n - x\| \to 0$ as $n \to \infty$. Applying (2.52) to $x - w_n$ and $x - w_m$, we get

$$\|(x - w_n) + (x - w_m)\|^2 + \|(x - w_n) - (x - w_m)\|^2$$
$$= 2\left(\|x - w_n\|^2 + \|x - w_m\|^2\right);$$

which simplifies to

$$\|w_n - w_m\|^2 = 2\left(\|x - w_n\|^2 + \|x - w_m\|^2\right) - 4\left\|x - \frac{w_n + w_m}{2}\right\|^2$$
$$\leq 2\left(\|x - w_n\|^2 + \|x - w_m\|^2\right) - 4d. \tag{2.53}$$

Notice here all we require is $\frac{1}{2}(w_n + w_m) \in \mathscr{W}$, hence the argument carries over if we simply assume $\mathscr{W}$ is a closed convex subset in $\mathscr{H}$. We conclude from (2.53) that $\|w_n - w_m\| \to 0$, and so $\{w_n\}$ is a Cauchy sequence. Since $\mathscr{H}$ is complete, there is a unique limit,

$$Px := \lim_{n \to \infty} w_n \in \mathscr{W} \tag{2.54}$$

and

$$d = \|x - Px\| \left(= \inf_{w \in \mathscr{W}} \|x - w\|\right). \tag{2.55}$$

See Figure 2.11.

Set $P^{\perp}x := x - Px$. We proceed to verify that $P^{\perp}x \in \mathscr{W}^{\perp}$. By the minimizing property in (2.55), we have

$$\left\|P^{\perp}x\right\|^2 \leq \left\|P^{\perp}x + tw\right\|^2$$
$$= \left\|P^{\perp}x\right\|^2 + |t|^2 \|w\|^2 + t\left\langle P^{\perp}x, w\right\rangle + \bar{t}\left\langle w, P^{\perp}x\right\rangle \tag{2.56}$$

for all $t \in \mathbb{C}$, and all $w \in \mathscr{W}$. Assuming $w \neq 0$ (the non-trivial case), and setting

$$t = -\frac{\left\langle w, P^{\perp}x\right\rangle}{\|w\|^2}$$

in (2.56), it follows that

$$0 \leq -\frac{\left|\left\langle w, P^{\perp}x\right\rangle\right|^2}{\|w\|^2} \implies \left\langle w, P^{\perp}x\right\rangle = 0, \ \forall w \in \mathscr{W}.$$

This shows that $P^\perp x \in \mathscr{W}^\perp$, for all $x \in \mathscr{H}$.

For uniqueness, suppose P_1 and P_2 both have the stated properties, then for all $x \in \mathscr{H}$, we have

$$x = P_1 x + P_1^\perp x = P_2 x + P_2^\perp x; \text{ i.e.,}$$

$$P_1 x - P_2 x = P_2^\perp x - P_1^\perp x \in \mathscr{W} \cap \mathscr{W}^\perp = \{0\}$$

thus, $P_1 x = P_2 x$, $\forall x \in \mathscr{H}$.

We leave the rest to the reader. See, e.g., [Rud91], [Nel69, p.62].   □

Exercise 2.17 (Riesz). As a corollary to Theorem 2.7, prove the following version of Riesz' theorem. Let $\mathscr{H}$ be a fixed Hilbert space:

For every $l \in \mathscr{H}^*$, show that there is a unique $h \, (= h_l) \in \mathscr{H}$ such that

$$l(f) = \langle h, f \rangle, \; \forall f \in \mathscr{H}. \tag{2.57}$$

Exercise 2.18 (Lax-Milgram [Lax02]). Let $B : \mathscr{H} \times \mathscr{H} \to \mathbb{C}$ be sesquilinear, and suppose there is a finite constant c such that

$$|B(h, k)| \le c \, \|h\| \, \|k\| \, ;$$

and $b > 0$ such that

$$|B(h, h)| \ge b \, \|h\|^2, \; \forall h, k \in \mathscr{H}.$$

Then prove that, for every $h \in \mathscr{H}$, there is a unique $k \, (= k_h) \in \mathscr{H}$ such that

$$\langle h, f \rangle = B(k_h, f), \; \forall f \in \mathscr{H}. \tag{2.58}$$

Remark 2.14. In view of Riesz, Lax-Milgram[12] is an assertion about $\mathscr{H}^*$. The Lax-Milgram lemma was proved with view to solving elliptic PDEs, but in Chapter 9 we give an application to frame expansion.

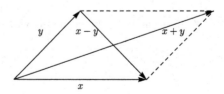

Figure 2.10: The parallelogram law.

---

[12]Appendix C includes a short biographical sketch of P. Lax.

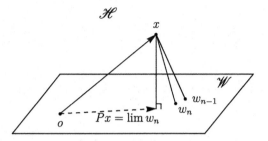

Figure 2.11: $\|x - Px\| = \inf\{\|x - w\| : w \in \mathscr{W}\}$. Projection from optimization in Hilbert space.

2.4.4 The Gram-Schmidt Process and Applications

Every Hilbert space has an ONB, but it does not mean in practice it is easy to select one that works well for a particular problem. The Gram-Schmidt orthogonalization process was developed a little earlier than von Neumann's formulation of abstract Hilbert space. It is an important tool to get an orthonormal set out of a set of linearly independent vectors.

Among the applications of the Gram-Schmidt algorithm we highlight below the theory of orthogonal polynomials: For the three families, the Legendre, Chebyshev, and Hermite polynomials, we outline the use of Gram-Schmidt in the derivation of recursive identities, of generating functions (applications to probability), and to spectral theory of ODEs.

Lemma 2.5 (Gram-Schmidt). *Let $\{u_n\}$ be a sequence of linearly independent vectors in $\mathscr{H}$, then there exists a sequence $\{v_n\}$ of unit vectors so that $\langle v_n, v_k \rangle = \delta_{n,k}$ and*

$$span\left\{u_k\right\}_{k=1}^{n} = span\left\{v_k\right\}_{k=1}^{n}$$

for all n; and therefore,

$$\overline{span}\left\{u_k\right\} = \overline{span}\left\{v_k\right\}.$$

Proof. Given $\{u_n\}$ as in the statement of the lemma, we set

$$v_1 = \frac{u_1}{\|u_1\|}.$$

$$v_2 = \frac{u_2 - \langle v_1, u_2 \rangle v_1}{\|u_2 - \langle v_1, u_2 \rangle v_1\|}, \cdots.$$

The inductive step: Suppose we have constructed the orthonormal set $F_n := \{v_1, \ldots, v_n\}$, and let P_{F_n} be the projection on F_n. For the induction step, we set

$$v_{n+1} := \frac{u_{n+1} - P_{F_n} u_{n+1}}{\|u_{n+1} - P_{F_n} u_{n+1}\|}, \quad n = 1, 2, \ldots \tag{2.59}$$

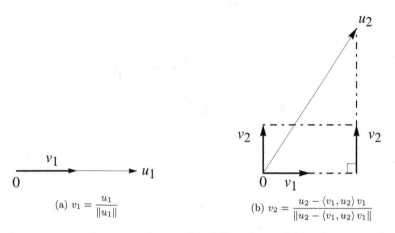

(a) $v_1 = \dfrac{u_1}{\|u_1\|}$

(b) $v_2 = \dfrac{u_2 - \langle v_1, u_2 \rangle\, v_1}{\|u_2 - \langle v_1, u_2 \rangle\, v_1\|}$

Figure 2.12: The first two steps in G-S.

See Figure 2.12. Note the LHS in (2.59) a unit vector, and orthogonal to $P_{F_n}\mathscr{H}$. The formula for P_{F_n}, the projection onto the span of F_n, is

$$P_{F_n} = \sum_{k=1}^{n} |v_k\rangle\langle v_k|.$$

$\square$

Remark 2.15. If $\mathscr{H}$ is *non-separable*, the standard induction does not work, and the transfinite induction is needed.

Example 2.5 (Legendre (see Table 2.2)). Let $\mathscr{H} = L^2(-1,1)$. The polynomials $\{1, x, x^2, \ldots\}$ are linearly independent in $\mathscr{H}$, for if

$$\sum_{k=1}^{n} c_k x^k = 0$$

then as an analytic function, the left-hand-side must be identically zero. By Stone-Weierstrass' theorem, $span\{1, x, x^2, \ldots\}$ is dense in $C([-1,1])$ under the $\|\cdot\|_\infty$ norm. Since $\|\cdot\|_{L^2} \le \|\cdot\|_\infty$, it follows that $span\{1, x, x^2, \ldots\}$ is also dense in $\mathscr{H}$.

By the Gram-Schmidt process, we get a sequence $\{V_n\}_{n=1}^{\infty}$ of finite dimensional subspaces in $\mathscr{H}$, where V_n has an orthonormal basis $\{h_0, \ldots, h_n\}$, so that

$$V_n = span\{1, x, \ldots, x^n\}$$
$$= span\{h_0, h_1, \ldots, h_n\}.$$

Details: Set $h_0 = \mathbb{1} = $ constant function, and

$$h_{n+1} := \frac{x^{n+1} - P_n x^{n+1}}{\|x^{n+1} - P_n x^{n+1}\|}, \ n \in \mathbb{N}.$$

Then the set $\{h_n : n \in \mathbb{N} \cup \{0\}\}$ is an ONB in $\mathscr{H}$. These are the Legendre polynomials, see Table 2.2.

The two important families of orthogonal polynomials on $(-1, 1)$, are in Table 2.2 and Figure 2.13. (See, e.g., [Sze59].)

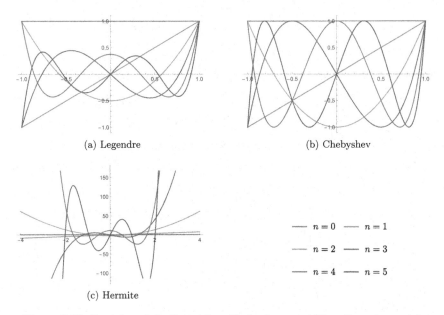

(a) Legendre

(b) Chebyshev

(c) Hermite

$$\text{---} \ n = 0 \quad \text{---} \ n = 1$$

$$\text{---} \ n = 2 \quad \text{---} \ n = 3$$

$$\text{---} \ n = 4 \quad \text{---} \ n = 5$$

Figure 2.13: Samples of the Legendre, Chebyshev, and Hermite polynomials.

Definition 2.21. Let $\{P_n(x)\}_{n \in \{0\} \cup \mathbb{N}}$ be a sequence of polynomials. We say that the expansion

$$G_P(x, t) = \sum_{n=0}^{\infty} P_n(x) t^n$$

is the corresponding *generating function*.

Exercise 2.19 (Generating functions). Show that the generating functions for the three cases of orthogonal polynomials in Table 2.2, *Legendre* (pg.58), *Chebyshev* (pg.58), and *Hermite* (pg.58) are as follows:

$$G_L(x, t) = \frac{1}{\sqrt{1 - 2xt + t^2}},$$

Name	Hilbert space	List
Legendre	$L^2\left(-1,1\right)$	$P_0\left(x\right)=1$
	$\|f\|_L^2=\int_{-1}^1|f\left(x\right)|^2\,dx,\ -1\le x\le 1$	$P_1\left(x\right)=x$
		$P_2\left(x\right)=\frac{1}{2}\left(3x^2-1\right)$
	orthogonal relation	$P_3\left(x\right)=\frac{1}{2}\left(5x^3-3x\right)$
	$\int_{-1}^1 P_n\left(x\right)P_k\left(x\right)\,dx=\delta_{n,k}\dfrac{2}{2n+1}$	$P_4\left(x\right)=\frac{1}{8}\left(35x^4-30x^2+3\right)$
		$\vdots$
Chebyshev	$L^2\left(\left(-1,1\right);\dfrac{dx}{\sqrt{1-x^2}}\right)$	$P_{n+1}\left(x\right)=2xP_n\left(x\right)-P_{n-1}\left(x\right)$
	$\|f\|_C^2=\int_{-1}^1|f\left(x\right)|^2\,\dfrac{dx}{\sqrt{1-x^2}}$	$P_n\left(\cos\theta\right)=\cos\left(n\theta\right)$
	$=\int_0^\pi|f\left(\theta\right)|^2\,d\theta,\ x\in\left[-1,1\right]$	$P_0\left(x\right)=1$
		$P_1\left(x\right)=x$
	orthogonal relation	$P_2\left(x\right)=2x^2-1$
	$\int_{-1}^1\dfrac{P_n\left(x\right)P_k\left(x\right)}{\sqrt{1-x^2}}\,dx$	$P_3\left(x\right)=4x^3-3x$
	$=\begin{cases}0 & n\ne k\\ \frac{\pi}{2} & n=k\ne 0\\ \pi & n=k=0\end{cases}$	$\vdots$
Hermite	$L^2\left(\mathbb{R},e^{-x^2}dx\right)$	(physics version)
	$\|f\|_H^2=\int_{-\infty}^\infty|f\left(x\right)|^2\,e^{-x^2}dx,\ x\in\mathbb{R}$	$P_n\left(x\right)=\left(-1\right)^n e^{x^2}\left(\frac{d}{dx}\right)^n e^{-x^2}$
		$P_0\left(x\right)=1$
	orthogonal relation	$P_1\left(x\right)=2x$
	$\int_{-\infty}^\infty P_n\left(x\right)P_m\left(x\right)e^{-x^2}dx$	$P_2\left(x\right)=4x^2-2$
	$=\ \sqrt{\pi}\,2^n\,n!\,\delta_{n,m}$	$P_3\left(x\right)=8x^3-12x$
		$\vdots$

Table 2.2: ORTHOGONAL POLYNOMIALS. Legendre, Chebyshev, Hermite.

$$G_C(x,t) = \frac{1 - xt}{1 - 2xt + t^2}, \text{ and}$$

$$G_H(x,t) = \sum_{\substack{n=0 \\ \text{(modified)}}}^{\infty} P_n^{(H)}(x) \frac{t^n}{n!} = \exp\left(2xt - t^2\right).$$

Make use of Taylor's expansion; see also [Akh65].

There is an ordinary differential equation (ODE) associated with each of the three families of orthogonal polynomials, those named after Legendre, Chebyshev, and Hermite, see Table 2.2. Each arises as solutions to an eigenvalue problem for accompanying ordinary differential equations; i.e., an ODE for each of the three families of orthogonal polynomials. These ODEs in turn are frequently encountered in numerical analysis, in physics, and in many other applications. They are as follows:

For the Legendre polynomials, the ODE is

$$\frac{d}{dx}\left[\left(1 - x^2\right)\frac{d}{dx}P_n(x)\right] + n(n+1)P_n(x) = 0.$$

For the Chebyshev polynomials (of the first kind), the ODE is

$$\left(1 - x^2\right)\frac{d^2}{dx^2}P_n(x) - x\frac{d}{dx}P_n(x) + n^2 P_n(x) = 0.$$

And for the Hermite polynomials, the ODE is as follows:

$$y'' - 2xu' = -2\lambda u.$$

Exercise 2.20 (Recursive identities). Verify the following recursive identities for the three classes of orthogonal polynomials:
Legendre:

$$(n+1)P_{n+1}(x) = (2n+1)xP_n(x) - nP_{n-1}(x)$$

Chebyshev:

$$P_{n+1}(x) = 2xP_n(x) - P_{n-1}(x),$$
$$2P_m(x)P_n(x) = P_{m+n}(x) + P_{m-n}(x)$$

Hermite:

$$P_{n+1}(x) = 2xP_n(x) - P_n'(x) \text{ (derivative)}$$
$$= 2xP_n(x) - 2nP_{n-1}(x),$$
$$P_n(x+y) = 2^{-\frac{n}{2}}\sum_{k=0}^{n}\binom{n}{k}P_{n-k}(x\sqrt{2})P_k(y\sqrt{2}).$$

(For details regarding the essential properties of the Hermite polynomials, see e.g., [AAR99, pages 278-282].)

Exercise 2.21 (Legendre, Chebyshev, Hermite, and Jacobi). Find the three $\infty \times \infty$ Jacobi matrices J associated with the three systems of polynomials in Table 2.2.

Hint: Before writing down the three respective matrices J, you must first normalize the polynomials with respect to the respective Hilbert norms. See also [Sho36].

We shall return to the Hermite polynomials, and a corresponding system, the Hermite functions, in Examples 4.6 and 4.7, where they are used in a detailed analysis of the canonical commutation relations, see also Section 2.1 above; as well as the corresponding harmonic oscillator Hamiltonian H. With the use of raising and lowering operators, we show that the Hermite functions are eigenfunctions for H, and we derive the spectrum for H this way.

Exercise 2.22 (The orthogonality rules). Verify the orthogonality rules contained in Table 2.2.

Hint: You can use direct computations, a clever system of recursions, or Fourier transform (generating function).

Exercise 2.23 (M_x in Jacobi form). Let $J \subset \mathbb{R}$ be an interval (finite, or infinite), and let $\{p_n(x)\}_{n=0}^{\infty}$ be a system of polynomial functions on J; then show that there is a positive Borel measure μ on J, with infinite support, but moments of all orders, such that $\{p_n(x)\}_{n=0}^{\infty}$ is an ONB in $L^2(J, \mu)$, if and only if, the *multiplication operator* M_x has an $\infty \times \infty$ matrix representation, with $\alpha_n \in \mathbb{R}$, $\beta_n \in \mathbb{C}$, and

$$\beta_1 p_1(x) = (x - \alpha_0) p_0(x)$$
$$\beta_2 p_2(x) = (x - \alpha_1) p_1(x) - \overline{\beta_1} p_0(x)$$
$$\vdots$$
$$\beta_{n+1} p_{n+1}(x) = (x - \alpha_n) p_n(x) - \overline{\beta_n} p_{n-1}(x);$$

i.e., with Jacobi matrix given in Figure 2.14.

$$
J := \begin{pmatrix}
\alpha_0 & \overline{\beta_1} & 0 & & & & \mathbf{0} & \\
\beta_1 & \alpha_1 & \overline{\beta_2} & 0 & & & & \\
0 & \beta_2 & \alpha_2 & \overline{\beta_3} & \ddots & & & \\
& \ddots & \ddots & \ddots & \ddots & \ddots & & \\
& & 0 & \beta_{n-1} & \alpha_{n-1} & \overline{\beta_n} & 0 & \\
& & & 0 & \beta_n & \alpha_n & \overline{\beta_{n+1}} & \ddots \\
\mathbf{0} & & & & 0 & \beta_{n+1} & \alpha_{n+1} & \ddots \\
& & & & & \ddots & \ddots & \ddots
\end{pmatrix}
$$

Figure 2.14: An $\infty \times \infty$ matrix representation of M_x in Exercise 2.23.

Example 2.6 (Fourier basis). Let $\mathscr{H} = L^2[0,1]$. Consider the set of complex exponentials

$$\left\{ e^{i2\pi nx} : n \in \mathbb{Z} \right\},$$

or equivalently, one may also consider

$$\{1, \; \cos 2\pi nx, \; \sin 2\pi nx \; : \; n \in \mathbb{N}\}.$$

This is already an ONB in $\mathscr{H}$ and leads to Fourier series.

In the next example we construct the *Haar wavelet.*

Definition 2.22. A function $\psi \in L^2(\mathbb{R})$ is said to generate a wavelet if

$$\psi_{j,k}(x) = 2^{j/2}\psi\left(2^j x - k\right), \; j, k \in \mathbb{Z} \tag{2.60}$$

is an ONB in $L^2(\mathbb{R})$.

Note with the normalization in (2.60) we get

$$\int_{\mathbb{R}} |\psi_{j,k}(x)|^2 \, dx = \int_{\mathbb{R}} |\psi(x)|^2 \, dx, \; \forall j, k \in \mathbb{Z}.$$

Example 2.7 (Haar wavelet and its orthogonality relations). Let $\mathscr{H} = L^2(0,1)$, and let φ_0 be the characteristic function of the unit interval $[0,1]$. φ_0 is called a scaling function. Define

$$\begin{aligned} \varphi_1 &:= \varphi_0(2x) - \varphi_0(2x - 1), \text{ and} \tag{2.61} \\ \psi_{j,k} &:= 2^{j/2}\varphi_1(2^j x - k), \; j, k \in \mathbb{Z}. \tag{2.62} \end{aligned}$$

Claim: $\{\psi_{j,k} : j, k \in \mathbb{Z}\}$ is an orthonormal set (in fact, an ONB) in $L^2(0,1)$, when j, k are restricted as follows:

$$k = 0, 1, \dots, 2^j - 1, \; j \in \mathbb{N} \cup \{0\}.$$

Proof. Fix k, if $j_1 \neq j_2$, then $\psi_{j_1,k}$ and $\psi_{j_2,k}$ are orthogonal since their supports are nested. For fixed j, and $k_1 \neq k_2$, then ψ_{j,k_1} and ψ_{j,k_2} have disjoint supports, and so they are also orthogonal (see Figure 2.15)[13]. $\qquad\square$

Exercise 2.24 (The Haar wavelet, and multiplication by t). Let $M := M_t$ be the multiplication operator in L^2 of the unit interval, $L^2[0,1] \xrightarrow{M} L^2[0,1]$. Now compute the $\infty \times \infty$ matrix of M relative to the orthogonal Haar wavelet basis in Example 2.7.

---

[13]A summary of some highpoints from multiresolutions is contained in Appendix B. See also Sections 2.4 (pg. 39), 6.6 (pg. 241); and Chapter 6 (pg. 223).

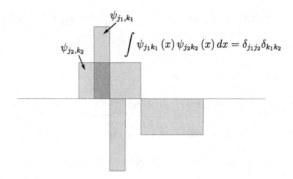

Figure 2.15: Haar wavelet. Scaling properties, multiresolutions.

Remark 2.16. Even though M_t has continuous spectrum $[0,1]$, uniform multiplicity, it is of interest to study the diagonal part in an $\infty \times \infty$ matrix representation of M_t. Indeed, in the wavelet ONB in $L^2(0,1)$ we get the following $\infty \times \infty$ matrix representation

$$(M_t)_{(j_1,k_1)(j_2,k_2)} = \int_0^1 \psi_{j_1,k_1}(t)\, t\psi_{j_2,k_2}(t)\, dt.$$

The diagonal part D consists of the sequence

$$D(jk) = \int_0^1 t\, (\psi_{j,k}(t))^2\, dt.$$

Anderson's theorem [And79b] states that $M_t - D \in \mathscr{K}$ (the compact operators in $L^2(0,1)$.) Indeed, Anderson computed the variance

$$V_{jk} = \int_0^1 t^2 \psi_{j,k}^2(t)\, dt - \left(\int_0^1 t\psi_{j,k}^2(t)\, dt\right)^2 = \frac{1}{12} 2^{-2j} \qquad (2.63)$$

for all $j \in \mathbb{N} \cup \{0\}$, and all $k \in \{0,1,2,\cdots,2^j - 1\}$.

Generally: It is known that if A is a selfadjoint operator acting in a separable Hilbert space, then $A = D + K$, where D is a diagonal operator, and K is a compact perturbation [vN35]. (In fact, there is even a representation $A = D + K$, where K is a Hilbert-Schmidt operator. See Section 2.5.1 and Chapter 4.)

Note that the multiplication operator M_t in Exercise 2.24 is bounded and selfadjoint in $L^2(0,1)$. Different ONBs will yield different diagonal representations D. But the wavelet basis is of special interest.

For more details on compact perturbation of linear operators in Hilbert space, we refer to [And74].

The conclusion from Anderson is of special interests as the function t on $[0,1]$ is as "nice" as can be, while the functions from the wavelet ONB (2.62) are wiggly, and in fact get increasingly more wiggly as the scaling degree j in the wavelet ONB tends to infinity. The scaling degree j is log to the base 2 of the frequency applied to the mother wavelet function (2.61). The conclusion from Anderson is that the variance numbers (2.63) fall off as the inverse square of the frequency.

Remark 2.17. It is of interest to ask the analogous questions for other functions than t, and for other wavelet bases, other than the Haar wavelet basis.

Exercise 2.25 (A duality). Let z be a complex number, and P be a selfadjoint projection. Show that $U(z) = zP + (I - P)$ is unitary if and only if $|z| = 1$.

Hint: $U(z)U(z)^* = U(z)^*U(z) = |z|^2 P + (I - P)$, so $U(z)$ is unitary iff $|z| = 1$. In block matrix form,

$$U(z) = \begin{pmatrix} z & 0 \\ 0 & 1 \end{pmatrix} \begin{matrix} P \\ P^\perp \end{matrix} \quad .$$

$$\begin{matrix} P & P^\perp \end{matrix}$$

2.5 Dirac's Notation

> "There is a great satisfaction in building good tools for other people to use."
> — Freeman Dyson

P.A.M. Dirac[14] was very efficient with notation, and he introduced the "bra-ket" vectors [Dir35, Dir47].

This Dirac formalism has proved extraordinarily efficient, and it is widely used in physics. It deserves to be better known in the math community.

Definition 2.23. Let $\mathscr{H}$ be a Hilbert space with inner product $\langle \cdot, \cdot \rangle$. We denote by "bra" for vectors $\langle x|$ and "ket" for vectors $|y\rangle$, for $x, y \in \mathscr{H}$.

With Dirac's notation, our first observation is the following lemma.

Lemma 2.6. *Let $v \in \mathscr{H}$ be a unit vector. The operator $x \mapsto \langle v, x \rangle v$ can be written as $P_v = |v\rangle\langle v|$, i.e., a "ket-bra" vector (Definition 2.23). And P_v is a rank-one selfadjoint projection.*

Proof. First, we see that

$$P_v^2 = (|v\rangle\langle v|)(|v\rangle\langle v|) = |v\rangle\langle v, v\rangle\langle v| = |v\rangle\langle v| = P_v.$$

Also, if $x, y \in \mathscr{H}$ then

$$\langle x, P_v y \rangle = \langle x, v \rangle \langle v, y \rangle = \left\langle \overline{\langle x, v \rangle} v, y \right\rangle = \langle \langle v, x \rangle v, y \rangle = \langle P_v x, y \rangle$$

so $P_v = P_v^*$. □

---

[14]Appendix C includes a short biographical sketch of P. Dirac.

Corollary 2.5. *Let $F = span\{v_i\}$ with $\{v_i\}$ a finite set of orthonormal vectors in $\mathscr{H}$, then*

$$P_F := \sum_{v_i \in F} |v_i\rangle\langle v_i|$$

is the selfadjoint projection onto F.

Proof. Indeed, we have

$$P_F^2 = \sum_{v_i, v_j \in F} (|v_i\rangle\langle v_i|)(|v_j\rangle\langle v_j|) = \sum_{v_i \in F} |v_i\rangle\langle v_i| = P_F$$

$$P_F^* = \sum_{v_i \in F} (|v_i\rangle\langle v_i|)^* = \sum_{v_i \in F} |v_i\rangle\langle v_i| = P_F,$$

and we have

$$P_F w = w \iff \sum_F |\langle v_i, w\rangle|^2 = \|w\|^2 \tag{2.64}$$

$$\iff w \in F.$$

Since we may take the limit in (2.64), it follows that the assertion also holds if F is infinite-dimensional, i.e., $F = \overline{span}\{v_i\}$, closure. $\qquad\square$

Remark 2.18. More generally, any rank-one operator can be written in Dirac notation as

$$|u\rangle\langle v| : \mathscr{H} \ni x \longmapsto \langle v, x\rangle u \in \mathscr{H}.$$

With the bra-ket notation, it is easy to verify that the set of rank-one operators forms an algebra, which easily follows from the fact that

$$(|v_1\rangle\langle v_2|)(|v_3\rangle\langle v_4|) = \langle v_2, v_3\rangle |v_1\rangle\langle v_4|.$$

The moment that an orthonormal basis is selected, the algebra of operators on $\mathscr{H}$ will be translated to the algebra of matrices (infinite). See Lemma 2.8.

Exercise 2.26 (Finite-rank reduction). Let $\mathscr{H}$ be a Hilbert space. For all $x, y \in \mathscr{H}$, let $|x\rangle\langle y|$ denote the corresponding (Dirac) rank-1 operator.

    1. Let $A, B \in \mathscr{B}(\mathscr{H})$. Verify that

$$A|x\rangle\langle y| = |Ax\rangle\langle y|, \text{ and} \tag{2.65}$$

$$|x\rangle\langle y| B = |x\rangle\langle B^*y|. \tag{2.66}$$

    In particular, $\mathscr{F}R(\mathscr{H})$ is a two-sided ideal in $\mathscr{B}(\mathscr{H})$.

    2. For all $x, y \in \mathscr{H}$, set

$$w_{x,y}(A) := \langle x, Ay\rangle, \quad \forall A \in \mathscr{B}(\mathscr{H}).$$

For $\|x\| = 1$, set $w_x(A) := \langle x, Ax \rangle$, i.e., a *pure state* on $\mathscr{B}(\mathscr{H})$.
Let $\{x_i\}_{i=1}^n \subset \mathscr{H}$, $\|x_i\| = 1$, and set T by

$$T = \sum_i |x_i\rangle\langle x_i|. \tag{2.67}$$

Show that then

$$TAT = \sum_i \sum_j w_{x_i, x_j}(A) \underbrace{|x_i\rangle\langle x_j|}_{\text{Dirac-rank-1}}. \tag{2.68}$$

In particular, if $n = 1$, and $T = |x_1\rangle\langle x_1|$, we have:

$$TAT = w_{x_1}(A) T. \tag{2.69}$$

3. Use part (1), and Theorem 2.8 below, to give a quick proof that the compact operators form an ideal in $\mathscr{B}(\mathscr{H})$.

Exercise 2.27 (Numerical Range and Toeplitz-Hausdorff). The set

$$\{w_x(A) : \|x\| = 1\} \subset \mathbb{C}$$

is called the *numerical range* of A, NR_A. Show that NR_A is convex.

Hint: Difficult! It is called the *Toeplitz-Hausdorff theorem*; see e.g., [Hal82, Hal64]. (There are few assertions that are true for *all* bounded operators. The Toeplitz-Hausdorff theorem is one on a short list.)

Lemma 2.7. *Let* $\{u_\alpha\}_{\alpha \in J}$ *be an ONB in* $\mathscr{H}$, *then we may write*

$$I_{\mathscr{H}} = \sum_{\alpha \in J} |u_\alpha\rangle\langle u_\alpha|.$$

Proof. This is equivalent to the decomposition

$$v = \sum_{\alpha \in J} \langle u_\alpha, v \rangle u_\alpha, \ \forall v \in \mathscr{H}.$$

$\square$

A selection of ONB makes a representation of the algebra of operators acting on $\mathscr{H}$ by infinite matrices. We check that, using Dirac's notation (Definition 2.23), the algebra of operators really becomes the algebra of *infinite matrices*.

Definition 2.24. For $A \in \mathscr{B}(\mathscr{H})$, and $\{u_i\}$ an ONB, set

$$(M_A)_{i,j} = \langle u_i, Au_j \rangle_{\mathscr{H}}.$$

Most of the operators we use in mathematical physics problems are unbounded, so it is a big deal that the conclusion about matrix product is valid for unbounded operators subject to the condition that the chosen ONB is in the domain of such operators.

Lemma 2.8 (Matrix product). *Assume some ONB* $\{u_i\}_{i \in J}$ *satisfies*

$$u_i \in dom\,(A^*) \cap dom\,(B)\,;$$

then $M_{AB} = M_A M_B$, *i.e.*, $(M_{AB})_{ij} = \sum_k (M_A)_{ik}\,(M_B)_{kj}$.

Proof. By AB we mean the operator given by

$$(AB)\,(u) = A\,(B\,(u))\,.$$

Pick an ONB $\{u_i\}$ in $\mathscr{H}$, and the two operators as stated. We denote by $M_A = A_{ij} := \langle u_i, Au_j \rangle$ the matrix of A under the ONB. We compute $\langle u_i, ABu_j \rangle$.

$$
\begin{aligned}
(M_A M_B)_{ij} = \sum_k A_{ik} B_{kj} &= \sum_k \langle u_i, Au_k \rangle \langle u_k, Bu_j \rangle \\
&= \sum_k \langle A^* u_i, u_k \rangle \langle u_k, Bu_j \rangle \\
&= \langle A^* u_i, Bu_j \rangle \quad \text{[by Parseval]} \\
&= \langle u_i, ABu_j \rangle \\
&= (M_{AB})_{ij}
\end{aligned}
$$

where we used that $I = \sum |u_i \rangle \langle u_i|$. $\qquad\qquad\qquad\qquad\qquad\qquad\qquad\square$

Exercise 2.28 (Matrix product of $\infty \times \infty$ banded matrices). Consider two linear operators A and B both defined on a dense subspace $\mathscr{D}$ in a fixed Hilbert space $\mathscr{H}$. Suppose $\mathscr{D}$ contains an ONB $\{e_i\}_{i \in \mathbb{N}}$, and that the corresponding matrices M_A and M_B with respect to $\{e_i\}$ are both banded. Then show that the matrix-product

$$M_{AB} = M_A M_B \tag{2.70}$$

is well defined, and is again banded. See Figure 2.16.

Hint: Use Lemma 2.8, and the equation

$$(M_{AB})_{i,j} = \langle e_i, ABe_j \rangle = \langle A^* e_i, Be_j \rangle\,,$$

and note that $\mathscr{D} \subset dom\,(A^*)$.

Remark 2.19. Here are two *open questions* regarding *banded* operators/matrices.

1. Fix a separable Hilbert space $\mathscr{H}$. Is there an intrinsic geometric characterization of the linear operators A (with dense domain) in $\mathscr{H}$ which admit an ONB $\{e_i\}_{i \in \mathbb{N}}$ such that the matrix

$$(M_A)_{i,j} := \langle e_i, Ae_j \rangle \tag{2.71}$$

   is banded?

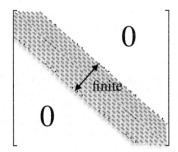

Figure 2.16: $\infty \times \infty$ *banded matrix*. Supported on a band around the diagonal.

2. Given an ONB $\{e_i\}_{i\in\mathbb{N}}$, what is the $*$-algebra $\mathfrak{A}$ of unbounded operators with dense domain

$$\mathscr{D} = span\,\{e_i\} \tag{2.72}$$

such that every $A \in \mathfrak{A}$ is banded with respect to $\{e_i\}_{i\in\mathbb{N}}$?

Exercise 2.29 (A transform). Let $\mathscr{H}$ be a Hilbert space, and A a set which indexes a fixed ONB $\{v_\alpha\}_{\alpha\in A}$. Now define $T : \mathscr{H} \to l^2\,(A)$, by

$$(Th)\,(\alpha) := (\langle v_\alpha, h\rangle)\,,\ \forall \alpha \in A,\ h \in \mathscr{H}.$$

Show that T is unitary, and onto $l^2\,(A)$, i.e.,

$$TT^* = I_{l^2(A)},\ \text{and}$$
$$T^*T = I_{\mathscr{H}}.$$

T is called the analysis transformation, and T^* the synthesis transformation.

2.5.1 Three Norm-Completions

Let $\mathscr{H}$ be a fixed Hilbert space, infinite-dimensional in the discussion below. Let

$$\mathscr{F}R\,(\mathscr{H}) = \{\text{all finite-rank operators } \mathscr{H} \longrightarrow \mathscr{H}\}$$
$$= span\,\{|v\,\rangle\langle\,w| \ :\ v, w \in \mathscr{H}\}$$

where $|v\,\rangle\langle\,w|$ denotes the Dirac ket-bra-operator.

Definition 2.25. On $\mathscr{F}R\,(\mathscr{H})$ we introduce the following three norms: the *uniform norm* (UN), the *trace-norm* (TN), and the *Hilbert-Schmidt* norm as follows:

- (UN) For all $T \in \mathscr{F}R\,(\mathscr{H})$, set

$$\|T\|_{UN} := \sup\,\{\|Tv\| \ :\ \|v\| = 1\}.$$

- (TN) Set

$$\|T\|_{TN} := \text{trace}\left(\sqrt{T^*T}\right);$$

- (HSN) Set

$$\|T\|_{HSN} = (\text{trace}\,(T^*T))^{\frac{1}{2}}.$$

Theorem 2.8. *The completion of $\mathscr{F}R(\mathscr{H})$ with respect to $\|\cdot\|_{UN}$, $\|\cdot\|_{TN}$, and $\|\cdot\|_{HSN}$ are respectively the compact operators, the trace-class operators, and the Hilbert-Schmidt operators. (See Definition 2.25 above.)*

We will prove, as a consequence of the Spectral Theorem (Chapter 4), that the $\|\cdot\|_{UN}$- completion agrees with the usual definition of the compact operators.

Remark 2.20. Note that Theorem 2.8 is for Hilbert space, and it is natural to ask "what carries over to Banach space?" Not everything by a theorem of Per Enflo [Enf73]. In detail: The assertion in the first part of Theorem 2.8 (Hilbert space) is that every compact operator is the norm limit of finite-rank operators; referring to the uniform norm (UN) in Definition 2.25. But there are Banach spaces where this is false; — although the easy implication is true, i.e., that operators in the norm closure of finite-rank operators are compact.

Definition 2.26. Let $T \in \mathscr{B}(\mathscr{H})$, then we say that T is compact iff (Def.) $T(\mathscr{H}_1)$ is relatively compact in $\mathscr{H}$, where $\mathscr{H}_1 := \left\{v \in \mathscr{H} \mid \|v\| \leq 1\right\}$.

A similar remark applies to the other two Banach spaces of operators. The Hilbert-Schmidt operators forms a Hilbert space.

Exercise 2.30 (The identity-operator in infinite dimension). If $\dim \mathscr{H} = \infty$, show that the identity operator $I_{\mathscr{H}}$ is *not* compact.

 Hint: Use an ONB.

The following is useful in working through the arguments above.

Lemma 2.9. *Let $v, w \in \mathscr{H}$, then*

$$trace\,(|v\rangle\langle w|) = \langle v, w\rangle_{\mathscr{H}}.$$

Proof. Introduce an ONB $\{u_\alpha\}$ in $\mathscr{H}$, and compute:

$$\text{trace}\,(|v\rangle\langle w|) = \sum_\alpha \langle u_\alpha, |v\rangle\langle w|\, u_\alpha\rangle_{\mathscr{H}}$$

$$= \sum_\alpha \langle u_\alpha, v\rangle_{\mathscr{H}} \langle w, u_\alpha\rangle_{\mathscr{H}} = \langle v, w\rangle_{\mathscr{H}}$$

where we used Parseval in the last step of the computation. $\qquad\square$

Exercise 2.31 (Comparing norms). Let $T \in \mathscr{F}R(\mathscr{H})$, and let the three norms be as in Definition 2.25. Then show that

$$\|T\|_{UN} \leq \|T\|_{HSN} \leq \|T\|_{TN}; \tag{2.73}$$

and conclude the following contractive inclusions:

$$\{\text{trace-class operators}\} \subset \{\text{Hilbert-Schmidt operators}\}$$
$$\subset \{\text{compact operators}\} .$$

Remark 2.21. In the literature, the following notation is often used for the three norms in Definition 2.25:

$$\begin{cases} \|T\|_{UN} = \|T\|_\infty \\ \|T\|_{TN} = \|T\|_1 \\ \|T\|_{HSN} = \|T\|_2 . \end{cases} \tag{2.74}$$

Note that if T is a diagonal operator, $T = \sum_k x_k |u_k\rangle\langle u_k|$ in some ONB $\{u_k\}$, then the respective norms are $\|x\|_\infty$, $\|x\|_1$, and $\|x\|_2$.

With the notation in (2.74), the inequalities (2.73) now take the form

$$\|T\|_\infty \leq \|T\|_2 \leq \|T\|_1 , \ T \in \mathscr{F}R(\mathscr{H}) . \tag{2.75}$$

Exercise 2.32 (Matrix entries in infinite dimensions). Let $\mathscr{H}$ be a separable Hilbert space. Pick an ONB $\{e_j\}_{j\in J}$, and set $E_{ij} := |e_i\rangle\langle e_j|$, $(i,j) \in J^2$.

1. Show that this is an ONB in $\mathscr{H}S(\mathscr{H})$ (= Hilbert Schmidt operators) with respect to the inner product

$$\langle A, B\rangle_{\mathscr{H}S} := trace\,(A^*B) , \ A, B \in \mathscr{H}S(\mathscr{H}) . \tag{2.76}$$

2. Show that the corresponding orthogonal expansion for $A \in \mathscr{H}S(\mathscr{H})$ is

$$A = \sum_{(i,j)\in J^2}\sum \langle e_i, Ae_j\rangle_{\mathscr{H}}\, E_{ij}. \tag{2.77}$$

Exercise 2.33 (The three steps). Let $\mathscr{H}$ be a separable Hilbert space, and let

1. $\mathscr{B}(\mathscr{H})$: all bounded operators in $\mathscr{H}$; with the uniform norm.

2. $\mathscr{T}_1(\mathscr{H})$: all trace-class operators, with the trace-norm; see Definition 2.25.

Use the three steps from Section 2.2 to show that

$$(\mathscr{T}_1(\mathscr{H}))^* = \mathscr{B}(\mathscr{H}) ;$$

i.e., that $\mathscr{B}(\mathscr{H})$ is the dual Banach space where the respective norms are specified as in (1)-(2). (For more about this duality, see also Theorem 5.7.)

2.5.2   Connection to Quantum Mechanics

One of the powerful applications of the theory of operators in Hilbert space, and more generally of functional analysis, is in quantum mechanics (QM) [Pol02, PK88, CP82]. Even the very formulation of the central questions in QM entails the mathematics of *unbounded selfadjoint operators*, of projection valued measures (PVM), and *unitary one-parameter groups* of operators[15]. The latter usually abbreviated to "unitary one-parameter groups."

By contrast to what holds for the more familiar case of bounded operators, we stress that for *unbounded* selfadjoint operators, mathematical precision necessitates making a sharp distinction between the following three notions: *selfadjoint*, *essentially selfadjoint*, and *formally selfadjoint* (also called Hermitian, or symmetric). See [RS75, Nel69, vN32a, DS88b]. We define them below; see especially the appendix at the end of Chapter 3. One reason for this distinction is that *quantum mechanical observables*, momentum P, position Q, energy etc, become *selfadjoint operators* in the axiomatic language of QM. What makes it even more subtle is that these operators are both unbounded and non-commuting (take the case of P and Q which was pair from Heisenberg's pioneering paper on uncertainty.) Another subtle point entails the relationship between selfadjoint operators, projection-valued measures, and unitary one-parameter groups (as used in the dynamical description of states in QM, i.e., describing the solution of the wave equation of Schrödinger.) Unitary one-parameter groups are also used in the study of other partial differential equations, especially hyperbolic PDEs.

The discussion which follows below will make reference to this setting from QM, and it serves as motivation. The more systematic mathematical presentation of *selfadjoint operators*, *projection-valued measures*, and *unitary one-parameter groups* will be postponed to later in the book. We first need to develop a number of technical tools. However we have included an outline of the bigger picture in the appendix (Stone's Theorem), to the present chapter. Stone's theorem shows that the following three notions, (i) selfadjoint operator, (ii) projection-valued measure, and (iii) unitary one-parameter group, are incarnations of one and the same; i.e., when one of the three is known, anyone of the other two can be computed from it.

We emphasize that there is a host of other applications of this, for example to harmonic analysis, to statistics, and to PDE. These will also play an important role in later chapters. See Chapters 7, 8, 11, and 12.

Much of the motivation for the axiomatic approach to the theory of linear operators in Hilbert space dates back to the early days of quantum mechanics (Planck, Heisenberg [Hei69], and Schrödinger [Sch32, Sch40, Sch99]), but in the form suggested by J. von Neumann [vN31, vN32a, vN32c, vN35]. (von Neumann's formulation is the one now adopted by most books on functional analysis.) Here we will be brief, as a systematic and historical discussion is far beyond our present scope. Suffice it to mention here that what is known as the

---

[15]A summary of some highpoints from quantum mechanics is contained in Appendix B. See also Sections 2.5.2 (pg. 70), 3.1 (pg. 90), 4.1 (pg. 120), and Chapter 9 (pg. 333).

"matrix-mechanics" of Heisenberg takes the form infinite by infinite matrices with entries representing, in turn, *transition probabilities*, where "transition" refers to "jumps" between energy levels. See (2.80)-(2.81) below, and Exercises 2.34-2.35. By contrast to matrices, in Schrödinger's wave mechanics, the Hilbert space represents wave solutions to Schrödinger's equation. Now this entails the study of one-parameter groups of unitary operators in $\mathscr{H}$. In modern language, with the two settings we get the dichotomy between the case when the Hilbert space is an l^2-space (i.e., a l^2-sequence space), vs the case of Schrödinger when $\mathscr{H}$ is an L^2-space of functions on phase-space.

In both cases, the observables are represented by families of selfadjoint operators in the respective Hilbert spaces. For the purpose here, we pick the pair of selfadjoint operators representing momentum (denoted P) and position (denoted Q). In one degree of freedom, we only need a single pair. The canonical commutation relation is

$$PQ - QP = -i\,I;\ \text{or}\ PQ - QP = -i\,\hbar\,I$$

where $\hbar = \frac{h}{2\pi}$ is Planck's constant, and $i = \sqrt{-1}$.

A few years after the pioneering work of Heisenberg and Schrödinger, J. von Neumann and M. Stone proved that the two approaches are *unitarily equivalent*, hence they produce the same "measurements." In modern lingo, the notion of measurement takes the form of projection valued measures, which in turn are the key ingredient in the modern formulation of the spectral theorem for selfadjoint, or normal, linear operators in Hilbert space. (See [Sto90, Yos95, Nel69, RS75, DS88b].) Because of dictates from physics, the "interesting" operators, such as P and Q are *unbounded*.

The first point we will discuss about the pair of operators P and Q is non-commutativity. As is typical in mathematical physics, non-commuting operators will satisfy conditions on the resulting commutators. In the case of P and Q, the commutation relation is called the canonical commutation relation; see below. For reference, see [Dir47, Hei69, vN31, vN32c].

Quantum mechanics was born during the years from 1900 to 1933. It was created to explain phenomena in black body radiation, and for the low energy levels of the hydrogen atom, where a discrete pattern occurs in the frequencies of waves in the radiation. The radiation energy turns out to be $E = \nu\hbar$, with $\hbar$ being the Plank's constant, and ν is frequency. Classical mechanics ran into trouble in explaining experiments at the atomic level.

In response to these experiments, during the years of 1925 and 1926, Heisenberg found a way to represent the energy E as a matrix (spectrum = energy levels), so that the matrix entries $\langle v_j, Ev_i \rangle$ represent transition probability for transitions from energy level i to energy level j. (See Figure 2.19 below.)

A fundamental relation in quantum mechanics is the *canonical commutation relation* satisfied by the momentum operator P and the position operator Q, see (2.78).

Definition 2.27. Given a Hilbert space $\mathscr{H}$, we say two operators P and Q in $\mathscr{L}(\mathscr{H})$ satisfy the *canonical commutation relation*, if the following identity holds:

$$PQ - QP = -i\,I, \quad i = \sqrt{-1}. \tag{2.78}$$

Heisenberg represented the operators P, Q by infinite matrices, although his solution to (2.78) is not really matrices, and not finite matrices.

Lemma 2.10. *Eq. (2.78) has no solutions for finite matrices, in fact, not even for bounded operators.*

Proof. The reason is that for matrices, there is a trace operation where

$$trace(AB) = trace(BA). \tag{2.79}$$

This implies the trace on the left-hand-side is zero, while the trace on the RHS is not. □

This shows that there is no finite dimensional solution to the commutation relation above, and one is forced to work with infinite dimensional Hilbert space and operators on it. Notice that P, Q do not commute, and the above commutation relation leads to the uncertainty principle (Hilbert, Max Born, von Neumann worked out the mathematics). It states that the statistical variance $\triangle P$ and $\triangle Q$ satisfy $\triangle P \triangle Q \geq \hbar/2$ (see Theorem 4.5). We will come back to this later in Exercise 4.8.

We will show that non-commutativity always yields "uncertainty."

However, Heisenberg [vN31, Hei69] found his "matrix" solutions by tri-diagonal $\infty \times \infty$ matrices, where

$$P = \frac{1}{\sqrt{2}} \begin{bmatrix} 0 & 1 & & & \\ 1 & 0 & \sqrt{2} & & \\ & \sqrt{2} & 0 & \sqrt{3} & \\ & & \sqrt{3} & 0 & \ddots \\ & & & \ddots & \ddots \end{bmatrix} \tag{2.80}$$

and

$$Q = \frac{1}{i\sqrt{2}} \begin{bmatrix} 0 & 1 & & & \\ -1 & 0 & \sqrt{2} & & \\ & -\sqrt{2} & 0 & \sqrt{3} & \\ & & -\sqrt{3} & 0 & \ddots \\ & & & \ddots & \ddots \end{bmatrix} \tag{2.81}$$

the complex i in front of Q is to make it selfadjoint.

Exercise 2.34 (The canonical commutation relation). Using matrix multiplication for $\infty \times \infty$ matrices verify directly that the two matrices P and Q satisfy $PQ - QP = -i\,I$, where I is the identity matrix in $l^2(\mathbb{N}_0)$, i.e., $(I)_{ij} = \delta_{ij}$. Hint: Use the rules in Lemma 2.8.

Exercise 2.35 (Raising and lowering operators (non-commutative complex variables)). Set

$$A_{\mp} := P \pm iQ; \tag{2.82}$$

and show that the *matrix representation* for these operators is as follows:

$$A_- = \sqrt{2} \begin{bmatrix} 0 & 1 & 0 & 0 & 0 & \cdots \\ \vdots & 0 & \sqrt{2} & 0 & 0 & \cdots \\ \vdots & \vdots & 0 & \sqrt{3} & 0 & \cdots \\ \vdots & \vdots & \vdots & 0 & \sqrt{4} & \cdots \\ \vdots & \vdots & \vdots & \vdots & \ddots & \ddots \end{bmatrix}$$

and

$$A_+ = \sqrt{2} \begin{bmatrix} 0 & 0 & \cdots & \cdots & \cdots & \cdots \\ 1 & 0 & 0 & \cdots & \cdots & \cdots \\ 0 & \sqrt{2} & 0 & 0 & \cdots & \cdots \\ \vdots & 0 & \sqrt{3} & 0 & 0 & \cdots \\ \vdots & \vdots & 0 & \sqrt{4} & \ddots & \ddots \end{bmatrix}.$$

In other words, the raising operator A_+ is a sub-banded matrix, while the lowering operator A_- is a supper-banded matrix. Both A_+ and A_- have 0s down the diagonal.

Further, show that

$$A_- A_+ = 2 \begin{bmatrix} 1 & 0 & \cdots & \cdots & \cdots & \cdots \\ 0 & 2 & 0 & \cdots & \cdots & \cdots \\ \vdots & 0 & 3 & 0 & \cdots & \cdots \\ \vdots & \vdots & 0 & 4 & 0 & \cdots \\ \vdots & \vdots & \vdots & \vdots & \ddots & \ddots \end{bmatrix}$$

i.e., a diagonal matrix, with the numbers $\mathbb{N}$ down the diagonal inside $\begin{bmatrix} \ddots \end{bmatrix}$; and

$$A_+ A_- = 2 \begin{bmatrix} 0 & 0 & \cdots & \cdots & \cdots & \cdots & \cdots \\ 0 & 1 & 0 & \cdots & \cdots & \cdots & \cdots \\ \vdots & 0 & 2 & 0 & \cdots & \cdots & \cdots \\ \vdots & & 0 & 3 & 0 & \cdots & \cdots \\ \vdots & & \vdots & 0 & 4 & 0 & \cdots \\ \vdots & & \vdots & \vdots & \vdots & \ddots & \ddots \end{bmatrix};$$

so that

$$\frac{1}{2} [A_-, A_+] = I.$$

Remark 2.22 (Raising and lowering in an ONB). In the canonical ONB $\{e_n\}_{n=0}^{\infty}$ in $l^2(\mathbb{N}_0)$, we have the following representations of the two operators $A_{\pm}$ (see (2.82)):

$$A_+ e_n = \sqrt{2}\sqrt{n+1}e_{n+1}; \ n = 0, 1, 2, \ldots, \text{ and}$$
$$A_- e_n = \sqrt{2}\sqrt{n}e_{n-1}; \ n = 1, 2, \ldots,$$
$$A_- e_0 = 0, \text{ see Fig. 2.17.}$$

The vector e_0 is called the *ground state*, or *vacuum vector*.

Remark 2.23. The conclusion of the discussion above is that the Heisenberg commutation relation (2.78) for pairs of selfadjoint operators has two realizations, one in $L^2(\mathbb{R})$, and the other in $l^2(\mathbb{N})$.

In the first one we have

$$(Pf)(x) = \frac{1}{i}\frac{d}{dx}f$$
$$(Qf)(x) = xf(x)$$

for all $f \in \mathcal{S} \subset L^2(\mathbb{R})$, where $\mathcal{S}$ denotes the Schwartz test-function subspace in $L^2(\mathbb{R})$.

The second realization is by $\infty \times \infty$ matrices, and it is given in detail above. In Section 8.5 we shall return to the first realization.

The *Stone-von Neumann uniqueness theorem* (see [vN32b, vN31]) implies the two solutions are unitarily equivalent; see Chapter 8.

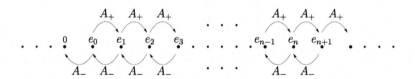

Figure 2.17: The raising and lowering operators $A_{\pm}$. The lowering operator A_- kills e_0.

Exercise 2.36 (Infinite banded matrices). Give an example of two sequences

$$d_1, d_2, \ldots \in \mathbb{R}, \ a_1, a_2, \ldots \in \mathbb{C}$$

such that the corresponding Hermitian symmetric $\infty \times \infty$ tri-diagonal (banded) matrix A in Figure 2.18 satisfies $A \subset A^*$, but $\overline{A} \neq A^*$, i.e., A is *not* essentially selfadjoint when realized as a Hermitian operator in l^2.

Remark 2.24 (Matrices vs operators). Every bounded linear operator (and many unbounded operators too) in separable Hilbert space (and, in particular, in l^2) can be realized as a well-defined *infinite "square" matrix*. In l^2 we pick the canonical ONB, but in a general Hilbert space, a choice of ONB must be made. We saw that most rules for finite matrices carry over to the case of infinite matrices; sums, products, and adjoints.

For instance, in order to find the matrix of the sum of two bounded operators, just find the sum of the matrices of these operators. And the matrix of the adjoint operator A^* (of a bounded operator A in Hilbert space) is the adjoint matrix (conjugate transpose) of the matrix of the operator A^{16}.

So while it is "easy" to go from bounded operators to infinite "square" matrices, the converse is much more subtle.

Exercise 2.37 (The Hilbert matrix).

1. Show that the Hilbert matrix

$$H = \left(\frac{1}{1+j+k} \right)_{j,k \in \mathbb{N}}$$

   defines a bounded selfadjoint operator T_H in $l^2(\mathbb{N})$.

2. Show that

$$\|T_H\|_{UN} = \sqrt{\pi}$$

   where $\|\cdot\|_{UN}$ denotes the uniform operator norm

$$\|Tx\| = \sup \left\{ \|Tx\|, \; x \in l^2, \; \|x\| = 1 \right\}.$$

3. Show that T_H is positive definite.
   <u>Hint:</u>

$$\int_0^1 x^n dx = \frac{1}{1+n}, \; n \in \mathbb{N}.$$

Remark 2.25. Note that the Hilbert matrix H is not banded; in fact every entry in H is positive. Nonetheless, it follows from an application of Corollary 2.6 (Section 2.7) that the Hilbert matrix H in Exercise 2.37 is equivalent to a banded matrix J; and there is a choice of J to be tri-diagonal; a Jacobi matrix; see Figure 2.18. Since H yields a bounded selfadjoint operator in l^2, it follows from Corollary 2.6 that J is in fact a bounded Jacobi-matrix with the same norm as H.

2.5.3 Probabilistic Interpretation of Parseval in Hilbert Space

Case 1. Let $\mathscr{H}$ be a complex Hilbert space, and let $\{u_k\}_{k \in \mathbb{N}}$ be an ONB, then Parseval's formula reads:

$$\langle v, w \rangle_{\mathscr{H}} = \sum_{k \in \mathbb{N}} \langle v, u_k \rangle \langle u_k, w \rangle, \; \forall v, w \in \mathscr{H}. \tag{2.83}$$

---

[16] The reader will find the precise definition of adjoint operators in Definition 2.6 (pg. 33), see also Sections 2.4 (pg. 2.4), 3.1 (pg. 3.1).

$$A = \begin{bmatrix} d_1 & a_1 & 0 & \cdots & & \cdots & & & \\ \overline{a_1} & d_2 & a_2 & 0 & \cdots & & \cdots & & \\ 0 & \overline{a_2} & d_3 & a_3 & 0 & & \cdots & & \cdots \\ \vdots & 0 & \overline{a_3} & \ddots & \ddots & \ddots & \cdots & & \cdots \\ \vdots & \vdots & \ddots & \ddots & \ddots & \ddots & \ddots & \cdots & \cdots \\ \vdots & \vdots & 0 & \overline{a_{n-2}} & d_{n-1} & a_{n-1} & 0 & \cdots \\ \vdots & \vdots & & 0 & \overline{a_{n-1}} & d_n & a_n & \ddots \\ \vdots & \vdots & & & 0 & \overline{a_n} & d_{n+1} & \ddots \\ \vdots & \vdots & & & & \ddots & & \ddots & \ddots \end{bmatrix}$$

Figure 2.18: $A \subset A^*$ (a Hermitian Jacobi matrix).

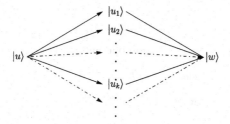

Figure 2.19: Transition of quantum-states.

Translating this into a statement about "transition probabilities" for quantum states, $v, w \in \mathscr{H}$, with $\|v\|_{\mathscr{H}} = \|w\|_{\mathscr{H}} = 1$, we get

$$\text{Prob}\,(v \to w) = \sum_{k \in \mathbb{N}} \text{Prob}\,(v \to u_k)\,\text{Prob}\,(u_k \to w). \qquad (2.84)$$

See Figure 2.19. The states v and w are said to be *uncorrelated* iff (Def.) they are orthogonal.

Fix a state $w \in \mathscr{H}$, then

$$\|w\|^2 = \sum_{k \in \mathbb{N}} |\langle u_k, w \rangle|^2 = 1.$$

The numbers $|\langle u_i, w \rangle|^2$ represent a probability distribution over the index set, where $|\langle u_k, w \rangle|^2$ is the probability that the quantum system is in the state $|u_k\rangle$.

Disclaimer: The notation "transition probability" in (2.84) and Figure 2.19 is a stretch since the inner products $\langle v, u_k \rangle$ are not positive. Nonetheless, it is justified by

$$\sum_k |\langle v, u_k \rangle|^2 = 1$$

when $v \in \mathscr{H}$ (is a state vector).

Case 2. If $P : \mathcal{B}(\mathbb{R}) \to \mathrm{Proj}(\mathscr{H})$ is a projection valued measure (Appendix 3.A), we get the analogous assertions, but with integration, as opposed to summation. In this case (2.83) holds in the following form:

$$\langle v, w \rangle_{\mathscr{H}} = \int_{\mathbb{R}} \langle v, P(d\lambda) w \rangle_{\mathscr{H}} ; \qquad (2.85)$$

and for $v = w$, it reads:

$$\|v\|_{\mathscr{H}}^2 = \int_{\mathbb{R}} \|P(d\lambda) v\|_{\mathscr{H}}^2 . \qquad (2.86)$$

Definition 2.28. We say $P : \mathcal{B}(\mathbb{R}) \to \mathrm{Proj}(\mathscr{H})$ is a projection valued measure, if the following conditions are satisfied:

1. $P(A) = P(A)^* = P(A)^2$, $\forall A \in \mathcal{B}(\mathbb{R})$.

2. $P(\cdot)$ is countably additive on the Borel subsets of $\mathbb{R}$, i.e.,

$$\sum_j P(A_j) = P(\cup_j A_j)$$

where $A_j \in \mathcal{B}(\mathbb{R})$, $A_i \cap A_j = \emptyset$, $i \neq j$.

3. $P(A \cap B) = P(A) P(B)$, $\forall A, B \in \mathcal{B}(\mathbb{R})$.

Remark 2.26. It is customary, when only axioms (1) and (2) hold, to refer to P as a projection valued measure (PVM). But if a particular projection valued measure P should also satisfy condition (3), then we say that P is an *orthogonal* PVM. We will show in Chapter 6 that every PVM in a fixed Hilbert space has an orthogonal realization in a "bigger" Hilbert space; we say that P has a (so called) orthogonal dilation.

2.6 The Lattice Structure of Projections

A lattice is a partially ordered set in which every two elements have a supremum (also called a least upper bound or join) and an infimum (also called a greatest lower bound or meet).

The purpose of the discussion below is twofold; one to identify two cases: (i) the (easy) *lattice of subsets* of a fixed total set; and (ii) the *lattice of projections* in a fixed Hilbert space. Secondly we point out how non-commutativity of projections makes the comparison of (i) and (ii) subtle; even though there are some intriguing correspondences; see Table 2.4 for illustration.

Notation. In this section we will denote projections P, Q, etc.

von Neumann invented the notion of abstract Hilbert space in 1928 as shown in one of the earliest papers.[17] His work was greatly motivated by quantum

---

[17]Earlier authors, Schmidt and Hilbert, worked with infinite bases, and $\infty \times \infty$ matrices.

mechanics. In order to express quantum mechanics logic operations, he created lattices of projections, so that everything we do in set theory with set operation has a counterpart in the operations of projections. See Table 2.4.

For example, if P and Q are two projections in $\mathscr{B}(\mathscr{H})$, then

$$P\mathscr{H} \subset Q\mathscr{H} \tag{2.87}$$
$$\Updownarrow$$
$$P = PQ \tag{2.88}$$
$$\Updownarrow$$
$$P \leq Q. \tag{2.89}$$

This is similar to the following equivalence relation in set theory

$$A \subset B \text{ (containment of sets)} \tag{2.90}$$
$$\Updownarrow$$
$$A = A \cap B. \tag{2.91}$$

In general, product and sum of projections are not projections. But if $P\mathscr{H} \subset Q\mathscr{H}$ then the product PQ is in fact a projection. Taking adjoint in (2.88) yields

$$P^* = (PQ)^* = Q^*P^* = QP.$$

It follows that $PQ = QP = P$, i.e., *containment of subspaces implies the corresponding projections commute.*

Two decades before von Neumann developed his Hilbert space theory, Lebesgue developed his integration theory [Leb05] which extends the classical Riemann integral. The monotone sequence of sets $A_1 \subset A_2 \subset \cdots$ in Lebesgue's integration theory also has a counterpart in the theory of Hilbert space.

Lemma 2.11. *Let P_1 and P_2 be orthogonal projections acting on $\mathscr{H}$, then*

$$P_1 \leq P_2 \Longleftrightarrow \|P_1 x\| \leq \|P_2 x\|, \ \forall x \in \mathscr{H} \tag{2.92}$$
$$\Updownarrow$$
$$P_1 = P_1 P_2 = P_2 P_1 \tag{2.93}$$

(see Table 2.4.)

Proof. Indeed, for all $x \in \mathscr{H}$, we have

$$\|P_1 x\|^2 = \langle P_1 x, P_1 x \rangle = \langle x, P_1 x \rangle = \langle x, P_2 P_1 x \rangle \leq \|P_1 P_2 x\|^2 \leq \|P_2 x\|^2.$$

$\square$

Theorem 2.9. *For every monotonically increasing sequence of projections*

$$P_1 \leq P_2 \leq \cdots,$$

and setting

$$P := \vee P_k = \lim_k P_k,$$

then P defines a projection, the limit.

SETS	CHAR	PROJ	DEF
$A \cap B$	$\chi_A \chi_B$	$P \wedge Q$	$P\mathscr{H} \cap Q\mathscr{H}$
$A \cup B$	$\chi_{A \cup B}$	$P \vee Q$	$\overline{span}\{P\mathscr{H} \cup Q\mathscr{H}\}$
$A \subset B$	$\chi_A \chi_B = \chi_A$	$P \leq Q$	$P\mathscr{H} \subset Q\mathscr{H}$
$A_1 \subset A_2 \subset \cdots$	$\chi_{A_i} \chi_{A_{i+1}} = \chi_{A_i}$	$P_1 \leq P_2 \leq \cdots$	$P_i\mathscr{H} \subset P_{i+1}\mathscr{H}$
$\bigcup_{k=1}^{\infty} A_k$	$\chi_{\cup_k A}$	$\vee_{k=1}^{\infty} P_k$	$\overline{span}\{\bigcup_{k=1}^{\infty} P_k\mathscr{H}\}$
$\bigcap_{k=1}^{\infty} A_k$	$\chi_{\cap_k A_k}$	$\wedge_{k=1}^{\infty} P_k$	$\bigcap_{k=1}^{\infty} P_k\mathscr{H}$
$A \times B$	$(\chi_{A \times X})(\chi_{X \times B})$	$P \otimes Q$	$P \otimes Q \in proj(\mathscr{H} \otimes \mathscr{K})$

Table 2.4: Lattice of projections in Hilbert space, Abelian vs non-Abelian.

Proof. The assumption $P_1 \leq P_2 \leq \cdots$ implies that $\{\|P_k x\|\}_{k=1}^{\infty}$, $x \in \mathscr{H}$, is a monotone increasing sequence in $\mathbb{R}$, and the sequence is bounded by $\|x\|$, since $\|P_k x\| \leq \|x\|$, for all $k \in \mathbb{N}$. Therefore the sequence $\{P_k\}_{k=1}^{\infty}$ converges to $P \in \mathscr{B}(\mathscr{H})$ (strongly), and P really defines a selfadjoint projection. (We use "$\leq$" to denote the lattice operation on projection.) Note the convergence refers to the strong operator topology, i.e., for all $x \in \mathscr{H}$, there exists a vector, which we denote by Px, so that $\lim_k \|P_k x - Px\| = 0$. $\qquad\square$

The examples in Section 2.4 using Gram-Schmidt process can now be formulated in the lattice of projections.

Definition 2.29. Recall (Lemma 2.5) that for a linearly independent subset $\{u_k\} \subset \mathscr{H}$, the Gram-Schmidt process yields an orthonormal set $\{v_k\} \subset \mathscr{H}$, with $v_1 := u_1/\|u_1\|$, and

$$v_{n+1} := \frac{u_{n+1} - P_n u_{n+1}}{\|u_{n+1} - P_n u_{n+1}\|}, \ n = 1, 2, \ldots;$$

where P_n is the orthogonal projection on the n-dimensional subspace

$$V_n := span\{v_1, \ldots, v_n\}.$$

See Figure 2.20.

Note that

$$V_n \subset V_{n+1} \rightarrow \bigcup_n V_n \quad \sim \quad P_n \leq P_{n+1} \rightarrow P$$

$$P_n^\perp \geq P_{n+1}^\perp \to P^\perp.$$

Assume $\bigcup_n V_n$ is dense in $\mathscr{H}$, then $P = I$ and $P^\perp = 0$. In lattice notations, we may write

$$\begin{aligned} \vee P_n = \sup P_n &= I \\ \wedge P_n^\perp = \inf P_n^\perp &= 0. \end{aligned}$$

Lemma 2.12. *Let $P, Q \in Proj(\mathscr{H})$, then*

$$P + Q \in Proj(\mathscr{H}) \iff PQ = QP = 0, \ i.e., \ P \perp Q.$$

Proof. Notice that
$$(P + Q)^2 = P + Q + PQ + QP \tag{2.94}$$

and so

$$(P + Q)^2 = P + Q \tag{2.95}$$
$$\Updownarrow$$
$$PQ + QP = 0. \tag{2.96}$$

Suppose $PQ = QP = 0$ then

$$(P + Q)^2 = P + Q = (P + Q)^*, \text{ i.e., } P + Q \in Proj(\mathscr{H}).$$

Conversely, if $P + Q \in Proj(\mathscr{H})$, then $(P+Q)^2 = P+Q \implies PQ+QP = 0$ by (2.96). Also, $(PQ)^* = Q^*P^* = QP$, combining with (2.96) yields

$$(PQ)^* = QP = -PQ. \tag{2.97}$$

Then,

$$(PQ)^2 = P(QP)Q \underset{(2.97)}{=} -P(PQ)Q = -PQ$$

which implies $PQ(I + PQ) = 0$. Hence,

$$PQ = 0 \quad \text{or} \quad PQ = I. \tag{2.98}$$

But by (2.97), PQ is skew-adjoint, it follows that $PQ = 0$, and so $QP = 0$. $\square$

Lemma 2.13. *Let P and Q be two orthogonal projections (i.e., $P = P^* = P^2$) on a fixed Hilbert space $\mathscr{H}$, and consider the operator-norm $\|P + Q\|$. Suppose $\|P + Q\| \leq 1$; then the two projections are orthogonal, $PQ = 0$. As a result (Lemma 2.12), the sum $P + Q$ is itself an orthogonal projection.*

Proof. Let $u \in P\mathscr{H}$, i.e., $Pu = u$ holds. Then

$$\begin{aligned} \|u\|^2 &\geq \|(P + Q)u\|^2 = \|u\|^2 + \|Qu\|^2 + \langle u, Qu \rangle + \langle Qu, u \rangle \\ &= \|u\|^2 + 3\|Qu\|^2, \end{aligned}$$

since $\langle Qu, u \rangle = \langle u, Qu \rangle = \|Qu\|^2$. Hence $Qu = 0$. Since

$$P\mathscr{H} = \{u \in \mathscr{H} \ ; \ Pu = u\},$$

we get $QP = 0$; equivalently, $PQ = 0$; equivalently, $P\mathscr{H} \perp Q\mathscr{H}$. $\square$

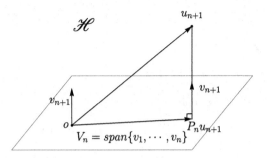

Figure 2.20: Gram-Schmidt: $V_n \longrightarrow V_{n+1}$

Exercise 2.38 (Sums of projections).

1. In the setting of Lemma 2.13, show that every number in the closed interval $[1,2]$ arises as $\|P+Q\|$ for a pair of orthogonal projections P and Q.

2. Give examples of each configuration in a 2-dimensional Hilbert space.

3. General: For each case, characterize the associated geometric configuration of the two given projections P and Q.

4. Formulate and prove the appropriate generalization of Lemma 2.13 to the case of a sum of any finite number of orthogonal projections.

5. Discuss extensions of your result in part 4 to the analogous case of an infinite sum of orthogonal projections.

Remark 2.27. Eq. (2.94) is analogous to the following identity for characteristic functions:
$$\chi_A + \chi_B = \chi_{A\cup B} - \chi_{A\cap B}$$
Therefore, $\left(\chi_A + \chi_B\right)^2 = \chi_A + \chi_B$ iff $\chi_{A\cap B} = 0$, i.e., iff $A \cap B = \emptyset$.

The set of projections in a Hilbert space $\mathscr{H}$ is partially ordered according to the corresponding closed subspaces partially ordered by inclusion. Since containment implies commuting, the chain of projections

$$P_1 \le P_2 \le \cdots$$

is a family of commuting selfadjoint operators. By the spectral theorem (Chapter 4), $\{P_i\}$ may be simultaneously diagonalized, so that P_i is unitarily equivalent to the operator of multiplication by χ_{E_i} on the Hilbert space $L^2(X,\mu)$, where X is compact and Hausdorff. Therefore the lattice structure of projections in $\mathscr{H}$ is precisely the lattice structure of χ_E, or equivalently, the lattice structure of measurable sets in X.

Lemma 2.14. *Consider $L^2(X,\mu)$. The following are equivalent.*

1. $E \subset F$;

2. $\chi_E \chi_F = \chi_F \chi_E = \chi_E$;

3. $\|\chi_E f\| \leq \|\chi_F f\|$, for any $f \in L^2$;

4. $\chi_E \leq \chi_F$, in the sense that

$$\langle f, \chi_E f \rangle \leq \langle f, \chi_F f \rangle, \ \forall f \in L^2(X).$$

Proof. The proof is trivial. Note that

$$\langle f, \chi_E f \rangle = \int \chi_E |f|^2 \, d\mu$$

$$\|\chi_E f\|^2 = \int |\chi_E f|^2 \, d\mu = \int \chi_E |f|^2 \, d\mu$$

where we used that fact that

$$\chi_E = \overline{\chi_E} = \chi_E^2.$$

$\square$

What makes $Proj\,(\mathcal{H})$ intriguing is the non-commutativity. For example, if $P, Q \in Proj\,(\mathcal{H})$ are given, it does not follow (in general) that $P + Q \in Proj\,(\mathcal{H})$; nor that $PQP \in Proj\,(\mathcal{H})$. These two conclusions only hold if it is further assumed that P and Q commute; see Lemmas 2.12 and 2.15.

Lemma 2.15. *Let $P, Q \in Proj\,(\mathcal{H})$; then the following conditions are equivalent:*

1. $PQP \in Proj\,(\mathcal{H})$;

2. $PQ = QP$.

Proof. First note that the operator $A = PQ - QP$ is skew-symmetric, i.e., $A^* = -A$, and so its spectrum is contained in the imaginary line $i\mathbb{R}$.

The implication (2)$\Rightarrow$(1) above is immediate so assume (1), i.e., that

$$(PQP)^2 = PQP.$$

And using this, one checks by a direct computation that $A^3 = 0$. But with $A^* = -A$, and the spectral theorem, we therefore conclude that $A = 0$, in other words, (2) holds. $\square$

2.7 Multiplication Operators

Exercise 2.39 (Multiplication operators). Let $(X, \mathcal{F}, \mu)$ be a measure space, assume μ is σ-finite. Let $L^2(\mu)$ be the corresponding Hilbert space. Let φ be a locally integrable function on X, and set

$$M_\varphi f := \varphi f,$$

pointwise product, defined for

$$f \in dom(M_\varphi) = \left\{ f \in L^2(\mu) \ : \ \varphi f \in L^2(\mu) \right\}.$$

M_φ is called a *multiplication operator*.

1. Show that M_φ is normal.

2. Show that M_φ is selfadjoint iff φ is μ-a.e. real-valued.

3. Show that M_φ is bounded in $L^2(\mu)$ iff $\varphi \in L^\infty(\mu)$; and, in this case,

$$\|M_\varphi\|_{UN} = \|\varphi\|_{L^\infty(\mu)}. \tag{2.99}$$

4. Show that if $\varphi \in L^\infty(\mu)$, then $dom(M_\varphi) = L^2(\mu)$.

5. Discuss the converse.

Exercise 2.40 (Moment theory [Akh65]). Let μ be a positive Borel measure on $\mathbb{R}$ such that

$$\int_{\mathbb{R}} x^{2n} d\mu(x) < \infty \tag{2.100}$$

for all $n \in \mathbb{N}$, i.e., μ has finite moments of all orders. Let $\varphi(x) = x$, and $M = M_\varphi$ the corresponding multiplication operator in $L^2(\mu) = L^2(\mathbb{R}, \mathcal{B}, \mu)$, i.e.,

$$(Mf)(x) = (Qf)(x) = xf(x), \tag{2.101}$$

for all $f \in L^2(\mu)$ such that $xf \in L^2(\mu)$.

1. Using Gram-Schmidt (Lemma 2.5), show that M has a matrix representation by an $\infty \times \infty$ tri-diagonal (banded) matrix as in Figure 2.18.

   Akhiezer calls these infinite banded matrices *Jacobi matrices*. They define formally selfadjoint (alias symmetric) operators in l^2; unbounded of course. And these operators can only attain von Neumann indices $(0,0)$ or $(1,1)$. Both are possible. For details of von Neumann's deficiency indices, see Chapter 10 and [DS88b].

2. Work out a recursive formula for the two sequences $(a_n)_{n \in \mathbb{N}}$ and $(d_n)_{n \in \mathbb{N}}$ in the expression for M by the matrix of Figure 2.18 in terms of the moments $(s_n)_{n \in \mathbb{N} \cup \{0\}}$:

$$s_n := \int_{\mathbb{R}} x^n d\mu(x).$$

3. Same question as in (2) but for the special case when μ is

$$d\mu(x) = \frac{1}{\sqrt{2\pi}} e^{-\frac{x^2}{2}} dx$$

i.e., the $N(0,1)$ Gaussian measure on $\mathbb{R}$.

Hint: Show first that the moments $\{s_k\}_{k \in \{0\} \cup \mathbb{N}}$ of $\mu_{N(0,1)}$ are as follows (the Gaussian moments):

$s_{2n+1} = 0$, and

$$s_{2n} = \frac{(2n)!}{2^n \cdot n!} = (2n-1)!! \, (= (2n-1)(2n-3)\cdots 5 \cdot 3).$$

4. General: Give necessary and sufficient conditions for *essential selfadjointness* of the associated Jacobi matrix as an operator in $L^2(\mu)$, expressed in terms μ and of the moments (s_n) in (2).

Corollary 2.6. *Let A be a selfadjoint operator in a separable Hilbert space, and suppose there is a cyclic vector u_0, $\|u_0\| = 1$, such that*

$$u_0 \in \bigcap_{k \in \mathbb{N}} dom\left(A^k\right). \tag{2.102}$$

Then there is an ONB $\{e_i\}_{i \in \mathbb{N}}$ in $\mathscr{H}$, contained in $dom(A)$ such that the corresponding $\infty \times \infty$ matrix

$$(M_A)_{i,j} = \langle e_i, Ae_j \rangle, \ i,j \in \mathbb{N} \tag{2.103}$$

is banded; what is more, it is a Jacobi-matrix, see Exercise 2.40.

Proof. Using the Spectral Theorem, we conclude that there is a measure μ_0 on $\mathbb{R}$ and a unitary transform $W : L^2(\mathbb{R}, \mu_0) \longrightarrow \mathscr{H}$ such that

1. $Wf = f(A)u_0$, $\forall f \in L^2(\mu_0)$;

2. $WM_t = AW$, where M_t denotes multiplication by t in $L^2(\mu_0)$; and

3. $\langle u_0, f(A)u_0 \rangle = \int_{\mathbb{R}} f(t) d\mu_0(t)$, $\forall f \in L^2(\mu_0)$.

We refer the reader to Chapter 4 for more details.

Now apply Gram-Schmidt to the monomials $\{t^k\}_{k \in \{0\} \cup \mathbb{N}}$, to get orthogonal polynomials $\{p_k(t)\}$ such that

$$span_{k \leq n}\{p_k(t)\} = span_{k \leq n}\{t^k\}$$

holds for all $n \in \mathbb{N}$.

Set

$$e_k := p_k(A)u_0, \ k \in \{0\} \cup \mathbb{N}, \tag{2.104}$$

$e_0 = u_0$, and this is then the desired ONB. To see this, use the conclusion from Exercise 2.40, together with the following:

$$\begin{aligned}
\langle e_j, e_k \rangle &= \langle p_j(A) u_0, p_k(A) u_0 \rangle \\
&= \langle u_0, p_j(A) p_k(A) u_0 \rangle \\
&= \langle u_0, (p_j p_k)(A) u_0 \rangle \\
&= \int_{\mathbb{R}} p_j(t) p_k(t) \, d\mu_0(t), \text{ by condition 3} \\
&= \delta_{j,k} \text{ (by Gram-Schmidt.)}
\end{aligned}$$

$\square$

Historical Note. In [GIS90], Lax relates an account of von Neumann and F. Rellich speaking in Hilbert's seminar, in Göttingen (around 1930). When they came to "selfadjoint operator in Hilbert space," Erhard Schmidt (of Gram-Schmidt) would interrupt: "Please, young man, say *infinite matrix*."

Ironically, von Neumann invented numerical methods for "large" matrices toward the end of his career [GvN51].

A summary of relevant numbers from the Reference List

For readers wishing to follow up sources, or to go in more depth with topics above, we suggest:
[Tay86, Arv72, Ban93, BR79, Con90, DM72, DS88b, Lax02, RS75, RSN90, Rud91, Rud87, AJS14, AJLM13, AJL13, AJ12, ARR13, BM13, Hid80, HØUZ10, Itô06, Loè63, Jør14, KL14a, BJ02, Jor06, CW14, Gro64, Joh88, KF75, JM80, Hel13, KW12, Con07, AH09, Hut81, BHS05, CDR01, BDM+97, Dop95].

While there are existing books dealing with diverse areas in mathematics where non-commutativity plays an essential role, our present point of view is more general. Nonetheless, our book will serve as an introduction to such more specialized research monographs as: [CCG+04, Vár06] covering non-commutative geometry, [Tay86, Kir88] on non-commutative harmonic analysis, [BGV12] on non-commutative probability, and non-commutativity in particle physics, see e.g., [vS15].

2.A Hahn-Banach Theorems

Version 1. Let S be a subspace of a real vector space X. Let $l : S \to \mathbb{R}$ be a linear functional, and let $p : X \to \mathbb{R}$ satisfy

$$p(x+y) \le p(x) + p(y), \quad x, y \in X; \tag{2.105}$$

$$p(tx) = t p(x), \quad t \in \mathbb{R}_+, \, x \in X. \tag{2.106}$$

Theorem 2.10 (HB1). *Let X, S, p, and l be as above, and assume:*

$$l(x) \le p(x), \quad x \in S. \tag{2.107}$$

Then there is a linear functional $\tilde{l} : X \to \mathbb{R}$, extending l on the subspace, and satisfying

$$\tilde{l}(x) \le p(x), \quad \forall x \in X. \tag{2.108}$$

Hint: Introduce a partially ordered set (p.o.s.) (T, m) where $S \subset T \subset X$, T is a subspace, $m : T \to X$ is a linear functional extending (S, l) and satisfying

$$m(x) \le p(x), \quad \forall x \in T. \tag{2.109}$$

Define the order $(T, m) \le (T', m')$ to mean that $T \subseteq T'$ and m' agrees with m on T. (Both satisfying (2.109).) Apply Zorn's lemma to this p.o.s., and show that every maximal element must be a solution to (2.108).

2.B   Banach-Limit

Consider $X := l_{\mathbb{R}}^{\infty}(\mathbb{N}) =$ all bounded real sequences $x = (x_1, x_2, \cdots)$, and the shift $\sigma : l^{\infty} \to l^{\infty}$, defined by

$$\sigma(x_1, x_2, x_3, \cdots) = (x_2, x_3, \cdots).$$

Consider the subspace $S \subset X$ consisting of all convergent sequences, and for $x \in S$, set $l(x) = \lim_{k \to \infty} x_k$, i.e., it holds that, for $\forall \varepsilon \in \mathbb{R}_+$, $\exists n$ such that

$$|x_k - l(x)| < \varepsilon \quad \text{for} \quad \forall k \ge n.$$

Theorem 2.11 (Banach). *There is a linear functional, called $LIM : X \to \mathbb{R}$, (a Banach limit) having the following properties:*

1. *LIM is an extension of l on S.*

2.
$$\liminf_{k} x_k \le LIM(x) \le \limsup_{k} x_k \quad \text{and}$$

3.
$$LIM \circ \sigma = LIM,$$
   i.e., LIM is shift-invariant.

Proof. This is an application of (HB1) but with a modification; we set

$$q(x) = \limsup x_k, \quad \text{and}$$

$$p(x) = \inf_{n} \frac{1}{n+1} \sum_{k=0}^{n} q(\sigma^k(x)), \quad \forall x \in X,$$

and we note that $l(x) \le p(x)$, $\forall x \in S$. The rest of the proof follows that of HB1 *mutatis mutandis*. $\qquad\square$

Remark 2.28. Note that LIM is *not* unique.

It follows by the theorem above that LIM is in $(l^\infty\,(\mathbb{N}))^*$, and that it is *not* represented by any $y \in l^1\,(\mathbb{N})$, see Table 2.1. As a result, we have

$$(l^\infty\,(\mathbb{N}))^* \supsetneq l^1\,(\mathbb{N})\,; \qquad (2.110)$$

i.e., the dual $(l^\infty)^*$ is (much) bigger than l^1.

Theorem 2.12 (HB2). *Let X be a normed space, $S \subset X$ a closed subspace, $l : S \to \mathbb{R}$ a linear functional such that*

$$|l\,(x)| \leq \|x\|\,, \quad \forall x \in S. \qquad (2.111)$$

Then there is a $\widetilde{l} \in X^*$ such that

$$|\widetilde{l}\,(x)\,| \leq \|x\|\,, \quad \forall x \in X,$$

$\widetilde{l}$ extending l from S, and

$$\|\widetilde{l}\|_{X^*} = \|l\|_{S^*}\,. \qquad (2.112)$$

Theorem 2.13 (HB3 - separation). *Let X be a real vector space. Assume X is equipped with a topology making the two vector-operations continuous. Let $K \neq \emptyset$ be an open convex subset of X. Let $y \in X\backslash K$ (in the complement). Then there is a linear functional $l : X \to \mathbb{R}$ such that*

$$l\,(x) < l\,(y)\,, \quad \forall x \in K. \qquad (2.113)$$

(In fact, there exists $c \in \mathbb{R}$ such that $l\,(x) < c$, $\forall x \in K$, and $l\,(y) > c$, and the hyperplane $H_c = \{z \in X\,:\,l\,(z) = c\}$ satisfies $K \subset H_c^-$, $y \in H_c^+$.)

Hint: Assume (by translation) that $0 \in K$; and set

$$p_K\,(x) := \inf\left\{a\,:\,a \in \mathbb{R}_+, \frac{x}{a} \in K\right\}; \qquad (2.114)$$

and then apply version 2 to p_K, which can be shown to be sub-additive. For the separation property, see Figure 2.21.

Theorem 2.14 (HB4). *Let $\mathfrak{A}$ be a C^*-algebra and let $\mathfrak{B} \subset \mathfrak{A}$ be a $*$-subalgebra. Let $l : \mathfrak{B} \to \mathbb{C}$ satisfy*

$$l\,(b^*b) \geq 0,\ \forall b \in \mathfrak{B}, \quad and \quad \|l\| = 1\ (positivity),$$

then there is a positive linear functional $\widetilde{l} : \mathfrak{A} \to \mathbb{C}$, such that

1. $\widetilde{l}$ extends l on $\mathfrak{B}$;

2. $\widetilde{l}\,(a^*a) \geq 0$, $\forall a \in \mathfrak{A}$; and

3. $\|\widetilde{l}\| = 1$.

Remark 2.29. This version is due to M. Krein, but its proof uses the same ideas which we sketched above in versions 1-2.

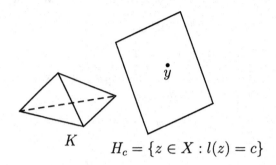

$$H_c = \{z \in X : l(z) = c\}$$

Figure 2.21: Separation of K and y by the hyperplane H_c.

Chapter 3

Unbounded Operators in Hilbert Space

"We were [initially] entirely in Heisenberg's footsteps. He had the idea that one should take matrices, although he did not know that his dynamical quantities were matrices.... And when one had such a programme of formulating everything in matrix language, it takes some effort to get rid of matrices. Though it seemed quite natural for me to represent perturbation theory in the algebraic way, this was not a particularly new way."
— Max Born

"...practical methods of applying quantum mechanics should be developed, which can lead to an explanation of the main features of complex atomic systems without too much computation."
— Paul Adrien Maurice Dirac

"...Mathematics ... are not only part of a special science, but are also closely connected with our general culture ..., a bridge to the Arts and Sciences, and the seemingly so non-exact sciences ...Our purpose is to help build such a bridge. Not for the sake of history but for the genesis of problems, facts and proofs, ... By going back to the roots of these conceptions, back through the dust of times past, the scars of long use would disappear, and they would be reborn to us as creatures full of life."
— Otto Toeplitz, 1926

Outline Quantum physics is one of the sources of problems in Functional Analysis, in particular the study of operators in Hilbert space. In the dictionary translating between Quantum physics and operators in Hilbert space we already saw that "quantum observables" are "selfadjoint operators." (See also Chapter 9, and especially Figure 9.1.)

As noted, even in a finite number of degrees of freedom, the relevant operators such as momentum, position, and energy are unbounded. Most Functional Analysis books stress the bounded case, and below we identify questions and theorems related to key issues for unbounded linear operators. (See, e.g., Appendix 3.A.)

In this chapter, we review the basic theory of unbounded operators in Hilbert space. For general notions, we refer to [DS88a, DS88b].

In the details below, the following concepts and terms will play an important role; (for reference, see also Appendix F): Hilbert space, Itō-isometry, wavelets, the canonical commutation relations, selfadjoint, and essentially selfadjoint linear operators, the adjoint operator to a densely defined operator, projection valued measure, unitary one-parameter groups.

3.1   Domain, Graph, and Adjoints

Among the classes of operators in Hilbert space, the family of selfadjoint linear operators is crucially important for a host of applications, e.g., to mathematical physics, and to the study of partial differential equations (PDE). For a study of each of the three classes of linear PDOs, elliptic, hyperbolic, and parabolic, the Spectral Theorem (see [Sto90, Yos95, Nel69, RS75, DS88b]) for associated unbounded selfadjoint operators is a "workhorse."

Every selfadjoint operator is densely defined, is closed, and it is necessarily (Hermitian) symmetric. For *unbounded operators*, the converse fails; although it does hold for bounded operators. It follows that selfadjointness is a much more restricting property than the related three properties. Moreover we will see that the distinction (between "symmetric" and "selfadjoint") lies at the heart of key issues from applications. We will further see that symmetric operators (also called "Hermitian" and "Hermitian symmetric" operators) with dense domain are automatically closable; but they may, or may not, have selfadjoint extensions; — again an issue of importance in physics.

The Spectral Theorem holds for selfadjoint operators, and for normal operators. But the case of normal operators reduces to the Spectral Theorem for two commuting selfadjoint operators.

On account of *Stone's theorem* (Appendix 3.A, Theorem 3.17) for one-parameter unitary groups we know that the class of selfadjoint operators coincides precisely with the infinitesimal generators of *strongly continuous* one-parameter groups of unitary operators acting on Hilbert space; — hence applications to the Schrödinger equation[1], and to wave equations.

Definition 3.1. Let $\mathcal{H}$ be a complex Hilbert space. An operator A is a linear mapping whose domain $dom\,(A)$ and range $ran\,(A)$ are subspaces in $\mathcal{H}$. The

---

[1]The reader will find the precise definition of Schrödinger equation in Section 10.1 (pg. 350); see also Appendix B (pg. 451).

kernel $ker\,(A)$ of A consists of all $a \in dom\,(A)$ such that $Aa = 0$.

Definition 3.2. The operator A is uniquely determined by its *graph*

$$\mathscr{G}(A) = \{(a, Aa) : a \in dom\,(A)\}. \tag{3.1}$$

(The parentheses in (3.1) denote an ordered pair, rather than an inner product.) Thus, $\mathscr{G}(A)$ is a subspace in $\mathscr{H} \oplus \mathscr{H}$ equipped with the inherited *graph inner product*

$$\langle a, b \rangle_A = \langle a, b \rangle + \langle Aa, Ab \rangle, \text{ and} \tag{3.2}$$

$$\|a\|_A^2 = \langle a, a \rangle_A, \ \forall a, b \in dom\,(A) \tag{3.3}$$

Proposition 3.1. *A subspace $K \subset \mathscr{H} \oplus \mathscr{H}$ is the graph of an operator if and only if $(0, a) \in K$ implies $a = 0$.*

Given two operators A and B, we say B is an extension of A, denoted by $B \supset A$, if $\mathscr{G}(B) \supset \mathscr{G}(A)$ in $\mathscr{H} \oplus \mathscr{H}$. The operator A is *closable* if $\overline{\mathscr{G}(A)}$ is the graph of an operator $\overline{A}$, namely, the closure of A. We say A is *closed* if $A = \overline{A}$.

Let A be a closed operator. A dense subspace $K \subset \mathscr{H}$ is called a *core* of A, if the closure of the restriction $A\big|_K$ is equal to A.

Let G be the *group* of all 3×3 real matrices $\begin{bmatrix} 1 & a & c \\ 0 & 1 & b \\ 0 & 0 & 1 \end{bmatrix}$ with *Lie algebra* $\mathfrak{g}$

consisting of matrices $\begin{bmatrix} 0 & a & c \\ 0 & 0 & b \\ 0 & 0 & 0 \end{bmatrix}$, $(a, b, c) \in \mathbb{R}^3$. Let $\mathcal{U}$ be a strongly continuous unitary representation of G acting on a Hilbert space $\mathscr{H}$. Then for every $X \in \mathfrak{g}$, $t \longmapsto \mathcal{U}\left(e^{tX}\right)$ defines a strongly continuous one-parameter group; and hence its infinitesimal generator, denoted $d\mathcal{U}(X)$ is a *skew-adjoint* operator with dense domain in $\mathscr{H}$.

For a more systematic account of the interplay between the Lie algebra and the corresponding Lie group, as it relates to representations, see also Chapter 8 below, especially Exercise 8.10 for the present setting.

Definition 3.3 (Schrödinger representation). The group of real 3×3 upper triangular matrices with ones in the diagonal places is called the Heisenberg group. It is a 3-dimensional nilpotent Lie group. The equivalence classes of its non-trivial irreducible unitary representations are indexed by $\mathbb{R} \setminus \{0\}$. The parameter in $\mathbb{R} \setminus \{0\}$ is usually denoted by h. If h is taken to be Plank's constant, we then say that the corresponding irreducible representation is the Schrödinger representation. As we note in formula (3.4) below, it is easy to realize this representation in the Hilbert space $L^2(\mathbb{R})$. (See Chapter 8, especially Section 8.5 and Theorem 8.4.)

Example 3.1. In this example, we consider $\mathscr{H} = L^2(\mathbb{R})$, and the unitary representation $\mathcal{U}$ of G is defined as follows:

For $f \in \mathcal{H} = L^2(\mathbb{R})$, set

$$(\mathcal{U}(g)f)(x) = e^{ih(c+bx)}f(x+a), \quad x \in \mathbb{R}; \tag{3.4}$$

called the *Schrödinger representation*.

Setting $h = 1$, and differentiating in the three directions in the Lie algebra, we get

$$(d\mathcal{U}(X_1)f)(x) = f'(x) = \frac{d}{dx}f, \tag{3.5}$$

$$(d\mathcal{U}(X_2)f)(x) = ix\,f(x), \text{ and} \tag{3.6}$$

$$(d\mathcal{U}(X_3)f)(x) = i\,f(x). \tag{3.7}$$

The first two operators in (3.5)-(3.6) are often written as follows:

$$d\mathcal{U}(X_1) = iP$$

where P is the *momentum operator* of a single quantum mechanical particle (wave function); and

$$d\mathcal{U}(X_2) = iQ$$

where Q is the corresponding *position operator* (in a single degree of freedom.) These two operators may be realized as follows:

Remark 3.1. It would appear that the function $f_\lambda(x) = e^{i\lambda x}$, λ fixed is an eigenfunction for $P = \frac{1}{i}\frac{d}{dx}$. For all λ,

$$Pf_\lambda = \lambda f_\lambda$$

holds pointwise, but "spectrum" depends on the ambient Hilbert space $\mathcal{H}$, in this case $\mathcal{H} = L^2(\mathbb{R})$; and $f_\lambda \notin L^2(\mathbb{R})$, so λ is not an eigenvalue. Nonetheless, if we allow an intervals for the λ variable, e.g., $a < \lambda < b$, with a and b being finite, then

$$F_{a,b}(x) = \int_a^b e^{i\lambda x}d\lambda = \frac{e^{ibx} - e^{iax}}{ix}$$

is in $L^2(\mathbb{R})$; and hence P has *continuous* spectrum. The functions $F_{a,b}(\cdot)$ are examples of wave-packets in quantum mechanics.

Example 3.2. The two operators d/dx and M_x in QM, are acting on $L^2(\mathbb{R})$ with common dense domain = the *Schwartz space* $\mathcal{S}$.

An alternative way to get a dense common domain, a way that works for all representations, is to use *Gårding*[2] *space*, or C^∞-*vectors*.

Let $u \in \mathcal{H}$ and define

$$u_\varphi := \int_G \varphi(g)\mathcal{U}_g u\,dg$$

---

[2]Appendix C includes a short biographical sketch of L. Gårding.

where $\varphi \in C_c^\infty$, and $\mathcal{U} \in Rep(G, \mathcal{H})$. Let φ_ϵ be an approximation of identity. Then for functions on G, $\varphi_\epsilon \star \psi \to \psi$ as $\epsilon \to 0$; and for C^∞ vectors, $u_{\varphi_\epsilon} \to u$, as $\epsilon \to 0$ in $\mathcal{H}$, i.e., in the $\|\cdot\|_{\mathcal{H}}$- norm.

The set $\{u_\varphi\}$ is dense in $\mathcal{H}$. It is called the Gårding space, or C^∞ vectors, or Schwartz space. Notice that not only u_φ is dense in $\mathcal{H}$, their derivatives are also dense in $\mathcal{H}$.

Differentiating $\mathcal{U}_g$, we then get a Lie algebra representation

$$\rho(X) := \frac{d}{dt}\big|_{t=0} \mathcal{U}\left(e^{tX}\right) = d\mathcal{U}(X).$$

Lemma 3.1. $\|u_{\varphi_\epsilon} - u\| \to 0$, as $\epsilon \to 0$.

Proof. Since $u - u_{\varphi_\epsilon} = \int \varphi_\epsilon(g)(u - \mathcal{U}_g u)dg$, we have

$$\|u_{\varphi_\epsilon} - u\| = \left\|\int_G \varphi_\epsilon(g)(u - \mathcal{U}_g u)dg\right\|$$
$$\leq \int_G \varphi_\epsilon(g)\|u - \mathcal{U}_g u\|\,dg$$

where the integration on G is with respect to Haar measure, and where we used the fact that $\int_G \varphi_\epsilon = 1$. Notice that we always assume the representations are norm continuous in the g variable, otherwise it is almost impossible to get anything interesting. i.e., assume $\mathcal{U}$ being strongly continuous. So for all $\delta > 0$, there is a neighborhood $\mathcal{O}$ of $e \in G$ so that $\|u - \mathcal{U}_g u\| < \delta$ for all $g \in \mathcal{O}$. Choose ϵ_δ so that φ_ϵ is supported in $\mathcal{O}$ for all $\epsilon < \epsilon_\delta$. Then the statement is proved.

$\square$

Corollary 3.1. *For all $X \in \mathfrak{g}$, $d\mathcal{U}(X)$ is essentially skew-adjoint on the Gårding domain.*

Proof. Let $\mathcal{U} \in Rep(G, \mathcal{H})$ be a unitary strongly continuous representation, where G is a Lie group with Lie algebra $\mathfrak{g}$. Let $X \in \mathfrak{g}$; we claim that $d\mathcal{U}(X)$ is essentially skew-adjoint (see [RS75, Nel69, vN32a, DS88b]) on the Gårding space, i.e., the span of vectors

$$v_\varphi = \int_G \varphi(g)\mathcal{U}_g v\,dg \tag{3.8}$$

where $\varphi \in C_c^\infty(G)$.

Hence we must show that

$$\ker\left(d\mathcal{U}(X)^* \pm I\right) = 0. \tag{3.9}$$

The argument is the same in both cases of (3.9). Hence we must show that, if $w \in \mathcal{H}$ satisfies

$$\langle d\mathcal{U}(X) v_\varphi - v_\varphi, w \rangle = 0 \tag{3.10}$$

for all v_φ in (3.8), then $w = 0$. Now if we view X as an invariant vector field on G, then (3.10) states that the continuous function

$$f_w(g) := \mathcal{U}(g)\,w \tag{3.11}$$

is a weak solution to the ODE

$$X f_w = f_w;$$

equivalently

$$f_{w,X}(t) := \mathcal{U}(\exp(tX))\,w \tag{3.12}$$

satisfies

$$\frac{d}{dt} f_{w,X}(t) = f_{w,X}(t), \quad t \in \mathbb{R}; \tag{3.13}$$

and so

$$f_{w,X}(t) = \text{const} \cdot e^t, \; t \in \mathbb{R}. \tag{3.14}$$

But, since $\mathcal{U}$ is unitary, $f_{w,X}$ in (3.12) is bounded; so the constant in (3.14) is zero. Hence $f_{w,X}(t) \equiv 0$. But $f_{w,X}(0) = w$, and so $w = 0$. $\square$

Now, let A be an arbitrary linear operator in a Hilbert space $\mathscr{H}$ with $dom\,(A)$ dense in $\mathscr{H}$.

Theorem 3.1. *The following are equivalent.*

1. *$A = \bar{A}$.*

2. *$\mathscr{G}(A) = \overline{\mathscr{G}(A)}$.*

3. *$dom\,(A)$ is a Hilbert space with respect to the graph inner product $\langle \cdot, \cdot \rangle_A$.*

4. *If $\{(a_n, Aa_n)\}_{n=1}^\infty$ is a sequence in $\mathscr{G}(A)$, and $(a_n, Aa_n) \to (a, b)$ as $n \to \infty$, then $(a, b) \in \mathscr{G}(A)$. In particular, $b = Aa$. (The round braces $(\cdot, \cdot)$ mean "pair-of vectors.")*

Proof. All follow from definitions. $\square$

Let X be a vector space over $\mathbb{C}$. Suppose there are two norms defined on X, such that

$$\|\cdot\|_1 \le \|\cdot\|_2. \tag{3.15}$$

Let $\overline{X_i}$ be the completion of X with respect to $\|\cdot\|_i$, $i = 1, 2$. The ordering (3.15) implies the identify map

$$\varphi : (X, \|\cdot\|_2) \to (X, \|\cdot\|_1)$$

is continuous, hence it has a unique continuous extension $\tilde{\varphi}$ to $\overline{X_2}$; and (3.15) passes to the closure $\overline{X_2}$. If $\tilde{\varphi}$ is injective, $\overline{X_2}$ is embedded into $\overline{X_1}$ as a dense subspace. In that case, $\|\cdot\|_1$ and $\|\cdot\|_2$ are said to be *topologically consistent*.

Lemma 3.2. $\|\cdot\|_1$ and $\|\cdot\|_2$ are topologically equivalent if and only if

$$\left\{ \begin{array}{c} \{x_n\} \subset X \text{ is Cauchy under } \|\cdot\|_2 \\ \text{(hence Cauchy under } \|\cdot\|_1) \\ \|x_n\|_1 \to 0 \end{array} \right\} \implies \|x_n\|_2 \to 0.$$

Proof. Note $\widetilde{\varphi}$ is linear, and

$$\ker \widetilde{\varphi} = \left\{ \begin{array}{c} x \in \overline{X_2} \mid \exists\, (x_n) \subset X,\ \|x_n - x\|_2 \to 0,\ \widetilde{\varphi}(x) = 0 \\ \text{(note } \widetilde{\varphi}(x) = \lim_n \varphi(x_n) = \lim_n x_n \text{ in } \overline{X_1}) \end{array} \right\}.$$

The lemma follows from this. □

Lemma 3.3. *The graph norm of A is topologically equivalent to $\|\cdot\| + \|A\cdot\|$.*

Proof. This follows from the estimate

$$\frac{1}{2} \left(\|x\| + \|Ax\| \right)^2 \leq \|x\|^2 + \|Ax\|^2 \leq \left(\|x\| + \|Ax\| \right)^2, \ \forall x \in dom\,(A).$$

□

Theorem 3.2. *An operator A is closable if and only if $\|\cdot\|$ and $\|\cdot\|_A$ are topologically equivalent. (When they are, the completion of $dom\,(A)$ with respect to $\|\cdot\|_A$ is identified as a subspace of $\mathscr{H}$.)*

Proof. First, assume A is closable. Let $\{x_n\}$ be a sequence in $dom\,(A)$. Suppose $\{x_n\}$ is a Cauchy sequence with respect to $\|\cdot\|_A$, and $\|x_n\| \to 0$. We need to show $\{x_n\}$ converges to 0 under the A-norm, i.e., $\|x_n\|_A \to 0$. Since $\{(x_n, Ax_n)\} \subset \mathscr{G}(A)$, and A is closable, it follows that $(x_n, Ax_n) \to (0,0) \in \mathscr{G}(A)$. Therefore, $\|Ax_n\| \to 0$, and (see Lemma 3.3)

$$\|x_n\|_A = \|x_n\| + \|Ax_n\| \to 0.$$

Conversely, assume $\|\cdot\|$ and $\|\cdot\|_A$ are topologically consistent. Let $\{x_n\} \subset dom\,(A)$, such that

$$(x_n, Ax_n) \to (0, b) \text{ in } \mathscr{H} \oplus \mathscr{H}. \tag{3.16}$$

We proceed to show that $b = 0$, which implies that A is closable.

By (3.16), $\{x_n\} \subset dom\,(A)$ is a Cauchy sequence with respect to the $\|\cdot\|_A$-norm, and $\|x_n\| \to 0$. Since the two norms are topologically consistent, then $\|x_n\|_A \to 0$ and so $\|Ax_n\| \to 0$. We conclude that $b = 0$. □

Corollary 3.2. *An operator A with dense domain is closable if and only if its adjoint A^* has dense domain.*

We will focus on unbounded operators. In the sequel, we will consider densely defined Hermitian (symmetric) operators. Such operators are necessarily closable.

The following result is usually applied to operators whose inverses are bounded.

Proposition 3.2. *Let A be a bounded operator with domain $dom\,(A)$, and acting in $\mathscr{H}$. Then $dom\,(A)$ is closed in $\|\cdot\|_A$ if and only if it is closed in $\|\cdot\|$. (That is, for bounded operators, $\|\cdot\|$ and $\|\cdot\|_A$ are topologically equivalent.)*

Proof. This is the result of the following estimate:

$$\|x\| \leq \|x\|_A = \|x\| + \|Ax\| \leq (1 + \|A\|)\,\|x\|\,,\ \forall x \in dom\,(A)\,.$$

$\square$

Corollary 3.3. *If A is a closed operator in $\mathscr{H}$ and A^{-1} is bounded, then $ran\,(A)$ is closed in both $\|\cdot\|$ and $\|\cdot\|_{A^{-1}}$.*

Proof. Note the $\mathscr{G}\,(A)$ is closed iff $\mathscr{G}\,(A^{-1})$ is closed; and $ran\,(A) = dom\,(A^{-1})$. Now, apply Proposition 3.2 to A^{-1}. $\square$

Let A be an operator in a Hilbert space $\mathscr{H}$. The set $\mathscr{G}\,(A)^{\perp}$ consists of $(-b^*, b)$ such that $(-b^*, b) \perp \mathscr{G}\,(A)$ in $\mathscr{H} \oplus \mathscr{H}$.

Proposition 3.3. *The following are equivalent.*

1. $\mathscr{D}\,(A)$ is dense in $\mathscr{H}$.

2. $(b, 0) \perp \mathscr{G}\,(A) \Longrightarrow b = 0$.

3. If $(b, -b^*) \perp \mathscr{G}\,(A)$, the map $b \mapsto b^*$ is well-defined.

Proof. Let $a \in \mathscr{D}\,(A)$, and $b, b^* \in \mathscr{H}$; then

$$(-b^*, b) \perp (a, Aa) \text{ in } \mathscr{H} \oplus \mathscr{H} \Longleftrightarrow \langle b^*, a \rangle = \langle b, Aa \rangle$$

and the desired results follow from this. $\square$

If any of the conditions is satisfied, $A^* : b \mapsto b^*$ defines an operator, called the adjoint of A, such that

$$\langle b, Aa \rangle = \langle A^*b, a \rangle \tag{3.17}$$

for all $a \in \mathscr{D}\,(A)$. $\mathscr{G}\,(A)^{\perp}$ is the inverted graph of A^*. The adjoints are only defined for operators with dense domains in $\mathscr{H}$.

Example 3.3. $A = d/dx$ on $L^2[0, 1]$ with dense domain

$$\mathscr{D} = \left\{ f \in C^1 \mid f\,(0) = f\,(1) = 0 \right\}.$$

Integration by parts shows that $A \subset -A^*$.

For unbounded operators, $(AB)^* = B^*A^*$ does not hold in general. The situation is better if one of them is bounded.

Theorem 3.3 ([Rud90, Theorem 13.2]). *If S, T, ST are densely defined operators then $(ST)^* \supset T^*S^*$. If, in addition, S is bounded then $(ST)^* = T^*S^*$.*

The next theorem follows directly from the definition of the adjoint operators.

Theorem 3.4. *If A is densely defined then $\mathscr{H} = \overline{\mathscr{R}(A)} \oplus \mathscr{K}(A^*)$.*

Finally, we recall some definitions.

Definition 3.4. Let A be a linear operator acting in $\mathscr{H}$. A is said to be

- selfadjoint if $A = A^*$.

- essentially selfadjoint if $\overline{A} = A^*$.

- normal if $A^*A = AA^*$.

- regular if $\mathscr{D}(A)$ is dense in $\mathscr{H}$, and closed in $\|\cdot\|_A$.

Definition 3.5. Let A be a linear operator on a Hilbert space $\mathscr{H}$. The resolvent $R(A)$ is defined as

$$R(A) = \left\{ \lambda \in \mathbb{C} : (\lambda - A)^{-1} \text{ exists} \right\} \text{ (the resolvent set)}$$

and the *spectrum* of A is the complement of $R(A)$, and it is denoted by $sp(A)$ or $\sigma(A)$.

Exercise 3.1 (The resolvent identity). Let A be a linear operator in a Hilbert space $\mathscr{H}$, and, for $\lambda_i \in R(A)$, $i = 1, 2$ consider two operators $(\lambda_i - A)^{-1}$. Show that

$$(\lambda_1 - A)^{-1} - (\lambda_2 - A)^{-1} = (\lambda_2 - \lambda_1)(\lambda_1 - A)^{-1}(\lambda_2 - A)^{-1}.$$

This formula is called the resolvent identity.

3.2 Characteristic Matrix

The method of characteristic matrix was developed by M.H. Stone [Sto51]. It is extremely useful in operator theory, but has long been overlooked in the literature. We recall some of its applications in normal operators[3].

If $\mathscr{H}$ is a fixed Hilbert space, and A a given liner operator, then its graph

$$\mathscr{G}(A) = \left\{ \begin{bmatrix} u \\ Au \end{bmatrix} : u \in dom(A) \right\}$$

is a linear subspace in $\mathscr{H} \oplus \mathscr{H}$, represented as column vectors $\begin{bmatrix} u \\ v \end{bmatrix}$, $u, v \in \mathscr{H}$.

In the case where A is assumed closed, we now compute the projection onto $\mathscr{G}(A) = $ the $\mathscr{H} \oplus \mathscr{H}$-closure of the graph.

---

[3]Appendix C includes a short biographical sketch of M. H. Stone. See also [Sto90].

Let A be an operator in a Hilbert space $\mathcal{H}$. Let $P = (P_{ij})$ be the projection from $\mathcal{H} \oplus \mathcal{H}$ onto $\overline{\mathcal{G}(A)}$. The 2×2 operator matrix (P_{ij}) of bounded operators in $\mathcal{H}$ is called the characteristic matrix of A. (Also called the characteristic projection. See the discussion in Section 3.3 below.)

Since $P^2 = P^* = P$, the following identities hold

$$P_{ij}^* = P_{ji} \tag{3.18}$$

$$\sum_k P_{ik}P_{kj} = P_{ij}. \tag{3.19}$$

In particular, P_{11} and P_{22} are selfadjoint.

Theorem 3.5. *Let $P = (P_{ij})$ be the projection from $\mathcal{H} \oplus \mathcal{H}$ onto a closed subspace $\mathcal{K}$. The following are equivalent.*

1. $\mathcal{K}$ is the graph of an operator.

2.

$$\begin{bmatrix} P_{11} & P_{12} \\ P_{21} & P_{22} \end{bmatrix} \begin{bmatrix} 0 \\ a \end{bmatrix} = \begin{bmatrix} 0 \\ a \end{bmatrix} \implies a = 0.$$

3.

$$\left(P_{12}a = 0, P_{22}a = a \right) \implies a = 0.$$

If any of these conditions is satisfied, let A be the operator with $\mathcal{G}(A) = \mathcal{K}$, then for all $a, b \in \mathcal{H}$,

$$\begin{bmatrix} P_{11} & P_{12} \\ P_{21} & P_{22} \end{bmatrix} \begin{bmatrix} a \\ b \end{bmatrix} = \begin{bmatrix} P_{11}a + P_{12}b \\ P_{21}a + P_{22}b \end{bmatrix} \in \mathcal{G}(A);$$

i.e.,

$$A : (P_{11}a + P_{12}b) \mapsto P_{21}a + P_{22}b. \tag{3.20}$$

In particular,

$$AP_{11} = P_{21} \tag{3.21}$$

$$AP_{12} = P_{22}. \tag{3.22}$$

Proof. Let $v := (a, b) \in \mathcal{H} \oplus \mathcal{H}$. Then $v \in \mathcal{K}$ if and only if $Pv = v$; and the theorem follows from this.   □

The next theorem describes the adjoint operators.

Theorem 3.6. *Let A be an operator with characteristic matrix $P = (P_{ij})$. The following are equivalent.*

1. $\mathcal{D}(A)$ is dense in $\mathcal{H}$.

2. $\begin{bmatrix} b \\ 0 \end{bmatrix} \perp \mathscr{G}(A) \Longrightarrow b = 0.$

3. If $\begin{bmatrix} -b^* \\ b \end{bmatrix} \in \mathscr{G}(A)^\perp$, the map $A^* : b \mapsto b^*$ is a well-defined operator.

4. $\begin{bmatrix} 1 - P_{11} & -P_{12} \\ -P_{21} & 1 - P_{22} \end{bmatrix} \begin{bmatrix} b \\ 0 \end{bmatrix} = \begin{bmatrix} b \\ 0 \end{bmatrix} \Longrightarrow b = 0.$

5. $\left(P_{11}b = 0, P_{21}b = 0 \right) \Longrightarrow b = 0.$

If any of the above conditions is satisfied, then

$$\begin{bmatrix} 1 - P_{11} & -P_{12} \\ -P_{21} & 1 - P_{22} \end{bmatrix} \begin{bmatrix} a \\ b \end{bmatrix} = \begin{bmatrix} (1 - P_{11})a - P_{12}b \\ (1 - P_{22})b - P_{21}a \end{bmatrix} \in \mathscr{G}(A)^\perp$$

that is,

$$A^* : P_{21}a - (1 - P_{22})b \mapsto (1 - P_{11})a - P_{12}b. \tag{3.23}$$

In particular,

$$A^* P_{21} = 1 - P_{11} \tag{3.24}$$
$$A^*(1 - P_{22}) = P_{12}. \tag{3.25}$$

Proof. $(1) \Leftrightarrow (2) \Leftrightarrow (3)$ is a restatement of Proposition 3.3. Note the projection from $\mathscr{H} \oplus \mathscr{H}$ on $\mathscr{G}(A)^\perp = \overline{\mathscr{G}(A)}^\perp$ is

$$1 - P = \begin{bmatrix} 1 - P_{11} & -P_{12} \\ -P_{21} & 1 - P_{22} \end{bmatrix}$$

and so

$$\begin{bmatrix} b \\ 0 \end{bmatrix} \perp \mathscr{G}(A) \Longleftrightarrow (1 - P) \begin{bmatrix} b \\ 0 \end{bmatrix} = \begin{bmatrix} b \\ 0 \end{bmatrix}.$$

Therefore, $(2) \Leftrightarrow (4) \Leftrightarrow (5)$. Finally, (3.23)-(3.25) follow from the definition of A^*. $\qquad\square$

Theorem 3.7. *Let A be a regular operator (i.e., densely defined, closed) with characteristic matrix $P = (P_{ij})$.*

1. *The matrix entries P_{ij} are given by*

$$\begin{array}{llll} P_{11} &=& (1 + A^*A)^{-1} & \qquad P_{12} &=& A^*(1 + AA^*)^{-1} \\ P_{21} &=& A(1 + A^*A)^{-1} & \qquad P_{22} &=& AA^*(1 + AA^*)^{-1} \end{array} \tag{3.26}$$

2. $1 - P_{22} = (1 + AA^*)^{-1}.$

3. $1 + A^*A,\ 1 + AA^*$ *are selfadjoint operators.*

4. The following containments hold

$$A^*(1 + AA^*)^{-1} \supset (1 + A^*A)^{-1}A^* \qquad (3.27)$$

$$A(1 + A^*A)^{-1} \supset (1 + AA^*)^{-1}A. \qquad (3.28)$$

Proof. By (3.21) and (3.24), we have

$$\begin{bmatrix} AP_{11} = P_{21} \\ A^*P_{21} = 1 - P_{11} \end{bmatrix} \Longrightarrow A^*AP_{11} = 1 - P_{11}, \text{ i.e., } (1 + A^*A)P_{11} = 1.$$

That is, $1 + A^*A$ is a Hermitian extension of P_{11}^{-1}. By (3.18), P_{11} is selfadjoint and so is P_{11}^{-1}. Therefore, $1 + A^*A = P_{11}^{-1}$, or

$$P_{11} = (1 + A^*A)^{-1}.$$

By (3.21),

$$P_{21} = AP_{11} = A(1 + A^*A)^{-1}.$$

Similarly, by (3.22) and (3.25), we have

$$\begin{bmatrix} AP_{12} = P_{22} \\ A^*(1 - P_{22}) = P_{12} \end{bmatrix} \Longrightarrow AA^*(1 - P_{22}) = P_{22}, \text{ i.e.,}$$

$$(1 + AA^*)(1 - P_{22}) = 1.$$

This means $1 + AA^* \supset (1 - P_{22})^{-1}$ is a Hermitian extension of the selfadjoint operator $(1 - P_{22})^{-1}$ (note P_{22} is selfadjoint), hence $1 + AA^*$ is selfadjoint, and

$$1 - P_{22} = (1 + AA^*)^{-1}.$$

By (3.25),

$$P_{12} = A^*(1 - P_{22}) = A^*(1 + AA^*)^{-1}.$$

By (3.22),

$$P_{22} = AP_{12} = AA^*(1 + AA^*)^{-1}.$$

We have proved (1), (2) and (3).

Finally,

$$P_{12} = P_{21}^* = (AP_{11})^* \supset P_{11}A^*$$

yields (3.27); and

$$P_{21} = P_{12}^* = (A^*(1 - P_{22}))^* \supset (1 - P_{22})A$$

gives (3.28).   □

Corollary 3.4. *Let A be a regular operator in a fixed Hilbert space $\mathcal{H}$, and let $P = (P_{ij})_{i,j=1}^2$ be the characteristic projection in $\mathcal{H} \oplus \mathcal{H}$. Then A is normal in $\mathcal{H}$, i.e., $AA^* = A^*A$, if and only if the four bounded operator $P_{ij} : \mathcal{H} \longrightarrow \mathcal{H}$ form a commutative family.*

Exercise 3.2 $(A^{**} = \overline{A})$. Let A be a *regular* operator in a Hilbert space (i.e., we assume that A has dense domain and is closable.) Then show that

$$A^{**} = \overline{A} \tag{3.29}$$

where $\overline{A}$ denotes the *closure* of A; i.e., $\mathscr{G}\left(\overline{A}\right) = \overline{\mathscr{G}\left(A\right)}$.

Hint: Establish the desired identity (3.29) by justifying the following steps: Set $\chi : \mathscr{H}^2 \longrightarrow \mathscr{H}^2$,

$$\chi \begin{pmatrix} x \\ y \end{pmatrix} = \begin{pmatrix} -y \\ x \end{pmatrix}, \; \forall \begin{pmatrix} x \\ y \end{pmatrix} \in \mathscr{H}^2.$$

Then

$$\begin{aligned}
\mathscr{G}\left(A^{**}\right) &= \left(\chi \mathscr{G}\left(A^*\right)\right)^{\perp} \\
&= \left(\chi \left(\chi \mathscr{G}\left(A\right)\right)^{\perp}\right)^{\perp} \\
&= \left(\chi^2 \mathscr{G}\left(A\right)\right)^{\perp\perp} \\
&= \left(\mathscr{G}\left(A\right)\right)^{\perp\perp} = \overline{\mathscr{G}\left(A\right)} = \mathscr{G}\left(\overline{A}\right).
\end{aligned}$$

3.2.1 Commutants

Let A, B be operators in a Hilbert space $\mathscr{H}$, and suppose B is bounded. The operator B is said to commute (strongly) with A if $BA \subset AB$.

Lemma 3.4. *Assume that $\overline{A}$ exists. Then B commutes with A if and only if B commutes with $\overline{A}$.*

Proof. Suppose $BA \subset AB$, and we check that $B\overline{A} \subset \overline{A}B$. The converse is trivial. For $(a, \overline{A}a) \in \mathscr{G}(\overline{A})$, choose a sequence $(a_n, Aa_n) \in \mathscr{G}(A)$ such that $(a_n, Aa_n) \to (a, \overline{A}a)$. By assumption, $(Ba_n, ABa_n) = (Ba_n, BAa_n) \in \mathscr{G}(A)$. Thus,

$$(Ba_n, ABa_n) \to \left(Ba, B\overline{A}a\right) \in \mathscr{G}(\overline{A}).$$

That is, $Ba \in \mathscr{D}(\overline{A})$ and $\overline{A}Ba = B\overline{A}a$. □

Lemma 3.5. *Let A be a closed operator with characteristic matrix $P = (P_{ij})$. Let B be a bounded operator, and*

$$Q_B := \begin{bmatrix} B & 0 \\ 0 & B \end{bmatrix}.$$

1. *B commutes with $A \Leftrightarrow B$ leaves $\mathscr{G}(A)$ invariant $\Leftrightarrow Q_B P = P Q_B P$.*

2. *B commutes with $P_{ij} \Leftrightarrow Q_B P = P Q_B \Leftrightarrow Q_{B^*} P = P Q_{B^*} \Leftrightarrow B^*$ commutes with P_{ij}.*

3. *If B, B^* commute with A, then B, B^* commute with P_{ij}.*

Proof. Obvious.   □

A closed operator is said to be *affiliated* with a von Neumann algebra $\mathfrak{M}$ if it commutes with every unitary operator in $\mathfrak{M}'$. By [KR97a, Thm 4.1.7], every operator in $\mathfrak{M}'$ can be written as a finite linear combination of unitary operators in $\mathfrak{M}'$. Thus, A is affiliated with $\mathfrak{M}$ if and only if A commutes with every operator in $\mathfrak{M}'$.

Remark 3.2. Let $\mathfrak{M}$ be a von Neumann algebra. Let $x \in \mathfrak{M}$ such that $\|x\| \leq 1$ and $x = x^*$. Set $y := x + i\sqrt{1 - x^2}$. Then, $y^*y = yy^* = x^2 + 1 - x^2 = 1$, i.e., y is unitary. Also, $x = (y + y^*)/2$.

Theorem 3.8. *Let A be a closed operator with characteristic matrix $P = (P_{ij})$. Let $\mathfrak{M}$ be a von Neumann algebra, and*

$$Q_B := \begin{bmatrix} B & 0 \\ 0 & B \end{bmatrix}, \quad \forall B \in \mathfrak{M}'.$$

The following are equivalent:

1. *A is affiliated with $\mathfrak{M}$.*

2. *$PQ_B = Q_BP$, for all $B \in \mathfrak{M}'$.*

3. *$P_{ij} \in \mathfrak{M}$.*

4. *If $\mathscr{D}(A)$ is dense, then A^* is affiliated with $\mathfrak{M}$.*

Proof. Notice that $\mathfrak{M}$ is selfadjoint. The equivalence of $1, 2, 3$ is a direct consequence of Lemma 3.5.

$P^\perp := 1 - P$ is the projection onto the inverted graph of A^*, should the latter exists. $PQ_B = Q_BP$ if and only if $P^\perp Q_B = Q_B P^\perp$. Thus, 1 is equivalent to 4.   □

3.3   Unbounded Operators Between Different Hilbert Spaces

While the theory of unbounded operators has been focused on spectral theory where it is then natural to consider the setting of linear *endomorphisms* with dense domain in a fixed Hilbert space; many applications entail operators between distinct Hilbert spaces, say $\mathscr{H}_1$ and $\mathscr{H}_2$. Typically the facts given about the two differ greatly from one Hilbert space to the next.

Let $\mathscr{H}_i$, $i = 1, 2$ be two complex Hilbert spaces. The respective inner products will be written $\langle \cdot, \cdot \rangle_i$, with the subscript to identify the Hilbert space in question.

Definition 3.6. A linear operator T from $\mathscr{H}_1$ to $\mathscr{H}_2$ is a pair $\mathscr{D} \subset \mathscr{H}_1$, T, where $\mathscr{D}$ is a linear subspace in $\mathscr{H}_1$, and $T\varphi \in \mathscr{H}_2$ is well-defined for all $\varphi \in \mathscr{D}$. We require linearity on $\mathscr{D}$; so

$$T(\varphi + c\psi) = T\varphi + cT\psi, \quad \forall \varphi, \psi \in \mathscr{D}, \forall c \in \mathbb{C}. \tag{3.30}$$

Notation. When T is given as in (3.30), we say that $\mathscr{D} = dom\,(T)$ is the domain of T, and

$$\mathscr{G}\,(T) = \left\{ \begin{pmatrix} \varphi \\ T\varphi \end{pmatrix} \; ; \; \varphi \in \mathscr{D} \right\} \subset \begin{pmatrix} \mathscr{H}_1 \\ \oplus \\ \mathscr{H}_2 \end{pmatrix} \qquad (3.31)$$

is the graph. By closure, we shall refer to closure in the norm of $\mathscr{H}_1 \oplus \mathscr{H}_2$, i.e.,

$$\left\| \begin{pmatrix} h_1 \\ h_2 \end{pmatrix} \right\|^2 = \|h_1\|_1^2 + \|h_2\|_2^2, \quad h_i \in \mathscr{H}_i. \qquad (3.32)$$

If the closure $\overline{\mathscr{G}\,(T)}$ is the graph of a linear operator, we say that T is *closable*.

If $dom\,(T)$ is dense in $\mathscr{H}_1$, we say that T is densely defined. The abbreviated notation $\mathscr{H}_1 \xrightarrow{\;T\;} \mathscr{H}_2$ will be used when the domain of T is understood from the context.

Definition 3.7. Let $\mathscr{H}_1 \xrightarrow{\;T\;} \mathscr{H}_2$ be a densely defined operator, and consider the subspace $dom\,(T^*) \subset \mathscr{H}_2$ defined as follows:

$$dom\,(T^*) = \Big\{ h_2 \in \mathscr{H}_2 \; ; \; \exists C = C_{h_2} < \infty \text{ s.t.}$$

$$|\langle T\varphi, h_2 \rangle_2| \le C \,\|\varphi\|_1 \,, \; \forall \varphi \in dom\,(T) \Big\}. \qquad (3.33)$$

Then there is a unique $h_1 \in \mathscr{H}_1$ s.t.

$$\langle T\varphi, h_2 \rangle_2 = \langle \varphi, h_1 \rangle_1, \text{ and} \qquad (3.34)$$

we set $T^* h_2 = h_1$.

It is immediate that T^* is an operator from $\mathscr{H}_2$ to $\mathscr{H}_1$; we write

$$\mathscr{H}_1 \overset{T}{\underset{T^*}{\rightleftarrows}} \mathscr{H}_2 \qquad (3.35)$$

The following holds:

Lemma 3.6. *Given a densely defined operator* $\mathscr{H}_1 \xrightarrow{\;T\;} \mathscr{H}_2$, *then* T *is closable if and only if* $dom\,(T^*)$ *is dense in* $\mathscr{H}_2$.

Proof. Set $\chi : \mathscr{H}_1 \oplus \mathscr{H}_2 \longrightarrow \mathscr{H}_2 \oplus \mathscr{H}_1$, $\chi \begin{pmatrix} h_1 \\ h_2 \end{pmatrix} = \begin{pmatrix} -h_2 \\ h_1 \end{pmatrix}$, then

$$\mathscr{G}\,(T^*) = (\chi\,(\mathscr{G}\,(T)))^{\perp}. \qquad (3.36)$$

For a general vector $\begin{pmatrix} h_2 \\ h_1 \end{pmatrix} \in \begin{pmatrix} \mathscr{H}_2 \\ \oplus \\ \mathscr{H}_1 \end{pmatrix}$ we have the following:

$$\begin{pmatrix} h_2 \\ h_1 \end{pmatrix} \in (\chi \left(\mathscr{G} \left(T \right) \right))^{\perp}$$

$$\Updownarrow$$

$$\left\langle \begin{pmatrix} h_2 \\ h_1 \end{pmatrix}, \begin{pmatrix} -T\varphi \\ \varphi \end{pmatrix} \right\rangle = 0, \; \forall \varphi \in \mathscr{D} \; (\text{dense in } \mathscr{H}_1)$$

$$\Updownarrow$$

$$\langle h_2, T\varphi \rangle_2 = \langle h_1, \varphi \rangle_1, \; \forall \varphi \in \mathscr{D}$$

$$\Updownarrow$$

$$h_2 \in dom \left(T^* \right), \text{ and } T^* h_2 = h_1$$

$$\Updownarrow$$

$$\begin{pmatrix} h_2 \\ h_1 \end{pmatrix} \in \mathscr{G} \left(T^* \right).$$

Notation. The symbol "$\perp$" denotes ortho-complement.

We must show that

$$\begin{pmatrix} 0 \\ h_2 \end{pmatrix} \in \mathscr{G} \left(T \right)^{\perp\perp} = \overline{\mathscr{G} \left(T \right)}$$

$$\Updownarrow$$

$$h_2 \in dom \left(T^* \right)^{\perp} = \mathscr{H}_2 \ominus dom \left(T^* \right).$$

But this is immediate from

$$\mathscr{G} \left(T \right)^{\perp\perp} = (\chi \left(\mathscr{G} \left(T^* \right) \right))^{\perp}. \tag{3.37}$$

$$\square$$

Remark 3.3 (Notation and Facts).

1. Let T be an operator $\mathscr{H}_1 \xrightarrow{T} \mathscr{H}_2$ and $\mathscr{H}_i$, $i = 1, 2$, two given Hilbert spaces. Assume $\mathscr{D} := dom \left(T \right)$ is dense in $\mathscr{H}_1$, and that T is *closable*. Then there is a unique *closed* operator, denoted $\overline{T}$ such that

$$\mathscr{G} \left(\overline{T} \right) = \overline{\mathscr{G} \left(T \right)} \tag{3.38}$$

where "—" on the RHS in (3.38) refers to norm closure in $\mathscr{H}_1 \oplus \mathscr{H}_2$, see (3.32).

2. We say that an operator $\mathscr{H}_1 \xrightarrow{T} \mathscr{H}_2$ is bounded iff $dom \left(T \right) = \mathscr{H}_1$, and $\exists C < \infty$ s.t.,

$$\|T h_1\|_2 \leq C \|h_1\|_1, \quad \forall h_1 \in \mathscr{H}_1, \tag{3.39}$$

where the subscripts refer to the respective Hilbert spaces. The infimum of the set of constants C in (3.39) is called the *norm* of T, written $\|T\|$.

3. If T is bounded, then so is T^*, in particular, $dom\,(T)^* = \mathcal{H}_2$, and the following identities hold:

$$\|T^*\| = \|T\|, \tag{3.40}$$

$$\|T^*T\| = \|T\|^2. \tag{3.41}$$

Note that formulas like (3.40) and (3.41) refer to the underlying Hilbert spaces, and that (3.40) writes out in more detail as follows:

$$\|T^*\|_{\mathcal{H}_2 \to \mathcal{H}_1} = \|T\|_{\mathcal{H}_1 \to \mathcal{H}_2}. \tag{3.42}$$

Example 3.4. An operator $T : \mathcal{H}_1 \longrightarrow \mathcal{H}_2$ with dense domain such that $dom\,(T^*) = 0$, i.e., "extremely" non-closable.

Let $J = [-\pi, \pi]$ be the 2π-period interval, $d\lambda_1 = dx$ the normalized Lebesgue measure, $\mu_3 = $ the middle-third Cantor measure. Recall the transformation of μ_3 is given by

$$\widehat{\mu}_3\,(\xi) = \int_{-\pi}^{\pi} e^{i\xi x} d\mu_3\,(x) = \prod_{n=1}^{\infty} \cos\left(\xi/3^n\right), \quad \xi \in \mathbb{R}. \tag{3.43}$$

Recall, the support E of μ_3 is the standard Cantor set, and that $\lambda_1\,(E) = 0$.

Now set $\mathcal{H}_1 = L^2\,(\lambda_1)$, and $\mathcal{H}_2 = L^2\,(\mu_3)$. The space $\mathcal{D} := C_c\,(-\pi, \pi)$ is dense in both $\mathcal{H}_1$ and in $\mathcal{H}_2$ with respect to the two L^2-norms. Hence, setting

$$T\varphi = \varphi, \quad \forall \varphi \in \mathcal{D}; \tag{3.44}$$

the identity operator, then becomes a Hilbert space operator $\mathcal{H}_1 \xrightarrow{T} \mathcal{H}_2$.

Using Definition 3.7, we see that $h_2 \in L^2\,(\mu_3)$ is in $dom\,(T^*)$ iff $\exists h_1 \in L^2\,(\lambda_1)$ such that

$$\int \varphi\,h_1\,d\lambda_1 = \int \varphi\,h_2\,d\mu_3, \quad \forall \varphi \in \mathcal{D}. \tag{3.45}$$

Since $\mathcal{D}$ is dense in both L^2-spaces, we get

$$\int_E h_1\,d\lambda_1 = \int_E h_2\,d\mu_3. \tag{3.46}$$

Now suppose $h_2 \neq 0$ in $L^2\,(\mu_3)$, then there is a subset $A \subset E$ s.t. $h_2 > 0$ on A, $\mu_3\,(A) > 0$, and $\int_A h_2\,d\mu_3 > 0$. But $\int_A h_1\,d\lambda_1 = \int_A h_2\,d\mu_3$, and $\int_A h_1\,d\lambda_1 = 0$ since $\lambda_1\,(A) = 0$. This contradiction proves that $dom\,(T^*) = 0$; and in particular T in (3.44) is unbounded and non-closable as claimed.

Remark 3.4. In Example 3.4, i.e., T is given by (3.44), and $\mathcal{H}_1 = L^2\,(\lambda_1)$, $\mathcal{H}_2 = L^2\,(\mu_3)$, $\lambda_1 = $ Lebesgue, $\mu_3 = $ Cantor measure, we have

$$\overline{\mathcal{G}\,(T)} = \begin{pmatrix} \mathcal{H}_1 \\ \oplus \\ \mathcal{H}_2 \end{pmatrix} = L^2\,(\lambda_1) \oplus L^2\,(\mu_3). \tag{3.47}$$

The proof of this fact (3.47) uses basic approximation properties of iterated function system (IFS) measures . Both λ_1 and μ_3 are IFS-measures. See, e.g., [Hut81, BHS05].

Theorem 3.9 (The Polar Decomposition/Factorization). *Let $\mathscr{H}_1 \xrightarrow{T} \mathscr{H}_2$ be a densely defined operator, and assume that $dom(T^*)$ is dense in $\mathscr{H}_2$, i.e., T is closable, then both of the operators $T^*\overline{T}$ and $\overline{T}T^*$ are densely defined, and both are selfadjoint.*

Moreover, there is a partial isometry *$U : \mathscr{H}_1 \longrightarrow \mathscr{H}_2$ with initial space in $\mathscr{H}_1$ and final space in $\mathscr{H}_2$ such that*

$$T = U \left(T^*\overline{T}\right)^{\frac{1}{2}} = \left(\overline{T}T^*\right)^{\frac{1}{2}} U. \tag{3.48}$$

(Eq. (3.48) is called the polar decomposition.)

The proof of Theorem 3.9 will follow from the following more general considerations.

Remark 3.5. If $\mathscr{H}_i$, $i = 1, 2, 3$, are three Hilbert spaces with operators $\mathscr{H}_1 \xrightarrow{A} \mathscr{H}_2 \xrightarrow{B} \mathscr{H}_3$, then the domain of BA is as follows:

$$dom(BA) = \{x \in dom(A) \ ; \ Ax \in dom(B)\}, \text{ and} \tag{3.49}$$

on $x \in dom(BA)$, we have $(BA)(x) = B(Ax)$. In general, $dom(BA)$ may be 0, even if the two operators A and B have dense domains; see Example 3.4.

Definition 3.8 (The characteristic projection; see [Jør80, Sto51]). Let $\mathscr{H}_i$, $i = 1, 2$ be Hilbert spaces and $\mathscr{H}_1 \xrightarrow{T} \mathscr{H}_2$ a fixed operator; the characteristic projection P of T is the projection in $\mathscr{H}_1 \oplus \mathscr{H}_2$ onto $\mathscr{G}(T)$. We shall write

$$P = \begin{pmatrix} P_{11} & P_{12} \\ P_{21} & P_{22} \end{pmatrix}$$

where the components are bounded operators as follows:

$$P_{11} : \mathscr{H}_1 \longrightarrow \mathscr{H}_1, \qquad P_{12} : \mathscr{H}_2 \longrightarrow \mathscr{H}_1,$$
$$P_{21} : \mathscr{H}_1 \longrightarrow \mathscr{H}_2, \qquad P_{22} : \mathscr{H}_2 \longrightarrow \mathscr{H}_2.$$

Since

$$P = P^* = P^2 \tag{3.50}$$

we have

$$P_{11} = P_{11}^* \geq 0, \qquad P_{12} = P_{21}^*,$$
$$P_{21} = P_{12}^*, \text{ and} \qquad P_{22} = P_{22}^* \geq 0,$$

where "$\geq$" refers to the natural order as selfadjoint operators.

From (3.50), we further get

$$P_{ij} = \sum_{k=1}^{2} P_{ik}P_{kj}, \quad \forall i, j = 1, 2. \tag{3.51}$$

Lemma 3.7. *Let T be as above, and let $P = P(T) = (P_{i,j})_{i,j=1}^2$ be its characteristic projection, then T is closable if and only if*

$$\ker\left(I_{\mathscr{H}_2} - P_{22}\right) = 0; \tag{3.52}$$

i.e., iff the following implication holds:

$$\boxed{y \in \mathscr{H}_2,\ P_{22}y = y} \Longrightarrow \boxed{y = 0} \tag{3.53}$$

Proof. If $P = (P_{ij})$ is the characteristic projection for T, it follows from (3.36) and Lemma 3.6, that

$$P^{\vee} := \begin{pmatrix} I_2 - P_{22} & P_{21} \\ P_{12} & I_1 - P_{11} \end{pmatrix} \text{ in } \begin{pmatrix} \mathscr{H}_2 \\ \oplus \\ \mathscr{H}_1 \end{pmatrix} \tag{3.54}$$

is the characteristic projection matrix for the operator $T^* : \mathscr{H}_2 \longrightarrow \mathscr{H}_1$.
    Hence

$$T^*\left(y - P_{22}y\right) = P_{12}y, \quad \forall y \in \mathscr{H}_2, \text{ and} \tag{3.55}$$
$$T^* P_{21}x = x - P_{11}x, \quad \forall x \in \mathscr{H}_1. \tag{3.56}$$

Note that T is closable iff

$$\begin{pmatrix} 0 \\ y \end{pmatrix} \in \overline{\mathscr{G}(T)} \Longrightarrow y = 0. \tag{3.57}$$

But by (3.51), $\begin{pmatrix} 0 \\ y \end{pmatrix} \in \overline{\mathscr{G}(T)}$ holds iff $P_{12}y = 0$ and $P_{22}y = y$.

    But by (3.55), we have $P_{22}y = y \Longrightarrow P_{12}y = 0$, and the desired conclusion follows. $\square$

    We now turn to the proof of Theorem 3.9. Given is a closable operator $\mathscr{H}_1 \xrightarrow{T} \mathscr{H}_2$ with dense domain. To simplify notation, we shall denote $\overline{T}$ also by T.

Proof. (Theorem 3.9) By passing to the closure $\overline{T}$ of the given operator T, we get $\mathscr{G}(T) = \mathscr{G}(\overline{T})$, where $\overline{T}$ is then a closed operator. It is easy to see that

$$\left(\overline{T}\right)^* = T^* \tag{3.58}$$

on its dense domain in $\mathscr{H}_2$. Below, we shall write T for the associated closured operator $\overline{T}$.
    In Lemma 3.7 we formed the characteristic projection for both T and T^*; see especially (3.54). As a result we get

$$\begin{cases} TP_{11} = P_{21} & \text{on } \mathscr{H}_1, \text{ and} \\ TP_{12} = P_{22} & \text{on } \mathscr{H}_2. \end{cases} \tag{3.59}$$

Similarly (see (3.54))

$$\begin{cases} T^* \left(I - P_{22} \right) = P_{12} & \text{on } \mathscr{H}_2, \text{ and} \\ T^* P_{21} = I - P_{11} & \text{on } \mathscr{H}_1. \end{cases} \tag{3.60}$$

Note that part of the conclusion in (3.59) is that $ran\,(P_{11}) = \{P_{11}x \,;\, x \in \mathscr{H}_1\}$ is contained in $dom\,(T)$.

Further note that

$$\ker\,(P_{11}) = ran\,(P_{11})^{\perp} = 0, \quad \text{and} \tag{3.61}$$

$$\ker\,(I - P_{22}) = (ran\,(I - P_{22}))^{\perp} = 0. \tag{3.62}$$

We already proved (3.62), and we now turn to (3.61).

Combining (3.59) and (3.60), we get $T^*T P_{11} = I - P_{11}$, and so

$$(1 + T^*T)\,P_{11} = I; \tag{3.63}$$

equivalently,

$$P_{11}x + T^*T P_{11}x = x, \quad \forall x \in \mathscr{H}_1.$$

So $ran\,(P_{11})$ is dense in $\mathscr{H}_1$, and $ran\,(P_{11}) \subset dom\,(T^*T)$; so $dom\,(T^*T)$ is dense in $\mathscr{H}_1$. By (3.63) therefore,

$$T^*T = (P_{11})^{-1} - I \tag{3.64}$$

is a selfadjoint operator with dense domain $ran\,(P_{11})$ in $\mathscr{H}_1$.

The conclusions in Theorem 3.9 now follow.   □

Corollary 3.5. *Let $\mathscr{H}_1 \xrightarrow{T} \mathscr{H}_2$ be a closed operator with dense domain in $\mathscr{H}_1$, and let T^*T and TT^* be the selfadjoint operators from Theorem 3.9. Then the entries of the characteristic matrix are as follows:*

$$\begin{pmatrix} P_{11} = (I_1 + T^*T)^{-1} & P_{12} = T^* (I_2 + TT^*)^{-1} \\ P_{21} = T(I_1 + T^*T)^{-1} & P_{22} = I_2 - (I_2 + TT^*)^{-1} \end{pmatrix} \tag{3.65}$$

where I_i denote the identity operators in the respective Hilbert spaces, $i = 1, 2$.

Corollary 3.6. *Let Let $\mathscr{H}_1 \xrightarrow{T} \mathscr{H}_2$ be as above (T densely defined but not assumed closable), and let $(P_{ij})_{ij=1}^2$ be the characteristic projection referring to the sum Hilbert space $\mathscr{H}_1 \oplus \mathscr{H}_2$. Let $Q =$ the projection onto $\overline{(I_2 - P_{22})\,\mathscr{H}_2} = (\ker\,(I_2 - P_{22}))^{\perp}$, then*

$$\begin{array}{cc} & \mathscr{H}_1 \quad\; Q\mathscr{H}_2 \\ \begin{matrix} \mathscr{H}_1 \\ Q\mathscr{H}_2 \end{matrix} & \begin{pmatrix} P_{11} & P_{12}Q \\ QP_{21} & P_{22}Q \end{pmatrix} \end{array} \tag{3.66}$$

is a characteristic projection of a closable operator T_{clo}, now referring to the sum Hilbert space $\mathcal{H}_1 \oplus Q\mathcal{H}_2$.

Moreover, T_{clo} is given by

$$T_{clo}x = \lim_{n \to \infty} \frac{1}{n+1} \sum_{k=1}^{n} k\, P_{22}^{n-k} P_{21}x, \tag{3.67}$$

with $dom(T_{clo}) = dom(T)$ dense in $\mathcal{H}_1$.

Proof. (Sketch) Let T, (P_{ij}), Q and x be as given above; then

$$\frac{1}{n+1} \sum_{k=0}^{n} P_{22}^{k} T x = Tx - \frac{1}{n+1} \sum_{k=1}^{n} k\, P_{22}^{n-k} P_{21}x. \tag{3.68}$$

Note that the LHS in (3.68) converges in the norm of $\mathcal{H}_1$ by virtue of the ergodic theorem (see [Yos95]) applied to the selfadjoint contraction P_{22}. Moreover, the limit is $Q^\perp Tx$ where $Q^\perp := I - Q = \text{proj}\,(\ker(I_2 - P_{22}))$. □

3.3.1 An application to the Malliavin derivative

Below we give an application of the closability criterion for linear operators T between different Hilbert spaces $\mathcal{H}_1$ and $\mathcal{H}_2$, but having dense domain in the first Hilbert space. In this application, we shall take for T to be the so called Malliavin derivative. The setting for it is that of the Wiener process; see Example 2.3 above. For the Hilbert space $\mathcal{H}_1$ we shall take the L^2 space, $L^2(\Omega, \mathbb{P})$ where $\mathbb{P}$ is Wiener measure. Below we shall outline the rudimentary basics of the Malliavin derivative, and we shall specify the two Hilbert spaces corresponding to the setting of Theorem 3.9. We also stress that the literature on Malliavin calculus and its applications is vast, see e.g., [BØSW04, AØ15].

Let $(\Omega, \mathcal{F}_\Omega, \mathbb{P}, \Phi)$ be as specified in Example 2.3, i.e., we consider the Wiener process. Set $\mathcal{L} := L^2(0, \infty)$, $\mathcal{H}_1 = L^2(\Omega, \mathbb{P})$, and $\mathcal{H}_2 = L^2(\Omega \to \mathcal{L}, \mathbb{P}) = L^2(\Omega, \mathbb{P}) \otimes \mathcal{L}$, i.e., vector valued random variables.

For $h \in \mathcal{L} = L^2(0, \infty)$, we set $\Phi(h) = \int_0^\infty h(t)\, d\Phi_t =$ the Itō-integral introduced in Exercise 2.10. Recall we then have

$$\mathbb{E}\left(|\Phi(h)|^2\right) = \int_0^\infty |h(t)|^2\, dt, \tag{3.69}$$

or equivalently (the Itō-isometry),

$$\|\Phi(h)\|_{L^2(\Omega, \mathbb{P})} = \|h\|_{\mathcal{L}}, \quad \forall h \in \mathcal{L}. \tag{3.70}$$

In the discussion below we shall assume for convenience that h from $\mathcal{L}$ is real valued.

Definition 3.9. The operator T ($=$ Malliavin derivative), $T : \mathcal{H}_1 \longrightarrow \mathcal{H}_2$, and its dense domain will now be introduced:

Let $n \in \mathbb{N}$, and $p(x_1, \cdots, x_n)$ be a polynomial in n real variables, i.e., $p : \mathbb{R}^n \longrightarrow \mathbb{R}$, and let $h_1, h_2, \cdots, h_n \in \mathscr{L}$. Then set

$$F = F_n = p(\Phi(h_1), \cdots, \Phi(h_n)) \in L^2(\Omega, \mathbb{P}), \text{ and} \qquad (3.71)$$

$$T(F) = \sum_{j=1}^{n} \left(\frac{\partial}{\partial x_j} p \right) (\Phi(h_1), \cdots, \Phi(h_n)) \otimes h_j \in \mathscr{H}_2. \qquad (3.72)$$

We shall resume our discussion of the Malliavin derivative in the appendix to Chapter 5.

Theorem 3.10. $T : \mathscr{H}_1 \longrightarrow \mathscr{H}_2$ *is an unbounded closable operator with dense domain consisting of the span of all the functions F from (3.71).*
Moreover, for all $F \in dom(T)$, and $k \in \mathscr{L}$, we have

$$\mathbb{E}(\langle T(F), k \rangle_{\mathscr{L}}) = \mathbb{E}(F\Phi(k)). \qquad (3.73)$$

Proof. (Sketch) We shall refer to the literature [BØSW04, AØ15] for details. But hint: In order to establish (3.73), it is convenient to first apply the Gram-Schmidt algorithm to $\{h_j\}_{j=1}^{n} \subset \mathscr{L} = L^2(0, \infty)$. (See Section 2.4.4.) Once (3.73) is established, then there is a recursive argument which yields a dense subspace in $\mathscr{H}_2$, contained in $dom(T^*)$; and so T is closable.

Moreover, formula (3.73) yields directly the evaluation of $T^* : \mathscr{H}_2 \longrightarrow \mathscr{H}_1$ as follows: If $k \in \mathscr{L}$, set $\mathbb{1} \otimes k \in \mathscr{H}_2$ where $\mathbb{1}$ denotes the constant function "one" on Ω. We get

$$T^*(\mathbb{1} \otimes k) = \Phi(k) = \int_0^\infty k(t) \, d\Phi_t \, (= \text{the Itō-integral.}) \qquad (3.74)$$

$\square$

Exercise 3.3 (Leibniz-Malliavin). Let $\mathscr{H}_1 \xrightarrow{T} \mathscr{H}_2$ be the Malliavin derivative from (3.71)-(3.72).

1. Show that $dom(T) =: \mathscr{D}$, given by (3.71), is an *algebra* of functions on Ω under pointwise product, i.e., $FG \in \mathscr{D}$, $\forall F, G \in \mathscr{D}$.

2. Show that $\mathscr{H}_2$ is a *module* over $\mathscr{D}$ where $\mathscr{H}_2 = L^2(\Omega, \mathbb{P}) \otimes \mathscr{L}$ (= vector valued L^2-random variables.)

3. Show that
$$T(FG) = T(F)G + FT(G), \quad \forall F, G \in \mathscr{D}, \qquad (3.75)$$
i.e., T is a module-derivation.

Notation. The eq. (3.75) is called the Leibniz-rule. By the Leibniz, we refer to the traditional rule of Leibniz for the derivative of a product. And the Malliavin derivative is thus an infinite-dimensional extension of Leibniz calculus.

There is an extensive literature on the theory of densely defined unbounded derivations in C^*-algebras. This includes both the cases of abelian and non-abelian *-algebras. And moreover, this study includes both derivations in these algebras, as well as the parallel study of module derivations. So the case of the Malliavin derivative is in fact a special case of this study. Readers interested in details are referred to [Sak98], [BJKR84], [BR79], and [BR81b].

Corollary 3.7. *Let* $\mathscr{H}_1$, $\mathscr{H}_2$, *and* $\mathscr{H}_1 \xrightarrow{T} \mathscr{H}_2$ *be as in Theorem 3.10, i.e.,* T *is the Malliavin derivative. Then, for all* $h, k \in \mathscr{L} = L^2(0, \infty)$, *we have for the closure* $\overline{T}$ *of* T *the following:*

$$\overline{T}\left(e^{\Phi(h)}\right) = e^{\Phi(h)} \otimes h.$$

(Here $\overline{T}$ *denotes the graph-closure of* T = *the Malliavin derivative.)*
  Moreover,

$$\mathbb{E}\left(\left\langle \overline{T}\left(e^{\Phi(h)}\right), k\right\rangle_{\mathscr{L}}\right) = e^{\frac{1}{2}\|h\|_{\mathscr{L}}^2} \langle h, k\rangle_{\mathscr{L}}.$$

Proof. Follows immediately from (3.73) and a polynomial approximation to

$$e^x = \lim_{n \to \infty} \sum_0^n \frac{x^j}{j!}, \quad x \in \mathbb{R};$$

see (3.71). In particular, $e^{\Phi(h)} \in dom\left(\overline{T}\right)$, and $\overline{T}\left(e^{\Phi(h)}\right)$ is well defined.
    We first establish the formula (3.73) from a direct computation where we use the Itō rules (3.69)-(3.70). □

3.4 Normal Operators

We now return to the general setting from Section 3.2. In other words, we resume our study the axiomatic case of densely defined unbounded closable operators acting in a fixed Hilbert space $\mathscr{H}$. Our aim is to give necessary and sufficient conditions for when the given operator A is normal, as an operator in $\mathscr{H}$. In Theorem 3.13, we give a set of necessary and sufficient conditions for A to be normal, and expressed in terms of the associated characteristic matrix. Since the initial operator A is unbounded, there are subtle issues having to do with domains. We first address these issues in Theorems 3.11 and 3.12.
    Theorem 3.11 below concerning operators of the form A^*A is an application of Stone's characteristic matrix.

Theorem 3.11 (von Neumann). *If* A *is a regular operator in a Hilbert space* $\mathscr{H}$, *then*

1. A^*A *is selfadjoint;*

2. $\mathscr{D}(A^*A)$ *is a core of* A, *i.e.,*

$$\overline{A\big|_{\mathscr{D}(A^*A)}} = A;$$

3. *In particular,* $\mathscr{D}(A^*A)$ *is dense in* $\mathscr{H}$.

Proof. By Theorem 3.7, $A^*A = P_{11}^{-1} - 1$. Since P_{11} is selfadjoint, so is P_{11}^{-1}. Thus, AA^* is selfadjoint.

Suppose $(a, Aa) \in \mathscr{G}(A)$ such that

$$(a, Aa) \perp \mathscr{G}(A|_{\mathscr{D}(A^*A)}); \text{ i.e.,}$$

$$\langle a, b \rangle + \langle Aa, Ab \rangle = \langle a, (1 + A^*A) b \rangle = 0, \ \forall b \in \mathscr{D}(A^*A).$$

Since $1 + A^*A = P_{11}^{-1}$, and P_{11} is a bounded operator, then

$$\mathscr{R}(1 + A^*A) = \mathscr{D}(P_{11}) = \mathscr{H}.$$

It follows that $a \perp \mathscr{H}$, and so $a = 0$. $\square$

Theorem 3.12 (von Neumann). *Let A be a regular operator in a Hilbert space $\mathscr{H}$. Then A is normal if and only if $\mathscr{D}(A) = \mathscr{D}(A^*)$ and $\|Aa\| = \|A^*a\|$, for all $a \in \mathscr{D}(A)$.*

Proof. Suppose A is normal. Then for all $a \in \mathscr{D}(A^*A) (= \mathscr{D}(AA^*))$, we have

$$\|Aa\|^2 = \langle Aa, Aa \rangle = \langle a, A^*Aa \rangle = \langle a, AA^*a \rangle = \langle Aa, A^*a \rangle = \|A^*a\|^2;$$

i.e., $\|Aa\| = \|A^*a\|$, for all $a \in \mathscr{D}(A^*A)$. It follows that

$$\mathscr{D}\left(A|_{\mathscr{D}(A^*A)}\right) = \mathscr{D}\left(A^*|_{\mathscr{D}(AA^*)}\right).$$

By Theorem 3.11, $\mathscr{D}(A) = \mathscr{D}(\overline{A|_{\mathscr{D}(A^*A)}})$ and $\mathscr{D}(A^*) = \mathscr{D}(\overline{A^*|_{\mathscr{D}(AA^*)}})$. Therefore, $\mathscr{D}(A) = \mathscr{D}(A^*)$ and $\|Aa\| = \|A^*a\|$, for all $a \in \mathscr{D}(A)$.

Conversely, the map $Aa \mapsto A^*a$, $a \in \mathscr{D}(A)$, extends uniquely to a partial isometry V with initial space $\mathscr{R}(A)$ and final space $\mathscr{R}(A^*)$, such that $A^* = VA$. By Theorem 3.3, $A = A^*V^*$. Then $A^*A = A^*(V^*V)A = (A^*V^*)(VA) = AA^*$. Thus, A is normal. $\square$

The following theorem is due to M.H. Stone.

Theorem 3.13. *Let A be a regular operator in a Hilbert space $\mathscr{H}$. Let $P = (P_{ij})$ be the characteristic matrix of A. The following are equivalent.*

1. *A is normal.*

2. *P_{ij} are mutually commuting.*

3. *A is affiliated with an Abelian von Neumann algebra.*

Remark 3.6. The equivalence of (1) and (2) is straightforward; see Corollary 3.4, and the original paper of Stone. The more interesting part is (1)⇔(3). The idea of characteristic matrix gives rise to an elegant proof without reference to the spectral theorem.

Proof. Assuming (1)⇔(2), we prove that (1)⇔(3).

Suppose A is normal, i.e. P_{ij} are mutually commuting. Then A is affiliated with the Abelian von Neumann algebra $\{P_{ij}\}''$. For if $B \in \{P_{ij}\}'$, then B commutes P_{ij}, and so B commutes with A by Lemma 3.5.

Conversely, if A is affiliated with an Abelian von Neumann algebra $\mathfrak{M}$, then by Theorem 3.8, $P_{ij} \in \mathfrak{M}$. This shows that P_{ij} are mutually commuting, and A is normal. □

3.5 Polar Decomposition

We show that the intuition behind the familiar polar decomposition (or polar factorization) for complex numbers carries over remarkably well to operators in Hilbert space. Indeed (Theorem 3.15) the operators that admit a polar decomposition are precisely the regular operators, meaning closable and with dense domain.

Let A be a regular operator in a Hilbert space $\mathscr{H}$. By Theorem 3.11, A^*A is a positive selfadjoint operator and it has a unique positive square root $|A| := \sqrt{A^*A}$.

Theorem 3.14.

1. $|A| := \sqrt{A^*A}$ is the unique positive selfadjoint operator T satisfying $\mathscr{D}(T) = \mathscr{D}(A)$, and $\|Ta\| = \|Aa\|$ for all $a \in \mathscr{D}(A)$.

2. $\ker(|A|) = \ker(A)$, $\overline{\mathscr{R}(|A|)} = \overline{\mathscr{R}(A^*)}$.

Proof. Suppose $T = \sqrt{A^*A}$, i.e. $T^*T = A^*A$. Let $\mathscr{D} := \mathscr{D}(T^*T) = \mathscr{D}(A^*A)$. By Theorem 3.11, $\mathscr{D}$ is a core of both T and A. Moreover, $\|Ta\| = \|Aa\|$, for all $a \in \mathscr{D}$. We conclude from this norm identity that $\mathscr{D}(T) = \mathscr{D}(A)$ and $\|Ta\| = \|Aa\|$, for all $a \in \mathscr{D}(A)$.

Conversely, suppose T has the desired properties. For all $a \in \mathscr{D}(A) = \mathscr{D}(T)$, and $b \in \mathscr{D}(A^*A)$,

$$\langle Tb, Ta \rangle = \langle Ab, Aa \rangle = \langle A^*Ab, a \rangle .$$

This implies that $Tb \in \mathscr{D}(T^*) = \mathscr{D}(T)$, $T^2b = A^*Ab$, for all $b \in \mathscr{D}(A^*A)$. That is, T^2 is a selfadjoint extension of A^*A. Since A^*A is selfadjoint, $T^2 = A^*A$.

The second part follows from Theorem 3.4. □

Consequently, the map $|A|\, a \mapsto Aa$ extends to a unique partial isometry V with initial space $\overline{\mathscr{R}(A^*)}$ and final space $\overline{\mathscr{R}(A)}$ (the over-bar means "norm-closure"), such that

$$A = V\,|A| . \tag{3.76}$$

Equation (3.76) is called the *polar decomposition* of A. It is clear that such decomposition is unique.

We have proved:

Theorem 3.15 (Polar decomposition). *Let A, V and $|A|$ be as described; then*

$$A = V|A|.$$

Taking adjoints in (3.76) yields $A^* = |A|V^*$, so that

$$AA^* = VA^*AV^*. \tag{3.77}$$

Proof. The proof follows from the considerations above. Specifically, restrict AA^* to $\overline{\mathscr{R}(A)}$, and restrict A^*A restricted to $\overline{\mathscr{R}(A^*)}$. Then the two restrictions are unitarily equivalent. It follows that A^*A, AA^* have the same spectrum, aside from possibly the point 0.

By (3.77), $|A^*| = V|A|V^* = VA^*$, where $|A^*| = \sqrt{AA^*}$. Apply V^* on both sides gives

$$A^* = V^*|A^*|. \tag{3.78}$$

By uniqueness, (3.78) is the polar decomposition of A^*.   □

Theorem 3.16. *A is affiliated with a von Neumann algebra $\mathfrak{M}$ if and only if $|A|$ is affiliated with $\mathfrak{M}$ and $V \in \mathfrak{M}$.*

Proof. Let U be a unitary operator in $\mathfrak{M}'$. The operator UAU^* has polar decomposition

$$UAU^* = (UVU^*)(U|A|U^*).$$

By uniqueness, $A = UAU^*$ if and only if $V = UVU^*$, $|A| = U|A|U^*$. Since U is arbitrary, we conclude that $V \in \mathfrak{M}$, and A is affiliated with $\mathfrak{M}$.   □

A summary of relevant numbers from the Reference List

For readers wishing to follow up sources, or to go in more depth with topics above, we suggest:

Of these, refs [vN32a] and [DS88b] are especially central. A more comprehensive list is: [BR81a, DS88b, Jor08, Kat95, KR97b, Sto51, Sto90, Wei03, Yos95, JL01, Die75, Emc84, Jor88, Jor94, RS75, Akh65, BN00, BR79, Con90, dBR66, FL28, Fri80, GJ87, JM84, Kre46, Nel69, vN32a, Hel13].

3.A   Stone's Theorem

The gist of the result (Theorem 3.17) is as follows: Given a fixed Hilbert space, there is then a 1-1 correspondence between any two in pairs from the following three: (i) strongly continuous unitary one-parameter groups $\mathcal{U}(t)$; (ii) selfadjoint operators H (generally unbounded) with dense domain; and (iii) projection valued measures $P(\cdot)$, abbreviated PVM. Starting with $\mathcal{U}(t)$, we say that the corresponding selfadjoint operator H is its generator, and then the PVM $P(\cdot)$ will be from the Spectral Theorem applied to H.

Definition 3.10 (Projection valued measure (PVM)). Let $\mathcal{B}(\mathbb{R})$ be the Borel sigma algebra of subsets of $\mathbb{R}$. Let $\mathcal{H}$ be a Hilbert space.

A function $P : \mathcal{B}(\mathbb{R}) \longrightarrow \text{Proj}(\mathcal{H})$ is called a projection valued measure (PVM) iff (Def):

1. $P(\emptyset) = 0$, $P(\mathbb{R}) = I_{\mathcal{H}}$;

2. $P(A \cap B) = P(A)P(B)$, $\forall A, B \in \mathcal{B}(\mathbb{R})$; and

3. for all $(E_i)_{i=1}^{\infty}$ such that $E_i \cap E_j = \emptyset$ $(i \neq j)$, we have:

$$P\left(\bigcup_i E_i\right) = \sum_i P(E_i). \tag{3.79}$$

Remark 3.7. Suppose $P : \mathcal{B}(\mathbb{R}) \longrightarrow \text{Proj}(\mathcal{H})$ is given as in Definition 3.10, *but* only conditions 1 and 3 hold (*not* 2); then we will show in Chapter 5 that there is a system $\mathcal{K}$, V, $Q(\cdot)$ where:

$\mathcal{K}$ is a Hilbert space,

$V : \mathcal{H} \longrightarrow \mathcal{K}$ is isometric,

$Q : \mathcal{B}(\mathbb{R}) \longrightarrow \text{Proj}(\mathcal{K})$

where now Q is a PVM in the stronger sense, i.e., all three conditions 1–3 hold for $Q(\cdot)$; and

$$P(B) = V^*Q(B)V, \quad \forall B \in \mathcal{B}(\mathbb{R}).$$

We say that Q is a *dilation* of P, or that Q is an orthogonal realization of P.

Definition 3.11. A unitary one-parameter group is a function:

$$\mathcal{U} : \mathbb{R} \longrightarrow \left(\text{unitary operators in } \mathcal{H}\right)$$

such that:

$$\mathcal{U}(s+t) = \mathcal{U}(s)\mathcal{U}(t), \quad \forall s, t \in \mathbb{R}; \tag{3.80}$$

and for $\forall h \in \mathcal{H}$,

$$\lim_{t \to 0} \mathcal{U}(t)h = h \text{ (strong continuity)}. \tag{3.81}$$

Theorem 3.17 (Stone's Theorem [Lax02, RS75, Rud91]). *There is a sequence of bijective correspondences between (1)-(3) below, i.e., (1)⇒(2)⇒(3)⇒(1):*

1. *PVMs $P(\cdot)$;*

2. *unitary one-parameter groups $\mathcal{U}$; and*

3. *selfadjoint operators H with dense domain in $\mathcal{H}$.*

The correspondence is given explicitly as follows:
   (1) $\Rightarrow$ (2): *Given P, a PVM, set*

$$\mathcal{U}(t) = \int_{\mathbb{R}} e^{i\lambda t} P(d\lambda) \tag{3.82}$$

where the integral on the RHS in (3.82) is the limit of finite sums of

$$\sum_k e^{i\lambda_k t} P(E_k), \ t \in \mathbb{R}; \tag{3.83}$$

$E_i \cap E_j = \emptyset \ (i \neq j), \ \bigcup_k E_k = \mathbb{R}.$
   (2) $\Rightarrow$ (3): *Given $\{\mathcal{U}(t)\}_{t\in\mathbb{R}}$, set*

$$dom(H) = \left\{ f \in \mathcal{H}, \ s.t. \ \lim_{t\to 0_+} \frac{1}{it}(\mathcal{U}(t)f - f) \ exists \right\}$$

and

$$iHf = \lim_{t\to 0_+} \frac{\mathcal{U}(t)f - f}{t}, \quad f \in dom(H), \tag{3.84}$$

then $H^* = H$.
   (3) $\Rightarrow$ (1): *Given a selfadjoint operator H with dense domain in $\mathcal{H}$; then by the spectral theorem (Section 3.4) there is a unique PVM, $P(\cdot)$ such that*

$$H = \int_{\mathbb{R}} \lambda P(d\lambda); \quad and \tag{3.85}$$

$$dom(H) = \left\{ f \in \mathcal{H}; \ s.t. \ \int_{\mathbb{R}} \lambda^2 \|P(d\lambda)f\|^2 < \infty \right\}. \tag{3.86}$$

Remark 3.8. We state Stone's theorem already now even though the proof details will require a number of technical tools to be developed systematically only in Chapters 4 and 5 below.

Remark 3.9. Note that the selfadjointness condition on H in (3) in Theorem 3.17 is stronger than merely Hermitian symmetry, i.e., the condition

$$\langle Hu, v \rangle = \langle u, Hv \rangle \tag{3.87}$$

for all pairs of vectors u and $v \in dom(H)$. We shall discuss this important issue in much detail in Part IV of the book, both in connection with the theory, and its applications. The applications are in physics, statistics, and infinite networks.

   Here we limit ourselves to comments and some definitions; a full discussion will follow in Part IV below.

   Observations. Introducing the adjoint operator H^*, we note that (3.87) is equivalent to

$$H \subset H^*, \text{ or} \tag{3.88}$$

$$\mathscr{G}(H) \subset \mathscr{G}(H^*), \tag{3.89}$$

where $\mathscr{G}$ denotes the graph of the respective operators and where (3.88) & (3.89) mean that $dom\,(H) \subset dom\,(H^*)$, and $Hu = H^*u$ for $\forall u \in dom\,(H)$.

If (3.88) holds, then it may, or may not, have selfadjoint extensions. We introduce the two indices $d_\pm$ (deficiency-indices)

$$d_\pm = \dim\,(H^* \pm i\,I)\,. \tag{3.90}$$

The following will be proved in part 4:

Theorem 3.18. *(i) Suppose $H \subset H^*$, then H has selfadjoint extensions iff $d_+ = d_-$.*

(ii) If H has selfadjoint extensions, say K (i.e., $K^* = K$,) so $H \subset K$, then it follows that

$$H \subset K \subset H^*. \tag{3.91}$$

So, if there are selfadjoint extensions, they lie between H and H^*.

Definition 3.12. If $H \subset H^*$, and if the closure $\overline{H} = H^{**}$ is selfadjoint, we say that H is *essentially selfadjoint.*

Chapter 4

Spectral Theory

"As far as the laws of mathematics refer to reality, they are not certain, and as far as they are certain, they do not refer to reality."
— Albert Einstein

"A large part of mathematics which becomes useful developed with absolutely no desire to be useful, and in a situation where nobody could possibly know in what area it would become useful; and there were no general indications that it ever would be so. By and large it is uniformly true in mathematics that there is a time lapse between a mathematical discovery and the moment when it is useful; and that this lapse of time can be anything from 30 to 100 years, in some cases even more; and that the whole system seems to function without any direction, without any reference to usefulness, and without any desire to do things which are useful."
— John von Neumann

"The spectral theorem together with the multiplicity theory is one of the pearls of mathematics."
— M. Reed and B. Simon [RS75]

Most Functional Analysis books, when covering the Spectral Theorem, stress the bounded case. Because of dictates from applications (especially quantum physics), below we stress questions directly related to key issues for unbounded linear operators. These themes will be taken up again in Chapters 10 and 11. In a number of applications, some operator from physics may only be "formally selfadjoint" also called Hermitian; and in such cases, one asks for selfadjoint extensions (if any), Chapter 10. Chapter 11 is a particular case in point, arising in the study of infinite graphs.

In the details below, the following concepts and terms will play an important role; (for reference, see also Appendix F): Unitary equivalence of operators, disintegration of measures.

4.1 An Overview

von Neumann's spectral theorem (see [Sto90, Yos95, Nel69, RS75, DS88b]) states that an operator A acting in a Hilbert space $\mathscr{H}$ is normal if and only if there exits a projection-valued measure on $\mathbb{C}$ so that

$$A = \int_{sp(A)} z P_A(dz) \tag{4.1}$$

i.e., A is represented as an integral against the projection-valued measure P_A over its spectrum.

In quantum mechanics, an *observable* is represented by a selfadjoint operator. Functions of observables are again observables. This is reflected in the spectral theorem as the functional calculus, where we may define

$$\varphi(A) = \int_{sp(A)} \varphi(z) P_A(dz) \tag{4.2}$$

using the spectral representation of A.

When P is a selfadjoint projection, $\langle f, Pf \rangle_{\mathscr{H}} = \|Pf\|_{\mathscr{H}}^2$ is a real number and it represents the expected value of the observable P prepared in the state f, unit vector in $\mathscr{H}$. Hence, in view of (4.2), $\|P_A(\cdot) f\|_{\mathscr{H}}^2$ is a Borel probability measure on $sp(A)$, and

$$\langle f, \varphi(A) f \rangle_{\mathscr{H}} = \int_{sp(A)} \varphi(z) \|P(dz) f\|_{\mathscr{H}}^2 \tag{4.3}$$

is the expected value of the observable $\varphi(A)$.

Remark 4.1. Let $\varphi : \mathbb{R} \to \mathbb{R}$ be measurable and let $A = A^*$ be given; then, for every $f \in \mathscr{H} \backslash \{0\}$, set $d\mu_f^{(A)}(\lambda) := \|P_A(d\lambda) f\|^2 \in \mathcal{M}_+(\mathbb{R})$ (the finite positive Borel measures on $\mathbb{R}$.) Then the transformation formula (4.3) takes the following equivalent form:

$$d\mu_f^{(\varphi(A))} = d\mu_f^{(A)} \circ \varphi^{-1}, \text{ i.e.,} \tag{4.4}$$

$$d\mu_f^{(\varphi(A))}(\triangle) = d\mu_f^{(A)}\left(\varphi^{-1}(\triangle)\right), \ \forall \triangle \in \mathcal{B}(\mathbb{R}), \tag{4.5}$$

where $\varphi^{-1}(\triangle) = \{x \ : \ \varphi(x) \in \triangle\}$.

Corollary 4.1. Let $A = A^*$, and $f \in \mathscr{H} \backslash \{0\}$ be given, and let μ_f and φ be as in Remark 4.1, then $\varphi(A)^* = \varphi(A)$, and

$$f \in dom(\varphi(A)) \Longleftrightarrow \varphi \in L^2(\mathbb{R}, \mu_f),$$

where "dom" is short for "domain."

Proof. This is immediate from (4.3)-(4.5). Indeed, setting

$$d\mu_f(\lambda) := \|P(d\lambda)f\|_{\mathscr{H}}^2,$$

we get

$$\int_{\mathbb{R}} |\varphi(\lambda)|^2 \, d\mu_f(\lambda) = \|\varphi(A)f\|_{\mathscr{H}}^2.$$

$\square$

Remark 4.2. The standard diagonalization of Hermitian matrices in linear algebra is a special case of the spectral theorem. Recall that any Hermitian matrix A can be decomposed as $A = \sum_k \lambda_k P_k$, where $\lambda_k's$ are the eigenvalues of A and $P_k's$ are the selfadjoint projections onto the eigenspaces associated with $\lambda_k's$. The projection-valued measure in this case can be written as $P(E) = \sum_{\lambda_k \in E} P_k$, for all $E \in \mathcal{B}(\mathbb{R})$; i.e., the counting measure supported on $\lambda_k's$.

Quantum mechanics is stated using an abstract Hilbert space as the state space. In practice, one has the freedom to choose exactly which Hilbert space to use for a particular problem. Physical measurements remain unchanged when choosing different realizations of a Hilbert space. The concept needed here is unitary equivalence.

Definition 4.1. Let $A : \mathscr{H}_1 \to \mathscr{H}_1$ and $B : \mathscr{H}_2 \to \mathscr{H}_2$ be operators. A is said to be unitarily equivalent to B if there exists a unitary operator $U : \mathscr{H}_1 \to \mathscr{H}_2$ such that $B = UAU^*$.

Suppose $U : \mathscr{H}_1 \to \mathscr{H}_2$ is a unitary operator, $P : \mathscr{H}_1 \to \mathscr{H}_1$ is a selfadjoint projection. Then $UPU^* : \mathscr{H}_2 \to \mathscr{H}_2$ is a selfadjoint projection on $\mathscr{H}_2$. In fact,

$$(UPU^*)(UPU^*) = UPU^*$$

where we used $UU^* = U^*U = I$, since U is unitary. Let $|f_1\rangle$ be a state in $\mathscr{H}_1$ and $|f_2\rangle = |Uf_1\rangle$ be the corresponding state in $\mathscr{H}_2$. Then

$$\langle f_2, UPU^*f_2 \rangle_{\mathscr{H}_2} = \langle U^*f_2, PU^*f_2 \rangle_{\mathscr{H}_1} = \langle f_1, Pf_1 \rangle_{\mathscr{H}_1}$$

i.e., the observable P has the same expectation value. Since every selfadjoint operator is, by the spectral theorem, decomposed into selfadjoint projections, it follows that the expectation value of any observable remains unchanged under unitary transformations.

We will also consider family of selfadjoint operators. Heisenberg's commutation relation $PQ - QP = -iI$, $i = \sqrt{-1}$, is an important example of two non-commuting selfadjoint operators.

Example 4.1. The classical Fourier transform $\mathcal{F} : L^2(\mathbb{R}) \to L^2(\mathbb{R})$ is unitary, so $\mathcal{F}^*\mathcal{F} = \mathcal{F}\mathcal{F}^* = I_{L^2(\mathbb{R})}$, and in particular, the Parseval identity

$$\|\mathcal{F}f\|_{L^2(\mathbb{R})}^2 = \|f\|_{L^2(\mathbb{R})}^2$$

holds for all $f \in L^2(\mathbb{R})$.

Example 4.2. Let Q and P be the position and momentum operators in quantum mechanics. That is, $Q = M_x =$ multiplication by x, and $P = -id/dx$ both defined on the Schwartz space $\mathcal{S}(\mathbb{R})$, i.e., space of rapidly decreasing functions on $\mathbb{R}$, which is dense in the Hilbert space $L^2(\mathbb{R})$. On $\mathcal{S}(\mathbb{R})$, the operators P and Q satisfy the canonical commutation relation: $PQ - QP = -i\,I_{L^2(\mathbb{R})}$.

Example 4.3. Denote $\mathcal{F}$ the Fourier transform on $L^2(\mathbb{R})$ as before. Specifically, setting

$$(\mathcal{F}\varphi)(x) = \widehat{\varphi}(x) = \frac{1}{\sqrt{2\pi}} \int_{\mathbb{R}} \varphi(\xi) e^{-i\xi x} d\xi, \text{ and}$$

$$(\mathcal{F}^*\psi)(\xi) = \psi^{\vee}(\xi) = \frac{1}{\sqrt{2\pi}} \int_{\mathbb{R}} \psi(x) e^{i\xi x} dx, \ \xi \in \mathbb{R}.$$

Note that $\mathcal{F}$ is an automorphism in $\mathcal{S}(\mathbb{R})$, continuous with respect to the standard l.c. topology. Moreover,

$$(\mathcal{F}^*Q\mathcal{F}\varphi)(\xi) = \mathcal{F}^*(x\widehat{\varphi}(x)) = \frac{1}{i}\frac{d}{d\xi}\varphi(\xi), \ \forall \varphi \in \mathcal{S}.$$

Therefore,

$$P = \mathcal{F}^*Q\mathcal{F} \tag{4.6}$$

and so P and Q are unitarily equivalent.

A multiplication operator version of the spectral theorem is also available. It works especially well in physics. It says that A is a normal operator in $\mathscr{H}$ if and only if A is unitarily equivalent to the operator of multiplication by a measurable function f on $L^2(X, \mu)$, where X is locally compact and Hausdorff. The two versions are related via a measure transformation.

Example 4.4. Eq. (4.6) says that P is diagonalized by Fourier transform in the following sense.

Let ψ be any Borel function on $\mathbb{R}$, and set $M_\psi =$ multiplication by $\psi(x)$ in $L^2(\mathbb{R})$, with

$$dom(M_\psi) = \left\{ f \mid f, \psi f \in L^2(\mathbb{R}) \right\} \tag{4.7}$$

$$= \left\{ f \mid \int_{-\infty}^{\infty} \left(1 + |\psi(x)|^2 \right) |f(x)|^2 \, dx < \infty \right\};$$

then we define, via eq. (4.6),

$$\psi(P) := \mathcal{F}^*\psi(Q)\mathcal{F}.$$

In particular, given any $\triangle \in \mathcal{B}(\mathbb{R})$, let $\psi = \chi_\triangle =$ characteristic function, then

$$E(\triangle) = \mathcal{F}^* M_{\chi_\triangle} \mathcal{F}.$$

One checks directly that $E(\triangle)^2 = E(\triangle) = E(\triangle)^*$, so $E(\triangle)$ is a selfadjoint projection. Set $\triangle = [a, b]$. Indeed, $E(\cdot)$ is a convolution operator, where

$$(E(\triangle) f)(x) = \int_a^b e^{i\xi x} \widehat{f}(\xi) \, d\xi = f * (\chi_{[a,b]})^\wedge (x), \ \forall f \in L^2(\mathbb{R}).$$

Thus,

$$E(\triangle)(L^2(\mathbb{R})) = \left\{ f \in L^2(\mathbb{R}) \mid \mathrm{supp}\left(\widehat{f}\right) \subset \triangle \right\};$$

i.e., the space of "band-limited" functions, with the "pass-band" being $\triangle$.

Example 4.5. Below, it helps to denote the Fourier transformed space (or frequency space) by $L^2(\widehat{\mathbb{R}})$. Fix any $f \in L^2(\mathbb{R})$, $\triangle \in \mathcal{B}(\mathbb{R})$, then

$$\begin{aligned}
\mu_f(\triangle) := \|E(\triangle) f\|_{L^2(\mathbb{R})}^2 &= \langle f, E(\triangle) f \rangle_{L^2(\mathbb{R})} \\
&= \langle \mathcal{F}f, M_{\chi_\triangle} \mathcal{F}f \rangle_{L^2(\widehat{\mathbb{R}})} \\
&= \int_\triangle \left| \widehat{f}(x) \right|^2 dx
\end{aligned}$$

which is a Borel measure on $\mathbb{R}$, such that

$$\mu_f(\mathbb{R}) = \int_{-\infty}^\infty \left| \widehat{f}(x) \right|^2 dx = \int_{-\infty}^\infty |f(x)|^2 \, dx = \|f\|_{L^2(\mathbb{R})}^2$$

by the Parseval identity (see Corollary 2.3).

Now, let $\psi(x) = x$ be the identity function, and note that it is approximated pointwisely by simple functions of the form $\sum_{\text{finite}} c_i \chi_{\triangle_i}$, where $\triangle_i \in \mathcal{B}(\mathbb{R})$, and $\triangle_i$'s are mutually disjoint, i.e., $x = \lim_{n \to \infty} \sum_{i=1}^n c_i \chi_{\triangle_i}$.

Fix $f \in dom(M_x)$, see (4.7), it follows from Lebesgue dominated convergence theorem, that

$$\begin{aligned}
\langle f, Pf \rangle_{L^2(\mathbb{R})} &= \int_{-\infty}^\infty \overline{\widehat{f}(x)} x \widehat{f}(x) \, dx \\
&= \lim_{n \to \infty} \int_{-\infty}^\infty \overline{\widehat{f}(x)} \left(\sum_{i=1}^n c_i \chi_{\triangle_i}(x) \right) \widehat{f}(x) \, dx \\
&= \lim_{n \to \infty} \sum_{i=1}^n c_i \langle f, E(\triangle_i) f \rangle_{L^2(\mathbb{R})} \\
&= \lim_{n \to \infty} \sum_{i=1}^n c_i \|E(\triangle_i) f\|_{L^2(\mathbb{R})}^2 \\
&= \int_{-\infty}^\infty x \|E(dx) f\|_{L^2(\mathbb{R})}^2.
\end{aligned}$$

The last step above yields the projection-valued measure (PVM) version of the spectral theorem for P, where we write

$$P = \int_{-\infty}^{\infty} x \, dE(x). \tag{4.8}$$

Consequently, we get two versions of the spectral theorem for $P = \frac{1}{i} \frac{d}{dx} \Big|_{\mathcal{S}(\mathbb{R})}$:

1. Multiplication operator version, i.e., $P \simeq M_x = $ multiplication by x in $L^2(\widehat{\mathbb{R}})$; and

2. PVM version, as in (4.8).

This example illustrates the main ideas of the spectral theorem of a single selfadjoint operator in Hilbert space. We will develop the general theory in this chapter, and construct both versions of the spectral decomposition.

Example 4.6. Applying the Gram-Schmidt process to all polynomials against the measure $d\mu = e^{-x^2/2}dx$, one gets orthogonal polynomials P_n in $L^2(\mu)$. These are called the *Hermite polynomials,* and the associated Hermite functions are given by

$$h_n := e^{-x^2/2}P_n = e^{-x^2}\left(\frac{d}{dx}\right)^n e^{x^2/2}.$$

(See Table 2.2 and [Sze59].) The Hermite functions (after normalization) form an ONB in $L^2(\mathbb{R})$, which transforms P and Q to Heisenberg's infinite matrices in (2.80)-(2.81).

Example 4.7 (The harmonic oscillator Hamiltonian). Let P, Q be as in the previous example. We consider the quantum Hamiltonian

$$H := \frac{1}{2}(Q^2 + P^2 - 1).$$

It can be shown that

$$Hh_n = nh_n$$

or equivalently,

$$(P^2 + Q^2)h_n = (2n+1)h_n$$

$n = 0, 1, 2, \ldots$. That is, H is diagonalized by the Hermite functions.

H is called the energy operator in quantum mechanics (more precisely, the harmonic oscillator Hamiltonian.) This explains mathematically why the energy levels are discrete (in quanta), being a multiple of the Plank's constant $\hbar$.

Example 4.8 (Purely discrete spectrum v.s. purely continuous spectrum). The two operators $P^2 + Q^2$ and $P^2 - Q^2$ acting in $L^2(\mathbb{R})$; see Figure 4.1.

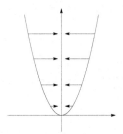

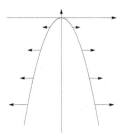

(a) Harmonic oscillator $P^2 + Q^2$ (bound-states).

(b) Repulsive potential $P^2 - Q^2$. This operator has purely continuous spectrum.

Figure 4.1: Illustration of forces: attractive vs repulsive. The case of "only bound states" (a), vs continuous Lebesgue spectrum (b).

Remark 4.3. Note that both of the two operators $H_\pm := P^2 \pm Q^2$ in Example 4.8 are essentially selfadjoint as operators in $L^2(\mathbb{R})$ (see [RS75, Nel69, vN32a, DS88b]), and with common dense domain equal to the Schwartz space. The potential in H_- is repulsive, see Figure 4.1 (b).

By comparison, the operator $H_4 := P^2 - Q^4$ is not essentially selfadjoint. (It can be shown that it has deficiency indices $(2, 2)$.) The following argument from physics is illuminating: For $E \in \mathbb{R}_+$, consider a classical particle $x(t)$ on the energy surface

$$S_E := \left\{ x(t) \; : \; (x'(t))^2 - (x(t))^4 = E \right\}.$$

The travel time to $\pm\infty$ is finite; in fact, it is

$$t_\infty = \int_0^\infty \frac{dx}{\sqrt{E + x^4}} < \infty.$$

There is a principle from quantum mechanics which implies that the quantum mechanical particle must be assigned conditions at $\pm\infty$, which translates into non-zero deficiency indices. (A direct computation, which we omit, yields indices $(2, 2)$.)

In the following sections, we present some main ideas of the spectral theorem for single normal operators acting in Hilbert space. Since every normal operator N can be written as $N = T_1 + iT_2$, where T_1 and T_2 are strongly commuting and selfadjoint, the presentation will be focused on selfadjoint operators.

4.2   Multiplication Operator Version

Together, the results below serve to give a spectral representation (by multiplication operators) for the most general case: an arbitrary given selfadjoint

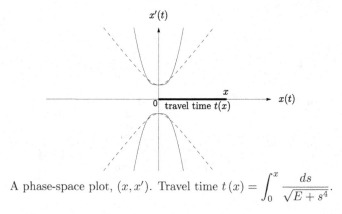

A phase-space plot, (x, x'). Travel time $t\left(x\right) = \int_0^x \dfrac{ds}{\sqrt{E + s^4}}.$

Figure 4.2: The energy surface S_E for the quantum mechanical $H_4 = P^2 - Q^4$, with $P \rightsquigarrow x'\left(t\right)$.

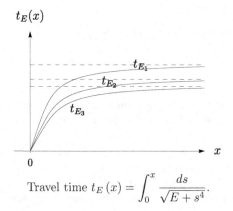

Travel time $t_E\left(x\right) = \int_0^x \dfrac{ds}{\sqrt{E + s^4}}.$

Figure 4.3: Fix $E_1 < E_2 < E_3$, then $0 < t_{E_3}\left(\infty\right) < t_{E_2}\left(\infty\right) < t_{E_1}\left(\infty\right)$.

(or normal) operator with dense domain in a Hilbert space. It applies both to the bounded, and unbounded cases; and it even applies to arbitrary families of strongly commuting selfadjoint operators. Caution: There is a number of subtle points in such representations. Since we aim for realizations up to unitary equivalence, care must be exercised in treating "multiplicity" for the most general spectral types. What we present below may be thought of as a modern version of what is often called the *Hahn-Hellinger theory of spectral multiplicity*.

This version of the spectral theory states that every selfadjoint operator A is unitarily equivalent to the operator of multiplication by a measurable function on some L^2-space.

Theorem 4.1 (The Spectral Representation Theorem). *Let A be a linear operator acting in the Hilbert space $\mathscr{H}$, then $A = A^*$ iff there exists a measure space (X, μ) and a unitary operator $U : L^2(X, \mu) \to \mathscr{H}$ such that*

$$M_\varphi = U^* A U; \tag{4.9}$$

where X is locally compact and Hausdorff, φ is a real-valued μ-measurable function, and

$$M_\varphi f \ := \ \varphi f, \ \forall f \in dom(M_\varphi), \ where \tag{4.10}$$
$$dom(M_\varphi) \ := \ \{h \in L^2(X, \mu) \ : \ \varphi h \in L^2(X, \mu)\}. \tag{4.11}$$

Hence, the following diagram commutes.

$$
\begin{array}{ccc}
\mathscr{H} & \xrightarrow{\ A\ } & \mathscr{H} \\
U \uparrow & & \uparrow U \\
L^2(X, \mu) & \xrightarrow{\ M_\varphi\ } & L^2(X, \mu)
\end{array}
$$

If $A \in \mathscr{B}(\mathscr{H})$, then $\varphi \in L^\infty(X, \mu)$ and $dom(M_\varphi) = \mathscr{H}$.

We postpone the detailed proof till Section 4.2.3 below.

Exercise 4.1 (Multiplication operators, continued). Prove that M_φ in (4.10)-(4.11) is selfadjoint.

Exercise 4.2 (Continuous spectrum). Let $M_t : L^2[0, 1] \to L^2[0, 1]$, $f(t) \longmapsto t f(t)$. Show that M_t has no eigenvalues in $L^2[0, 1]$.

Before giving a proof of Theorem 4.1, we show below that one can go one step further and get that A is unitarily equivalent to the operator of multiplication by the independent variable on some L^2-space. This is done by a transformation of the measure μ in (4.11).

4.2.1   Transformation of Measures

Definition 4.2. Let $\varphi : X \to Y$ be a measurable function, $\mathcal{T}_X$ and $\mathcal{T}_Y$ be the respective sigma-algebras. Fix a measure μ on $\mathcal{T}_X$, the measure

$$\mu_\varphi := \mu \circ \varphi^{-1} \tag{4.12}$$

defined on $\mathcal{T}_Y$ is called the transformation of μ under φ. Note that

$$\chi_E \circ \varphi(x) = \chi_{\varphi^{-1}(E)}(x) \tag{4.13}$$

for all $E \in \mathcal{T}_Y$ and $x \in X$.

Lemma 4.1. *For all $\mathcal{T}$-measurable functions f, we have*

$$\int_X f \circ \varphi \, d\mu = \int_Y f \, d\left(\mu \circ \varphi^{-1}\right). \tag{4.14}$$

(This is a generalization of the substitution formula in calculus.)

Proof. For any simple function $s = \sum c_i \chi_{E_i}$, $E_i \in \mathcal{T}_Y$, it follows from (4.13) that

$$\int_X s \circ \varphi d\mu = \sum c_i \int_X \chi_{E_i} \circ \varphi d\mu$$

$$= \sum c_i \int_X \chi_{\varphi^{-1}(E_i)} d\mu$$

$$= \sum c_i \mu\left(\varphi^{-1}(E_i)\right)$$

$$= \int_Y s \, d\left(\mu \circ \varphi^{-1}\right).$$

Note all the summations in the above calculation are finite.

Since any measurable function $f : X \to Y$ is approximated pointwisely by simple functions, eq. (4.14) follows.   □

Remark 4.4. If φ is nasty, then even if μ is a nice measure (say the Lebesgue measure), the transformation measure $\mu \circ \varphi^{-1}$ in (4.14) can still be nasty, e.g., it could even be singular.

To simplify the discussion, we consider *bounded* selfadjoint operators below.

Corollary 4.2. *Let $\varphi : X \to X$ be any measurable function. Then the operator $Uf := f \circ \varphi$ in $L^2(X, \mu)$ is isometric iff $\mu \circ \varphi^{-1} = \mu$. Moreover, $M_\varphi U = U M_t$. In particular, U is unitary iff φ is invertible.*

Proof. Follows immediately from Lemma 4.1.   □

Lemma 4.2. *In Theorem 4.1, assume A is bounded selfadjoint, so that $\varphi \in L^\infty(X, \mu)$, and real-valued. Let $\mu_\varphi := \mu \circ \varphi^{-1}$ (eq. (4.12)), supported on the essential range of φ. Then the operator $W : L^2(\mathbb{R}, \mu_\varphi) \longrightarrow L^2(X, \mu)$, by*

$$(Wf)(x) = f(\varphi(x)), \ \forall f \in L^2(\mu_\varphi) \tag{4.15}$$

is isometric, and

$$WM_t = M_\varphi W, \tag{4.16}$$

where $M_t : L^2(Y, \mu_f) \longrightarrow L^2(Y, \mu_f)$, given by

$$(M_t f)(t) = t f(t); \tag{4.17}$$

i.e., multiplication by the identify function.

Proof. For all $f \in L^2(Y, \mu_\varphi)$, we have

$$\|f\|^2_{L^2(Y,\mu_\varphi)} = \int_Y |f|^2 \, d\mu_\varphi = \int_X |f \circ \varphi|^2 \, d\mu = \|Wf\|^2_{L^2(X,\mu)}$$

so W is isometric. Moreover,

$$M_\varphi W f = \varphi(x) f(\varphi(x))$$
$$WM_t f = W(tg(t)) = \varphi(x) f(\varphi(x))$$

hence (4.16)-(4.17) follows. □

Corollary 4.3. *Let φ be as in Lemma 4.2. Assume φ is invertible, we get that W in (4.15)-(4.17) is unitary. Set $\mathcal{F} = UW : L^2(\mathbb{R}, \mu_\varphi) \longrightarrow \mathcal{H}$, then $\mathcal{F}$ is unitary and*

$$M_t = \mathcal{F}^* A \mathcal{F}; \tag{4.18}$$

i.e., the following diagram commutes.

Remark 4.5.

1. Eq. (4.18) is a vast extension of *diagonalizing hermitian matrices* in linear algebra, or a generalization of Fourier transform.

2. Given a selfadjoint operator A in the Hilbert space $\mathscr{H}$, what's involved are two algebras: the algebra of measurable functions on X, treated as multiplication operators, and the algebra of operators generated by A (with identity). The two algebras are $*$-isomorphic. The Spectral Theorem offers two useful tools:

   (a) Representing the algebra generated by A by the algebra of functions. In this direction, it helps us understand A.

   (b) Representing the algebra of functions by the algebra of operators generated by A. In this direction, it reveals properties of the function algebra and the underlying space X.

3. Let $\mathfrak{A}$ be the algebra of functions. We say that π is a representation of $\mathfrak{A}$ on the Hilbert space $\mathscr{H}$, denoted by $\pi \in Rep\,(\mathfrak{A}, \mathscr{H})$, if $\pi : \mathfrak{A} \to \mathscr{B}\,(\mathscr{H})$ is a $*$-homomorphism, i.e., $\pi\,(gh) = \pi\,(g)\,\pi\,(h)$, and $\pi\,(\overline{g}) = \pi\,(g)^*$, for all $g, h \in \mathfrak{A}$. Given $\mathcal{F}$ as in (4.18), then

$$\pi(\psi) = \mathcal{F}M_\psi\mathcal{F}^* \in Rep\,(\mathfrak{A}, \mathscr{H})\,; \qquad (4.19)$$

where the LHS in (4.19) defines the operator

$$\psi\,(A) := \pi\,(\psi)\,, \ \psi \in \mathfrak{A}. \qquad (4.20)$$

To see that (4.20) is an algebra isomorphism, one checks that

$$\begin{aligned}
(\psi_1\psi_2)\,(A) &= \mathcal{F}M_{\psi_1\psi_2}\mathcal{F}^* \\
&= \mathcal{F}M_{\psi_1}M_{\psi_2}\mathcal{F}^* \\
&= (\mathcal{F}M_{\psi_1}\mathcal{F}^*)\,(\mathcal{F}M_{\psi_2}\mathcal{F}^*) \\
&= \psi_1\,(A)\,\psi_2\,(A)
\end{aligned}$$

using the fact that

$$M_{\psi_1\psi_2} = M_{\psi_1}M_{\psi_2}$$

i.e., multiplication operators always commute.

4. Eq. (4.19) is called the *spectral representation* of A. In particular, the spectral theorem for A implies that the following substitution rule

$$\sum c_k x^k \longmapsto \sum c_k A^k$$

is well-defined, and it extends to all bounded measurable functions.

Let φ be as in Lemma 4.2. For the more general case when φ is not necessarily invertible, so W in (4.16) may not be unitary, we may still diagonalize A, i.e., get that A is unitarily equivalent to multiplication by the independent variable in some L^2-space; but now the corresponding L^2-space is vector-valued, and we get a direct integral representation. This approach is sketched in the next section.

4.2.2   Direct Integral Representation

Throughout, we assume all the Hilbert spaces are separable.

The multiplication operator version of the spectral theorem states that $A = A^* \iff A \simeq M_\varphi$, where

$$M_\varphi : L^2(X, \mu) \longrightarrow L^2(X, \mu) \tag{4.21}$$

$$f \longmapsto \varphi f. \tag{4.22}$$

Note that φ is real-valued. Moreover, A is bounded iff $M_\varphi \in L^\infty(X, \mu)$. When the Hilbert space $\mathscr{H}$ is separable, we may further assume that μ is finite, or even a probability measure.

To further diagonalize M_φ in the case when φ is "nasty", we will need the following tool from measure theory.

Definition 4.3. Let X be a locally compact Hausdorff space, and μ a Borel probability measure on X. Let $\varphi : X \to Y$ be a measurable function, and set

$$\nu := \mu \circ \varphi^{-1}.$$

A *disintegration* of μ with respect to φ is a system of probability measures $\{\mu_y : y \in Y\}$ on X, satisfying

1. $\mu_y \left(X \backslash \varphi^{-1}(\{y\}) \right) = 0$, ν-a.e, i.e., μ_y is supported on the "fiber" $\varphi^{-1}(\{y\})$.

2. For all Borel set E in X, the function $y \mapsto \mu_y(E)$ is ν-measurable, and

$$\mu(E) = \int_Y \mu_y(E) \, d\nu(y). \tag{4.23}$$

Now, back to the Spectral Theorem.

Let M_φ be as in (4.21)-(4.22), and let $\nu := \mu \circ \varphi^{-1}$, i.e., a Borel probability measure on $\mathbb{R}$. In fact, ν is supported on the essential range of φ.

It is well-known that, in this case, there exists a unique (up to measure zero sets) disintegration of μ with respect to φ. See, e.g., [Par05]. Therefore, we get:

Definition 4.4. The direct integral decomposition

$$L^2(\mu) \simeq \int^\oplus L^2(\mu_y) d\nu(y) \quad \text{(unitarily equivalent)} \tag{4.24}$$

where $\{\mu_y : y \in \text{essential range of } \varphi \subset \mathbb{R}\}$ is the system of probability measures as in Definition 4.3.

The RHS in (4.24) is the Hilbert space consisting of measurable cross-sections $f : \mathbb{R} \to \bigcup L^2(\mu_y)$, where $f(y) \in L^2(\mu_y)$, $\forall y$, and with the inner product given by

$$\langle f, g \rangle_{L^2(\nu)} := \int_Y \langle f(y), g(y) \rangle_{L^2(\mu_y)} \, d\nu(y). \tag{4.25}$$

Exercise 4.3 (Direct integral Hilbert space). Let the setting be as in (4.23)-(4.24); let $(Y, \mathcal{F}_Y, \nu)$ be a fixed measure space; and let $\{\mu_y\}_{y \in Y}$ be a field of Borel measures. Show that the space of all functions f specified as follows (i)-(iii) form a Hilbert space:

(i) $f : \mathbb{R} \longrightarrow \bigcup_{y \in Y} L^2(\mu_y)$;

(ii) $y \longmapsto \|f(y, \cdot)\|^2_{L^2(\mu_y)}$ is measurable, and in $L^1(Y, \nu)$; with

(iii) $\int_Y \|f(y, \cdot)\|^2_{L^2(\mu_y)} d\nu(y) < \infty$.

Set
$$\|f\|^2_{\text{Dir. sum}} = \text{RHS in (iii)},$$

and define the corresponding inner product by the RHS in (4.25).

Theorem 4.2. *Let $M_\varphi : L^2(X, \mu) \to L^2(X, \mu)$ be the multiplication operator in (4.21)-(4.22), $\nu := \mu \circ \varphi^{-1}$ as before. Then M_φ is unitarily equivalent to multiplication by the independent variable on $\int^\oplus L^2(\mu_y) d\nu(y)$.*

For additional details, see, e.g., [Dix81, Seg50].

4.2.3   Proof of Theorem 4.1 continued:

We try to get the best generalization of diagonalizing Hermitian matrices in finite dimensional linear algebra.

Nelson's idea [Nel69] is to get from selfadjoint operators $\to$ cyclic representation of function algebra $\to$ measure $\mu \to L^2(\mu)$.

Sketch.

1. Start with a single selfadjoint operator A acting in an abstract Hilbert space $\mathcal{H}$. Assume A is bounded.

2. Fix $u \in \mathcal{H}$. The set $\{f(A)u\}$, as f runs through some function algebra, generates a subspace $\mathcal{H}_u \subset \mathcal{H}$. $\mathcal{H}_u$ is called a *cyclic subspace*, and u the corresponding *cyclic vector*. The function algebra might be taken as the algebra of polynomials, then later it is extended to a much bigger algebra containing polynomials as a dense sub-algebra.

3. Break up $\mathcal{H}$ into a direct sum of mutually disjoint cyclic subspaces,
$$\mathcal{H} = \oplus_j \mathcal{H}_j, \tag{4.26}$$
with the family of cyclic vectors $u_j \in \mathcal{H}_j$.

4. Each $\mathcal{H}_j$ leaves A invariant, and the restriction of A to each $\mathcal{H}_j$ is unitarily equivalent to M_x on $L^2(sp(A), \mu_j)$, where $sp(A)$ denotes the spectrum of A.

5. Piecing together all the cyclic subspace: set

$$X = \bigsqcup_j sp(A), \quad \mu = \bigsqcup_j \mu_j$$

i.e., taking disjoint union as u_j runs through all the cyclic vectors. When $\mathscr{H}$ is separable, we get $\mathscr{H} = \oplus_{j \in \mathbb{N}} \mathscr{H}_j$, and we may set $\mu := \sum_{j=1}^{\infty} 2^{-j} \mu_j$.

Details below.

Lemma 4.3. *There exists a family of cyclic vector $\{u_\alpha\}$ such that $\mathscr{H} = \oplus_\alpha \mathscr{H}_{u_\alpha}$, orthogonal sum of cyclic subspaces.*

Proof. An application of Zorn's lemma. See Theorem 5.2. $\square$

Lemma 4.4. *Set $K := [-\|A\|, \|A\|]$. For each cyclic vector u, there exists a Borel measure μ_u such that $supp(\mu_u) \subset K$; and $\mathscr{H}_{u_\alpha} \simeq L^2(K, \mu_u)$.*

Proof. The map

$$f \mapsto w_u(f) := \langle u, f(A)u \rangle_{\mathscr{H}}$$

is a positive, bounded linear functional on polynomials over K; the latter is dense in $C(K)$ by Stone-Weierstrass theorem. Hence w_u extends uniquely to $C(K)$. (w_u is a *state* of the C^*-algebra $C(K)$. See Chapter 5.) By Riesz, there exists a unique Borel measure μ_u on K, such that

$$w_u(f) = \langle u, f(A)u \rangle_{\mathscr{H}} = \int_K f d\mu_u. \tag{4.27}$$

Therefore we get $L^2(K, \mu_u)$, a Hilbert space containing polynomials as a dense subspace. Let

$$\mathscr{H}_u = \overline{span}\{f(A)u : f \in \text{polynomials}\}.$$

Define $W : \mathscr{H}_u \longrightarrow L^2(K, \mu_u)$, by

$$W : f(A)u \longmapsto f \in L^2(\mu_u) \tag{4.28}$$

for polynomials f, which then extends to $\mathscr{H}_u$ by density. $\square$

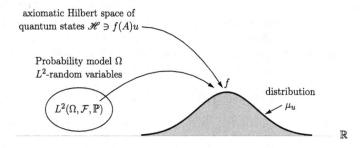

Figure 4.4: Application of the transform (4.28).

Lemma 4.5. *Let W be the operator in (4.28), then*

1. *W is an isometric isomorphism; and*

2. *$WA = M_t W$, i.e., W intertwines A and M_t. Hence W diagonalizes A.*

Remark 4.6. $WA = M_t W \iff WAW^* = M_t$. In finite dimension, it is less emphasized that the adjoint W^* equals the inverse W^{-1}. For finite dimensional case, $M_t = diag(\lambda_1, \lambda_2, \ldots, \lambda_n)$ where the measure $\mu = \sum_{\text{finite}} \delta_{\lambda_i}$, where δ_{λ_i} is the Dirac measure at the eigenvalue λ_i of A.

Proof. For the first part, let $f \in L^2(\mu_u)$, then

$$
\begin{aligned}
\|f\|_{L^2(K,\mu_u)}^2 = \int_{\mathbb{R}} |f|^2 \, d\mu_u &= \left\langle u, |f|^2(A)u \right\rangle_{\mathcal{H}} \\
&= \left\langle u, \bar{f}(A)f(A)u \right\rangle_{\mathcal{H}} \\
&= \left\langle u, f(A)^* f(A)u \right\rangle_{\mathcal{H}} \\
&= \left\langle f(A)u, f(A)u \right\rangle_{\mathcal{H}} \\
&= \|f(A)u\|_{\mathcal{H}}^2 .
\end{aligned}
$$

Notice that strictly speaking, $f(A^*)^* = \bar{f}(A)$. Since $A^* = A$, we then get

$$
f(A)^* = \overline{f}(A).
$$

Also, $f \xrightarrow{\pi} f(A)$ is a $*$-representation of the algebra $C(K)$; i.e., π is a homomorphism, and $\pi(\bar{f}) = f(A)^*$.

For the second part, let $f(t) = a_0 + a_1 t + \cdots a_n t^n$ be any polynomial, then

$$
\begin{aligned}
WAf(A) &= WA(a_0 + a_1 A + a_2 A^2 + \cdots + a_n A^n) \\
&= W(a_0 A + a_1 A^2 + a_2 A^3 + \cdots + a_n A^{n+1}) \\
&= a_0 t + a_1 t^2 + a_2 t^3 + \cdots + a_n t^{n+1} \\
&= tf(t) \\
&= M_t W f(A)
\end{aligned}
$$

thus $WA = M_t W$. The assertion then follows from standard approximation.

It remains to show that the isometry $\mathcal{H}_u \xrightarrow{W} L^2(\mu_u)$ in (4.28) maps <u>onto</u> $L^2(\mu_u)$. But this follows from (4.27). Indeed if $f \in L^2(\mu_u)$, then $f(A)u \in \mathcal{H}_u$ is well defined by the reasoning above. As a result, for the adjoint operator $L^2(\mu_u) \xrightarrow{W^*} \mathcal{H}_u$, we have

$$
W^*(f) = f(A)u, \ \forall f \in L^2(\mu_u). \tag{4.29}
$$

$\square$

Remark 4.7. Let $(\mathcal{H}, A, u)$ be as in Lemma 4.4 and in Figure 4.4, and let $u \in \mathcal{H}$, $\|u\| = 1$. Then μ_u from (4.27) is always a Borel probability measure

on $\mathbb{R}$, but it may not have finite moments. As a consequence of Lemma 4.5, we get the following: Given $n \in \mathbb{N}$, then TFAE:
(i) The vector u satisfies $u \in dom\,(A^n)$, and
$\Updownarrow$
(ii) the measure μ_u has finite moments up to order $2n$.
When the conditions are satisfied, we get:

$$\|A^n u\|^2 = \int_{\mathbb{R}} x^{2n} d\mu_u\,(x)\,.$$

Finally we piece together all the cyclic subspaces.

Lemma 4.6. *There exists a locally compact Hausdorff space X and a Borel measure μ, a unitary operator $\mathcal{F} : \mathscr{H} \longrightarrow L^2\,(X,\mu)$, such that*

$$A = \mathcal{F}^* M_\varphi \mathcal{F}$$

where $\varphi \in L^\infty\,(\mu)$.

Proof. Recall that we get a family of states w_j, with the corresponding measures μ_j, and Hilbert spaces $\mathscr{H}_j = L^2(\mu_j)$. Note that all the L^2-spaces are on $K = sp(A)$. So it's the same underlying set, but with possibly different measures. $\square$

To get a single measure space with μ, Nelson [Nel69] suggested taking the disjoint union

$$X := \bigcup_j K \times \{j\}$$

and $\mu :=$ the disjoint union of $\mu_j's$. The existence of μ follows from Riesz. Then we get

$$\mathscr{H} = \oplus \mathscr{H}_j \xrightarrow{\ \mathcal{F}\ } L^2(X,\mu).$$

Remark 4.8. Note the representation of $L^\infty\,(X,\mu)$ onto $\mathscr{H} = \sum^{\oplus} \mathscr{H}_j$ is highly non unique. There we enter into the multiplicity theory, which starts with breaking up each $\mathscr{H}_j$ into irreducible components.

Remark 4.9. A representation $\pi \in Rep\,(\mathfrak{A}, \mathscr{B}\,(\mathscr{H}))$ is said to be multiplicity free if and only if $\pi\,(\mathfrak{A})'$ is Abelian. We say π has multiplicity equal to n if and only if $(\pi(\mathfrak{A}))' \simeq M_n(\mathbb{C})$. This notation of multiplicity generalizes the one in finite dimensional linear algebra. See Section 5.11.

Exercise 4.4 (Multiplicity free). Prove that $\pi \in Rep(L^\infty(\mu), L^2(\mu))$ is multiplicity free. Conclude that each cyclic representation is multiplicity free (i.e., it is maximal Abelian.)
   <u>Hint</u>: Suppose $B \in \mathscr{B}(L^2\,(\mu))$ commutes with all M_φ, $\varphi \in L^2\,(\mu)$. Define $g = B\mathbb{1}$, where $\mathbb{1}$ is the constant function. Then, for all $\psi \in L^2\,(\mu)$, we have

$$B\psi = B\psi\mathbb{1} = BM_\psi \mathbb{1} = M_\psi B\mathbb{1} = M_\psi g = \psi g = g\psi = M_g\psi$$

thus $B = M_g$.

Corollary 4.4. $T \in \mathscr{B}(\mathscr{H})$ *is unitary iff there exists* $\mathscr{F} : \mathscr{H} \to L^2(X, d\mu)$, *unitary, such that*

$$T = \mathscr{F}^* M_z \mathscr{F},$$

where $|z| \in \mathbb{T}^1 = \{z \in \mathbb{C} : |z| = 1\}$.

Exercise 4.5 (The Cayley transform). Finish the proof of Theorem 4.1 for the case when A is unbounded and selfadjoint.

Hint: Suppose $A = A^*$, unbounded. The Cayley transform

$$C_A := (A - i)(A + i)^{-1}$$

is then unitary. See Section 10.2. Apply Corollary 4.4 to C_A, and convert the result back to A. See, e.g., [Nel69, Rud91].

4.3 Projection-Valued Measure (PVM)

A projection valued measure (PVM) P (see Definition 3.10) satisfies the usual axioms of measures (here Borel measures) but with the main difference:

1. $P(\triangle)$ is a projection for all $\triangle \in \mathcal{B}$, i.e., $P(\triangle) = P(\triangle)^* = P(\triangle)^2$.

2. We assume that $P(\triangle_1 \cap \triangle_2) = P(\triangle_1) P(\triangle_2)$ for all $\triangle_1, \triangle_2 \in \mathcal{B}$; this property is called "orthogonality".

Remark 4.10. The notion of PVM extends the familiar notion of an ONB:

Let $\mathscr{H}$ be a separable Hilbert space, and suppose $\{u_k\}_{k \in \mathbb{N}}$ is a ONB in $\mathscr{H}$; then for $\triangle \in \mathcal{B}(\mathbb{R})$, set

$$P(\triangle) := \sum_{k \in \triangle} |u_k\rangle\langle u_k|. \tag{4.30}$$

Exercise 4.6 (A concrete PVM). Show that P is a PVM.

Note that, under the summation on the RHS in (4.30), we used Dirac's notation $|u_k\rangle\langle u_k|$ for the rank-one projection onto $\mathbb{C}u_k$ (see Definition 2.23). Further note that the summation is over k from $\triangle$; so it varies as $\triangle$ varies.

The PVM version of the spectral theorem says that $A = A^*$ iff

$$A = \int x P_A(dx)$$

where P is a projection-valued measure defined on the Borel σ-algebra of $\mathbb{R}$.

Definition 4.5 (See also Definition 3.10). Let $\mathcal{B}(\mathbb{C})$ be the Borel σ-algebra of $\mathbb{C}$. $\mathscr{H}$ is a Hilbert space.

$$P : \mathcal{B}(\mathbb{C}) \longrightarrow \mathrm{Proj}(\mathscr{H})$$

is a *projection-valued measure* (PVM), if

1. $P(\emptyset) = 0$, $P(\mathbb{C}) = I$, $P(A)$ is a projection for all $A \in \mathcal{B}$;

2. $P(A \cap B) = P(A)P(B)$, for all $A, B \in \mathcal{B}(\mathbb{C})$;

3. $P(\cup_k E_k) = \sum P(E_k)$, $E_k \cap E_j = \phi$ if $k \neq j$. The convergence is in terms of the strong operator topology. By assumption, the sequence of projections $\sum_{k=1}^{N} P(E_k)$ is monotone increasing, hence it has a limit, and

$$\lim_{N \to \infty} \sum_{k=1}^{N} P(E_k) = P(\cup E_k).$$

The standard Lebesgue integration extends to a PVM.

$$\langle \varphi, P(E)\varphi \rangle = \langle P(E)\varphi, P(E)\varphi \rangle = \|P(E)\|^2 \geq 0$$

since P is countably additive, the map $E \mapsto P(E)$ is also countably additive. Therefore, each $\varphi \in \mathcal{H}$ induces a regular Borel measure μ_φ on the Borel σ-algebra of $\mathbb{R}$.

For a measurable function ψ,

$$\begin{aligned} \int \psi d\mu_\varphi &= \int \psi(x) \langle \varphi, P(dx)\varphi \rangle \\ &= \left\langle \varphi, \left(\int \psi P(dx) \right) \varphi \right\rangle \end{aligned}$$

hence we may define

$$\int \psi P(dx)$$

as the operator so that for all $\varphi \in \mathcal{H}$,

$$\left\langle \varphi, \left(\int \psi P(dx) \right) \varphi \right\rangle.$$

Remark 4.11. $P(E) = F\chi_E F^{-1}$ defines a PVM. In fact all PVMs come from this way. In this sense, the M_t version of the spectral theorem is better, since it implies the PVM version. However, the PVM version facilitates some formulations in quantum mechanics, so physicists usually prefer this version.

Remark 4.12. Suppose we start with the PVM version of the spectral theorem. How to prove $(\psi_1 \psi_2)(A) = \psi_1(A)\psi_2(A)$? i.e. how to check we do have an algebra isomorphism? Recall in the PVM version, $\psi(A)$ is defined as the operator so that for all $\varphi \in \mathcal{H}$, we have

$$\int \psi d\mu_\varphi = \langle \varphi, \psi(A)\varphi \rangle.$$

As a standard approximation technique, once starts with simple or even step functions. Once it is worked out for simple functions, the extension to any measurable functions is straightforward. Hence let's suppose (WLOG) that the functions are simple.

Lemma 4.7. *We have* $\psi_1(A)\psi_2(A) = (\psi_1\psi_2)(A)$.

Proof. Let

$$
\begin{aligned}
\psi_1 &= \sum \psi_1(t_i)\chi_{E_i} \\
\psi_2 &= \sum \psi_2(t_j)\chi_{E_j}
\end{aligned}
$$

then

$$
\begin{aligned}
\int \psi_1 P(dx) \int \psi_2 P(dx) &= \sum_{i,j} \psi_1(t_i)\psi_2(t_j)P(E_i)P(E_j) \\
&= \sum_i \psi_1(t_i)\psi_2(t_i)P(E_i) \\
&= \int \psi_1\psi_2 P(dx)
\end{aligned}
$$

where we used the fact that $P(A)P(B) = 0$ if $A \cap B = \phi$. $\qquad\square$

Remark 4.13. As we delve into Nelson's lecture notes [Nel69], we notice that on page 69, there is another unitary operator. By piecing these operators together is precisely how we get the spectral theorem. This "piecing" is a vast generalization of Fourier series.

Lemma 4.8. *Pick* $\varphi \in \mathscr{H}$, *get the measure* μ_φ *where*

$$
\mu_\varphi(\cdot) = \|P(\cdot)\varphi\|^2
$$

and we have the Hilbert space $L^2(\mu_\varphi)$. *Take* $\mathscr{H}_\varphi := \overline{span}\{\psi(A)\varphi : \psi \in L^2(\mu_\varphi)\}$. *Then the map*

$$
\mathscr{H} \ni \psi(A)\varphi \mapsto \psi \in L^2(\mu_\varphi)
$$

is an isometry, and it extends uniquely to a unitary operator from $\mathscr{H}_\varphi$ *to* $L^2(\mu_\varphi)$.

Proof. We have

$$
\begin{aligned}
\|\psi(A)\varphi\|^2 &= \langle \psi(A)\varphi, \psi(A)\varphi \rangle \\
&= \langle \varphi, \bar{\psi}(A)\psi(A)\varphi \rangle \\
&= \int_{\mathbb{R}} |\psi(\lambda)|^2 \|P(d\lambda)\varphi\|^2 \\
&= \langle \varphi, |\psi|^2(A)\varphi \rangle \\
&= \int |\psi|^2 \, d\mu_\varphi.
\end{aligned}
$$

$\qquad\square$

Remark 4.14. $\mathscr{H}_\varphi$ is called the cyclic space generated by φ. Before we can construct $\mathscr{H}_\varphi$, we must make sense of $\psi(A)\varphi$.

Lemma 4.9 ([Nel69, p.67]). *Let $p = a_0 + a_1 x + \cdots + a_n x^n$ be a polynomial. Then $\|p(A)u\| \leq \max |p(t)|$, where $\|u\| = 1$ i.e. u is a state.*

Proof. $M := span\{u, Au, \ldots, A^n u\}$ is a finite dimensional subspace in $\mathscr{H}$ (automatically closed). Let E be the orthogonal projection onto M. Then

$$p(A)u = Ep(A)Eu = p(EAE)u.$$

sSince EAE is a Hermitian matrix on M, we may apply the spectral theorem for finite dimensional space and get

$$EAE = \sum \lambda_k P_{\lambda_k}$$

where $\lambda_k's$ are eigenvalues associated with the projections P_{λ_k}. It follows that

$$p(A)u = p(\sum \lambda_k P_{\lambda_k})u = \left(\sum p(\lambda_k) P_{\lambda_k}\right) u$$

and

$$
\begin{aligned}
\|p(A)u\|^2 &= \sum |p(\lambda_k)|^2 \|P_{\lambda_k} u\|^2 \\
&\leq \max |p(t)|^2 \sum \|P_{\lambda_k} u\|^2 \\
&= \max |p(t)|^2
\end{aligned}
$$

since

$$\sum \|P_{\lambda_k} u\|^2 = \|u\|^2 = 1.$$

Notice that $I = \sum P_{\lambda_k}$. $\qquad \square$

Remark 4.15. How to extend this? polynomials - continuous functions - measurable functions. $[-\|A\|, \|A\|] \subset \mathbb{R}$,

$$\|EAE\| \leq \|A\|$$

is a uniform estimate for all truncations. Apply Stone-Weierstrass' theorem to the interval $[-\|A\|, \|A\|]$ we get that any continuous function ψ is uniformly approximated by polynomials. i.e. $\psi \sim p_n$. Thus

$$\|p_n(A)u - p_m(A)u\| \leq \max |p_n - p_m| \|u\| = \|p_n - p_m\|_\infty \to 0$$

and $p_n(A)u$ is a Cauchy sequence, hence

$$\lim_n p_n(A)u =: \psi(A)u$$

where we may define the operator $\psi(A)$ so that $\psi(A)u$ is the limit of $p_n(A)u$.

4.4   Convert M_φ to a PVM (projection-valued measure)

Theorem 4.3 (Spectral Theorem, PVM version). *Let $A : \mathcal{H} \to \mathcal{H}$ be a selfadjoint operator. Then A is unitarily equivalent to the operator M_t of multiplication by the independent variable on the Hilbert space $L^2(\mu)$. There exists a unique projection-valued measure P so that*

$$A = \int t P(dt)$$

i.e. for all $h, k \in \mathcal{H}$,

$$\langle k, Ah \rangle_{\mathcal{H}} = \int t \langle k, P(dt) h \rangle_{\mathcal{H}}.$$

Proof. The uniqueness part follows from a standard argument. We will only prove the existence of P.

Let $\mathcal{F} : L^2(\mu) \to \mathcal{H}$ be the unitary operator so that $A = \mathcal{F} M_t \mathcal{F}^*$. Define

$$P(E) := \mathcal{F} \chi_E \mathcal{F}^*$$

for all E in the Borel σ-algebra $\mathfrak{B}$ of $\mathbb{R}$. Then $P(\emptyset) = 0$, $P(\mathbb{R}) = I$; and for all $E_1, E_2 \in \mathfrak{B}$,

$$
\begin{aligned}
P(E_1 \cap E_2) &= \mathcal{F} \chi_{E_1 \cap E_2} \mathcal{F}^{-1} \\
&= \mathcal{F} \chi_{E_1} \chi_{E_2} \mathcal{F}^{-1} \\
&= \left(\mathcal{F} \chi_{E_1} \mathcal{F}^{-1} \right) \left(\mathcal{F} \chi_{E_2} \mathcal{F}^{-1} \right) \\
&= P(E_1) P(E_2).
\end{aligned}
$$

Suppose $\{E_k\}$ is a sequence of mutually disjoint elements in $\mathfrak{B}$. Let $h \in \mathcal{H}$ and write $h = \mathcal{F} \widehat{h}$ for some $\widehat{h} \in L^2(\mu)$. Then

$$
\begin{aligned}
\langle h, P(\cup_k E_k) h \rangle_{\mathcal{H}} &= \left\langle \mathcal{F} \widehat{h}, P(\cup E_k) \mathcal{F} \widehat{h} \right\rangle_{\mathcal{H}} = \left\langle \widehat{h}, \mathcal{F}^{-1} P(\cup E_k) \mathcal{F} \widehat{h} \right\rangle_{L^2(\mu)} \\
&= \left\langle \widehat{h}, \chi_{\cup E_k} \widehat{h} \right\rangle_{L^2(\mu)} = \int_{\cup E_k} |\widehat{h}|^2 d\mu \\
&= \sum_k \int_{E_k} |\widehat{h}|^2 d\mu = \sum_k \langle h, P(E_k) h \rangle_{\mathcal{H}}.
\end{aligned}
$$

Therefore, P is a projection-valued measure.

For any $h, k \in \mathcal{H}$, write $h = \mathcal{F} \widehat{h}$ and $k = \mathcal{F} \widehat{k}$. Then

$$
\begin{aligned}
\langle k, Ah \rangle_{\mathcal{H}} &= \left\langle \mathcal{F} \widehat{k}, A \mathcal{F} \widehat{h} \right\rangle_{\mathcal{H}} = \left\langle \widehat{k}, \mathcal{F}^* A \mathcal{F} \widehat{h} \right\rangle_{\mathcal{H}} \\
&= \left\langle \widehat{k}, M_t \widehat{h} \right\rangle_{L^2(\mu)} = \int \overline{t \widehat{k}(t)} \widehat{h}(t) d\mu(t)
\end{aligned}
$$

$$= \int t \langle k, P(dt) h \rangle_{\mathscr{H}}.$$

Thus $A = \int t P(dt)$. $\qquad\qquad\square$

Remark 4.16. In fact, A is in the closed (under norm or strong topology) span of $\{P(E) : E \in \mathfrak{B}\}$. Equivalently, since $M_t = \mathcal{F}^* A \mathcal{F}$, the function $f(t) = t$ is in the closed span of the set of characteristic functions; the latter is again a standard approximation in measure theory. It suffices to approximate $t\chi_{[0,\infty]}$.

The wonderful idea of Lebesgue is not to partition the domain, as was the case in Riemann integral over $\mathbb{R}^n$, but instead the range. Therefore integration over an arbitrary set is made possible. Important examples include analysis on groups.

Proposition 4.1. *Let* $f : [0,\infty] \to \mathbb{R}$, $f(x) = x$, *i.e.* $f = x\chi_{[0,\infty]}$. *Then there exists a sequence of step functions* $s_1 \le s_2 \le \cdots \le f(x)$ *such that* $\lim_{n\to\infty} s_n(x) = f(x)$.

Proof. For $n \in \mathbb{N}$, define

$$s_n(x) = \begin{cases} i2^{-n} & x \in [i2^{-n}, (i+1)2^{-n}) \\ n & x \in [n, \infty] \end{cases}$$

where $0 \le i \le n2^{-n} - 1$. Equivalently, s_n can be written using characteristic functions as

$$s_n = \sum_{i=0}^{n2^n - 1} i2^{-n}\chi_{[i2^{-n},(i+1)2^{-n})} + n\chi_{[n,\infty]}.$$

Notice that on each interval $[i2^{-n}, (i+1)2^{-n})$,

$$\begin{aligned} s_n(x) &\equiv i2^{-n} \le x \\ s_n(x) + 2^{-n} &\equiv (i+1)2^{-n} > x \\ s_n(x) &\le s_{n+1}(x). \end{aligned}$$

Therefore, for all $n \in \mathbb{N}$ and $x \in [0,\infty]$,

$$x - 2^{-n} < s_n(x) \le x \qquad\qquad (4.31)$$

and $s_n(x) \le s_{n+1}(x)$.

It follows from (4.31) that

$$\lim_{n\to\infty} s_n(x) = f(x)$$

for all $x \in [0,\infty]$. $\qquad\qquad\square$

Corollary 4.5. *Let* $f(x) = x\chi_{[0,M]}(x)$. *Then there exists a sequence of step functions* s_n *such that* $0 \le s_1 \le s_2 \le \cdots \le f(x)$ *and* $s_n \to f$ *uniformly, as* $n \to \infty$.

Proof. Define s_n as in Proposition 4.1. Let $n > M$, then by construction

$$f(x) - 2^{-n} < s_n(x) \leq f(x)$$

for all $s \in [0, M]$. Hence $s_n \to f$ uniformly as $n \to \infty$. □

Proposition 4.1 and its corollary immediate imply the following.

Corollary 4.6. *Let $(X, \mathcal{S}, \mu)$ be a measure space. A function (real-valued or complex-valued) is measurable if and only if it is the point-wise limit of a sequence of simple functions. A function is bounded measurable if and only if it is the uniform limit of a sequence of simple functions. Let $\{\tilde{s}_n\}$ be an approximation sequence of simple functions. Then $\tilde{s}_n$ can be chosen such that $|\tilde{s}_n(x)| \leq |f(x)|$ for all $n = 1, 2, 3 \ldots$.*

Proof. Let (s_n) be the functions from Proposition 4.1. Each s_n is a linear combination of indicator functions χ_J where J is a dyadic interval. Now set $\tilde{s}_n := s_n \circ f$, where $f : X \longrightarrow \mathbb{R}$ is given and assumed measurable. Then $f^{-1}(J) \in \mathcal{S}$ (the σ-algebra). Since $\chi_J \circ f = \chi_{f^{-1}(J)}$, it follows that the sequence $(\tilde{s}_n)$ of simple functions has the desired properties. □

Theorem 4.4. *Let $M_f : L^2(X, \mathcal{S}, \mu) \to L^2(X, \mathcal{S}, \mu)$ be the operator of multiplication by f. Then,*

1. *if $f \in L^\infty$, M_f is a bounded operator, and M_f is in the closed span of the set of selfadjoint projections under norm topology.*

2. *if f is unbounded, M_f is an unbounded operator. M_f is in the closed span of the set of selfadjoint projections under the strong operator topology.*

Proof. If $f \in L^\infty$, then by Corollary 4.6 there exists a sequence of simple functions s_n so that $s_n \to f$ uniformly. Hence $\|f - s_n\|_\infty \to 0$, as $n \to \infty$.

Suppose f is unbounded. By Proposition 4.1 and its corollaries, there exists a sequence of simple functions s_n such that $|s_n(x)| \leq |f(x)|$ and $s_n \to f$ pointwisely, as $n \to \infty$. Let h be any element in the domain of M_f, i.e.

$$\int \left(|h| + |fh|^2 \right) d\mu < \infty.$$

Then

$$\lim_{n \to \infty} |(f(x) - s_n(x))h(x)|^2 = 0$$

and

$$|(f(x) - s_n(x))h(x)|^2 \leq \text{const} \cdot |h(x)|^2.$$

Hence by the dominated convergence theorem,

$$\lim_{n \to \infty} \int |(f(x) - s_n(x))h(x)|^2 \, d\mu = 0$$

or equivalently,

$$\|(f - s_n)h\|^2 \to 0$$

as $n \to \infty$. i.e. M_{s_n} converges to M_f in the strong operator topology. □

Exercise 4.7 (An application to numerical range). Let A be a bounded normal operator in a separable Hilbert space $\mathscr{H}$; then prove that

$$NR_A \subseteq \overline{conv}\left(spec\left(A\right)\right); \qquad (4.32)$$

i.e., that the numerical range of A is contained in the closed convex hull of the spectrum of A. (We refer to Exercise 2.27 for details on "numerical range.")

Hint: Since A is normal, by the Spectral Theorem, it is represented in a PVM $P_A\left(\cdot\right)$, i.e., taking valued in $\mathrm{Proj}\left(\mathscr{H}\right)$. For $x \in \mathscr{H}$, $\|x\| = 1$, we have

$$w_x\left(A\right) = \langle x, Ax \rangle = \int_{spec(A)} \lambda \left\| P_A\left(d\lambda\right) x \right\|^2, \text{ and} \qquad (4.33)$$

$$\int_{spec(A)} \left\| P_A\left(d\lambda\right) x \right\|^2 = \|x\|^2 = 1, \qquad (4.34)$$

so $d\mu_x\left(\lambda\right) = \left\| P_A\left(d\lambda\right) x \right\|^2$ is a regular Borel probability measure on $spec\left(A\right)$. Now approximate (4.33) with simple functions on $spec\left(A\right)$.

Definition 4.6 (Quantum States). Let $\mathscr{H}$ be a Hilbert space (corresponding to some quantum system), and let A be a selfadjoint operator in $\mathscr{H}$, possibly unbounded. Vectors $f \in \mathscr{H}$ represent quantum states if $\|f\| = 1$.

The *mean* of A in the state f is

$$\langle f, Af \rangle = \int_{\mathbb{R}} \lambda \left\| P_A\left(d\lambda\right) f \right\|^2$$

where $P_A\left(\cdot\right)$ denotes the spectral resolution of A.

The *variance* of A in the state f is

$$\begin{aligned} v_f\left(A\right) &= \|Af\|^2 - \left(\langle f, Af \rangle\right)^2 \qquad (4.35) \\ &= \int_{\mathbb{R}} \lambda^2 \left\| P_A\left(d\lambda\right) f \right\|^2 - \left(\int_{\mathbb{R}} \lambda \left\| P_A\left(d\lambda\right) f \right\|^2\right)^2 \\ &= \int_{\mathbb{R}} \left(\lambda - \langle f, Af \rangle\right)^2 \left\| P_A\left(d\lambda\right) f \right\|^2. \end{aligned}$$

Theorem 4.5 (Uncertainty Principle). *Let $\mathscr{D}$ be a dense subspace in $\mathscr{H}$, and A, B be two Hermitian operators such that $A, B : \mathscr{D} \hookrightarrow \mathscr{D}$ (i.e., $\mathscr{D}$ is assumed invariant under both A and B.)*
Then,

$$\|Ax\| \, \|Bx\| \geq \frac{1}{2} \left| \langle x, [A, B] x \rangle \right|, \ \forall x \in \mathscr{D}; \qquad (4.36)$$

where $[A, B] := AB - BA$ is the commutator of A and B.
In particular, setting

$$\begin{aligned} A_1 &:= A - \langle x, Ax \rangle \\ B_1 &:= B - \langle x, Bx \rangle \end{aligned}$$

then A_1, B_1 are Hermitian, and

$$[A_1, B_1] = [A, B].$$

Therefore,

$$\|A_1 x\| \, \|B_1 x\| \geq \frac{1}{2} |\langle x, [A, B] x \rangle|, \ \forall x \in \mathscr{D}.$$

Proof. By the Cauchy-Schwarz inequality (Lemma 2.2), and for $x \in \mathscr{D}$, we have

$$
\begin{aligned}
\|Ax\| \, \|Bx\| \ &\geq \ |\langle Ax, Bx \rangle| \\
&\geq \ |\Im\{\langle Ax, Bx \rangle\}| \\
&= \ \frac{1}{2} \left| \langle Ax, Bx \rangle - \overline{\langle Ax, Bx \rangle} \right| \\
&= \ \frac{1}{2} |\langle Ax, Bx \rangle - \langle Bx, Ax \rangle| \\
&= \ \frac{1}{2} |\langle x, [AB - BA] x \rangle| .
\end{aligned}
$$

$\square$

Corollary 4.7. *If $[A, B] = ihI$, $h \in \mathbb{R}_+$, and $\|x\| = 1$ (i.e., is a state); then*

$$w_x \left(A^2\right)^{\frac{1}{2}} w_x \left(B^2\right)^{\frac{1}{2}} \geq \frac{h}{2}. \tag{4.37}$$

Exercise 4.8 (Heisenberg's uncertainty principle)**.** Let $\mathscr{H} = L^2(\mathbb{R})$, and $f \in L^2(\mathbb{R})$ given, with $\|f\|^2 = \int |f(x)|^2 \, dx = 1$. Suppose $f \in dom(P) \cap dom(Q)$, where P and Q are the momentum and position operators, respectively. (See eq. (3.5)-(3.6) in Section 3.1.)

Show that

$$v_f(P) \, v_f(Q) \geq \frac{1}{4}. \tag{4.38}$$

Inequality (4.38) is the mathematically precise form of Heisenberg's uncertainty relation, often written in the form

$$\sigma_f(P) \, \sigma_f(Q) \geq \frac{1}{2}$$

where $\sigma_f(P) = \sqrt{v_f(P)}$, and $\sigma_f(Q) = \sqrt{v_f(Q)}$.

4.5 The Spectral Theorem for Compact Operators

4.5.1 Preliminaries

The setting for the first of the two Spectral Theorems (direct integral vs representation) we will consider is as following: (Restricting assumptions will be relaxed in subsequent versions!)

Let $\mathscr{H}$ be a separable (typically infinite dimensional Hilbert space assumed here!) and let $A \in \mathscr{B}(\mathscr{H}) \setminus \{0\}$ be compact and selfadjoint, i.e., $A = A^*$, and A is in the $\|\cdot\|_{UN}$-closure of $\mathscr{F}R(\mathscr{H})$. See Section 2.5.1.

Theorem 4.6 (Spectral Theorem for compact operators). *With A as above, there is an orthonormal set $\{u_k\}_{k \in \mathbb{N}}$, and a sequence $\{\lambda_k\}_{k \in \mathbb{N}}$ in $\mathbb{R} \setminus \{0\}$, such that $|\lambda_1| \geq |\lambda_2| \geq \cdots \geq |\lambda_k| \geq |\lambda_{k+1}| \geq \cdots$, $\lim_{k \to \infty} \lambda_k = 0$, and*

1. $Au_k = \lambda_k u_k$, $k \in \mathbb{N}$,

2. $A = \sum_{k=1}^{\infty} \lambda_k |u_k\rangle\langle u_k|$,

3. $spec(A) = \{\lambda_k\} \cup \{0\}$,

4. $\dim\{v \in \mathscr{H} \mid Av = \lambda_k v\} < \infty$, for all $k \in \mathbb{N}$.

The set $\{u_k\}$ extends to an ONB, containing an ONB (possibly infinite) for the subspace

$$\ker(A) = \{v \in \mathscr{H} \mid Av = 0\}.$$

Proof. The proof is a guided tour through the technical points, collected into the next two exercises. $\qquad\square$

Exercise 4.9 (An eigenspace). Let $A \in \mathscr{B}(\mathscr{H})$ be compact, and let $\lambda \in \mathbb{C} \setminus \{0\}$. Show that the eigenspace

$$\mathscr{E}_\lambda := \mathrm{Ker}(\lambda I - A)$$

is finite-dimensional.

    Hint: Assume the contrary $\dim \mathscr{E}_\lambda = \infty$. Pick an ONB in $\mathscr{E}_\lambda$, say $\{u_i\}_{i \in \mathbb{N}}$. Since A is compact, the sequence $\{Au_i\}_{i \in \mathbb{N}}$ has a convergent subsequence, say $\{Au_{i_k}\}$. But then

$$\|Au_{i_k} - Au_{i_l}\| \xrightarrow[k,l \to \infty]{} 0.$$

On the other hand,

$$\|Au_{i_k} - Au_{i_l}\|^2 = |\lambda|^2 \|u_{i_k} - u_{i_l}\|^2 = 2|\lambda|^2;$$

so a contradiction.

Exercise 4.10 (Attaining the sup). Suppose $A \in \mathscr{B}(\mathscr{H})$ is compact, and $A^* = A$. Suppose further that

$$\lambda = \sup\{\langle x, Ax\rangle : \|x\| = 1\}$$

satisfies $\lambda > 0$, strict.

1. Show that $\|A\| = \lambda$.

2. Show that, if $\{x_i\}_{i\in\mathbb{N}}$ satisfies:

$$\|x_i\| = 1, \quad \langle x_i, Ax_i\rangle \xrightarrow[i\to\infty]{} \lambda,$$

then $\exists\, x \in \mathscr{H}$, $\|x\| = 1$, and a subsequence $\{x_{i_k}\}$ such that

$$\|Ax_{i_k} - Ax_{i_l}\| \xrightarrow[k,l\to\infty]{} 0, \text{ and}$$

$$\langle x_{i_k} - x, v\rangle \xrightarrow[k\to\infty]{} 0, \ \forall v \in \mathscr{H}.$$

3. Conclude from (1)-(2) that $Ax = \lambda x$.

4. Make (1)-(2) the first step in an induction; thus finishing the proof of the Spectral Theorem for compact selfadjoint operators.

Exercise 4.11 (A Green's function). Let $\mathscr{H} = L^2(0,1)$, and set

$$K(s,t) = s \wedge t - st, \quad \forall s,t \in [0,1]. \tag{4.39}$$

Let $A : \mathscr{H} \longrightarrow \mathscr{H}$ be the integral operator

$$(Af)(t) = \int_0^1 K(t,s)f(s)\,ds. \tag{4.40}$$

1. Show that the functions $u_k(t) = \sqrt{2}\sin(k\pi t)$, $k \in \mathbb{N} = \{1,2,\cdots\}$, satisfy the condition in Theorem 4.6 for the pair $(\mathscr{H}, A)$.

2. Use part 1 to establish the formula

$$K(t,s) = 2\sum_{k=1}^{\infty} \frac{\sin(k\pi t)\sin(k\pi s)}{(k\pi)^2}. \tag{4.41}$$

Note, we shall return to (4.39)-(4.41) in Chapter 7. In fact, K in (4.39) here is the covariance function for pinned Brownian motion $\{X_t\}_{t\in[0,1]}$ (Figure 4.5), i.e., pinned by $X_0 = X_1 = 0$. (To see that $X(t) = tB\left(\frac{1}{t}\right)$ is again a "copy" of Brownian motion, the reader may consult Definition 7.6, and the discussion after that.)

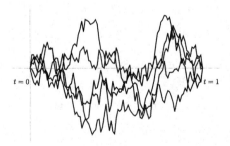

Figure 4.5: Pinned Brownian motion (with 5 sample paths).

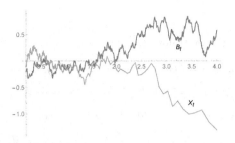

Figure 4.6: Time-reflection of Brownian motion: $X_t = tB\left(\frac{1}{t}\right)$, $t \in \mathbb{R}_+$, reflection $B \longrightarrow X$.

Exercise 4.12 (Pinned Brownian motion from reflection). Let $\{B_t\}_{t\in[0,\infty]}$ be standard Brownian motion, and let $X_t := tB\left(\frac{1}{t}\right)$, be its time-reflection. Set

$$Y(t) := \frac{B(t) - X(t)}{\sqrt{2}}, \quad t \in \mathbb{R}_+. \tag{4.42}$$

Show that then $Y(t)$ in (4.42) on $t \in [0,1]$, is a "copy" of pinned Brownian motion (also called the Brownian bridge between two points). See Figures 4.5-4.6, and 4.7 below.

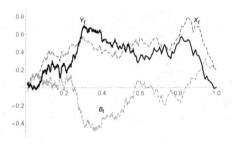

Figure 4.7: A simulation of pinned Brownian motion as $Y(t) = \frac{1}{\sqrt{2}}(B(t) - X(t))$ (black).

Readers not familiar with Brownian motion and its variants, may wish to consult Chapter 7; especially Definitions 7.6 and 7.8, Remark 7.1.

We begin with some preliminaries in the preparation for the proof of Theorem 4.6.

Lemma 4.10 (Polarization identity). *Let X be a complex vector space, and $f : X \times X \to \mathbb{C}$ a sesquilinear form, conjugate linear in the first variable and linear in the second variable. Then the following polarization identity holds:*

$$f(x,y) = \frac{1}{4} \sum_{k=0}^{3} i^k f\left(y + i^k x, y + i^k x\right) \tag{4.43}$$

for all $x, y \in X$.

Proof. A direct computation shows that

$$\begin{aligned}
f(y+x, y+x) - f(y-x, y-x) &= 2f(y,x) + 2f(x,y), \text{ and} \\
i(f(y+ix, y+ix) - f(y-ix, y-ix)) &= -2f(y,x) + 2f(x,y).
\end{aligned}$$

Adding the above two equations yields the desired result.   □

Corollary 4.8. *A bounded operator A in $\mathscr{H}$ is selfadjoint if and only if*

$$\langle x, Ax \rangle \in \mathbb{R}, \ \forall x \in \mathscr{H}.$$

Proof. Suppose A is selfadjoint, i.e., $\langle x, Ay \rangle = \langle Ax, y \rangle$, $\forall x, y \in \mathscr{H}$. Setting $x = y$, then

$$\langle x, Ax \rangle = \langle Ax, x \rangle = \overline{\langle x, Ax \rangle} \implies \langle x, Ax \rangle \in \mathbb{R}, \ \forall x \in \mathscr{H}.$$

Conversely, suppose $\langle x, Ax \rangle \in \mathbb{R}, \ \forall x \in \mathscr{H}$. Note that

$$(x, y) \longmapsto \langle x, Ay \rangle \quad \text{and} \quad (x, y) \longmapsto \langle Ax, y \rangle$$

are both sesquilinear forms defined on $\mathscr{H}$. It follows from Lemma 4.10, that

$$\langle x, Ay \rangle = \frac{1}{4} \sum_{k=0}^{3} i^k \langle y + i^k x, A(y + i^k x) \rangle, \text{ and} \tag{4.44}$$

$$\langle Ax, y \rangle = \frac{1}{4} \sum_{k=0}^{3} i^k \langle A(y + i^k x), y + i^k x \rangle. \tag{4.45}$$

But by the assumption $(\langle x, Ax \rangle \in \mathbb{R}, \ x \in \mathscr{H})$, the RHS in (4.44) and (4.45) are equal. Therefore, we conclude that $\langle x, Ay \rangle = \langle Ax, y \rangle, \ \forall x, y \in \mathscr{H}$, i.e., A is selfadjoint.   □

Theorem 4.7. *Let $\mathscr{H}$ be a Hilbert space over $\mathbb{C}$. Let $f : \mathscr{H} \times \mathscr{H} \to \mathbb{C}$ be a sesquilinear form, and set*

$$M := \sup\{|f(x,y)| : \|x\| = \|y\| = 1\} < \infty.$$

Then there exists a unique bounded operator A in $\mathscr{H}$, satisfying

$$\begin{aligned}
f(x,y) &= \langle Ax, y \rangle, \ \forall x, y \in \mathscr{H}; \text{ and} \tag{4.46} \\
\|A\| &= M. \tag{4.47}
\end{aligned}$$

Proof. Given $x, y \in \mathscr{H}$, nonzero, we have

$$\left| f\left(\frac{x}{\|x\|}, \frac{y}{\|y\|} \right) \right| \leq M \iff |f(x,y)| \leq M \|x\| \|y\|. \tag{4.48}$$

Thus, for each $x \in \mathcal{H}$, the map $y \mapsto f(x,y)$ is a bounded linear functional on $\mathcal{H}$. By Riesz's theorem, there exists a unique element $\xi_x \in \mathcal{H}$, such that

$$f(x,y) = \langle \xi_x, y \rangle.$$

Set $\xi_x := Ax$, $x \in \mathcal{H}$. (The uniqueness part of follows from Riesz.)

Note the map $x \longmapsto Ax$ is linear. For if $c \in \mathbb{C}$, then

$$
\begin{aligned}
f(x_1 + cx_2, y) &= \langle A(x_1 + cx_2), y \rangle, \text{ and} && (4.49) \\
f(x_1 + cx_2, y) &= f(x_1, y) + \bar{c}f(x_2, y) \\
&= \langle Ax_1, y \rangle + \bar{c}\langle Ax_2, y \rangle \\
&= \langle Ax_1 + cAx_2, y \rangle; && (4.50)
\end{aligned}
$$

where in (4.50), we used the fact that f is conjugate linear in the first variable. It follows that $A(x_1 + cx_2) = Ax_1 + cAx_2$, i.e., A is linear.

Finally,

$$\|A\| = \sup_{\|x\|=1} \|Ax\| = \sup_{\|x\|=1} \left(\sup_{\|y\|=1} |\langle Ax, y \rangle| \right)$$

and (4.47) follows. $\qquad \square$

Corollary 4.9. *Any bounded operator A in $\mathcal{H}$ is uniquely determined by the corresponding sesquilinear form $(x,y) \longmapsto \langle x, Ay \rangle$, $(x,y) \in \mathcal{H} \times \mathcal{H}$.*

Corollary 4.10. *For all $A \in \mathcal{B}(\mathcal{H})$, we have*

$$\|A\| = \sup \{|\langle x, Ay \rangle| : \|x\| = \|y\| = 1\}. \qquad (4.51)$$

Eq. (4.52) below is the key step in the proof of Theorem 4.6.

Corollary 4.11. *Let A be a bounded selfadjoint operator in $\mathcal{H}$, then*

$$\|A\| = \sup \{|\langle x, Ax \rangle| : \|x\| = 1\}. \qquad (4.52)$$

Proof. Set $M := \sup \{|\langle x, Ax \rangle| : \|x\| = 1\}$.

For all unit vector x in $\mathcal{H}$, we see that

$$|\langle x, Ax \rangle| \le \|x\| \|Ax\| \le \|x\| \|x\| \|A\| = \|A\|;$$

where the first step above uses the Cauchy-Schwarz inequality. Thus, $M \le \|A\|$. Conversely, by the polarization identity (4.43), we have

$$
\begin{aligned}
4\langle x, Ay \rangle = &\langle A(x+y), x+y \rangle - \langle A(-x+y), -x+y \rangle \\
&+ i\langle A(ix+y), ix+y \rangle - i\langle A(-ix+y), -ix+y \rangle. \quad (4.53)
\end{aligned}
$$

Sine A is selfadjoint, the four inner products on the RHS of (4.53) are all real-valued (Corollary 4.8). Therefore,

$$\Re\{\langle x, Ay \rangle\} = \frac{1}{4}\left(\langle A(x+y), x+y \rangle - \langle A(-x+y), -x+y \rangle \right).$$

Now, there exists a phase factor $e^{i\theta}$ (depending on x, y) such that

$$
\begin{aligned}
|\langle x, Ay \rangle| &= e^{i\theta} \langle x, Ay \rangle \\
&= |\Re \{\langle x, Ay \rangle\}| \\
&= \frac{1}{4} |\langle A(x+y), x+y \rangle - \langle A(-x+y), -x+y \rangle| \\
&\leq \frac{1}{4} M \left(\|x+y\|^2 + \|x-y\|^2 \right) \\
&= \frac{1}{4} \|M\| \left(2\|x\|^2 + 2\|y\|^2 \right) = M
\end{aligned}
$$

valid for all unit vectors x, y in $\mathcal{H}$, i.e., with $\|x\| = \|y\| = 1$. It follows from this and (4.51) that $\|A\| \leq M$.

Therefore, we have

$$
M = \|A\| = \sup_{\|x\|=1} |\langle x, Ax \rangle|
$$

which is the assertion. $\square$

4.5.2   Integral operators

Integral operators with continuous kernel form an important subclass of compact operators.

Setting. Let X be a compact space, μ a finite positive Borel measure on X, and

$$
K : X \times X \longrightarrow \mathbb{C} \tag{4.54}
$$

a given function, assumed *continuous* on $X \times X$. Define

$$
T_K : L^2(\mu) \longrightarrow L^2(\mu) \text{ by}
$$

$$
(T_K f)(x) = \int_X K(x,y) f(y) \, d\mu(y), \ \forall f \in L^2(\mu), \ \forall x \in X. \tag{4.55}
$$

Exercise 4.13 (An application of Arzelà-Ascoli). Prove that T_K is a compact operator in $L^2(\mu)$ subject to the stated assumptions above.

Hint:

Step 1. Show that, for $\forall x_1, x_2 \in X$, and $f \in L^2(\mu)$, we have:

$$
\begin{aligned}
&|T_K f(x_1) - T_K f(x_2)| \\
&\leq \sqrt{\mu(X)} \max_{y \in X} |K(x_1, y) - K(x_2, y)| \|f\|_{L^2(\mu)}.
\end{aligned}
$$

Step 2. Show that

$$
|(T_K f)(x)| \leq \sqrt{\mu(X)} \max_{y \in X} |K(x,y)| \|f\|_{L^2(\mu)}.
$$

Step 3. Conclude from steps 1-2, and an application of Arzelà-Ascoli's theorem that $T_K : L^2(\mu) \longrightarrow L^2(\mu)$ is a compact operator.

Exercise 4.14 (Powers-Størmer [PS70]). Let $\mathscr{H}$ be a Hilbert space, and let A and B be positive operators ($\in \mathscr{B}(\mathscr{H})$). Then show that

$$\left\| A^{\frac{1}{2}} - B^{\frac{1}{2}} \right\|_{HS}^2 \leq \|A - B\|_{TR}. \tag{4.56}$$

(Note $A^{\frac{1}{2}} = \sqrt{A}$ is defined via the Spectral Theorem.)

   Hint: Difficult. (The inequality (4.56) is called the Powers-Størmer inequality.) Set $S = A^{\frac{1}{2}} - B^{\frac{1}{2}}$, and $T = A^{\frac{1}{2}} + B^{\frac{1}{2}}$.

   Note that (4.56) is trivial if $A - B$ is not trace-class, so assume it has finite trace-norm. Then diagonalize S in an ONB (use the Spectral Theorem), i.e., pick an ONB $\{f_i\}$ of eigenvectors with eigenvalues $\{\lambda_i\}$, $Sf_i = \lambda_i f_i$. Then

$$
\begin{aligned}
Tr\left(|A - B|\right) &= \frac{1}{2} \sum_i \langle f_i, |ST + TS| f_i \rangle \\
&\geq \sum_i |\lambda_i \langle f_i, Tf_i \rangle| \\
&\geq \sum_i |\lambda_i|^2.
\end{aligned}
$$

Exercise 4.15 (Parseval frames). Let $\mathscr{H}$ be a Hilbert space. A system of vectors $\{\psi_i\}_{i \in I}$ in $\mathscr{H}$ is called a *Parseval frame* iff, for all $h \in \mathscr{H}$, we have

$$\|h\|^2 = \sum_{i \in I} |\langle \psi_i, h \rangle|^2. \tag{4.57}$$

Let $\{\psi_i\}_{i \in I}$ be a Parseval frame (i.e., assume (4.57) holds.)

1. Then prove that the following two conditions are equivalent:

   (a) $\{\psi_i\}_{i \in I}$ is an orthonormal basis (ONB) in $\mathscr{H}$; and

   (b) $\|\psi_i\| = 1$, $\forall i \in I$.

2. Given examples of Parseval frames which are *not* ONBs.

Exercise 4.16 (A dilation space). Let $\mathscr{H}$ be a Hilbert space, and let $\{\psi\}_{i \in I}$ be a Parseval frame in $\mathscr{H}$.

1. Define $V : \mathscr{H} \longrightarrow l^2(I)$ by $Vh = (\langle \psi_i, h \rangle_{\mathscr{H}})_{i \in I}$, and show that V is isometric, i.e.,

   $$\|Vh\|_{l^2(I)} = \|h\|_{\mathscr{H}}, \quad \forall h \in \mathscr{H}. \tag{4.58}$$

2. Show that the adjoint operator $V^* : l^2(I) \longrightarrow \mathscr{H}$ is given by

   $$V^*\xi = \sum_{i \in I} \xi_i \psi_i, \quad \forall (\xi_i) \in l^2(I); \tag{4.59}$$

and so the projection VV^* is as follows:

$$VV^*\xi = \sum_{j\in I} \langle \psi_i, \psi_j \rangle_{\mathscr{H}} \, \xi_j, \quad \forall \xi \in l^2(I). \tag{4.60}$$

It is a well-defined projection in $l^2(I)$.

3. Let ε_i (given by $\varepsilon_i(j) = \delta_{ij}$), $i \in I$ be the standard ONB in $l^2(I)$; then show that $V^*\varepsilon_i = \psi_i$, $\forall i \in I$.

A summary of relevant numbers from the Reference List

For readers wishing to follow up sources, or to go in more depth with topics above, we suggest: [Con90, FL28, Kat95, Kre55, Lax02, LP89, Nel69, RS75, Rud91, Sto51, Sto90, Yos95, HJL$^+$13, DHL09, HKLW07, Alp01, CZ07, Fan10, Jor06, AG93, DS88b, Hal82, Jor02, Mac52, vN35, Hel13, MJD$^+$15].

Part III

Applications

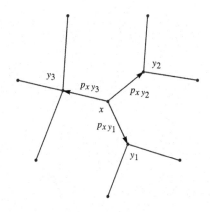

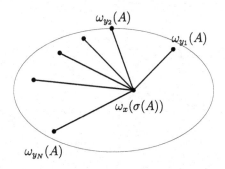

Chapter 5

GNS and Representations

"If one finds a difficulty in a calculation which is otherwise quite convincing, one should not push the difficulty away; one should rather try to make it the centre of the whole thing."
— Werner Heisenberg

"Mathematics is a way of thinking in everyday life ..."
— I.M. Gelfand

"First quantization is a mystery; – second quantization is a functor."
— Edward Nelson

Explanation:

A **category** is an algebraic structure that comprises "objects" linked by morphisms, also called "arrows." A category has two basic properties: one allows us to compose the arrows associatively; and the existence of an identity arrow for each object. A functor is a type of mapping, or transformation, between categories. Functors can be thought of as transformation between categories that transform the rules in the first category into those of the second: objects to objects, arrows to arrows, and diagrams to diagrams. In small categories, functors can be thought of as morphisms.

First quantization is the replacement of classical observables, such as energy or momentum, by operators; and of classical states by "wave functions". Second quantization usually refers to the introduction of field operators when describing quantum many-body systems. In second quantization, one passes from wave functions to an operators; hence the non-commutativity.

Expanding upon the more traditional interpretation, Ed Nelson suggested yet another second quantization functor; it goes from the category of Hilbert space (Hilb) to that of probability space (Prob). In this version, the objects in the category of Hilbert space "Hilb" are Hilbert spaces, and the morphisms are contractive linear operators.

In the category "Prob," the morphisms are point-transformation of measures.

Outline A state on a C^*-algebra $\mathfrak{A}$ is a positive linear (and normalized) functional on $\mathfrak{A}$. Given a C^*-algebra $\mathfrak{A}$, then there is a bijective correspondence between states of $\mathfrak{A}$, on one side, and cyclic representations of $\mathfrak{A}$ on the other; it is called the Gelfand-Naimark-Segal construction (abbreviated GNS), and it yields an explicit correspondence between the set of all cyclic $*$-representations of $\mathfrak{A}$, $Rep_{cyc}(\mathfrak{A})$; and the states of $\mathfrak{A}$, $S(\mathfrak{A})$. It is named for Israel Gelfand, Mark Naimark, and Irving Segal.

The importance of the GNS construction is that it offers answers to a host of questions about representations of algebras, and of groups; the natural question regarding elementary building blocks.

More precisely, in the case of unitary representations of groups, the first questions that present themselves are: "What is the "right" notion of decomposition of a given unitary representation in terms of irreducible unitary representations of G?" And "how to compute decompositions?" In this generality, there is not a precise answer. But if G is assumed locally compact and unimodular, I.E. Segal [Seg50] established precise answers. They entail direct integral theory for representations; see details below. The power of the GNS-construction (states vs representations) is that it allows us to answer a parallel question for states. Indeed, there is a precise notion of "building blocks" for states, they are the pure states (extreme points).

Segal's insight was twofold[1]: (i) Showing that the pure states correspond to irreducible representations via the GNS correspondence. And (ii), make the precise link between a direct integral decomposition for states, on one side, and on the other, direct integrals for unitary representations. Part (ii) in turn involves Krein-Milman and Choquet theory; see Sections 5.2, 5.8, and Chapter 9.

A corollary of the GNS construction is *the Gelfand-Naimark theorem*. The latter characterizes C^*-algebras as precisely the norm-closed $*$-algebras arising as $*$-subalgebras of $\mathscr{B}(\mathscr{H})$, the C^*-algebra of all bounded operators on a Hilbert space. By extreme-point theory [Phe01], one shows that every C^*-algebra has sufficiently many pure states (corresponding to irreducible representations under GNS). As a result, the representation of $\mathfrak{A}$ arising as a direct sum of these corresponding irreducible GNS-representations is faithful.

Since states in quantum physics are vectors (of norm one) in Hilbert space; two question arise: "Where does the Hilbert space come from?" And "What are the algebras of operators from which the selfadjoint observables must be selected?" In a general framework, we offer an answer below, it goes by the name "the Gelfand-Naimark-Segal (GNS) theorem," which offers a direct correspondence between states and cyclic representations.

---

[1]Appendix C includes a short biographical sketch of I. Segal. See also [Seg50].

Chapters 5 and 6 below form a pair; — to oversimplify, the theme in Chapter 6 is a generalization of that of the present chapter.

One could say that Chapter 5 is about scalar valued "states"; while the "states" in Chapter 6 are operator valued. In both cases, we must specify the appropriate notion of positivity, and this notion in the setting of Chapter 6 is more subtle; — it is called "complete positivity."

But the goal in both cases is to induce in order to create representation of some given non Abelian algebra $\mathfrak{A}$ coming equipped with a star-involution; for example a C^*-algebra. The representations, when induced from states, will be $*$-representations; i.e., will take the $*$-involution in $\mathfrak{A}$ to "adjoint operator", where "adjoint" refers to the Hilbert space of the induced representation.

In Chapters 5-6, this notion of induction is developed in detail; and its counterpart for the case of unitary representations of groups is discussed in detail in Chapter 8.

Historically, the two notion of *induction of representations* were used by researchers in parallel universes, for the case of operator algebras (Chapters 5-6), they were pioneered by Gelfand, Naimark, Segal, and brought to fruition by Stinespring and Arveson.

On the other side of the divide, in the study of unitary representations of groups, the names are Harish-Chandra, G.W. Mackey[2] (and more, see cited references in Chapter 8); — and this is the subject of Chapter 8 below. We caution the reader that the theory of unitary representations is a vast subject, and motivated by a number of diverse areas, such as quantum theory, ergodic theory, harmonic analysis, to mention just a few. And the theory of representations of groups, and their induction, is in turn developed by different researchers; and often with different groups G in mind; — continuous vs discrete; Lie groups vs the more general case of locally compact groups. The case when the group G is assumed locally compact is attractive because we then always will have left (or right-) Haar measure at our disposal. And there is an associated left-regular representation on the L^2 space of Haar measure. The left-invariance of Haar measure makes this representation unitary. The analogous hold of course for the constructions with right-invariant Haar measure; the two are linked by the *modular function* of G.

In the details below, the following concepts and terms will play an important role; (for reference, see also Appendix F): States on $*$-algebras, group representations, and operators in Hilbert space, representations of $*$-algebras, trace class operators, renormalization, Gelfand transform, Fredholm operators and their indices, multiplicity of a representation.

---

[2]Appendix C includes a short biographical sketch of G. Mackey. See also [Mac52, Mac85, Mac88].

5.1   Definitions and Facts: An Overview

Let $\mathfrak{A}$ be an algebra over $\mathbb{C}$, with an involution $\mathfrak{A} \ni a \mapsto a^* \in \mathfrak{A}$, and the unit-element $\mathbf{1}$. Let $\mathfrak{A}_+$ denote the set of positive elements in $\mathfrak{A}$; i.e.,

$$\mathfrak{A}_+ = \left\{ b^* b \mid b \in \mathfrak{A} \right\}.$$

Definition 5.1. We say $\mathfrak{A}$ is a C^*-algebra if it is complete in a norm $\|\cdot\|$, which satisfies:

1. $\|ab\| \leq \|a\| \, \|b\|$, $\forall a, b \in \mathfrak{A}$;

2. $\|\mathbf{1}\| = 1$;

3. $\|b^* b\| = \|b\|^2$, $\forall b \in \mathfrak{A}$.[3]

Example 5.1. Let X be a compact Hausdorff space, the algebra $C(X)$ of all continuous function on X is a C^*-algebra under the sup-norm.

Example 5.2. $\mathscr{B}(\mathscr{H})$: All bounded linear operators on a fixed Hilbert space $\mathscr{H}$ is a C^*-algebra.

Example 5.3. $\mathscr{O}_N$: The Cuntz-algebra, $N > 1$; it is the C^*-completion of N generators $s_1, s_2, \ldots, s_N$ satisfying the following relations [Cun77]:

1. $s_i^* s_j = \delta_{ij} \mathbf{1}$;

2. $\sum_{i=1}^N s_i s_i^* = \mathbf{1}$.

For the representations of $\mathscr{O}_N$, see [Gli60, Gli61, BJO04].

Remark 5.1. We denote $Rep(\mathfrak{A}, \mathscr{H})$ the representations of $\mathfrak{A}$ acting on some Hilbert space $\mathscr{H}$, i.e., $\pi \in Rep(\mathfrak{A}, \mathscr{H})$ iff $\pi : \mathfrak{A} \to \mathscr{B}(\mathscr{H})$ is a homomorphism of $*$-algebras, $\pi(\mathbf{1}) = I_{\mathscr{H}}$ = the identity operator in $\mathscr{H}$; in particular

$$\langle \pi(b) u, v \rangle_{\mathscr{H}} = \langle u, \pi(b^*) v \rangle_{\mathscr{H}}, \ \forall b \in \mathfrak{A}, \forall u, v \in \mathscr{H}. \tag{5.1}$$

See Definition 5.6 below.

Let $S(\mathfrak{A})$ be the states $\varphi : \mathfrak{A} \to \mathbb{C}$ on $\mathfrak{A}$; i.e., (axioms) $\varphi \in \mathfrak{A}^*$ = the dual of $\mathfrak{A}$, $\varphi(\mathbf{1}) = 1$, and

$$\varphi(b^* b) \geq 0, \ \forall b \in \mathfrak{A}. \tag{5.2}$$

Theorem 5.1 (Gelfand-Naimark-Segal (GNS)). *There is a bijection:*

$$S(\mathfrak{A}) \longleftrightarrow \boxed{\textit{cyclic representations, up to unitary equivalence}}$$

as follows:

---

[3]Kadison et al. in 1950's reduced the axioms of C^*-algebra from about 6 down to just one (3) on the C^*-norm. (See e.g., [KR97a].)

$\longleftarrow$ *(easy direction)*: Given $\pi \in Rep(\mathfrak{A}, \mathscr{H})$, $u_0 \in \mathscr{H}$, $\|u_0\| = 1$, set

$$\varphi(a) = \langle u_0, \pi(a) u_0 \rangle_{\mathscr{H}}, \ \forall a \in \mathfrak{A}. \tag{5.3}$$

$\longrightarrow$ *(non-trivial direction)*: Given $\varphi \in S(\mathfrak{A})$, there is a system $(\pi, \mathscr{H}, u_0)$ such that (5.3) holds. (Notation, we set $\pi = \pi_\varphi$ to indicate the state φ.)

Proof. ($\longrightarrow$) Given $\varphi \in S(\mathfrak{A})$, then on $\mathfrak{A} \times \mathfrak{A}$ consider the sesquilinear form

$$(a, b) \longmapsto \varphi(a^* b) \tag{5.4}$$

$$\mathscr{H}_\varphi = \left\{ \mathfrak{A} / \left\{ b \in \mathfrak{A} \mid \varphi(b^* b) = 0 \right\} \right\}^{\sim}$$

where $\sim$ refers to Hilbert completion in (5.4). Note

$$|\varphi(a^* b)|^2 \le \varphi(a^* a) \varphi(b^* b), \forall a, b \in \mathfrak{A}.$$

Set $\Omega = \mathrm{class}(1)$ in $\mathscr{H}_\varphi$, and

$$\pi_\varphi(a)(\mathrm{class}(b)) = \mathrm{class}(ab), \ \forall a, b \in \mathfrak{A}; \ (\text{Schwarz.})$$

Then it is easy to show that $(\mathscr{H}_\varphi, \Omega, \pi_\varphi)$ satisfies conclusion (5.3), i.e.,

$$\varphi(a) = \langle \Omega, \pi(a) \Omega \rangle_{\mathscr{H}_\varphi}, \ \forall a \in \mathfrak{A}.$$

$\square$

Definition 5.2. Let $\varphi \in S(\mathfrak{A})$, we say it is a *pure state* iff $\varphi \in \mathrm{ext} S(\mathfrak{A}) :=$ the extreme-points in $S(\mathfrak{A})$. (See [Phe01].)

That is, a state s is pure if it cannot be broken up into a convex combination of two distinct states. More specifically, for all states s_1 and s_2, we have:

$$s = \lambda s_1 + (1 - \lambda) s_2 \implies s = s_1 \text{ or } s = s_2.$$

Remark 5.2 (GNS-correspondence). 1. If $\mathfrak{A}$ is a C^*-algebra , then $S(\mathfrak{A})(\subset \mathfrak{A}^*)$ is convex and weak *-compact.

2. Given $\varphi \in S(\mathfrak{A})$, and let $\pi_\varphi \in Rep(\mathfrak{A}, \mathscr{H})$ be the GNS-representation, see (5.3); then

$$\boxed{\varphi \in \mathrm{ext} S(\mathfrak{A}), \text{ i.e., it is pure}}$$
$$\Updownarrow$$
$$\boxed{\pi_\varphi \text{ is an irreducible representation}}$$

3. If $\psi \in S(\mathfrak{A})$, $\exists$ a measure P_ψ on $\mathrm{ext}(S(\mathfrak{A}))$ such that $\psi = \int w \, dP_\psi(w)$, and then

$$\pi_\psi = \int^{\oplus} \pi_w \, dP_\psi(w).$$

$\mathfrak{A}$	$\mathrm{ext}S\,(\mathfrak{A})$
$C\,(X)$	points $x \in X$, and $\varphi = \delta_x$ (Dirac mass); $\varphi\,(f) = f\,(x)$, $\forall f \in C\,(X)$
$\mathscr{B}\,(\mathscr{H})$	$v \in \mathscr{H},\ \|v\| = 1,\ \varphi = \varphi_v;\ \varphi_v\,(A) = \langle v, Av\rangle,\ \forall A \in \mathscr{B}\,(\mathscr{H})$
$\mathscr{O}_N$	Partial list: $u = (u_1,\ldots,u_N) \in \mathbb{C}^N,\ \sum_1^N |u_j|^2 = 1,\ \varphi = \varphi_u,$ specified by $\varphi\,(s_i s_j^*) = u_i\overline{u_j},\ \forall i, j = 1,\ldots,N$; see 1-2.

Table 5.1: Examples of pure states.

Example 5.4 (Pure states, cases where the full list is known!). $\varphi \in S\,(\mathfrak{A})$:

Exercise 5.1 (Irreducible representations). Using GNS, write down explicitly the irreducible representations of the three C^*-algebras in Table 5.1 corresponding to the listed pure states.

Hint: In the case of $C\,(X)$, the representations are one-dimensional, but in the other cases, they are infinite-dimensional, i.e., $\dim \mathscr{H}_{\pi_\varphi} = \infty$.

Remark 5.3. It is probably impossible to list <u>all</u> pure states of $\mathscr{O}_N$; see [Gli60].

Exercise 5.2 (Infinite-product measures and representations of $\mathscr{O}_N$). Fix $N \in \mathbb{N}$, $N > 1$, and denote the cyclic group of order N,

$$\mathbb{Z}_N = \mathbb{Z}/N\mathbb{Z} = \{0, 1, 2\ldots, N - 1\},$$

residue classes mod N. Let $(z_i)_{i=0}^{N-1}$ be complex numbers such that $\sum_i |z_i|^2 = 1$, and assume $z_j \neq 0$ for all j; see line 3 in Table 5.1.

Let p be the probability measure on $\mathbb{Z}_N$ with weights $p_i = |z_i|^2$, and let $\mu = \mu_p$ be the infinite-product measure on $\Omega_N := \mathsf{X}_{\mathbb{N}}\mathbb{Z}_N = \mathsf{X}_{\mathbb{N}}\{0, 1, \cdots, N - 1\}$,

$$\mu_p := \mathsf{X}_{\mathbb{N}}p = \underbrace{p \times p \times \cdots}_{\aleph_0-\text{infinite}}. \tag{5.5}$$

Set $\Omega_N\,(j) = \{(x_i) \in \Omega_N\ ;\ x_1 = j\} = \{j\} \times \Omega_N$.

1. For all $x = (x_1, x_2, x_3, \ldots)$, and all $f \in L^2\,(\mu)$, set

$$(S_j f)\,(x) = \frac{1}{z_j}\chi_{\Omega_N(j)}\,(x)\,f\,(x_2, x_3, \ldots).$$

Show that the adjoint operator with respect to $L^2\,(\mu)$ is

$$(S_j^* f)\,(x) = z_j f\,(j, x_1, x_2, x_3, \ldots);$$

and that this system $\{S_j\}_{j=0}^{N-1}$ defines an irreducible representation of $\mathscr{O}_N$, i.e., is in $Rep_{irr}\,(\mathscr{O}_N, L^2\,(\mu))$.

2. Denote the representation in (1) $\pi_p^{(N)}$, and setting $\mathbb{1}$ to be the constant function in $L^2(\mu)$, show that we recover the pure state from line 3 in Table 5.1 corresponding to $u_j = z_j$; i.e., using the formula:

$$\varphi\left(s_j s_k^*\right) = \langle \mathbb{1}, S_j S_k^* \mathbb{1} \rangle_{L^2(\mu)} = z_j \overline{z_k}, \ \forall j, k \in \mathbb{Z}_N.$$

3. Show that $\pi_p^{(N)}$ is <u>not</u> irreducible when restricted to the Abelian subalgebra in $\mathcal{O}_N$ generated by $\{S_J S_J^*\}$, as J ranges over all finite words in the fixed alphabet $\mathbb{Z}_N$.

Exercise 5.3 (A representation of $\mathcal{O}_N$). What can you say about the representation of $\mathcal{O}_N$ corresponding to $(0, z_1, \ldots, z_{N-1})$, $\sum |z_j|^2 = 1$?

    Hint: Modify (1) from Exercise 5.2. (The state $\varphi\left(s_j s_k^*\right) = z_j \overline{z_k}$ then yields $\varphi\left(s_j s_0^*\right) = 0$, $\forall j \in \mathbb{Z}_N$.)

Groups

Case 1. Groups contained in $\mathcal{B}(\mathcal{H})$ where $\mathcal{H}$ is a fixed Hilbert space:

Definition 5.3. Set

- $\mathcal{B}(\mathcal{H})^{-1}$: all bounded linear operators in $\mathcal{H}$ with bounded inverse.

- $\mathcal{B}(\mathcal{H})_{uni}$: all unitary operators $u : \mathcal{H} \to \mathcal{H}$, i.e., u satisfies

$$uu^* = u^*u = I_{\mathcal{H}}.$$

Definition 5.4. Fix a group G, and set:

- $Rep(G, \mathcal{H})$: all homomorphisms $\rho \in G \to \mathcal{B}(\mathcal{H})^{-1}$

- $Rep_{uni}(G, \mathcal{H})$: all homomorphisms, $\rho : G \to \mathcal{B}_{uni}(\mathcal{H})$, i.e.,

$$\rho(g^{-1}) = \rho(g)^{-1} = \rho(g)^*, \ \forall g \in G$$

- $Rep_{cont}(G, \mathcal{H})$: Elements $\rho \in Rep(G, \mathcal{H})$ such that $\forall v \in \mathcal{H}$,

$$G \ni g \mapsto \rho(g)v$$

    is continuous from G into $\mathcal{H}$; called *strongly continuous*.

Remark 5.4. In the case of $Rep_{cont}(G, \mathcal{H})$ it is assumed that G is a continuous group, i.e., is equipped with a topology such that the following two operations are both continuous:

1. $G \times G \ni (g_1, g_2) \longmapsto g_1 g_2 \in G$

2. $G \ni g \longmapsto g^{-1} \in G$

Exercise 5.4 (The regular representation of G). Let G be a locally compact group with $\mu = $ a left-invariant Haar measure. Set

$$(\rho_L(g) f)(x) := f(g^{-1}x), \; g, x \in G, \; f \in L^2(G, \mu).$$

Then show that ρ_L is a strongly continuous unitary representation of G acting in $L^2(G, \mu)$.

The Group Algebra

Let G be a group, and set $\mathbb{C}[G] := $ all linear combinations, i.e., finite sums

$$A = \sum_g A_g g \tag{5.6}$$

where $A_g \in \mathbb{C}$, and making $\mathfrak{A} := \mathbb{C}[G]$ into a $*$-algebra with the following two operations on finite sums as in (5.6): $\mathfrak{A} \times \mathfrak{A} \longrightarrow \mathfrak{A}$, given by

$$\left(\sum_{g \in G} A_g g\right)\left(\sum_{h \in G} B_h h\right) := \sum_g \left(\sum_{hk=g} A_h B_k\right) g \tag{5.7}$$

and

$$\left(\sum_{g \in G} A_g g\right)^* := \sum_{g \in G} \overline{A_g} g^{-1}. \tag{5.8}$$

Lemma 5.1. *There is a bijection between $Rep_{uni}(G, \mathscr{H})$ and $Rep(\mathbb{C}[G], \mathscr{H})$ as follows: If $\pi \in Rep_{uni}(G, \mathscr{H})$, set $\widetilde{\pi} \in Rep(\mathbb{C}[G], \mathscr{H})$:*

$$\widetilde{\pi}\left(\sum_{g \in G} A_g g\right) := \sum_{g \in G} A_g \pi(g) \tag{5.9}$$

where the element $\sum_{g \in G} A_g g$ in (5.9) is a generic element in $\mathbb{C}[G]$, see (5.6), i.e., is a finite sum with $A_g \in \mathbb{C}$, for all $g \in G$.

Exercise 5.5 (Unitary representations). Fill in the proof details of the assertion in Lemma 5.1.

Example 5.5. Let G be a group, considered as a countable discrete group (the countability is not important). Set $\mathscr{H} = l^2(G)$, and

$$\pi(g) \delta_h := \delta_{gh}, \; \forall g, h \in G. \tag{5.10}$$

Exercise 5.6 (A proof detail). Show that π in (5.10) is in $Rep(G, l^2(G))$.

Definition 5.5. Let G, and $\pi \in Rep\left(G, l^2\left(G\right)\right)$ be as in (5.10), and let

$$\widetilde{\pi} \in Rep\left(\mathbb{C}[G], l^2\left(G\right)\right)$$

be the corresponding representation of $\mathbb{C}[G]$; see Lemma 5.1. Set

$$C^*_{red}\left(G\right) := \text{the norm closure of } \widetilde{\pi}\left(\mathbb{C}[G]\right) \subset \mathscr{B}\left(l^2\left(G\right)\right);$$

then $C^*_{red}\left(G\right)$ is called the reduced C^*-algebra of the group G.

Exercise 5.7 (Reduced C^*-algebra). Prove that $C^*_{red}\left(G\right)$ is a C^*-algebra.

Remark 5.5. It is known [Pow75] that $C^*_{red}\left(F_2\right)$ is simple, where F_2 is the free group on two generators. ("red" short for reduced; it is called the reduced C^*-algebra on the group.)

5.2 The GNS Construction

The GNS construction is a general principle for getting representations from given data in applications, especially in quantum mechanics [Pol02, PK88, CP82]. It was developed independently by I. Gelfand, M. Naimark, and I. Segal around the 1960s, see e.g., [GJ60, Seg50].

Definition 5.6. Let $\mathfrak{A}$ be a $*$-algebra with identity. A representation of $\mathfrak{A}$ is a map $\pi : \mathfrak{A} \to B(\mathscr{H}_\pi)$, where $\mathscr{H}_\pi$ is a Hilbert space, such that for all $A, B \in \mathfrak{A}$,

1. $\pi(AB) = \pi(A)\pi(B)$

2. $\pi(A^*) = \pi(A)^*$.

The $*$ operation (involution) is given on $\mathfrak{A}$ so that $A^{**} = A$, $(AB)^* = B^*A^*$, $(\lambda A)^* = \overline{\lambda}A^*$, for all $\lambda \in \mathbb{C}$.

Example 5.6. The multiplication version of the spectral theorem of a single selfadjoint operator, say A acting on $\mathscr{H}$, yields a representation of the algebra of $L^\infty\left(sp\left(A\right)\right)$ (or $C(sp\left(A\right))$) as operators on $\mathscr{H}$, where

$$L^\infty\left(sp\left(A\right)\right) \ni f \xrightarrow{\pi} f\left(A\right) \in \mathscr{B}(\mathscr{H})$$

such that $\pi(fg) = \pi(f)\pi(g)$ and $\pi(\bar{f}) = \pi(f)^*$.

The general question is given any $*$-algebra, where to get such a representation? The answer is given by *states*. One gets representations from algebras vis states. For Abelian algebras, the states are Borel measures, so the measures come out as a corollary of representations.

Definition 5.7. Let $\mathfrak{A}$ be a $*$-algebra. A state on $\mathfrak{A}$ is a linear functional $\varphi : \mathfrak{A} \to \mathbb{C}$ such that $\varphi(1_{\mathfrak{A}}) = 1$, and $\varphi(A^*A) \geq 0$, for all $A \in \mathfrak{A}$.

Example 5.7. Let $\mathfrak{A} = C(X)$, i.e., C^*-algebra of continuous functions on a compact Hausdorff space X. Note that there is a natural involution $f \mapsto f^* := \overline{f}$ by complex conjugation. Let μ_φ be a Borel probability measure on X, then

$$C(X) \ni f \mapsto \varphi(f) = \int_X f d\mu_\varphi$$

is a state. In fact, in the Abelian case, all states are Borel probability measures.

Because of this example, we say that the GNS construction is non-commutative measure theory.

Example 5.8. Let G be a discrete group, and let $\mathfrak{A} = \mathbb{C}[G]$ be the group-algebra, see Section 5.1.

If we make the assumption (defining φ first on points in G)

$$\varphi(g) = \begin{cases} 1 & \text{if } g = e \text{ (the unit element in } G) \\ 0 & \text{if } g \in G \backslash \{e\}, \end{cases} \tag{5.11}$$

then the argument from above shows that φ extends to a *linear functional* on $\mathfrak{A}$.

Exercise 5.8 (The trace state on $\mathbb{C}[G]$).

1. Show that φ as defined in (5.11), extended to $\mathfrak{A} = \mathbb{C}[G]$ is a state, and if $A = \sum_g A_g g$ (finite sum), then

$$\varphi(A^* A) = \sum_{g \in G} |A_g|^2; \tag{5.12}$$

and moreover (the trace property):

$$\varphi(AB) = \varphi(BA), \ \forall A, B \in \mathfrak{A}. \tag{5.13}$$

We are aiming at a proof of the GNS theorem (Theorem 5.1), and a way to get more general representations of $*$-algebras. Indeed, any representation is built up by the cyclic representations (Definition 5.8), and each cyclic representation is in turn given by a GNS construction.

Definition 5.8. A representation $\pi \in Rep(\mathfrak{A}, \mathscr{H})$ is called *cyclic*, with a *cyclic vector* $u \in \mathscr{H}$, if $\mathscr{H} = \overline{span} \{\pi(A) u \mid A \in \mathfrak{A}\}$.

Theorem 5.2. *Given any representation* $\pi \in Rep(\mathfrak{A}, \mathscr{H})$, *there exists an index set* J, *and closed subspaces* $\mathscr{H}_j \subset \mathscr{H}$ $(j \in J)$ *such that*

1. $\mathscr{H}_i \perp \mathscr{H}_j, \ \forall i \neq j$;

2. $\sum_{j \in J}^{\oplus} \mathscr{H}_j = \mathscr{H}$; *and*

3. *there exists cyclic vectors* $v_j \in \mathscr{H}_j$ *such that the restriction of* π *to* $\mathscr{H}_j$ *is cyclic.*

Remark 5.6. The proof of 5.2 is very similar to the construction of orthonormal basis (ONB) (use Zorn's lemma!); but here we get a family of mutually orthogonal subspaces.

Of course, if $\mathscr{H}_j$'s are all one-dimensional, then it is a decomposition into ONB. Note that not every representation is irreducible, but every representation can be decomposed into direct sum of cyclic representations.

Exercise 5.9 (Cyclic subspaces). Prove Theorem 5.2. Hint: pick $v_1 \in \mathscr{H}$, and let

$$\mathscr{H}_{v_1} := \overline{span}\left\{\pi\left(A\right)v_1 : A \in \mathfrak{A}\right\},$$

i.e., the cyclic subspace generated by v_1. If $\mathscr{H}_{v_1} \neq \mathscr{H}$, then $\exists v_2 \in \mathscr{H}\backslash\mathscr{H}_{v_1}$, and the cyclic subspace $\mathscr{H}_{v_2}$, so that $\mathscr{H}_{v_1}$ and $\mathscr{H}_{v_2}$ are orthogonal. If $\mathscr{H}_{v_1} \oplus \mathscr{H}_{v_2} \neq \mathscr{H}$, we then build $\mathscr{H}_{v_3}$ and so on. Now use transfinite induction or Zorn's lemma to show the family of direct sum of mutually orthogonal cyclic subspaces is total. The final step is exactly the same argument for the existence of an ONB of any Hilbert space.

Now we proceed to prove the Theorem 5.1 (GNS), which is restated below.

Theorem 5.3 (Gelfand-Naimark-Segal). *There is a bijection between states φ and cyclic representations $\pi \in Rep\left(\mathfrak{A}, \mathscr{H}, u\right)$, with $\|u\| = 1$; where*

$$\varphi(A) = \langle u, \pi(A)u \rangle, \ \forall A \in \mathfrak{A}. \tag{5.14}$$

Moreover, fix a state φ, the corresponding cyclic representation is unique up to unitary equivalence. Specifically, if $(\pi_1, \mathscr{H}_1, u_1)$ and $(\pi_2, \mathscr{H}_2, u_2)$ are two cyclic representations, with cyclic vectors u_1, u_2, respectively, satisfying

$$\varphi(A) = \langle u_1, \pi_1(A)u_1 \rangle = \langle u_2, \pi_2(A)u_2 \rangle, \ \forall A \in \mathfrak{A}; \tag{5.15}$$

then

$$W : \pi_1\left(A\right)u_1 \longmapsto \pi_2\left(A\right)u_2, \ A \in \mathfrak{A} \tag{5.16}$$

extends to a unitary operator from $\mathscr{H}_1$ onto $\mathscr{H}_2$, also denoted by W, and such that

$$\pi_2 W = W\pi_1, \tag{5.17}$$

i.e., W intertwines the two representations.

Remark 5.7. For the non-trivial direction, let φ be a given state on $\mathfrak{A}$, and we need to construct a cyclic representation $(\pi, \mathscr{H}_\varphi, u_\varphi)$. Note that $\mathfrak{A}$ is an algebra, and it is also a complex vector space. Let us try to turn $\mathfrak{A}$ into a Hilbert space and see what conditions are needed. There is a homomorphism $\mathfrak{A} \to \mathfrak{A}$ which follows from the associative law of $\mathfrak{A}$ being an algebra, i.e., $(AB)C = A(BC)$. To continue, $\mathfrak{A}$ should be equipped with an inner product. Using φ, we may set $\langle A, B \rangle_\varphi := \varphi\left(A^*B\right), \forall A, B \in \mathfrak{A}$. Then $\langle\cdot,\cdot\rangle_\varphi$ is linear in the second variable, and conjugate linear in the first variable. It also satisfies $\langle A, A \rangle_\varphi = \varphi\left(A^*A\right) \geq 0$. Therefore we take $\mathscr{H}_\varphi := [\mathfrak{A}/\{A : \varphi(A^*A) = 0\}]^{cl}$.

Proof. Given a cyclic representation $\pi \in Rep(\mathfrak{A}, \mathcal{H}, u)$, define φ as in (5.14). Clearly φ is linear, and

$$
\begin{aligned}
\varphi(A^*A) &= \langle u, \pi(A^*A)u \rangle \\
&= \langle u, \pi(A^*)\pi(A)u \rangle \\
&= \langle \pi(A)u, \pi(A)u \rangle \\
&= \|\pi(A)u\|^2 \geq 0.
\end{aligned}
$$

Thus φ is a state.

Conversely, fix a state φ on $\mathfrak{A}$. Set

$$
\mathcal{H}_0 := \left\{ \sum_{i=1}^n c_i A_i \mid c_i \in \mathbb{C},\ n \in \mathbb{N} \right\}
$$

and define the inner product

$$
\left\langle \sum c_i A_i, \sum d_i B_i \right\rangle_\varphi := \sum \sum \overline{c_i} d_j \varphi(A_i^* B_j).
$$

Note that, by definition,

$$
\left\| \sum c_i A_i \right\|_\varphi^2 = \left\langle \sum c_i A_i, \sum c_i A_i \right\rangle_\varphi = \sum \sum \overline{c_i} c_j \varphi(A_i^* A_j) \geq 0. \qquad (5.18)
$$

The RHS of (5.18) is positive since φ is a state. Recall that $\varphi(A^*A) \geq 0$, for all $A \in \mathfrak{A}$, and this implies that for all $n \in \mathbb{N}$, the matrix $(\varphi(A_i^* A_j))_{i,j=1}^n$ is positive definite, hence (5.18) holds.

Proof of (5.14): Now, let $\mathcal{H}_\varphi :=$ completion of $\mathcal{H}_0$ under $\langle \cdot, \cdot \rangle_\varphi$ modulo elements s such that $\|s\|_\varphi = 0$. See Lemma 5.2 below. $\mathcal{H}_\varphi$ is the desired cyclic space, consisting of equivalence classes $[A]$, $\forall A \in \mathfrak{A}$. Next, let $u_\varphi = [1_\mathfrak{A}] =$ equivalence class of the identity element, and set

$$
\pi(A) := [A] = [A1_\mathfrak{A}] = [A][1_\mathfrak{A}];
$$

then one checks that $\pi \in Rep(\mathfrak{A}, \mathcal{H}_\varphi)$, and therefore $\varphi(A) = \langle u_\varphi, \pi(A)u_\varphi \rangle_\varphi$, $\forall A \in \mathfrak{A}$.

For uniqueness, let $(\pi_1, \mathcal{H}_1, u_1)$ and $(\pi_2, \mathcal{H}_2, u_2)$ be as in the statement of the theorem, and let W be as in (5.16). By (5.15), we have

$$
\varphi(A^*A) = \|\pi_2(A)u_2\|^2 = \|\pi_1(A)u_1\|^2
$$

so that W is isometric. But since $\mathcal{H}_i = \overline{span}\{\pi_i(A)u_i : A \in \mathfrak{A}\}$, $i = 1, 2$, then W extends by density to a unitary operator from $\mathcal{H}_1$ to $\mathcal{H}_2$.

Proof of (5.17): Finally, for all $A, B \in \mathfrak{A}$, we have

$$
\begin{aligned}
W\pi_1(A)(\pi_1(B)u_1) &= W\pi_1(AB)u_1 \\
&= \pi_2(AB)u_2
\end{aligned}
$$

$$= \pi_2(A)(\pi_2(B)u_2)$$
$$= \pi_2(A)W\pi_1(B)u_1;$$

therefore, by the density argument again, we conclude that

$$\pi_2(A) = W\pi_1(A) \ \forall A \in \mathfrak{A}.$$

This is the intertwining property in (5.17). □

Lemma 5.2. $\{A \in \mathfrak{A} : \varphi(A^*A) = 0\}$ is a closed two-sided ideal in $\mathfrak{A}$.

Proof. This follows from the Schwarz inequality. Note that

$$\begin{bmatrix} \varphi(A^*A) & \varphi(A^*B) \\ \varphi(B^*A) & \varphi(B^*B) \end{bmatrix}$$

is a positive definite matrix, and so its determinant is positive, i.e.,

$$|\varphi(A^*B)|^2 \le \varphi(A^*A)\varphi(B^*B); \tag{5.19}$$

using the fact that $\varphi(C^*) = \varphi(C)^*$, $\forall C \in \mathfrak{A}$. The lemma follows from the estimate (5.19). □

Example 5.9. Let $\mathfrak{A} = C[0,1]$. Set $\varphi : f \mapsto f(0)$, so that $\varphi(f^*f) = |f(0)|^2 \ge 0$. Then,

$$\ker \varphi = \{f \in C[0,1] \mid f(0) = 0\}$$

and $C[0,1]/\ker\varphi$ is one dimensional. The reason is that if $f \in C[0,1]$ such that $f(0) \ne 0$, then we have $f(x) \sim f(0)$ since $f(x) - f(0) \in \ker\varphi$, where $f(0)$ represents the constant function $f(0)$ over $[0,1]$. This shows that φ is a pure state, since the representation has to be irreducible.

Exercise 5.10 (The GNS construction). Fill in the remaining details in the above proof of the GNS theorem.

Using GNS construction we get the following structure theorem for abstract C^*-algebras. As a result, all C^*-algebras are sub-algebras of $\mathscr{B}(\mathscr{H})$ for some Hilbert space $\mathscr{H}$.

Theorem 5.4 (The Gelfand-Naimark Theorem). *Every C^*-algebra (Abelian or non-Abelian) is isometrically isomorphic to a norm-closed sub-algebra of $\mathscr{B}(\mathscr{H})$, for some Hilbert space $\mathscr{H}$.*

Proof. Let $\mathfrak{A}$ be any C^*-algebra, no Hilbert space $\mathscr{H}$ is given from outside. Let $S(\mathfrak{A})$ be the states on $\mathfrak{A}$, which is a compact convex subset of the dual space $\mathfrak{A}^*$. Here, compactness refers to the weak $*$-topology.

We use Hahn-Banach theorem to show that there are plenty of states. Specifically, $\forall a \in \mathfrak{A}$, $\exists \varphi \in \mathfrak{A}^*$ such that $\varphi(a) > 0$. It is done first on the 1-dimensional subspace

$$tA \mapsto t \in \mathbb{R},$$

and then extends to $\mathfrak{A}$. (Note this is also a consequence of Krein-Milman, i.e., $S(\mathfrak{A}) = cl(\text{pure states})$. We will come back to this point later.)

For each state φ, one gets a *cyclic representation* $(\pi_\varphi, \mathscr{H}_\varphi, u_\varphi)$. Applying transfinite induction, one concludes that $\pi := \oplus \pi_\varphi$ is a representation on the Hilbert space $\mathscr{H} := \oplus \mathscr{H}_\varphi$. For details, see e.g., [Rud91]. $\qquad\square$

Theorem 5.5. *Let $\mathfrak{A}$ be an Abelian C^*-algebra. Then there is a compact Hausdorff space X, unique up to homeomorphism, such that $\mathfrak{A} \cong C(X)$.*

5.3   States, Dual and Pre-dual

Let V be a *Banach space*, i.e., (recall, Chapter 2):

- V is a vector space over $\mathbb{C}$;

- $\exists$ norm $\|\cdot\|$

- V is complete with respect to $\|\cdot\|$

The dual space V^* consists of linear functionals $l : V \to \mathbb{C}$ satisfying

$$\|l\| := \sup_{\|v\|=1} |l(v)| < \infty.$$

These are the continuous linear functionals.

The *Hahn-Banach Theorem* implies that for all $v \in V$, $\|v\| \neq 0$, there exists $l_v \in V^*$, of norm 1, such that $l(v) = \|v\|$. Recall the construction is to first define l_v on the one-dimensional subspace spanned by the vector v, then use transfinite induction to extend l_v to all of V. Notice that V^* is always complete, even if V is an incomplete normed space. In other words, V^* is always a Banach space.

Now V is embedded into V^{**} (as we always do this) via the mapping

$$V \ni v \mapsto \psi(v) \in V^{**}, \text{ where}$$
$$\psi(v)(l) := l(v), \ \forall l \in V^*. \tag{5.20}$$

Below we give a number of applications:

Exercise 5.11 (Identification by isometry). Show that $V \xrightarrow{\psi} V^{**}$ in (5.20) is isometric, i.e.,

$$\|\psi(v)\|_{**} = \|v\|, \ \forall v \in V.$$

Example 5.10. Let X be a compact Hausdorff space. The algebra $C(X)$ of all continuous functions on X with the sup norm, i.e., $\|f\|_\infty := \sup_{x \in X} |f(x)|$, is a Banach space.

Example 5.11. The classical L^p space: $(l^p)^* = l^q$, $(L^p)^* = L^q$, for $1/p + 1/q = 1$ and $1 \leq p < \infty$. If $1 < p < \infty$, then $(l^p)^{**} = l^p$, i.e., these spaces are *reflexive*. For $p = 1$, however, we have $(l^1)^* = l^\infty$, but $(l^\infty)^*$ is much bigger than l^1. Also note that $(l^p)^* \neq l^q$ except for $p = q = 2$. And l^p is a Hilbert space iff $p = 2$.

Let B be a Banach space and denote by B^* its dual space. B^* is a Banach space as well, where the norm is defined by

$$\|f\|_{B^*} = \sup_{\|x\|=1} \{|f(x)|\}.$$

Let $B_1^* = \{f \in B^* : \|f\| \leq 1\}$ be the unit ball in B^*.

Theorem 5.6 (Banach-Alaoglu). *B_1^* is weak $*$ compact in B^*.*

Proof. This is proved by showing B_1^* is a closed subset in $\Omega := \prod_{\|x\|=1} \mathbb{C}_1$, with $\mathbb{C}_1 = \{z \in \mathbb{C} : |z| \leq 1\}$; and Ω is given its product topology, and is compact and Hausdorff. $\qquad\square$

As an application, we have

Corollary 5.1. *Let B be a separable Banach space. Then every bounded sequence in B^* has a convergent subsequence in the weak $*$-topology.*

Corollary 5.2. *Every bounded sequence in $\mathscr{B}(\mathscr{H})$ contains a convergence subsequence in the weak $*$-topology.*

We show in Theorem 5.7 that $\mathscr{B}(\mathscr{H}) = \mathscr{T}_1(\mathscr{H})^*$, where $\mathscr{T}_1(\mathscr{H}) = $ trace-class operators.

Now we turn to Hilbert space, say $\mathscr{H}$:

- $\mathscr{H}$ is a vector space over $\mathbb{C}$;

- it has an inner product $\langle \cdot, \cdot \rangle$, and the norm $\|\cdot\| := \sqrt{\langle \cdot, \cdot \rangle}$;

- $\mathscr{H}$ is complete with respect to $\|\cdot\|$;

- $\mathscr{H}^* = \mathscr{H}$, i.e., $\mathscr{H}$ is reflexive;

- every Hilbert space has an orthonormal basis (by Zorn's lemma).

The identification $\mathscr{H} = \mathscr{H}^*$ is due to Riesz, and the corresponding map is given by

$$h \mapsto \langle h, \cdot \rangle \in \mathscr{H}^*.$$

This can also be seen by noting via an ONB that $\mathscr{H}$ is unitarily equivalent to $l^2(A)$, with some index set A, and $l^2(A)$ is reflexive.

The set of all bounded operators $\mathscr{B}(\mathscr{H})$ on $\mathscr{H}$ is a Banach space. We ask two questions:

1. What is the dual $\mathscr{B}(\mathscr{H})^*$?

2. Is $\mathscr{B}(\mathscr{H})$ the dual space of some Banach space?

The first question is extremely difficult and we will discuss that later.

For the present section, we show that

$$\mathscr{B}(\mathscr{H}) = \mathscr{T}_1(\mathscr{H})^*$$

where we denote by $\mathscr{T}_1(\mathscr{H})$ the trace-class operators in $\mathscr{B}(\mathscr{H})$. For more details, see Theorem 5.7, and Section 2.5.1.

Let $\rho : \mathscr{H} \to \mathscr{H}$ be a compact selfadjoint operator. Assume ρ is positive, i.e., $\langle x, \rho x \rangle \geq 0$ for all $x \in \mathscr{H}$. By the spectral theorem of compact operators, we get the following decomposition

$$\rho = \sum \lambda_k P_k \tag{5.21}$$

where $\lambda_1 \geq \lambda_2 \geq \cdots \to 0$, and P_k is the projection onto the finite dimensional eigenspace of λ_k.

In general, we want to get rid of the assumption that $\rho \geq 0$. This is done using the polar decomposition, which we will consider in Section 3.5 even for unbounded operators. It is much easier for bounded operators: If $A \in \mathscr{B}(\mathscr{H})$, A^*A is positive, selfadjoint, and so by the spectral theorem, we may take $|A| := \sqrt{A^*A}$. Then, one checks that

$$\|Ax\|^2 = \langle Ax, Ax \rangle = \langle x, A^*Ax \rangle = \left\langle \sqrt{A^*A}x, \sqrt{A^*A}x \right\rangle = \||A|\,x\|^2,$$

thus

$$\|A\| = \||A|\| \tag{5.22}$$

and there is a partial isometry $V : \text{range}(|A|) \to \text{range}(A)$, and the following polar decomposition holds:

$$A = V|A|. \tag{5.23}$$

We will come back to this point in Section 3.5 when we consider unbounded operators.

Corollary 5.3. *Let $A \in \mathscr{T}_1(\mathscr{H})$, then A has the following decomposition*

$$A = \sum_n \lambda_n |f_n\rangle\langle e_n| \tag{5.24}$$

where $\{e_n\}$ and $\{f_n\}$ are ONBs in $\mathscr{H}$.

Proof. Using the polar decomposition $A = V|A|$, we may first diagonalize $|A|$ with respect to some ONB $\{e_n\}$ as

$$|A| = \sum_n \lambda_n |e_n\rangle\langle e_n|, \text{ then}$$

$$A = V|A| = \sum_n \lambda_n |Ve_n\rangle\langle e_n| = \sum_n \lambda_n |f_n\rangle\langle e_n|$$

where $f_n := Ve_n$.   □

With the above discussion, we may work, instead, with compact operators $A : \mathcal{H} \to \mathcal{H}$ so that A is a trace class operator if $|A|$ (positive, selfadjoint) satisfies condition (5.25).

Definition 5.9. Let A be a compact operator with its polar decomposition $A = V|A|$, where $|A| := \sqrt{A^*A}$. Let $\{\lambda_k\}_{k=1}^{\infty}$ be the eigenvalues of $|A|$, and P_k the corresponding spectral projections, see (5.21). We say A is a trace class operator, if

$$\|A\|_1 := trace\,(|A|) = \sum_n \lambda_n \, \mathrm{rank}\,(P_k) < \infty. \tag{5.25}$$

Caution. In our consideration of eigenvalue lists, we may of course have multiplicity. But for compact operators, the multiplicity is automatically finite for each non-zero eigenvalue. And if we have sets of associated eigenvectors run through a local ONB in each of the finite-dimensional eigenspaces, then multiplicity is counted this way. But, alternatively, when computing a trace as a sum of eigenvalues, then the term in such a sum must be counted with multiplicity. Or each of the distinct numbers in an eigenvalue list can be multiplied with the respective multiplicity. This will be clear from the context.

We now continue the discussion from Section 2.5.1 on spaces of operators.

Definition 5.10. Let $A \in \mathscr{T}_1\,(\mathcal{H})$, and $\{e_n\}$ an ONB in $\mathcal{H}$. Set

$$trace\,(A) := \sum_n \langle e_n, Ae_n \rangle. \tag{5.26}$$

Note the RHS in (5.26) is independent of the choice of the ONB. For if $\{f_n\}$ is another ONB in $\mathcal{H}$, using the Parseval identity repeatedly, we have

$$
\begin{aligned}
\sum_n \langle f_n, Af_n \rangle &= \sum_n \sum_m \langle f_n, e_m \rangle \langle e_m, Af_n \rangle \\
&= \sum_m \sum_n \langle f_n, e_m \rangle \langle A^*e_m, f_n \rangle \\
&= \sum_m \langle A^*e_m, e_m \rangle = \sum_m \langle e_m, Ae_m \rangle.
\end{aligned}
$$

Corollary 5.4. *Let $A \in \mathscr{T}_1\,(\mathcal{H})$, then*

$$|trace\,(A)| \leq \|A\|_1.$$

Therefore, the RHS in (5.26) is absolutely convergent.

Proof. By Corollary 5.3, there exists ONBs $\{e_n\}$ and $\{f_n\}$, and A has a decomposition as in (5.24). Then,

$$
\begin{aligned}
|trace\,(A)| &\leq \sum_n |\langle e_n, Ae_n \rangle| = \sum_n \lambda_n \, |\langle e_n, f_n \rangle| \\
&\leq \sum_n \lambda_n \, \|e_n\| \, \|f_n\| = \sum_n \lambda_n = \|A\|_1 < \infty.
\end{aligned}
$$

We have used the fact that $trace(A)$ is independent of the choice of an ONBs. $\square$

Lemma 5.3. *Let $\mathscr{T}_1(\mathscr{H})$ be the trace class introduced above. Then,*

1. $\mathscr{T}_1(\mathscr{H})$ *is a two-sided ideal in* $\mathscr{B}(\mathscr{H})$.

2. $trace(AB) = trace(BA)$.

3. $\mathscr{T}_1(\mathscr{H})$ *is a Banach space with respect to the trace norm (5.25).*

Exercise 5.12 (A pre-dual). Prove Lemma 5.3.

Lemma 5.4. *Let $\rho \in \mathscr{T}_1(\mathscr{H})$, then the map $A \mapsto trace(A\rho)$ is a state on* $\mathscr{B}(\mathscr{H})$. *These are called the* normal *states.*

Proof. By Lemma 5.3, $A\rho \in \mathscr{T}_1(\mathscr{H})$ for all $A \in \mathscr{B}(\mathscr{H})$. The map $A \mapsto trace(A\rho)$ is in $\mathscr{B}(\mathscr{H})^*$ means that the pairing $(A, \rho) \mapsto trace(A\rho)$ satisfies

$$|trace(A\rho)| \le \|A\| \, \|\rho\|_1 \, .$$

By Corollary 5.4, it suffices to verify, instead, that

$$\|A\rho\|_1 \le \|A\| \, \|\rho\|_1 \, .$$

Indeed, if we choose an ONB $\{e_n\}$ in $\mathscr{H}$ that diagonalizes $|\rho|$, i.e.,

$$|\rho| = \sum \lambda_n |e_n\rangle\langle e_n| \, , \text{ where } \sum \lambda_n < \infty, \lambda_n > 0, \forall n;$$

then

$$\begin{aligned}
\|A\rho\|_1 &= trace\left(\sqrt{\rho^* A^* A\rho}\right) = trace\left(\sqrt{\rho^*\rho}\sqrt{A^*A}\right) \\
&= \sum_n \langle e_n, |A| \, |\rho| \, e_n\rangle = \sum_n \lambda_n \langle e_n, |A| \, e_n\rangle \\
&\le \|A\| \sum_k \lambda_k = \|A\| \, \|\rho\|_1 \, .
\end{aligned}$$

$\square$

Theorem 5.7. $\mathscr{T}_1^*(\mathscr{H}) = \mathscr{B}(\mathscr{H})$.

Proof. Let $l \in \mathscr{T}_1^*$. By $\mathscr{T}_1^* = (\mathscr{T}_1)^*$ we mean the dual Banach, duality with respect to the trace-norm.

How to get an operator A? The operator A must satisfy

$$l(\rho) = trace(\rho A), \quad \forall \rho \in \mathscr{T}_1. \tag{5.27}$$

How to pull an operator A out of the hat? The idea also goes back to Dirac. It is in fact not difficult to find A. Since A is determined by its matrix, it suffices to find $\langle f, Af\rangle$, the entries in the matrix of A.

For any $f_1, f_2 \in \mathcal{H}$, the rank-one operator $|f_1 \rangle\langle f_2|$ is in $\mathcal{T}_1$, hence we know what l does to it, i.e., we know the numbers $l(|f_1 \rangle\langle f_2|)$. But since $l(|f_1 \rangle\langle f_2|)$ is linear in f_1, and conjugate linear in f_2, by the Riesz theorem for Hilbert space, there exists a unique operator A such that

$$l(|f_1 \rangle\langle f_2|) = \langle f_2, Af_1 \rangle.$$

Now we check that $l(\rho) = trace(\rho A)$. By Corollary 5.3, any $\rho \in \mathcal{T}_1$ can be written as $\rho = \sum_n \lambda_n |f_n \rangle\langle e_n|$, where $\{e_n\}$ and $\{f_n\}$ are some ONBs in $\mathcal{H}$. Then,

$$
\begin{aligned}
trace(\rho A) &= trace\left(\sum_n \lambda_n |f_n \rangle\langle e_n| A\right) \\
&= trace\left(\sum_n \lambda_n |Af_n \rangle\langle e_n|\right) \\
&= \sum_m \sum_n \lambda_n \langle u_m, Af_n \rangle \langle e_n, u_m \rangle \\
&= \sum_n \lambda_n \left(\sum_m \langle u_m, Af_n \rangle \langle e_n, u_m \rangle\right) \\
&= \sum_n \lambda_n \langle e_n, Af_n \rangle \ (= l(\rho))
\end{aligned}
$$

where $\{u_n\}$ is an ONB in $\mathcal{H}$, and the last step follows from Parseval's identity. $\square$

Remark 5.8. If B is the dual of a Banach space, then we say that B has a pre-dual. For example $l^\infty = (l^1)^*$, hence l^1 is the pre-dual of l^∞.

Another example: Let $\mathbb{H}_1$ be hardy space of analytic functions on the disk [Rud87]. $(\mathbb{H}_1)^* = $ BMO, where BMO refers to bounded mean oscillation. It was developed by Charles Fefferman in 1974 who won the fields medal for this theory. See [Fef71]. (Getting hands on a specific dual space is often a big thing.)

Definition 5.11. Let $\mathbb{D}$ be the complex disk

$$\mathbb{D} = \{z \in \mathbb{C} : |z| < 1\}.$$

Consider functions f analytic on $\mathbb{D}$ such that

$$\sup_{0 < r < 1} \frac{1}{2\pi} \int_{-\pi}^{\pi} |f(re^{it})| \, dt < \infty. \tag{5.28}$$

This is the $\mathbb{H}_1$-Hardy space, and the $\mathbb{H}_1$-norm is the supremum in (5.28).

The version of (5.28) on the upper half-plane $\{z \in \mathbb{C}, \Im\{z\} > 0\} =: U_+$ is as follows: For functions f, now assumed analytic on U_+, consider the condition

$$\sup_{y>0} \int_{-\infty}^{\infty} |f(x + iy)| \, dx < 0. \tag{5.29}$$

The corresponding $\mathbb{H}_1$ is called the Hardy space on the line $\mathbb{R}$.

The literature on Hardy space is extensive, and we refer to [Rud87] for overview and details.

Theorem 5.8 (C. Fefferman). $\mathbb{H}_1^* = \text{BMO}$.

Proof. See [Fef71]. □

Definition 5.12. Let f be a locally integrable function on $\mathbb{R}^n$, and let Q run through all n-cubes $\subset \mathbb{R}^n$. Set

$$f_Q = \frac{1}{|Q|} \int_Q f(y)\, dy.$$

We say that $f \in BMO$ iff (Def.)

$$\sup_Q \frac{1}{|Q|} \int_Q |f(x) - f_Q|\, dx < \infty. \tag{5.30}$$

In this case the LHS in (5.30) is the BMO-norm of f. Moreover, BMO is a Banach space.

Exercise 5.13 (The Bohr compactification). From abstract harmonic analysis (see [Rud90]), we know that every locally Abelian (l.c.a.) group G has a Haar measure, unique up to scalar normalization. When G is a given l.c.a. group, we denote by $\widehat{G}$ its dual group (of all continuous unitary characters.) The duality theorem for l.c.a. groups G states the following:

$$G \text{ is compact} \iff \widehat{G} \text{ is discrete.} \tag{5.31}$$

Moreover, in general, $G \simeq \widehat{\widehat{G}}$ when G is l.c.a. Now consider the group $\mathbb{R}$ (the reals) with addition, but in its *discrete topology*, (usually denoted $\mathbb{R}_d$.) The corresponding dual group $\mathbb{R}_b = (\mathbb{R}_d)^\wedge$ is therefore compact by (5.31). It is called the *Bohr-compactification* of $\mathbb{R}$. Let $d\chi$ denote its Haar measure.

1. Show that $L^2(G_b, d\chi)$ is a *non-separable* Hilbert space.

2. Find an ONB in $L^2(G_b, d\chi)$ indexed by $\mathbb{R}$.

5.4 New Hilbert Spaces From "old"

Below we consider some cases of building new Hilbert spaces from given ones. Only sample cases are fleshed out; and they will be needed in the sequel.

An Overview:

5.4.1 GNS

See Section 5.2.

5.4.2 Direct sum $\bigoplus_\alpha \mathscr{H}_\alpha$

(a) Let $\mathscr{H}_i$, $i = 1, 2$ be two given Hilbert spaces, then the direct "orthogonal" sum $\mathscr{H} = \mathscr{H}_1 \oplus \mathscr{H}_2$ is as follows:

$$\mathscr{H} = \{\text{symbol pairs } h_1 \oplus h_2,\ h_i \in \mathscr{H}_i,\ i = 1, 2\}, \text{ and}$$

$$\|h_1 \oplus h_2\|_{\mathscr{H}}^2 = \|h_1\|_{\mathscr{H}_1}^2 + \|h_2\|_{\mathscr{H}_2}^2. \tag{5.32}$$

(b) Given an indexed family of Hilbert spaces $\{\mathscr{H}_\alpha\}_{\alpha \in A}$ where A is a set; then set $\mathscr{H} := \bigoplus_A \mathscr{H}_\alpha$ to be

$$\left\| \sum_{\alpha \in A}^{\oplus} h_\alpha \right\|_{\mathscr{H}}^2 = \sum_{\alpha \in A} \|h_\alpha\|_{\mathscr{H}_\alpha}^2 < \infty; \tag{5.33}$$

i.e., finiteness of the sum in (5.33) is part of the definition.

Exercise 5.14 (Unitary operators on a direct Hilbert sum). Let $\mathscr{H}_i$, $i = 1, 2$, be Hilbert spaces, and set $\mathscr{H} := \mathscr{H}_1 \oplus \mathscr{H}_2$.

(i) Let $G_{\mathscr{H}}$, and $G_{\mathscr{H}_i}$, $i = 1, 2$, be the respective groups of unitary operators. Show that $G_{\mathscr{H}_1} \times G_{\mathscr{H}_2}$ is a subgroup of $G_{\mathscr{H}}$.

(ii) Let $L \in \mathscr{B}(\mathscr{H})$, where $\mathscr{H} = \mathscr{H}_1 \oplus \mathscr{H}_2$, and suppose L commutes with the group $G_{\mathscr{H}_1} \times G_{\mathscr{H}_2}$ in (i); then show that L must have the following form

$$L = (\alpha I_{\mathscr{H}_1}) \times (\beta I_{\mathscr{H}_2})$$

where $\alpha, \beta \in \mathbb{C}$. (We say that the *commutant* of the group $G_{\mathscr{H}_1} \times G_{\mathscr{H}_2}$ has this form; it is two-dimensional.)

(iii) Let $A \in \mathscr{B}(\mathscr{H}_2, \mathscr{H}_1)$, and $B \in \mathscr{B}(\mathscr{H}_1, \mathscr{H}_2)$; show that the block-operator matrix $\begin{pmatrix} 0 & A \\ B & 0 \end{pmatrix}$ defines a unitary operator in $\mathscr{H} = \mathscr{H}_1 \oplus \mathscr{H}_2$ if and only if

$$AA^* = I_{\mathscr{H}_1}, \quad A^*A = I_{\mathscr{H}_2},$$

and

$$BB^* = I_{\mathscr{H}_2}, \quad B^*B = I_{\mathscr{H}_1},$$

i.e., the two operators are unitary between the respective Hilbert spaces.

5.4.3 Hilbert-Schmidt operators (continuing the discussion in 1)

Let $\mathscr{H}$ be a fixed Hilbert space, and set

$$\mathscr{H}S(\mathscr{H}) := \{T \in \mathscr{B}(\mathscr{H}) \mid T^*T \text{ is trace class}\} \tag{5.34}$$

and set
$$\|T\|_{\mathscr{H}S}^2 := trace\,(T^*T)\,; \tag{5.35}$$
similarly if $S, T \in \mathscr{H}S\,(\mathscr{H})$, set

$$\langle S, T \rangle_{\mathscr{H}S} := trace\,(S^*T)\,. \tag{5.36}$$

Note that finiteness on the RHS in (5.35) is part of the definition.

5.4.4   Tensor-Product $\mathscr{H}_1 \otimes \mathscr{H}_2$

Let $\mathscr{H}_1$ and $\mathscr{H}_2$ be two Hilbert spaces, and consider finite-rank operators (rank-1 in this case):

$$|h_1\rangle\langle h_2| \text{ (Dirac ket-bra)}$$
$$|h_1\rangle\langle h_2|\,(u) = \langle h_2, u \rangle_{\mathscr{H}_2}\, h_1,\ \forall u \in \mathscr{H}_2, \tag{5.37}$$

so $T = |h_1\rangle\langle h_2| : \mathscr{H}_2 \longrightarrow \mathscr{H}_1$ with the identification

$$h_1 \otimes h_2 \longleftrightarrow |h_1\rangle\langle h_2|\,. \tag{5.38}$$

Set
$$\|h_1 \otimes h_2\|^2 := trace\,(T^*T) = \|h_1\|_{\mathscr{H}_1}^2\,\|h_2\|_{\mathscr{H}_2}^2\,. \tag{5.39}$$

For the Hilbert space $\mathscr{H}_1 \otimes \mathscr{H}_2$ we take the $\mathscr{H}S$-completion of the space of finite rank operators spanned by the set in (5.37). The tensor product construction fits with composite system in quantum mechanics.

5.4.5   Contractive Inclusion

Let $\mathscr{H}_1$ and $\mathscr{H}_2$ be two Hilbert spaces, and let $T : \mathscr{H}_1 \to \mathscr{H}_2$ be a contractive linear operator, i.e.,
$$I_{\mathscr{H}_1} - T^*T \geq 0. \tag{5.40}$$

On the subspace
$$\mathscr{R}\,(T) = \left\{ Th_1 \mid h_1 \in \mathscr{H}_1 \right\} \tag{5.41}$$

(generally not closed in $\mathscr{H}_2$,) set

$$\|Th_1\|_{\text{new}} := \|h_1\|,\ h_1 \in \mathscr{H}_1; \tag{5.42}$$

then with $\|\cdot\|_{\text{new}}$, $\mathscr{R}\,(T)$ becomes a Hilbert space.

5.4.6   Inflation (Dilation)

Let $T : \mathscr{H}_1 :\longrightarrow \mathscr{H}_2$ be a contraction, and set

$$\mathcal{U} = \left[\begin{array}{c|c} T & (I_2 - TT^*)^{\frac{1}{2}} \\ \hline (I_1 - T^*T)^{\frac{1}{2}} & -T^* \end{array} \right]. \tag{5.43}$$

The two operators in the off-diagonal slots are called the "defect operators" for the contraction T. Reason: the pair of defect-operators are $(0,0)$ if and only if T is a unitary isomorphism of $\mathcal{H}_1$ onto $\mathcal{H}_2$.

Exercise 5.15 (The Julia operator). Show that the matrix-block (5.43) defines a *unitary* operator $\mathcal{U}$ in $\mathcal{H} = \mathcal{H}_1 \oplus \mathcal{H}_2$ (called the Julia operator); and that $P_1 \mathcal{U} P_1 = T$ where P_1 denotes the projection of $\mathcal{H}$ onto $\mathcal{H}_1$.

5.4.7 Quantum Information

Finite-dimensional Hilbert spaces and finite tensor factors are central to applications to quantum information (QI)[4]: *Qubits* refer to a choice of an ONB in a 2-dimensional Hilbert space, so $\mathbb{C}^2$; usually written as follows: $|0\rangle$, $|1\rangle$ with tensors

$$\left.\begin{array}{rcl} |00\rangle & \longleftrightarrow & e_0 \otimes e_0, \\ |10\rangle & \longleftrightarrow & e_1 \otimes e_0, \\ |01\rangle & \longleftrightarrow & e_0 \otimes e_1, \\ |11\rangle & \longleftrightarrow & e_1 \otimes e_1. \end{array}\right\} \in \mathbb{C}^2 \times \mathbb{C}^2.$$

These are the four basis vectors in $\mathbb{C}^2 \otimes \mathbb{C}^2$. If $U = \begin{bmatrix} x_{00} & x_{01} \\ x_{10} & x_{11} \end{bmatrix} \in U(2) =$ the group of all 2×2 matrices, it acts on $\mathbb{C}^2 \otimes \mathbb{C}^2$ in either one of the following ways: $U \otimes I_2$, $I_2 \otimes U$, $U \otimes U$, for example:

$$I_2 \otimes U : \begin{cases} |00\rangle \longrightarrow |00\rangle \\ |01\rangle \longrightarrow |01\rangle \\ |10\rangle \longrightarrow |1\rangle (x_{00} |0\rangle + x_{10} |1\rangle) \\ |11\rangle \longrightarrow |1\rangle (x_{01} |0\rangle + x_{11} |1\rangle). \end{cases}$$

(This is a controlled U-gate, see Figure 5.1.)

The following $U(2)$-gates are elementary, but also important (see, e.g., [NC00]):

---

[4]A summary of some highpoints from quantum information and quantum field theory is contained in Appendix B. See also Sections 1.6 (pg. 17), 2.5 (pg. 63), 4.4 (pg. 140), 5.4 (pg. 174), 8.2 (pg. 280); and Chapters 5 (pg. 155), 6 (pg. 223), 9 (pg. 333).

name	matrix	operation
Hadamard	$H = \frac{1}{\sqrt{2}} \begin{pmatrix} 1 & 1 \\ 1 & -1 \end{pmatrix}$	rotation about the axis $\dfrac{\hat{x} + \hat{z}}{\sqrt{2}}$ (see Figure 5.7)
Pauli X	$X = \begin{pmatrix} 0 & 1 \\ 1 & 0 \end{pmatrix}$	$\simeq$ a NOT-gate, rotation around the x-axis
Pauli Y	$Y = \begin{pmatrix} 0 & -i \\ i & 0 \end{pmatrix}$	rotation around the y-axis
Pauli Z	$Z = \begin{pmatrix} 1 & 0 \\ 0 & -1 \end{pmatrix}$	rotation around the z-axis
Swap-gate	$\text{SWAP} = \begin{bmatrix} 1 & 0 & 0 & 0 \\ 0 & 0 & 1 & 0 \\ 0 & 1 & 0 & 0 \\ 0 & 0 & 0 & 1 \end{bmatrix}$	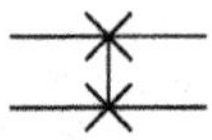
controlled NOT	$\text{CNOT} = \begin{bmatrix} 1 & 0 & 0 & 0 \\ 0 & 1 & 0 & 0 \\ 0 & 0 & 0 & 1 \\ 0 & 0 & 1 & 0 \end{bmatrix}$	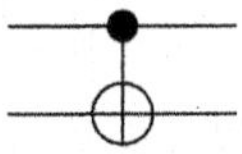

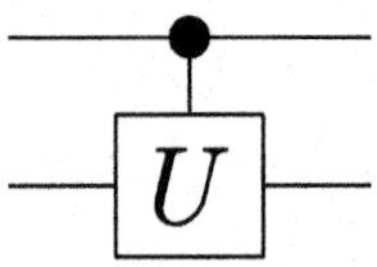

Figure 5.1: Circuit of a controlled U-gate, $U \in U(2)$.

The measurement problem. The Bloch sphere (Figure 5.7) representation of qubits as binary quantum states derives in part from the "notorious" measurement problem in quantum mechanics (QM): When an observation is made of a quantum state, say in the form of a wavefunction, the state then "collapses"? Direct observations in QM are limited this way, and we arrive at QM measurement interpretations, different from those in classical physics. In the Schrödinger picture, the wavefunction (quantum state) evolves according to the Schrödinger equation, and making use of linear superposition, but actual measurements always find the physical system in a definite state. As a result, and paradoxically, the Schrödinger wave equation determines the wavefunction at any "later time." In other words, how can one establish a correspondence between quantum and classical reality?

In the paradox of the Schrödinger's cat (see Figure 5.2), we have the binary states, |dead⟩ vs |alive⟩. A mechanism is arranged to kill a cat if the decay of a radioactive atom, occurs. The two quantum states are entangled with the outcome of a quantum object, the atom. Each of the resulting mixed state possibilities is associated with a specific probability amplitude; the cat seems to be in some kind of "combination" state called a "quantum superposition". But, a definite observation does not measure the probabilities: it always finds either a living cat, |alive⟩, or a dead cat, |dead⟩. How are the probabilities converted into an actual, sharply well-defined outcome?

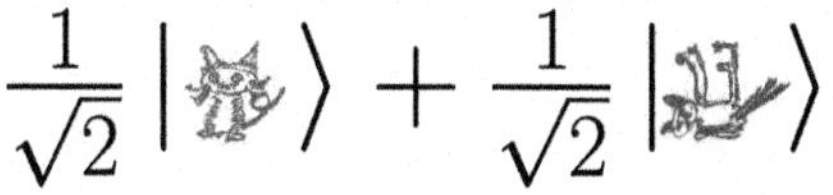

$$\frac{1}{\sqrt{2}}\,\left|\,\vphantom{x}\right\rangle + \frac{1}{\sqrt{2}}\,\left|\,\vphantom{x}\right\rangle$$

Figure 5.2: A mixed state. (With apologies to Schrödinger). The two Dirac kets represent the two states, the "alive" state and the "dead" state, the second. By QM, the cat would be both dead and alive at the same time, but of course, the cat is not a quantum system. In QM, the presence of an observer entails the "collapse" of the wavefunction $|\psi\rangle$, assuming the cat could be in a superposition, as opposed to in one of the two states (decoherence).

5.4.8 Reflection Positivity (or renormalization) $(\mathscr{H}_+/\mathscr{N})^{\sim}$

New Hilbert space from reflection positivity: Let $\mathscr{H}$ be a given Hilbert space, $\mathscr{H}_+ \subset \mathscr{H}$ a closed subspace, and let $\mathcal{U}, \mathcal{J} : \mathscr{H} \to \mathscr{H}$ be two unitary operators, $\mathcal{J}$ satisfying the idempotency condition

$$\mathcal{J}^2 = I, \text{ as well as} \tag{5.44}$$

$$\mathcal{J}\mathcal{U}\mathcal{J} = \mathcal{U}^*, \text{ and} \tag{5.45}$$

$$\mathcal{U}\mathscr{H}_+ \subset \mathscr{H}_+; \text{ and} \tag{5.46}$$

finally

$$\langle h_+, \mathcal{J}h_+ \rangle \geq 0, \ \forall h_+ \in \mathscr{H}_+. \tag{5.47}$$

Note that (5.45) states that $\mathcal{U}$ is unitarily equivalent to its adjoint $\mathcal{U}^*$.

Note 1. Set $P_+ := Proj\,\mathscr{H}_+$ (= the projection onto $\mathscr{H}_+$), then (5.47) is equivalent to

$$P_+ \mathcal{J} P_+ \geq 0$$

with respect to the usual ordering of operators.

Set

$$\begin{aligned}\mathscr{N} &= \text{Ker}\,(P_+ \mathcal{J} P_+) \tag{5.48}\\ &= \{h_+ \in \mathscr{H}_+ \ : \ \langle h_+, \mathcal{J}h_+ \rangle = 0\}.\end{aligned}$$

Set

$$\mathscr{K} = (\mathscr{H}_+/\mathscr{N})^{\sim} \tag{5.49}$$

where "∼" in (5.49) means Hilbert completion with respect to the sesquilinear form: $\mathscr{H}_+ \times \mathscr{H}_+ \to \mathbb{C}$, given by

$$\langle h_+, h_+ \rangle_{\mathscr{K}} := \langle h_+, \mathcal{J} h_+ \rangle, \qquad (5.50)$$

a renormalized inner product.

Exercise 5.16 (An induced operator). Let the setting be as above. Show that $\widetilde{U} : \mathscr{K} \to \mathscr{K}$, given by

$$\widetilde{U} (\text{class } h_+) = \text{class} (\mathcal{U} h_+), \quad h_+ \in \mathscr{H}_+ \qquad (5.51)$$

where class h_+ refers to the quotient in (5.49), is *selfadjoint* and *contractive* (see Figure 5.3).

Remark 5.9. The construction outlined above is called "reflection positivity"; see e.g., [JÓ00, PK88]. It has many applications in physics and in representation theory.

Proof of the assertions in Figure 5.3. Denote the "new" inner product in $\mathscr{K}$ by $\langle \cdot, \cdot \rangle_{\mathscr{K}}$, and the initial inner product in $\mathscr{H}$ by $\langle \cdot, \cdot \rangle$.

$\widetilde{U}$ *is symmetric*: Let $x, y \in \mathscr{H}_+$, then

$$\begin{aligned}
\langle x, \widetilde{U} y \rangle_{\mathscr{K}} &= \langle x, \mathcal{J} \mathcal{U} y \rangle = \langle x, \mathcal{U}^* \mathcal{J} y \rangle \\
&= \langle \mathcal{U} x, \mathcal{J} y \rangle = \langle \widetilde{U} x, y \rangle_{\mathscr{K}}
\end{aligned}$$

is the desired conclusion.

$\widetilde{U}$ *is contractive*: Let $x \in \mathscr{H}_+$, then

$$\begin{aligned}
\left\| \widetilde{U} x \right\|_{\mathscr{K}}^2 &= \langle \mathcal{U} x, \mathcal{J} \mathcal{U} x \rangle = \langle \mathcal{U} x, \mathcal{U}^* \mathcal{J} x \rangle \\
&= \langle \mathcal{U}^2 x, \mathcal{J} x \rangle = \langle \mathcal{U}^2 x, x \rangle_{\mathscr{K}} \\
&\leq \left\| \mathcal{U}^2 x \right\|_{\mathscr{K}} \cdot \| x \|_{\mathscr{K}} \qquad \text{(by Schwarz in } \mathscr{K}\text{)} \\
&\leq \left\| \mathcal{U}^4 x \right\|_{\mathscr{K}}^{\frac{1}{2}} \cdot \| x \|_{\mathscr{K}}^{1+\frac{1}{2}} \qquad \text{(by the first step)} \\
&\leq \left\| \mathcal{U}^{2^{n+1}} x \right\|_{\mathscr{K}}^{\frac{1}{2^n}} \cdot \| x \|_{\mathscr{K}}^{1+\frac{1}{2}+\cdots+\frac{1}{2^n}}. \qquad \text{(by iteration)}
\end{aligned}$$

By the spectral-radius formula,

$$\lim_{n \to \infty} \left\| \mathcal{U}^{2^n} x \right\|_{\mathscr{K}}^{\frac{1}{2^n}} = 1;$$

and we get $\left\| \widetilde{U} x \right\|_{\mathscr{K}}^2 \leq \| x \|_{\mathscr{K}}^2$, which is the desired contractivity. $\square$

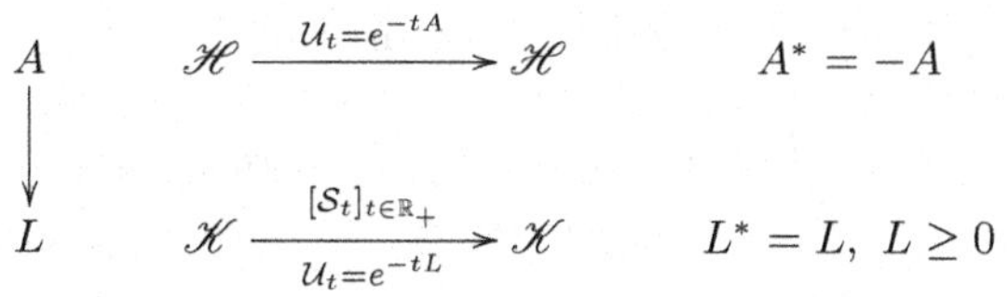

$$A \downarrow L \qquad \mathscr{H} \xrightarrow{\;\mathcal{U}_t = e^{-tA}\;} \mathscr{H} \qquad A^* = -A$$

$$\mathscr{K} \xrightarrow[\;\mathcal{U}_t = e^{-tL}\;]{\;[S_t]_{t \in \mathbb{R}_+}\;} \mathscr{K} \qquad L^* = L, \; L \geq 0$$

Figure 5.4: Transformation of skew-adjoint A into selfadjoint semibounded L.

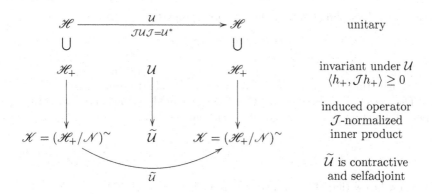

Figure 5.3: Reflection positivity. A unitary operator $\mathcal{U}$ transforms into a self-adjoint contraction $\widetilde{\mathcal{U}}$.

Exercise 5.17 (Time-reflection). Show that if $\{\mathcal{U}_t\}_{t \in \mathbb{R}}$ is a unitary one-parameter group in $\mathscr{H}$ such that

$$\mathcal{J}\mathcal{U}_t\mathcal{J} = \mathcal{U}_{-t}, \; t \in \mathbb{R}, \text{ and}$$

$$\mathcal{U}_t\mathscr{H}_+ \subset \mathscr{H}_+, \; t \in \mathbb{R}_+,$$

then

$$S_t = \widetilde{\mathcal{U}}_t : \mathscr{K} \to \mathscr{K}$$

is a *selfadjoint contraction semigroup*, $t \in \mathbb{R}_+$, i.e., there is a selfadjoint generator L in $\mathscr{K}$,

$$\langle k, Lk \rangle_{\mathscr{K}} \geq 0, \; \forall k \in dom(L), \tag{5.52}$$

where

$$S_t \left(= \widetilde{\mathcal{U}}_t \right) = e^{-tL}, \; t \in \mathbb{R}_+ \tag{5.53}$$

and

$$S_{t_1}S_{t_2} = S_{t_1+t_2}, \; t_1, t_2 \in \mathbb{R}_+. \tag{5.54}$$

Example 5.12 ([Jor02]). Fix $0 < \sigma < 1$, and let $\mathscr{H} (= \mathscr{H}_\sigma)$ be the Hilbert space of all locally integrable functions on $\mathbb{R}$ satisfying

$$\|f\|^2 = \int_\mathbb{R} \int_\mathbb{R} \overline{f(x)} f(y) |x - y|^{\sigma-1} \, dx dy < \infty. \tag{5.55}$$

Set

$$(\mathcal{U}(t) f)(x) = e^{(\sigma+1)t} f\left(e^{2t} x\right), \text{ and} \tag{5.56}$$

$$(\mathcal{J} f)(x) = |x|^{-\sigma-1} f\left(\frac{1}{x}\right). \tag{5.57}$$

Then $\{\mathcal{U}(t)\}_{t\in\mathbb{R}}$ and $\mathcal{J}$ satisfy the reflection property, i.e.,

$$\mathcal{J}\mathcal{U}(t) \mathcal{J} = \mathcal{U}(-t), \ t \in \mathbb{R} \tag{5.58}$$

as operators in $\mathscr{H}$; and $\{\mathcal{U}(t)\}_{t\in\mathbb{R}}$ is a unitary one-parameter group.

We now turn to the "reflected" version of the Hilbert norm (5.55):

The reflection Hilbert space $\mathscr{K}$ will be generated by the completion of the space of functions f supported in $(-1, 1)$, that satisfy

$$\|f\|_\mathscr{K}^2 = \int_{-1}^1 \int_{-1}^1 \overline{f(x)} f(y) |1 - xy|^{\sigma-1} \, dx dy < \infty. \tag{5.59}$$

We show below that this is a Hilbert space of *distributions*.

The selfadjoint contractive semigroup $\left\{\widetilde{\mathcal{U}}(t)\right\}_{t\in\mathbb{R}_+}$ acting in $\mathscr{K}$ is given by the same formula as in (5.56), but now acting in the Hilbert space $\mathscr{K}$ defined by (5.59). Note $\widetilde{\mathcal{U}}(t)$ is only defined for $t \in \mathbb{R}_+ \cup \{0\}$.

Exercise 5.18 (Renormalization).

1. Show that the distributions $\left\{\delta_0^{(n)}\right\}_{n\in\{0\}\cup\mathbb{N}}$ forms an orthogonal and total system in $\mathscr{K}_\sigma$ from (5.59), for all fixed $0 < \sigma < 1$.

2. Show that

$$\left\|\delta_0^{(n)}\right\|_{\mathscr{K}_\sigma}^2 = n! (1 - \sigma) (2 - \sigma) \cdots (n - \sigma). \tag{5.60}$$

The idea of reflection positivity originated in physics[5]. Now, when it is carried out in concrete cases, the initial function spaces change; but, more importantly, the inner product which produces the respective Hilbert spaces of quantum states changes as well.

What is especially intriguing is that before reflection we may have a Hilbert space of functions, but after the time-reflection is turned on, then, in the new

---

[5] A summary of some highpoints from quantum information and quantum field theory is contained in Appendix B. See also Sections 1.6 (pg. 17), 2.5 (pg. 63), 4.4 (pg. 140), 5.4 (pg. 174), 8.2 (pg. 280); and Chapters 5 (pg. 155), 6 (pg. 223), 9 (pg. 333).

inner product, the corresponding completion magically becomes a *Hilbert space of distributions*.

Now this is illustrated already in the simple examples above, Exercise 5.17, and Example 5.12. We include details below to stress the distinction between an abstract Hilbert-norm completion on the one hand, and a concretely realized Hilbert space on the other.

Constructing physical Hilbert spaces entail completions, often a completion of a suitable space of functions. What can happen is that the completion may fail to be a Hilbert space of *functions*, but rather a suitable Hilbert space of distributions.

Recall that a completion, say $\mathscr{H}$ is defined axiomatically, and the "real" secret is revealed only when the elements in $\mathscr{H}$ are identified.

To make the idea more clear we illustrate the point by considering functions on the interval $-1 < x < 1$.

Let $C_c^\infty(-1, 1)$ be the C^∞-functions with compact supports contained in $(-1, 1)$.

A linear functional φ on $C_c^\infty(-1, 1)$ is said to be a *distribution* if for $\forall$ $K \subset (-1, 1)$ compact, $\forall n \in \mathbb{N}$, $\exists C = C_{K,n}$ such that

$$|\varphi(f)| \leq C \sup_{x \in K} \max_{0 \leq j \leq n} \left| \left(\frac{d}{dx} \right)^j f(x) \right|, \ \forall f \in C_c^\infty(-1, 1). \tag{5.61}$$

Examples of distributions are Dirac "functions" δ_{x_0}, and the derivatives $\left(\frac{d}{dx} \right)^n \delta_{x_0}$, $x_0 \in (-1, 1)$, are defined by:

$$\left(\left(\frac{d}{dx} \right)^n \delta_{x_0} \right)(f) = (-1)^n f^{(n)}(x_0), \ f \in C_c^\infty(-1, 1). \tag{5.62}$$

(Note: Distributions are *not* functions, but in Gelfand's rendition of the theory [GS77] they are called "generalized functions.")

Now equip $C_c^\infty(-1, 1)$ with the sesquilinear form from (5.59) in Example 5.12, i.e.,

$$\langle f, g \rangle_{\mathscr{K}_\sigma} := \int_{-1}^1 \int_{-1}^1 \overline{f(x)} g(y) |1 - xy|^{\sigma-1} \, dx \, dy.$$

Exercise 5.19 (A Hilbert space of distributions).

1. Show that each of the distributions $\left(\frac{d}{dx} \right)^n \delta_{x_0}$, $n \in \{0\} \cup \mathbb{N}$, $x_0 \in (-1, 1)$ is in the completion $\mathscr{K}_\sigma$ with respect to (5.59).

2. Compute the Hilbert norm of $\left(\frac{d}{dx} \right)^n \delta_{x_0}$ in $\mathscr{K}_\sigma$, i.e., find

$$\left\| \left(\frac{d}{dx} \right)^n \delta_{x_0} \right\|_{\mathscr{K}_\sigma} \tag{5.63}$$

for all $n \in \{0\} \cup \mathbb{N}$, and $x_0 \in (-1, 1)$.

Hint: The answer to (2) (i.e., (5.63)) is as follows:

- $n = 0$:

$$\left\|\delta_{x_0}\right\|_{\mathscr{K}_\sigma}^2 = \left(1 - x_0^2\right)^{\sigma-1};$$

- $n = 1$ (one derivative):

$$\left\|\delta'_{x_0}\right\|_{\mathscr{K}_\sigma}^2 = (1 - \sigma)\left(1 - x_0^2\right)^{\sigma-3}\left(1 + (1 - \sigma)x_0^2\right).$$

Exercise 5.20 (Taylor for distributions). Fix $0 < \sigma < 1$, and let $\mathscr{K}_\sigma$ be the corresponding Hilbert space of distributions. As an identity in $\mathscr{K}_\sigma$, establish:

$$\delta_x = \sum_{n=0}^{\infty} \frac{(-x)^n}{n!} \delta_0^{(n)}, \tag{5.64}$$

valid for all x, $|x| < 1$.

Historical Note. Laurent Schwartz has developed a systematic study of Hilbert spaces of distributions; see [Sch64b].

5.5   A Second Duality Principle: A Metric on the Set of Probability Measures

Let (X, d) be a separable metric space, and denote by $\mathcal{M}_1(X)$ and $\mathcal{M}_1(X \times X)$ the corresponding sets of regular probability measures. Let π_i, $i = 1, 2$, denote the projections: $\pi_1(x_1, x_2) = x_1$, $\pi_2(x_1, x_2) = x_2$, for all $(x_1, x_2) \in X \times X$.
  For $\mu \in \mathcal{M}_1(X \times X)$, set

$$\mu^{\pi_i} := \mu \circ \pi_i^{-1}.$$

For $P_i \in \mathcal{M}_1(X)$, $i = 1, 2$, set

$$\mathcal{M}(P_1, P_2) = \{\mu \in \mathcal{M}_1(X \times X) : \mu^{\pi_i} = P_i, \ i = 1, 2\}.$$

Finally, let Lip_1 = the Lipchitz functions on (X, d), i.e., $f \in \mathrm{Lip}_1$ iff (Def.)

$$|f(x) - f(y)| \leq d(x, y), \ \forall x, y \in X.$$

Theorem 5.9 (Kantorovich-Rubinstein). *Setting*

$$dist_W(P_1, P_2) = \inf\left\{\int_{X \times X} d(x, y)\, d\mu(x, y) : \mu \in \mathcal{M}(P_1, P_2)\right\}$$

and

$$dist_K(P_1, P_2) = \sup\left\{\int_X f\, d(P_1 - P_2) : f \in Lip_1\right\}$$

then

$$dist_W(P_1, P_2) = dist_K(P_1, P_2)$$

for all $P_1, P_2 \in \mathcal{M}_1(X)$.

Proof. We omit the proof here, but refer to [Kan58, KR57, Rüs07]. □

Exercise 5.21 (A complete metric space). Show that $\mathcal{M}_1(X)$ is a complete metric space when equipped with the metric $dist_K$.

Exercise 5.22 (A distance formula). Let (X, d) be $\mathbb{R}$ with the usual distance $d(x, y) = |x - y|$. For $P \in \mathcal{M}_1(\mathbb{R})$ set $F_P(x) = P((-\infty, x])$. Show that then

$$dist(P_1, P_2) = \int_{\mathbb{R}} |F_{P_1}(x) - F_{P_2}(x)|\, dx.$$

Let (X, d) and $\mathcal{M}_1(X)$ be as above. Now apply Banach's Fixed point theorem to the complete metric space $(\mathcal{M}_1(X), dist_K)$ to get a solution to the following:

Exercise 5.23 (Iterated function systems). Let $N \in \mathbb{N}$, and let $\varphi_i : X \longrightarrow X$, $i = 1, \cdots, N$ be a system of strict contractions in (X, d). On $\mathcal{M}_1(X)$, set

$$T_\mu := \frac{1}{N} \sum_{i=1}^{N} d\mu \circ \varphi_i^{-1}. \tag{5.65}$$

Recall $\left(\mu \circ \varphi_i^{-1}\right)(\triangle) = \mu\left(\varphi_i^{-1}(\triangle)\right)$. (For IFS, see e.g., [Hut81, BHS05].)

1. Show that, if c is the smallest of the contractivity constant for $\{\varphi_i\}_{i=1}^{N}$, then

$$dist(T_\mu, T_\nu) \leq c\, dist(\mu, \nu), \ \forall \mu, \nu \in \mathcal{M}_1(X). \tag{5.66}$$

2. Show that there is a unique solution $\mu_L \in \mathcal{M}_1(X)$ to

$$T\mu_L = \mu_L, \text{ i.e.,} \tag{5.67}$$

$$\int_X f(x)\, d\mu_L(x) = \frac{1}{N} \sum_{i=1}^{N} \int_X f(\varphi_i(x))\, d\mu_L(x) \tag{5.68}$$

holds for $\forall f \in C_b(X)$ (= bounded continuous.)

Hint: The desired conclusion in (2), i.e., both existence and uniqueness of μ_L, follows from *Banach's fixed point theorem*: Every strict contraction in a complete metric space has a unique fixed-point.

Exercise 5.24 (The Middle-Third Cantor-measure). Set $X = [0, 1]$ = the unit interval with the usual metric, set $N = 2$, and

$$\varphi_1(x) = \frac{x}{3}, \quad \varphi_2(x) = \frac{x + 2}{3}, \tag{5.69}$$

and let μ_L be the corresponding measure, i.e.,

$$\int_0^1 f(x)\, d\mu_L(x) = \frac{1}{2} \left(\int_0^1 f\left(\frac{x}{3}\right) d\mu_L(x) + \int_0^1 f\left(\frac{x + 2}{3}\right) d\mu_L(x) \right). \tag{5.70}$$

Show that μ_L is supported on the Middle-Third Cantor set (see Figure 5.5).

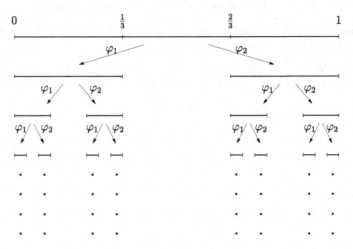

Figure 5.5: The Middle-Third Cantor set as a limit.

Remark 5.10. Starting with the Middle-Third Cantor measure μ_L; see (5.70), we get the cumulative distribution function F defined on the unit interval $[0, 1]$,

$$F(x) = \mu_L([0, x]). \tag{5.71}$$

It follows from Exercise 5.24 that the graph of F is the Devil's Staircase; see Figure 5.6. Endpoints: $F(0) = 0$, and $F(1) = 1$. The union $\mathcal{O}$ of all the omitted open intervals has total length:

$$\frac{1}{3} + \frac{2}{3^2} + \cdots = \frac{1}{3} \sum_{n=0}^{\infty} \left(\frac{2}{3} \right)^n = 1,$$

and $F'(x) = 0$ for all $x \in \mathcal{O}$.

Using the argument from Exercise 2.3 above, we get

$$\int_0^1 dF = \int_0^1 F'(x)\, dx = 0.$$

Since $F(1) - F(0) = 1$, it would seem that the conclusion in the Fundamental Theorem of Calculus fails. (Explain! See e.g., [Rud87, ch 7].)

Exercise 5.25 (Straightening out the Devil's staircase). Repeat the construction from the previous exercise, but now with the two functions φ_1, φ_2 modified as follows:

$$\varphi_1(x) = \frac{x}{2}, \quad \varphi_2(x) = \frac{x+1}{2}; \tag{5.72}$$

compare with (5.69) above.

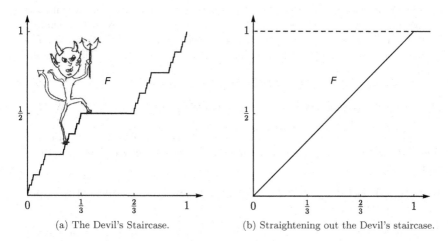

(a) The Devil's Staircase. (b) Straightening out the Devil's staircase.

Figure 5.6: The cumulative distribution of the middle-third Cantor measure.

Then rewrite formula (5.70), and show that the cumulative distribution F from (5.71) becomes (Figure 5.6)

$$F(x) = \begin{cases} 0 & x < 0 \\ x & 0 \le x \le 1 \\ 0 & x > 1. \end{cases}$$

Explain this!

5.6 Abelian C^*-algebras

Diagonalizing a commuting family of bounded selfadjoint operators may be formulated in the setting of Abelian C^*-algebras. By the structure theorem of Gelfand and Naimark, every Abelian C^*-algebra containing the identity element is isomorphic to the algebra $C(X)$ of continuous functions on some compact Hausdorff space X, which is unique up to homeomorphism. The classification of all the representations Abelian C^*-algebras, therefore, amounts to that of $C(X)$. This problem can be understood using the idea of σ-measures (square densities). It also leads to the multiplicity theory of selfadjoint operators. The best treatment on this subject can be found in [Nel69].

Here we discuss *Gelfand's theory on Abelian C^*-algebras*. Throughout, we assume all the algebras contain unit element.

Definition 5.13. $\mathfrak{A}$ is *Banach algebra* if it is a complex algebra and a Banach space such that the norm satisfies $\|ab\| \le \|a\| \, \|b\|$, for all $a, b \in \mathfrak{A}$.

Let $\mathfrak{A}$ be an Abelian Banach. Consider the closed ideals in $\mathfrak{A}$ (since $\mathfrak{A}$ is normed, so consider closed ideals) ordered by inclusion. By Zorn's lemma, there exists maximal ideals M, which are closed by maximality. Then $\mathfrak{A}/M$ is 1-dimensional, i.e., $\mathfrak{A}/M = \{tv\}$ for some $v \in \mathfrak{A}$, and $t \in \mathbb{R}$. Therefore the combined map

$$\varphi : \mathfrak{A} \to \mathfrak{A}/M \to \mathbb{C}, \ a \mapsto a/M \mapsto t_a$$

is a (complex) homomorphism. In particular, $\mathfrak{A} \ni 1_{\mathfrak{A}} \mapsto v := 1_{\mathfrak{A}}/M \in \mathfrak{A}/M \simeq \mathbb{C}$, and $\varphi(1_{\mathfrak{A}}) = 1$.

Conversely, the kernel of any homomorphism is a maximal ideal in $\mathfrak{A}$ (since the co-dimension $= 1$.) Therefore there is a bijection between maximal ideas and homomorphisms.

Lemma 5.5. *Let $\mathfrak{A}$ be an Abelian Banach algebra. If $a \in \mathfrak{A}$, and $\|a\| < 1$, then $1_{\mathfrak{A}} - a$ is invertible.*

Proof. It is easy to verify that $(1 - a)^{-1} = 1 + a + a^2 + \cdots$, and the RHS is norm convergent. $\qquad\square$

Corollary 5.5. *Any homomorphism $\varphi : \mathfrak{A} \to \mathbb{C}$ is a contraction.*

Proof. Let $a \in \mathfrak{A}$, $a \neq 0$. Suppose $\lambda := \varphi(a)$ such that $|\lambda| > \|a\|$. Then $\|a/\lambda\| < 1$ and so $1_{\mathfrak{A}} - a/\lambda$ is invertible by Lemma 5.5. Since φ is a homomorphism, it must map invertible element to invertible element, hence $\varphi(1_{\mathfrak{A}} - a/\lambda) \neq 0$, i.e., $\varphi(a) \neq \lambda$, which is a contradiction. $\qquad\square$

Let X be the set of all *maximal ideals*, identified with all homomorphisms in $\mathfrak{A}_1^*$, where $\mathfrak{A}_1^*$ is the unit ball in $\mathfrak{A}^*$. Since $\mathfrak{A}_1^*$ is compact (see Banach-Alaoglu, Theorem 5.6), and X is closed in it, therefore X is also compact. Here, compactness refers to the weak*-topology.

Definition 5.14. The *Gelfand transform* $\mathcal{F} : \mathfrak{A} \to C(X)$ is given by

$$\mathcal{F}(a)(\varphi) = \varphi(a), \ a \in \mathfrak{A}, \varphi \in C(X). \tag{5.73}$$

Hence $\mathfrak{A}/\ker\mathcal{F}$ is homomorphic to a closed subalgebra of $C(X)$. Note $\ker\mathcal{F} = \{a \in \mathfrak{A} : \varphi(a) = 0, \ \forall \varphi \in X\}$. It is called the *radical* of $\mathfrak{A}$.

The theory is takes a more pleasant form when $\mathfrak{A}$ is a C^*-algebra. So there is an involution, and the norm satisfies the C^* axiom: $\|aa^*\| = \|a\|^2$, for all $a \in \mathfrak{A}$.

Theorem 5.10 (Gelfand). *If $\mathfrak{A}$ is an Abelian C^*-algebra then the Gelfand transform (5.73) is an isometric $*$-isomorphism from $\mathfrak{A}$ onto $C(X)$, where X is the maximal ideal space of $\mathfrak{A}$.*

Example 5.13. Consider $l^1(\mathbb{Z})$, the convolution algebra:

$$(ab)_n = \sum_k a_k b_{n-k} \tag{5.74}$$

$$a_n^* = \overline{a_{-n}}$$

$$\|a\| = \sum_n |a_n|$$

$$1_{\mathfrak{A}} = \delta_0 \ (\text{Dirac mass at } 0);$$

the unit-element for the product (5.74) in $l^1(\mathbb{Z})$.

To identify X in practice, we always start with a guess, and usually it turns out to be correct. Since Fourier transform converts convolution to multiplication,

$$l^1(\mathbb{Z}) \ni a \xrightarrow{\ \varphi_z \ } \sum a_n z^n$$

is a complex homomorphism. To see φ_z is multiplicative, we have

$$
\begin{aligned}
\varphi_z(ab) &= \sum (ab)_n z^n \\
&= \sum_{n,k} a_k b_{n-k} z^n \\
&= \sum_k a_k z^k \sum_n b_{n-k} z^{n-k} \\
&= \left(\sum_k a_k z^k \right) \left(\sum_k b_k z^k \right) \\
&= \varphi_z(a)\, \varphi_z(b).
\end{aligned}
$$

Thus $\{z : |z| = 1\}$ is a subspace in the Gelfand space X. Note that we cannot use $|z| < 1$ since we are dealing with two-sided l^1 sequence. (If the sequences were truncated, so that $a_n = 0$ for $n < 0$ then we allow $|z| < 1$.)

φ_z is contractive: $|\varphi_z(a)| = |\sum a_n z^n| \le \sum_n |a_n| = \|a\|$.

Exercise 5.26 (The homomorphism of l^1). Prove that every homomorphism of $l^1(\mathbb{Z})$ is obtained as φ_z for some $|z| = 1$. Hence $X = \mathbb{T}^1 \, (= \{z \in \mathbb{C} : |z| = 1\})$.

Example 5.14. $l^\infty(\mathbb{Z})$, with $\|a\| = \sup_n |a_n|$. The Gelfand space in this case is $X = \beta\mathbb{Z}$, the Stone-Čech compactification of $\mathbb{Z}$, which can be realized as the set of all ultra-filters on $\mathbb{Z}$. (Caution, $\beta\mathbb{Z}$ is a much bigger compactification of $\mathbb{Z}$ than is, for example, the p-adic numbers.) From the discussion above, it follows that the pure states on the algebra of all bounded diagonal operators on $l^2(\mathbb{Z})$ correspond to the points in $\beta\mathbb{Z}$. See Chapter 9 for details.

5.7 States and Representations

Let $\mathfrak{A}$ be a $*$-algebra, a representation $\pi : \mathfrak{A} \to \mathscr{B}(\mathscr{H})$ generates a $*$-subalgebra $\pi(\mathfrak{A})$ in $\mathscr{B}(\mathscr{H})$. By taking norm closure, one gets a C^*-algebra, i.e., a Banach $*$-algebra with the axiom $\|a^*a\| = \|a\|^2$. On the other hand, by Gelfand and Naimark's theorem, all abstract C^*-algebras are isometrically isomorphic to closed subalgebras of $\mathscr{B}(\mathscr{H})$, for some Hilbert space $\mathscr{H}$ (Theorem 5.4). The construction of $\mathcal{H}$ comes down to states $S(\mathfrak{A})$ on $\mathfrak{A}$ and the GNS construction. Therefore, the GNS construction gives rise to a bijection between states and representations.

Let $\mathfrak{A}_+$ be the positive elements in $\mathfrak{A}$. $s \in S(\mathfrak{A})$, $s : \mathfrak{A} \to \mathbb{C}$ and $s(\mathfrak{A}_+) \subset [0, \infty)$. For C^*-algebra, positive elements can be written $f = (\sqrt{f})^2$ by the spectral theorem. In general, positive elements have the form a^*a. There is a bijection between states and GNS representations $Rep(\mathfrak{A}, \mathcal{H})$, where $s(A) = \langle \Omega, \pi(A)\Omega \rangle$.

Example 5.15. $\mathfrak{A} = C(X)$ where X is a compact Hausdorff space. s_μ given by $s_\mu(a) = \int a d\mu$ is a state. The GNS construction gives $\mathcal{H} = L^2(\mu)$, $\pi(f)$ is the operator of multiplication by f on $L^2(\mu)$. $\{\varphi 1 : \varphi \in C(X)\}$ is dense in L^2, where 1 is the cyclic vector. $s_\mu(f) = \langle \Omega, \pi(f)\Omega \rangle = \int 1 f 1 d\mu = \int f d\mu$, which is also seen as the expectation of f in case μ is a probability measure.

We consider decomposition of representations or equivalently states, i.e., breaking up representations corresponds to breaking up states.

The thing that we want to do with representations comes down to the smallest ones, i.e., the irreducible representations. Irreducible representations correspond to pure states which are *extreme points* in the states (see [Phe01]).

A representation $\pi : \mathfrak{A} \to \mathcal{B}(\mathcal{H})$ is *irreducible*, if whenever $\mathcal{H}$ breaks up into two pieces $\mathcal{H} = \mathcal{H}_1 \oplus \mathcal{H}_2$, where $\mathcal{H}_i$ is invariant under $\pi(\mathfrak{A})$, one of them is zero (the other is $\mathcal{H}$). Equivalently, if $\pi = \pi_1 \oplus \pi_2$, where $\pi_i = \pi\big|_{\mathcal{H}_i}$, then one of them is zero. This is similar to the decomposition of natural numbers into product of primes. For example, $6 = 2 \times 3$, but 2 and 3 are primes and they do not decompose further.

Hilbert spaces are defined up to unitary equivalence. A state φ may have equivalent representations on different Hilbert spaces (but unitarily equivalent), however φ does not see the distinction, and it can only detect equivalent classes of representations.

Example 5.16. Let $\mathfrak{A}$ be a $*$-algebra. Given two states s_1 and s_2, by the GNS construction, we get cyclic vectors ξ_i, and representations $\pi_i : \mathfrak{A} \to \mathcal{B}(\mathcal{H}_i)$, so that $s_i(A) = \langle \xi_i, \pi_i(A)\xi_i \rangle$, $i = 1, 2$. Suppose there is a unitary operator $W : \mathcal{H}_1 \to \mathcal{H}_2$, such that for all $A \in \mathfrak{A}$,

$$\pi_1(A) = W^*\pi_2(A)W.$$

Then

$$
\begin{aligned}
\langle \xi_1, \pi_1(A)\xi_1 \rangle_1 &= \langle \xi_1, W^*\pi_2(A)W\xi_1 \rangle_1 \\
&= \langle W\xi_1, \pi_2(A)W\xi_1 \rangle_2 \\
&= \langle \xi_2, \pi_2(A)\xi_2 \rangle_2, \quad \forall A \in \mathfrak{A};
\end{aligned}
$$

i.e., $s_2(A) = s_1(A)$. Therefore the same state $s = s_1 = s_2$ has two distinct (unitarily equivalent) representations.

Remark 5.11. A special case of states are measures when the algebra is Abelian. Recall that all Abelian C^*-algebras with identity are $C(X)$, where X is the corresponding Gelfand space. Two representations are mutually singular $\pi_1 \perp \pi_2$, if and only if the two measures are mutually singular, $\mu_1 \perp \mu_2$.

The theorem below is fundamental in representation theory. Recall that if M is a subset of $\mathscr{B}(\mathscr{H})$, the *commutant* M' consists of $A \in \mathscr{B}(\mathscr{H})$ that commutes with all elements in M.

Theorem 5.11 (Schur). *Let* $\pi : \mathfrak{A} \to \mathscr{B}(\mathscr{H})$ *be a representation. The following are equivalent.*

1. π is irreducible.

2. The commutant $(\pi(\mathfrak{A}))'$ is one-dimensional, i.e., $(\pi(\mathfrak{A}))' = cI_{\mathfrak{A}}$, $c \in \mathbb{C}$.

Proof. Suppose $(\pi(\mathfrak{A}))'$ has more than one dimension. Let $X \in (\pi(\mathfrak{A}))'$, then by taking adjoint, $X^* \in (\pi(\mathfrak{A}))'$. $X + X^*$ is selfadjoint, and $X + X^* \neq cI$ since by hypothesis $(\pi(\mathfrak{A}))'$ has more than one dimension. Therefore $X + X^*$ has a non trivial spectral projection $P(E)$, i.e., $P(E) \notin \{0, I\}$. Let $\mathscr{H}_1 = P(E)\mathscr{H}$ and $\mathscr{H}_2 = (I - P(E))\mathscr{H}$. $\mathscr{H}_1$ and $\mathscr{H}_2$ are both nonzero proper subspaces of $\mathscr{H}$. Since $P(E)$ commutes with $\pi(A)$, for all $A \in \mathfrak{A}$, it follows that $\mathscr{H}_1$ and $\mathscr{H}_2$ are both invariant under π.

Conversely, suppose $(\pi(\mathfrak{A}))'$ is one-dimensional. If π is not irreducible, i.e., $\pi = \pi_1 \oplus \pi_2$, then for

$$P_{\mathscr{H}_1} = \begin{bmatrix} I_{\mathscr{H}_1} & 0 \\ 0 & 0 \end{bmatrix}, \quad P_{\mathscr{H}_2} = 1 - P_{\mathscr{H}_1} = \begin{bmatrix} 0 & 0 \\ 0 & I_{\mathscr{H}_2} \end{bmatrix}$$

we have

$$P_{\mathscr{H}_i}\pi(A) = \pi(A)P_{\mathscr{H}_i}, \ i = 1, 2$$

for all $A \in \mathfrak{A}$. Hence $(\pi(\mathfrak{A}))'$ has more than one dimension. $\qquad \square$

Corollary 5.6. *π is irreducible if and only if the only projections in $(\pi(\mathfrak{A}))'$ are 0 or I.*

Thus to test invariant subspaces, one only needs to look at projections in the commutant.

Corollary 5.7. *If $\mathfrak{A}$ is Abelian, then π is irreducible if and only if $\mathscr{H}$ is one-dimensional.*

Proof. Obviously, if dim $\mathscr{H} = 1$, π is irreducible. Conversely, by Theorem 5.11, $(\pi(\mathfrak{A}))' = cI$. Since $\pi(\mathfrak{A})$ is Abelian, $\pi(\mathfrak{A}) \subset \pi(\mathfrak{A})'$. Thus for all $A \in \mathfrak{A}$, $\pi(A) = c_A I$, for some constant c_A. $\qquad \square$

If instead of taking the norm closure, but using the strong operator topology, ones gets a von Neumann algebra. von Neumann showed that the weak closure of $\mathfrak{A}$ is equal to $\mathfrak{A}''$.

Corollary 5.8. π *is irreducible* $\Longleftrightarrow$ $(\pi(\mathfrak{A}))'$ *is 1-dimensional* $\Longleftrightarrow$ $(\pi(\mathfrak{A}))'' = \mathscr{B}(\mathscr{H})$.

Remark 5.12. In matrix notation, we write $\pi = \pi_1 \oplus \pi_2$ as

$$\pi(A) = \begin{bmatrix} \pi_1(A) & 0 \\ 0 & \pi_2(A) \end{bmatrix}.$$

If

$$\begin{bmatrix} X & Y \\ U & V \end{bmatrix} \in (\pi(\mathfrak{A}))'$$

then

$$\begin{bmatrix} X & Y \\ U & V \end{bmatrix} \begin{bmatrix} \pi_1(A) & 0 \\ 0 & \pi_2(A) \end{bmatrix} = \begin{bmatrix} X\pi_1(A) & Y\pi_2(A) \\ U\pi_1(A) & V\pi_2(A) \end{bmatrix}$$

$$\begin{bmatrix} \pi_1(A) & 0 \\ 0 & \pi_2(A) \end{bmatrix} \begin{bmatrix} X & Y \\ U & V \end{bmatrix} = \begin{bmatrix} \pi_1(A)X & \pi_1(A)Y \\ \pi_2(A)U & \pi_2(A)V \end{bmatrix}.$$

Hence

$$\begin{aligned} X\pi_1(A) &= \pi_1(A)X \\ V\pi_2(A) &= \pi_2(A)V \\ U\pi_1(A) &= \pi_2(A)U \\ Y\pi_2(A) &= \pi_1(A)Y. \end{aligned}$$

Therefore,

$$X \in (\pi_1(\mathfrak{A}))', \ V \in (\pi_2(\mathfrak{A}))', \text{ and}$$

$$U, Y \in int(\pi_1, \pi_2) = \text{intertwining operators of } \pi_1, \pi_2.$$

This is illustrated by the diagram below.

We say π_1 and π_2 are *inequivalent* if and only if $int(\pi_1, \pi_2) = 0$. For $\pi_1 = \pi_2$, π has *multiplicity* 2. Multiplicity > 1 is equivalent to the commutant being non-Abelian. In the case $\pi = \pi_1 \oplus \pi_2$ where $\pi_1 = \pi_2$, $(\pi(\mathfrak{A}))' \simeq M_2(\mathbb{C})$.

Schur's lemma[6] addresses all representations. It says that a representation $\pi : \mathfrak{A} \to \mathscr{B}(\mathscr{H})$ is irreducible if and only if $(\pi(\mathfrak{A}))'$ is 1-dimensional. When specialize to the GNS representation of a given state s, this is also equivalent to saying that for all positive linear functional t, $t \leq s \Rightarrow t = \lambda s$ for some $\lambda \geq 0$.

---

[6]Appendix C includes a short biographical sketch of I. Schur.

This latter equivalence is obtained by using a more general result, which relates t and selfadjoint operators in the commutant $(\pi(\mathfrak{A}))'$.

We now turn to characterize the relation between state and its GNS representation, i.e., specialize to the GNS representation. Given a $*$ algebra $\mathfrak{A}$, the states $S(\mathfrak{A})$ forms a compact convex subset in the unit ball of the dual $\mathfrak{A}^*$.

Let $\mathfrak{A}_+$ be the set of positive elements in $\mathfrak{A}$. Given $s \in S(\mathfrak{A})$, let t be a positive linear functional. By $t \leq s$, we mean $t(A) \leq s(A)$ for all $A \in \mathfrak{A}_+$. We look for relation between t and the commutant $(\pi(\mathfrak{A}))'$.

Lemma 5.6 (Schur-Sakai-Nikodym). *Let t be a positive linear functional, and let s be a state. There is a bijection between t such that $0 \leq t \leq s$, and selfadjoint operator A in the commutant with $0 \leq A \leq I$. The relation is given by*

$$t(\cdot) = \langle \Omega, \pi(\cdot)A\Omega \rangle.$$

Remark 5.13. This is an extension of the classical Radon-Nikodym derivative theorem to the non-commutative setting. We may write $A = dt/ds$. The notation $0 \leq A \leq I$ refers to the partial order of selfadjoint operators. It means that for all $\xi \in \mathscr{H}$, $0 \leq$. See [Sak98, KR97b].

Proof. Easy direction, suppose $A \in (\pi(\mathfrak{A}))'$ and $0 \leq A \leq I$. As in many applications, the favorite functions one usually applies to selfadjoint operators is the square root function $\sqrt{\cdot}$. So let's take $\sqrt{A}$. Since $A \in (\pi(\mathfrak{A}))'$, so is $\sqrt{A}$. We need to show $t(a) = \langle \Omega, \pi(a)A\Omega \rangle \leq s(a)$, for all $a \geq 0$ in $\mathfrak{A}$. Let $a = b^2$, then

$$
\begin{aligned}
t(a) &= \langle \Omega, \pi(a)A\Omega \rangle \\
&= \langle \Omega, \pi(b^2)A\Omega \rangle \\
&= \langle \Omega, \pi(b)^*\pi(b)A\Omega \rangle \\
&= \langle \pi(b)\Omega, A\pi(b)\Omega \rangle \\
&\leq \langle \pi(b)\Omega, \pi(b)\Omega \rangle \\
&= \langle \Omega, \pi(a)\Omega \rangle \\
&= s(a).
\end{aligned}
$$

Conversely, suppose $t \leq s$. Then for all $a \geq 0$, $t(a) \leq s(a) = \langle \Omega, \pi(a)\Omega \rangle$. Again write $a = b^2$. It follows that

$$t(b^2) \leq s(b^2) = \langle \Omega, \pi(a)\Omega \rangle = \|\pi(b)\Omega\|^2.$$

By Riesz's theorem, there is a unique η, so that

$$t(a) = \langle \pi(b)\Omega, \eta \rangle.$$

Conversely, let $a = b^2$, then

$$t(b^2) \leq s(b^2) = \langle \Omega, \pi(a)\Omega \rangle = \|\pi(b)\Omega\|^2$$

i.e., $\pi(b)\Omega \mapsto t(b^2)$ is a bounded quadratic form. Therefore, there exists a unique $A \geq 0$ such that

$$t(b^2) = \langle \pi(b)\Omega, A\pi(b)\Omega \rangle.$$

It is easy to see that $0 \leq A \leq I$. Also, $A \in (\pi(\mathfrak{A}))'$, the commutant of $\pi(\mathfrak{A})$. $\quad\square$

Corollary 5.9. *Let s be a state. $(\pi, \Omega, \mathscr{H})$ is the corresponding GNS construction. The following are equivalent.*

1. *For all positive linear functional t, $t \leq s \Rightarrow t = \lambda s$ for some $\lambda \geq 0$.*

2. *π is irreducible.*

Proof. By use of the Sakai-Nikodym derivative, $t \leq s$ if and only if there is a selfadjoint operator $A \in (\pi(\mathfrak{A}))'$ so that

$$t(\cdot) = \langle \Omega, \pi(\cdot)A\Omega \rangle.$$

Therefore $t = \lambda s$ if and only if $A = \lambda I$.

Suppose $t \leq s \Rightarrow t = \lambda s$ for some $\lambda \geq 0$. Then π must be irreducible, since otherwise there exists $A \in (\pi(\mathfrak{A}))'$ with $A \neq cI$, hence $\mathfrak{A} \ni a \mapsto t(a) := \langle \Omega, \pi(a)A\Omega \rangle$ defines a positive linear functional, and $t \leq s$, however $t \neq \lambda s$. Thus a contradiction to the hypothesis.

Conversely, suppose π is irreducible. Then by Schur's lemma, $(\pi(\mathfrak{A}))'$ is 1-dimensional. i.e. for all $A \in (\pi(\mathfrak{A}))'$, $A = \lambda I$ for some λ. Therefore if $t \leq s$, by Sakai's theorem, $t(\cdot) = \langle \Omega, \pi(\cdot)A\Omega \rangle$. Thus $t = \lambda s$ for some $\lambda \geq 0$. $\quad\square$

The main theorem in this section is a corollary to Sakai's theorem.

Corollary 5.10. *Let s be a state. $(\pi, \Omega, \mathscr{H})$ is the corresponding GNS construction. The following are equivalent.*

1. *$t \leq s \Rightarrow t = \lambda s$ for some $\lambda \geq 0$.*

2. *π is irreducible.*

3. *s is a pure state.*

Proof. By Sakai-Nikodym derivative, $t \leq s$ if and only if there is a selfadjoint operator $A \in (\pi(\mathfrak{A}))'$ so that

$$t(a) = \langle \Omega, \pi(a)A\Omega \rangle, \ \forall a \in \mathfrak{A}.$$

Therefore $t = \lambda s$ if and only if $A = \lambda I$.

We show that $(1)\Leftrightarrow(2)$ and $(1)\Rightarrow(3)\Rightarrow(2)$.

$(1)\Leftrightarrow(2)$ Suppose $t \leq s \Rightarrow t = \lambda s$, then π must be irreducible, since otherwise there exists $A \in (\pi(\mathfrak{A}))'$ with $A \neq cI$, hence $t(\cdot) := \langle \Omega, \pi(\cdot)A\Omega \rangle$ defines a positive linear functional with $t \leq s$, however $t \neq \lambda s$. Conversely, suppose π is irreducible. If $t \leq s$, then $t(\cdot) = \langle \Omega, \pi(\cdot)A\Omega \rangle$ with $A \in (\pi(\mathfrak{A}))'$. By Schur's lemma, $(\pi(\mathfrak{A}))' = \{0, \lambda I\}$. Therefore, $A = \lambda I$ and $t = \lambda s$.

(1)$\Rightarrow$(3) Suppose $t \leq s \Rightarrow t = \lambda s$ for some $\lambda \geq 0$. If s is not pure, then $s = cs_1 + (1 - c)s_2$ where s_1, s_2 are states and $c \in (0, 1)$. By hypothesis, $s_1 \leq s$ implies that $s_1 = \lambda s$. It follows that $s = s_1 = s_2$.

(3)$\Rightarrow$(2) Suppose π is not irreducible, i.e. there is a non trivial projection $P \in (\pi(\mathfrak{A}))'$. Let $\Omega = \Omega_1 \oplus \Omega_2$ where $\Omega_1 = P\Omega$ and $\Omega_2 = (I - P)\Omega$. Then

$$
\begin{aligned}
s(a) &= \langle \Omega, \pi(a)\, \Omega \rangle \\
&= \langle \Omega_1 \oplus \Omega_2, \pi(a)\, \Omega_1 \oplus \Omega_2 \rangle \\
&= \langle \Omega_1, \pi(a)\, \Omega_1 \rangle + \langle \Omega_2, \pi(a)\, \Omega_2 \rangle \\
&= \|\Omega_1\|^2 \left\langle \frac{\Omega_1}{\|\Omega_1\|}, \pi(a)\, \frac{\Omega_1}{\|\Omega_1\|} \right\rangle + \|\Omega_2\|^2 \left\langle \frac{\Omega_2}{\|\Omega_2\|}, \pi(a)\, \frac{\Omega_2}{\|\Omega_2\|} \right\rangle \\
&= \|\Omega_1\|^2 \left\langle \frac{\Omega_1}{\|\Omega_1\|}, \pi(a)\, \frac{\Omega_1}{\|\Omega_1\|} \right\rangle + \left(1 - \|\Omega_1\|^2\right) \left\langle \frac{\Omega_2}{\|\Omega_2\|}, \pi(a)\, \frac{\Omega_2}{\|\Omega_2\|} \right\rangle \\
&= \lambda s_1(a) + (1 - \lambda)\, s_2(a).
\end{aligned}
$$

Hence s is not a pure state. $\qquad\square$

Corollary 5.11. *Let s be a state on $\mathscr{B}(\mathscr{H})$ (= the algebra of all bounded operators in $\mathscr{H}$).*

1. *Then the following two conditions are equivalent:*

   (a) s is pure, and

   (b) $\exists x \in \mathscr{H}$, $\|x\| = 1$ such that $s(A) = \langle x, Ax \rangle$, $\forall A \in \mathscr{B}(\mathscr{H})$.

2. *If $s(A) = \langle x_i, Ax_i \rangle$, $i = 1, 2$, $A \in \mathscr{B}(\mathscr{H})$, then $\exists \zeta \in \mathbb{T}$ such that $x_2 = \zeta x_1$.*

Exercise 5.27 (Pure states on $\mathbb{C}^n$)**.** Let $\mathscr{H}_n = \mathbb{C}^n$ be the finite-dimensional Hilbert space with its standard inner product, and let $U(n) =$ the unitary group in $\mathscr{H}_n$; i.e., all the unitary $n \times n$ matrices.

1. Show that the homogeneous space $U(n)/U(n-1) \times U(1)$ is naturally isomorphic to the pure states on $\mathscr{H}_n$.

2. For $n = 2$, show, as an application of part 1, that the pure states on $\mathscr{H}_2\,(\simeq \mathbb{C}^2)$ are the points on the Bloch sphere (see Figure 5.7)

$$
S^2 = \left\{ (x_1, x_2, x_3) \in \mathbb{R}^3 \; ; \; x_1^2 + x_2^2 + x_3^2 = 1 \right\},
$$

and that the mixed states are the points in the open ball $x_1^2 + x_2^2 + x_3^2 < 1$.

Hint: In $U(2)$, use the representation:

$$
x_1 \sigma_X + x_2 \sigma_Y + x_3 \sigma_Z = \begin{pmatrix} x_3 & x_1 - ix_2 \\ x_1 + ix_2 & -x_3 \end{pmatrix}
$$

where σ_X, σ_Y, and σ_Z are the Pauli spin matrices

$$
\sigma_X = \begin{pmatrix} 0 & 1 \\ 1 & 0 \end{pmatrix}, \quad \sigma_Y = \begin{pmatrix} 0 & -i \\ i & 0 \end{pmatrix}, \quad \sigma_Z = \begin{pmatrix} 1 & 0 \\ 0 & -1 \end{pmatrix}.
$$

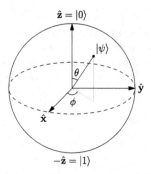

Figure 5.7: Bloch sphere

5.7.1   Normal States

More general states in physics come from the mixture of particle states, which correspond to composite system. These are called normal states in mathematics.

Let $\rho \in \mathscr{T}_1(\mathscr{H})$ = trace class operator, such that $\rho > 0$ and $tr(\rho) = 1$. Define state $s_\rho(A) := tr(A\rho)$, $A \in \mathscr{B}(\mathscr{H})$. Since ρ is compact, by spectral theorem of compact operators,

$$\rho = \sum_k \lambda_k P_k \tag{5.75}$$

such that $\lambda_1 > \lambda_2 > \cdots \to 0$; $\sum \lambda_k = 1$ and $P_k = |\xi_k\rangle\langle\xi_k|$, i.e., the rank-1 projections. (See Section 5.3.) We have

- $s_\rho(I) = tr(\rho) = 1$; and

- for all $A \in \mathscr{B}(\mathscr{H})$,

$$\begin{aligned}
s_\rho(A) = tr(A\rho) &= \sum_n \langle u_n, A\rho u_n \rangle \\
&= \sum_n \langle A^* u_n, \rho u_n \rangle \\
&= \sum_n \sum_k \lambda_k \langle A^* u_n, \xi_k \rangle \langle \xi_k, u_n \rangle \\
&= \sum_k \lambda_k \left(\sum_n \langle u_n, A\xi_k \rangle \langle \xi_k, u_n \rangle \right) \\
&= \sum_k \lambda_k \langle \xi_k, A\xi_k \rangle ;
\end{aligned}$$

where $\{u_k\}$ is any ONB in $\mathscr{H}$. Hence,

$$s_\rho = \sum_k \lambda_k s_{\xi_k} = \sum_k \lambda_k |\xi_k\rangle\langle\xi_k|$$

i.e., s_ρ is a convex combination of pure states $s_{\xi_k} := |\xi_k\rangle\langle\xi_k|$.

Remark 5.14. Notice that $tr(|\xi\rangle\langle\eta|) = \langle\eta, \xi\rangle$. In fact, take any ONB $\{e_n\}$ in $\mathcal{H}$, then

$$tr(|\xi\rangle\langle\eta|) = \sum_n \langle e_n\xi\rangle \langle\eta, e_n\rangle = \langle\eta, \xi\rangle$$

where the last step follows from Parseval identity. (If we drop the condition $\rho \geq 0$ then we get the duality $(\mathcal{T}_1\mathcal{H})^* = \mathcal{B}(\mathcal{H})$. See Theorem 5.7.)

5.7.2 A Dictionary of operator theory and quantum mechanics

- states - unit vectors $\xi \in \mathcal{H}$. These are all the pure (normal) states on $\mathcal{B}(\mathcal{H})$.

- observable - selfadjoint operators $A = A^*$

- measurement - spectrum

The spectral theorem was developed by J. von Neumann and later improved by Dirac and others. (See [Sto90, Yos95, Nel69, RS75, DS88b].) A selfadjoint operator A corresponds to a quantum observable, and result of a quantum measurement can be represented by the spectrum of A.

- simple eigenvalue: $A = \lambda|\xi_\lambda\rangle\langle\xi_\lambda|$,

$$s_{\xi_\lambda}(A) = \langle\xi_\lambda, A\xi_\lambda\rangle$$

- compact operator: $A = \sum_\lambda \lambda|\xi_\lambda\rangle\langle\xi_\lambda|$, such that $\{\xi_\lambda\}$ is an ONB of $\mathcal{H}$. If $\xi = \sum c_\lambda\xi_\lambda$ is a unit vector, then

$$s_\xi(A) = \sum_\lambda \lambda \langle\xi_\lambda, A\xi_\lambda\rangle$$

where $\{|c_\lambda|^2\}_\lambda$ is a probability distribution over the spectrum of A, and s_ξ is the expectation value of A.

- more general, allowing continuous spectrum:

$$A = \int \lambda E(d\lambda)$$
$$A\xi = \int \lambda E(d\lambda)\xi.$$

We may write the unit vector ξ as

$$\xi = \int \overbrace{E(d\lambda)\xi}^{\xi_\lambda}$$

so that

$$\|\xi\|^2 = \int \|E(d\lambda)\xi\|^2 = 1.$$

It is clear that $\|E(\cdot)\xi\|^2$ is a probability distribution on spectrum of A. $s_\xi(A)$ is again seen as the expectation value of A with respect to $\|E(\cdot)\xi\|^2$, since

$$s_\xi(A) = \langle \xi, A\xi \rangle = \int \lambda \|E(d\lambda)\xi\|^2.$$

5.8 Krein-Milman, Choquet, Decomposition of States

We study some examples of compact convex sets in locally convex topological spaces.[7] Typical examples include the set of positive semi-definite functions, taking values in $\mathbb{C}$ or $\mathscr{B}(\mathscr{H})$.

Definition 5.15. A vector space is locally convex if it has a topology which makes the vector space operators continuous, and if the neighborhoods

$$\{x + \mathrm{Nbh}_0\}$$

have a basis consisting of convex sets.

The context for Krein-Milman is locally convex topological spaces. It is in all functional analysis books. Choquet's[8] theorem however comes later, and it's not found in most books. A good reference is the book by R. Phelps [Phe01]. The proof of Choquet's theorem is not specially illuminating. It uses standard integration theory.

Theorem 5.12 (Krein-Milman). *Let K be a compact convex set in a locally convex topological space. Then K is the closed convex hull of its extreme points $E(X)$, i.e.,*

$$K = \overline{conv}(E(K)).$$

Proof. (sketch) If $K \supsetneq \overline{conv}(E(K))$, we get a linear functional w, such that w is zero on $\overline{conv}(E(K))$ and not zero on $w \in K \backslash \overline{conv}(E(K))$. Extend w by Hahn-Banach theorem to a linear functional to the whole space, and get a contradiction. $\square$

Note 2. The dual of a normed vector space is always a Banach space, so the theorem applies. The convex hull in an infinite dimensional space is not always closed, so close it. A good reference to locally convex topological space is the lovely book by F. Trèves [Trè06b].

---

[7] Almost all spaces one works with are locally convex.
[8] Appendix C includes a short biographical sketch of G. Choquet.

A convex combination of points (ξ_i) in K takes the form $v = \sum c_i \xi_i$, where $c_i > 0$ and $\sum c_i = 1$. Closure refers to taking limit, so we allow all limits of such convex combinations. Such a v is obviously in K, since K was assumed to be convex. The point of the Krein-Milman's theorem is the converse.

The decomposition of states into pure states was developed by Choquet et al; see [Phe01]. The idea goes back to Krein and Choquet.

Theorem 5.13 (Choquet). $K = S(\mathfrak{A})$ *is a compact convex set in a locally convex topological space. Let $E(K)$ be the set of extreme points on K. Then for all $p \in K$, there exists a Borel probability measure μ_p, supported on a Borel set $bE(K) \supset E(K)$, such that for all affine functions f, we have*

$$f(p) = \int_{bE(X)} f(\xi) \, d\mu_p(\xi). \tag{5.76}$$

The expression in Choquet's theorem is a generalization of convex combination. In stead of summation, it is an integral against a measure. Since there are some bizarre cases where the extreme points $E(K)$ do not form a Borel set, the measure μ_p is actually supported on $bE(K)$, such that $\mu_p(bE(K) - E(K)) = 0$.

Applications of Choquet theory and of Theorem 5.13 are manifold, and we shall discuss some of them in Chapter 8 below. Among them are applications to representations of C^*-algebras; e.g., the problem of finding "Borel-cross sections" for the set of equivalence classes of representations of a particular C^*-algebra. Equivalence here means "unitary equivalence." By a theorem of Glimm [Gli60, Gli61], we know that there are infinite simple C^*-algebras which do not admit such Borel parameterizations. Examples of this case include the Cuntz algebras $\mathcal{O}_N$, $N > 1$. Nonetheless we shall study subclasses of representations of $\mathcal{O}_N$ which correspond to sub-band filters in signal processing[9], and to pyramid algorithms for wavelet constructions. In Chapter 8 we shall also study representations of the C^*-algebra of the free group on 2 generators, as well as the C^*-algebra on two generators u and v, subject to the relation $uvu^{-1} = u^2$. It is called the Baumslag-Solitar algebra (BS_2), after Gilbert Baumslag and Donald Solitar; and it is of great importance in a more a systematic analysis of families of wavelets. It is the algebra of a Baumslag-Solitar group. In fact, there are indexed families of Baumslag-Solitar groups, given by their respective group presentation. They are examples of two-generator one-relator groups, and they play an important role in combinatorial group theory, and in geometric group theory as (counter) examples and test-cases.

Other examples of uses of Choquet theory in harmonic analysis and representation theory include such decompositions from classical analysis as Fourier transform, Laplace transform, as well as direct integral theory for representations [Sti59, Seg50].

Note 3. μ_p in (5.76) may not be unique. If it is unique, K is called a simplex. The unit disk has its boundary as extreme points. But representation of points

---

[9] A summary of some highpoints from signal processing is contained in Appendix B. See also Section 6.4 (pg. 233); and Chapters 6 (pg. 223), 9 (pg. 333).

in the interior using points on the boundary is not unique. Therefore the unit
disk is not a simplex. A tetrahedron is (Figure 5.8).

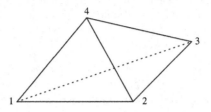

Figure 5.8: A simplex. The four extreme points are marked.

Example 5.17. Let $(X, \mathfrak{M}, \mu)$ be a measure space, where X is compact and
Hausdorff. The set of all probability measures $\mathcal{P}(X)$ is a convex set. To see
this, let $\mu_1, \mu_2 \in \mathcal{P}(X)$ and $0 \le t \le 1$, then $t\mu_1 + (1-t)\mu_2$ is a measure on X,
moreover $(t\mu_1 + (1-t)\mu_2)(X) = t + 1 - t = 1$, hence $t\mu_1 + (1-t)\mu_2 \in \mathcal{P}(X)$.
Usually we don't want all probability measures, but a closed subset.

Example 5.18. We compute extreme points in the previous example. $K =
\mathcal{P}(X)$ is compact convex in $C(X)^*$, which is identified as the set of all measures
due to Riesz. $C(X)^*$ is a Banach space hence is always convex. The importance
of being the dual of some Banach space is that the unit ball is always weak*-
compact (Banach-Alaoglu, Theorem 5.6). Note the weak*-topology is just the
cylinder/product topology. The unit ball B_1^* sits inside the infinite product
space (compact, Hausdorff) $\prod_{v \in B, \|v\|=1} D_1$, where $D_1 = \{z \in \mathbb{C} : |z| = 1\}$. The
weak $*$ topology on B_1^* is just the restriction of the product topology onto B_1^*.

Example 5.19. *Claim*: $E(K) = \{\delta_x : x \in X\}$, where δ_x is the Dirac measure
supported at $x \in X$. By Riesz, to know the measure is to know the linear
functional. $\int f d\delta_x = f(x)$. Hence we get a family of measures indexed by X.
If $X = [0, 1]$, we get a continuous family of measures. To see these really are
extreme points, we do the GNS construction on the algebra $\mathfrak{A} = C(X)$, with
the state $\mu \in \mathcal{P}(X)$. The Hilbert space so constructed is simply $L^2(\mu)$. It's
clear that $L^2(\delta_x)$ is 1-dimensional, hence the representation is irreducible. We
conclude that δ_x is a pure state, for all $x \in X$.

There is a bijection between state φ and Borel measure $\mu := \mu_\varphi$,

$$\varphi(a) = \int_X a \, d\mu_\varphi.$$

In $C(X)$, $1_{\mathfrak{A}} = \mathbb{1} = $ constant function. We check that

$$\varphi(1_{\mathfrak{A}}) = \varphi(1) = \int 1 d\mu = \mu(X) = 1$$

since $\mu \in \mathcal{P}(X)$ is a probability measure. Also, if $f \ge 0$ then $f = g^2$, with
$g := \sqrt{f}$; and

$$\varphi(f) = \int g^2 d\mu \ge 0.$$

Note 4. ν is an extreme point in $\mathcal{P}(X)$ if and only if

$$(\nu \in [\mu_1, \mu_2] = \text{convex hull of } \{\mu_1, \mu_2\}) \implies (\nu = \mu_1 \text{ or } \nu = \mu_2).$$

Example 5.20. Let $\mathfrak{A} = \mathcal{B}(\mathcal{H})$, and $S(\mathfrak{A}) = $ states of $\mathfrak{A}$. For each $\xi \in \mathcal{H}$, the map $A \mapsto w_\xi(A) := \langle \xi, A\xi \rangle$ is a state, called vector state.
Claim: $E(S) = $ vector states.
To show this, suppose W is a subspace of $\mathcal{H}$ such that $0 \subsetneq W \subsetneq \mathcal{H}$, and suppose W is invariant under the action of $\mathcal{B}(\mathcal{H})$. Then $\exists h \in \mathcal{H}, h \perp W$. Choose $\xi \in W$. The wonderful rank-1 operator (due to Dirac) $T : \xi \mapsto h$ given by $T := |h\rangle\langle\xi|$, shows that $h \in W$ (since $TW \subset W$ by assumption.) Hence $h \perp h$ and $h = 0$. Therefore $W = \mathcal{H}$. We say $\mathcal{B}(\mathcal{H})$ acts transitively on $\mathcal{H}$.

Note 5. In general, any C^*-algebra is a closed subalgebra of $\mathcal{B}(\mathcal{H})$ for some $\mathcal{H}$ (Theorem 5.4). All the pure states on $\mathcal{B}(\mathcal{H})$ are vector states.

Example 5.21. Let $\mathfrak{A}$ be a $*$-algebra, $S(\mathfrak{A})$ be the set of states on $\mathfrak{A}$. $w : \mathfrak{A} \to \mathbb{C}$ is a state on $\mathfrak{A}$ if $w(1_{\mathfrak{A}}) = 1$ and $w(A) \geq 0$, whenever $A \geq 0$. The set of completely positive (CP) maps is a compact convex set. CP maps are generalizations of states (Chapter 6).

Exercise 5.28 (Extreme measures). Take the two state sample space $\Omega = \prod_1^\infty \{0, 1\}$ with product topology. Assign probability measure, so that we might favor one outcome than the other. For example, let $s = x_1 + \cdots x_n$, $P_\theta(C_x) = \theta^s(1-\theta)^{n-1}$, i.e. s heads, $(n - s)$ tails. Notice that P_θ is invariant under permutation of coordinates. $x_1, x_2, \ldots, x_n \mapsto x_{\sigma(1)}x_{\sigma(2)}\ldots x_{\sigma(n)}$. P_θ is a member of the set of all such invariant measures (invariant under permutation) $P_{inv}(\Omega)$. Prove that

$$E(P_{inv}(\Omega)) = [0, 1]$$

i.e., P_θ are all the possible extreme points.

Remark 5.15. Let $\sigma : X \to X$ be a measurable transformation. A (probability) measure μ is *ergodic* if

$$[E \in \mathfrak{M}, \sigma E = E] \Rightarrow \mu(E) \in \{0, 1\}.$$

Intuitively, it says that the whole space X can't be divided non-trivially into parts where μ is invariant. The set X will be mixed up by the transformation σ.

Exercise 5.29 (Irrational rotation). Let $\theta > 0$ be a fixed irrational number, and set

$$\sigma_\theta(x) = \theta x \mod 1 \tag{5.77}$$

i.e., multiplication by θ modulo 1. Show that σ_θ in (5.77) is ergodic in the measure space $\mathbb{R}/\mathbb{Z} \simeq [0, 1)$ with Lebesgue measure (see Figure 5.9).

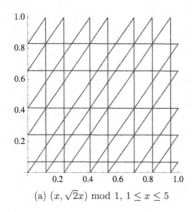
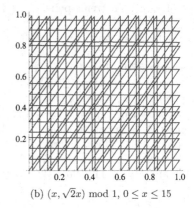

(a) $(x, \sqrt{2}x) \bmod 1$, $1 \leq x \leq 5$ (b) $(x, \sqrt{2}x) \bmod 1$, $0 \leq x \leq 15$

Figure 5.9: Irrational rotation.

5.8.1   Noncommutative Radon-Nikodym Derivative

Let w be a state on a C^*-algebra $\mathfrak{A}$, and let K be an operator in $\mathfrak{A}_+$. Set

$$w_K(A) = \frac{w(\sqrt{K}A\sqrt{K})}{w(K)}.$$

Then w_K is a state, and $w_K \ll w$, i.e., $w(A) = 0 \Rightarrow w_K(A) = 0$. We say that $K = \frac{dw}{dw_K}$ is a noncommutative Radon-Nikodym derivative.
   Check:

$$
\begin{aligned}
w_K(1) &= 1 \\
w_K(A^*A) &= \frac{w(\sqrt{K}A^*A\sqrt{K})}{w(K)} \\
&= \frac{w((A\sqrt{K})^*(A\sqrt{K}))}{w(K)} \geq 0
\end{aligned}
$$

The converse holds too [Sak98] and is called the noncommutative Radon-Nikodym theorem.

5.8.2   Examples of Disintegration

Example 5.22. $L^2(I)$ with Lebesgue measure. Let

$$F_x(t) = \begin{cases} 1 & t \geq x \\ 0 & t < x \end{cases}$$

F_x is a monotone increasing function on $\mathbb{R}$, hence by Riesz, we get the corresponding Riemann-Stieltjes measure dF_x.

$$d\mu = \int^{\oplus} dF_x(t)dx. \tag{5.78}$$

i.e.

$$\int f d\mu = \int dF_x(f)dx = \int f(x)dx.$$

Equivalently,

$$d\mu = \int \delta_x dx$$

i.e.

$$\int f d\mu = \int \delta_x(f)dx = \int f(x)dx.$$

μ is a state, $\delta_x = dF_x(t)$ is a pure state, $\forall x \in I$. This is a decomposition of state into direct integral of pure states. See [Sti59, Seg50].

Example 5.23. $\Omega = \prod_{t \geq 0} \bar{\mathbb{R}}$, $\Omega_x = \{w \in \Omega : w(0) = x\}$. Kolmogorov's inductive construction[10] of path-space measure gives rise to P_x by conditioning P with respect to "starting at x".

$$P = \int^{\oplus} P_x dx$$

i.e.,

$$P(\cdot) = \int P(\cdot \mid \text{start at } x) \, dx.$$

Example 5.24. Harmonic function on D

$$h \mapsto h(z) = \int_{\partial \mathbb{D}} \widehat{f} d\mu_z$$

Poisson integration.

5.9   Examples of C^*-algebras

Let $\mathscr{H}$ be an infinite-dimensional separable Hilbert space, and let $S : \mathscr{H} \to \mathscr{H}$ be an isometry; i.e., we have

$$S^*S = I_{\mathscr{H}}. \tag{5.79}$$

We shall be interested in the case when S is non-unitary, so the projection

$$P_S := SS^*$$

is not $I_{\mathscr{H}}$, i.e., $P_S \not\leq I_{\mathscr{H}}$.

---

[10] Appendix C includes a short biographical sketch of A. Kolmogorov.

Theorem 5.14 (Wold, see [Wol51, Con90]). *Let* $S : \mathscr{H} \to \mathscr{H}$ *be an isometry. Set*

$$\mathscr{H}_0 := \left\{ x \in \mathscr{H} : \lim_{n \to \infty} \|S^{*^n} x\| = 0 \right\}, \text{ and} \tag{5.80}$$

$$\mathscr{H}_1 := \left\{ x \in \mathscr{H} : \|S^{*^n} x\| = \|x\|, \forall n \in \mathbb{N} \right\}. \tag{5.81}$$

1. Then

$$\mathscr{H} = \mathscr{H}_0 \oplus \mathscr{H}_1, \tag{5.82}$$

where "$\oplus$" in (5.82) refers to orthogonal sum, i.e., $\mathscr{H}_0 \perp \mathscr{H}_1$.

2. $S\big|_{\mathscr{H}_0} : \mathscr{H}_0 \longrightarrow \mathscr{H}_0$ is a shift-operator;

3. $S\big|_{\mathscr{H}_1} : \mathscr{H}_1 \longrightarrow \mathscr{H}_1$ is a unitary operator in $\mathscr{H}_1$.

Exercise 5.30 (Wold's decomposition). Carry out the details in the proof of Wold's theorem.

In summary, associate to every isometry $\mathscr{H} \xrightarrow{S} \mathscr{H}$, there are three subspaces

$$
\begin{aligned}
\mathscr{H}_{\text{shift}} &= \mathscr{H}_0 \text{ in } (5.80) \\
\mathscr{H}_{\text{unit}} &= \mathscr{H}_1 \text{ in } (5.81), \text{ and} \\
\mathfrak{h} &= \ker (S^*) \text{ in } (5.84), \text{ the multiplicity space.}
\end{aligned}
$$

For the closed subspace $\mathscr{H}_0$ in the shift-part of the decomposition, it holds that $\mathscr{H}_0$ is the countable direct sum of $\mathfrak{h}$ with itself.

Exercise 5.31 (Substitution by z^N). Let $\mathscr{H} = \mathbb{H}_2 = \mathbb{H}_2(\mathbb{D})$ be the Hardy space of the disk, and let $N \in \mathbb{N}$, $N > 1$. Set

$$(Sf)(z) = f(z^N), \; f \in \mathbb{H}_2, \, z \in \mathbb{D}. \tag{5.83}$$

Show that the three closed subspaces for this isometry are as follows:

$$
\begin{aligned}
\mathscr{H}_{\text{unit}} &= \text{the constant functions on } \mathbb{D} \\
&= \mathbb{C}e_0, \; e_0(z) = z^0 = 1. \\
\mathscr{H}_{\text{shift}} &= \mathscr{H} \ominus \mathbb{C}e_0 \\
&= \{ f \in \mathbb{H}_2 : f(0) = 0 \} \\
\ker (S^*) &= \overline{span} \left\{ z^k : N \nmid k \text{ (not divisible by } N) \right\}, \text{ i.e.,} \\
&\quad \text{powers of } z^k, \; k \in (\{0\} \cup \mathbb{N}) \setminus N\mathbb{Z}, \text{ so } k \text{ not} \\
&\quad \text{divisible by } N.
\end{aligned}
$$

The isometry S in (5.83) is an example of an *isometry of infinite multiplicity*.

C^*-algebras generated by isometries. One isometric generator: With an oversimplification (see below), there is, up to C^*-isomorphism, only one C^*-algebra generated by a single non-trivial isometry. For more details see [Cob67]. By contrast, the case of multiple isometries is more subtle. Because of non-commutativity, we are then led to the study of multi-variable operator theory, and an analysis of the C^*-algebras of Toeplitz and of Cuntz. In addition to the issue of the C^*-algebras themselves, there is also the more important question of their representations. We now turn to these matters below.

Case 1. One Isometry Because of Wold's decomposition, if a C^*-algebra $\mathfrak{A}$ is generated by one isometry, we may "split off" the one generated by the unitary part; and then reduce the study to the case where $\mathfrak{A}$ is generated by a shift S.

Introduce $\mathfrak{h} := \ker(S^*)$, and

$$\mathfrak{h}^\infty = \oplus_\mathbb{N} \mathfrak{h} = \{(x, x_2, \cdots) \ : \ x_i \in \mathfrak{h}\} \tag{5.84}$$

$$\|(x_1, x_2, \cdots)\|_{\mathfrak{h}^\infty}^2 := \sum_{i=1}^\infty \|x_i\|_\mathfrak{h}^2 \ ; \tag{5.85}$$

and set

$$S_\infty(x_1, x_2, x_3, \cdots) = (0, x_1, x_2, x_3, \cdots) . \tag{5.86}$$

Exercise 5.32 (The backwards shift). Show that

$$S_\infty^*(x_1, x_2, x_3, \cdots) = (x_2, x_3, x_4, \cdots) , \text{ and that}$$

$$\|(S_\infty^*)^n x\| \xrightarrow[n \to \infty]{} 0.$$

Remark 5.16. It would seem like the backwards shift is an overly specialized example. Nonetheless it plays a big role in operator theory, see for example [AD03, MQ14, KLR09], and it is an example of a wider class of operators going by the name "the Cowen-Douglass class," playing an important role in complex geometry, see [CD78].

Exercise 5.33 (Infinite multiplicity). For the isometry $(Sf)(z) = f(z^N)$, $f \in \mathbb{H}_2$, $z \in \mathbb{D}$, write out the representation (5.84)-(5.86) above.

Exercise 5.34 (A shift is really a shift). Show that S and S_∞ are unitarily equivalent if and only if S is a shift.

Exercise 5.35 (Multiplication by z is a shift in $\mathbb{H}_2$). If $\dim \mathfrak{h} = 1$ (multiplicity one), show that S in (5.86) is unitarily equivalent to

$$\left(\widetilde{S}f\right)(z) = zf(z), \ z \in \mathbb{D}, \ f \in \mathbb{H}_2 = \text{the Hardy space.} \tag{5.87}$$

Hint: By $\mathbb{H}_2$, we mean the Hilbert space of all analytic functions f on the disk $\overline{\mathbb{D}} = \{z \in \mathbb{C} \ : \ |z| < 1\}$ such that

$$f(z) = \sum_{k=0}^\infty a_k z^k, \text{ and } (a_k) \in l^2. \tag{5.88}$$

We set $\|f\|_{\mathbb{H}_2} = \|(a_k)\|_{l^2}$.

Exercise 5.36 (The two shifts in $\mathbb{H}_2$). Show that the adjoint to the generator $\widetilde{S}$ (from (5.87)) is

$$\left(\widetilde{S}^* f\right)(z) = \frac{f(z) - f(0)}{z}, \ \forall f \in \mathbb{H}_2, \forall z \in \mathbb{D}\backslash\{0\}, \tag{5.89}$$

and

$$\left(\widetilde{S}^* f\right)(0) = f'(0), \ f \in \mathbb{H}_2.$$

Hint: Show that, if $f, g \in \mathbb{H}_2$, then the following holds:

$$\left\langle \widetilde{S}f, g \right\rangle_{\mathbb{H}_2} = \left\langle f, \widetilde{S}^* g \right\rangle_{\mathbb{H}_2} \tag{5.90}$$

where we use formula (5.89) in computing the $\mathbb{H}_2$-inner product on the RHS in (5.90).

Compare this with the result from Exercise 5.32.

Exercise 5.37 (A numerical range). Let $T := S_\infty^*$ be the backward shift (expressed in coordinates) in Exercise 5.32. Since $TT^* - T^*T$ is the rank-one projection $|e_1\rangle\langle e_1|$, of course T is not normal.

1. Show that $x_\lambda = \left(1, \lambda, \lambda^2, \lambda^3, \cdots\right)$ satisfies

$$Tx_\lambda = \lambda x_\lambda, \ \forall \lambda \in \mathbb{C}. \tag{5.91}$$

2. Since

$$x_\lambda \in l^2 \iff |\lambda| < 1, \tag{5.92}$$

conclude that the point-spectrum of T is $\mathbb{D} = \{\lambda \in \mathbb{C} : |\lambda| < 1\}$.

3. Combine (1) & (2) in order to conclude that

$$NR_T = \mathbb{D}.$$

Exercise 5.38 (The finite shift). Compare the infinite case above with the analogous matrix case

$$T_3 = \begin{bmatrix} 0 & 1 & 0 \\ 0 & 0 & 1 \\ 0 & 0 & 0 \end{bmatrix} \quad \text{and} \quad T_3' = \begin{bmatrix} 0 & 1 & 0 \\ 0 & 0 & 1 \\ 1 & 0 & 0 \end{bmatrix}.$$

A sketch of NR_{T_3} and $NR_{T_3'}$ are in Figure 5.10 below. See also Figure 5.11.

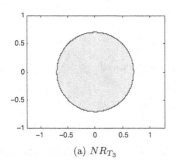

(a) NR_{T_3}

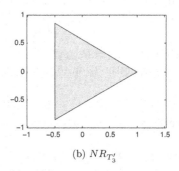

(b) $NR_{T_3'}$

Figure 5.10: The numerical range of T_3 vs T_3'.

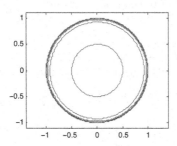

Figure 5.11: The numerical range (NR) of the truncated finite matrices: Expanding truncations of the infinite matrix T corresponding to the backward shift, and letting the size $\longrightarrow \infty$: $T_3, T_4, \cdots, T_n, T_{n+1}, \cdots$; the limit-NR fills the open disk of radius 1.

Exercise 5.39 (The Hardy space $\mathbb{H}_2$; a transform). Let $f(z) = \sum_{k=0}^{\infty} a_k z^k$, and set

$$\tilde{f}(t) = \sum_{k=0}^{\infty} a_k e^{i2\pi kt}, \; t \in \mathbb{R}. \tag{5.93}$$

Show that

$$f \in \mathbb{H}_2 \Longleftrightarrow \tilde{f} \in L^2(\mathbb{T}), \; \mathbb{T} = \partial \mathbb{D};$$

and that

$$\left\| \tilde{f} \right\|_{L^2(\mathbb{T})} = \|f\|_{\mathbb{H}_2} \tag{5.94}$$

holds.

Because of Exercise 5.39, we may identify $\mathbb{H}_2$ with a closed subspace in $L^2(\mathbb{T})$. Let P_+ denote the projection of $L^2(\mathbb{T})$ onto $\mathbb{H}_2$.

Definition 5.16. For $\varphi \in L^{\infty}(\mathbb{T})$, set

$$T_{\varphi}f = P_{+}(\varphi f), \ \forall f \in \mathbb{H}_{2}, \tag{5.95}$$

equivalently, $T_{\varphi} = P_{+}M_{\varphi}P_{+}$.

The operator T_{φ} in (5.95) is called a *Toeplitz-operator*; and

$$\mathscr{T} := C^{*}(\{T_{\varphi} \ : \ \varphi \in L^{\infty}(\mathbb{T})\}) \tag{5.96}$$

is called the *Toeplitz-algebra*.

Exercise 5.40 (Multiplicity-one and $\mathbb{H}_{2}$). Show that there is a short exact sequence (in the category of C^{*}-algebras):

$$0 \longrightarrow \mathscr{K} \longrightarrow \mathscr{T} \overset{\pi}{\longrightarrow} \mathscr{T}/\mathscr{K} \longrightarrow 0 \tag{5.97}$$

where $\mathscr{K}$ = the C^{*}-algebra of compact operators; and $\mathscr{T}/\mathscr{K}$ is the quotient; finally

$$\mathscr{T}/\mathscr{K} \simeq L^{\infty}(\mathbb{T}),$$

realized via the mapping $T_{\varphi} \overset{\pi}{\longrightarrow} \varphi$ (the symbol mapping) in (5.97), i.e., $\pi(T_{\varphi}) := \varphi$, is assigning the symbol φ to the Toeplitz operator T_{φ}.

Exercise 5.41 (The Toeplitz matrices). Suppose $\varphi \in L^{\infty}(\mathbb{T})$ has Fourier expansion

$$\varphi(t) = \sum_{n \in \mathbb{Z}} b_{n} e^{i2\pi nt}, \ t \in \mathbb{R}. \tag{5.98}$$

Then show that the $\infty \times \infty$ matrix of the corresponding Toeplitz operator T_{φ} is as follows w.r.t the standard ONB in $\mathbb{H}_{2}$, $\{z^{n} \ : \ n \in \{0\} \cup \mathbb{N}\}$.

$$Mat(T_{\varphi}) = \begin{bmatrix} b_{0} & b_{-1} & b_{-2} & b_{-3} & b_{-4} & \cdots & \cdots & \cdots & \cdots & \cdots \\ b_{1} & b_{0} & b_{-1} & b_{-2} & b_{-3} & \cdots & \cdots & \cdots & \cdots & \cdots \\ b_{2} & b_{1} & b_{0} & b_{-1} & b_{-2} & \cdots & \cdots & \cdots & \cdots & \cdots \\ b_{3} & b_{2} & b_{1} & b_{0} & b_{-1} & \cdots & \cdots & \cdots & \cdots & \cdots \\ b_{4} & b_{3} & b_{2} & b_{1} & b_{0} & \cdots & \cdots & \cdots & \cdots & \cdots \\ \ddots & \ddots & \ddots & \ddots & \ddots & \ddots & \ddots & \ddots & \cdots & \cdots \\ \ddots & \ddots & \ddots & \ddots & \ddots & \ddots & b_{0} & b_{-1} & b_{-2} & \cdots \\ \ddots & \ddots & \ddots & \ddots & \ddots & \ddots & b_{1} & b_{0} & b_{-1} & \cdots \\ \ddots & \ddots & \ddots & \ddots & \ddots & \ddots & b_{2} & b_{1} & b_{0} & \ddots \\ \ddots & \ddots & \ddots & \ddots & \ddots & \ddots & \ddots & \ddots & \ddots & \ddots \end{bmatrix} \tag{5.99}$$

Note 6. Matrices of the form given in (5.99) are called *Toeplitz matrices*, i.e., with the banded pattern, constant numbers down the diagonal lines, with b_{0} in the main diagonal.

Remark 5.17. Note that the mapping $\varphi \longrightarrow T_\varphi$ (Toeplitz), $L^\infty(\mathbb{T}) \to \mathscr{T}$ is not a homomorphism of the algebra $L^\infty(\mathbb{T})$ into $\mathscr{T} = C^*(\{T_\varphi\})$. Here we view $L^\infty(\mathbb{T})$ as an Abelian C^*-algebra under pointwise product, i.e.,

$$(\varphi_1\varphi_2)(t) := \varphi_1(t)\,\varphi_2(t), \ \forall t \in \mathbb{R}/\mathbb{Z}.$$

The point of mapping of the *short exact sequence* (lingo from homological algebra):

$$0 \longrightarrow \mathscr{K} \longrightarrow \mathscr{T} \longrightarrow L^\infty(\mathbb{T}) \longrightarrow 0 \tag{5.100}$$

is that $\varphi \longrightarrow T_\varphi$ is only a "homomorphism mod $\mathscr{K}$ (= the compact operators)", i.e., that we have

$$T_{\varphi_1}T_{\varphi_2} - T_{\varphi_1\varphi_2} \in \mathscr{K} \tag{5.101}$$

valid for all $\varphi_1, \varphi_2 \in L^\infty(\mathbb{T})$.

There is an extensive literature on (5.100) and (5.101), see especially [Dou80].

Exercise 5.42 (Homomorphism mod $\mathscr{K}$). Give a direct proof that the operator on the LHS in (5.101) is a compact operator in $\mathbb{H}_2$.

Remark 5.18. The subject of Toeplitz operators, and Toeplitz algebras is vast (see e.g., [AZ07]). The more restricted case where the symbol φ of $T_\varphi = P_+M_\varphi P_+$ is continuous (i.e., $\varphi \in C(S^1)$, $S^1 = $ the circle) is especially rich; starting with *Szegő's Index Theorem*:

Definition 5.17. Let X and Y be Banach spaces. A *Fredholm operator* is a bounded linear operator $T : X \to Y$, such that $\ker(T)$ and $\ker(T^*)$ are finite-dimensional, and $\mathrm{ran}(T)$ is closed. The *index* of T is given as

$$ind(T) := \dim(\ker(T)) - \dim(\ker(T^*)).$$

(The assumption on the range of T in the definition is redundant [AA02].)

Theorem 5.15 (Szegő [BS94]). *If $\varphi \in C(S^1)$ and φ does not vanish on S^1, then T_φ is Fredholm, and the index of T_φ computes as follows:*

$$ind(T_\varphi) = \dim(\ker(T_\varphi)) - \dim(\ker(T_\varphi^*)) = -\#w(\varphi) \tag{5.102}$$

where $\#w(\varphi)$ in (5.102) is the winding number

$$\#w(\varphi) = \frac{1}{2\pi i}\int_0^{2\pi} \frac{\varphi'(e^{i\theta})}{\varphi(e^{i\theta})}\,d\theta. \tag{5.103}$$

Note: $\#w(e^{in\theta}) = n$, *for* $n \in \mathbb{Z}$, *and* $\ker(T_\varphi^*) = (\mathrm{ran}(T_\varphi))^\perp$.

Case 2. Multiple Isometries Here we refer to the Cuntz-algebra $\mathscr{O}_N$ (see [Cun77]), the unique C^*-algebra $\mathscr{O}_N$, $N > 1$, generated by $\{S_i\}_{i=1}^N$ and the relations

$$S_i^*S_j = \delta_{ij}, \text{ and} \tag{5.104}$$

$$\sum_{i=1}^{N} S_i S_i^* = \mathbf{1}. \tag{5.105}$$

Cuntz showed ([Cun77]) that this is a simple C^*-algebra (i..e, no non-trivial closed two-sided ideals), purely infinite.

We shall return to the study of its *representation* in Chapter 8.

Exercise 5.43 (An element in $Rep\,(\mathcal{O}_N, \mathbb{H}_2)$). Fix $N \in \mathbb{N}$, $N > 1$, and consider the following operators $\{S_k\}_{k=0}^{N-1}$ acting in the Hardy space $\mathbb{H}_2 = \mathbb{H}_2\,(\mathbb{D})$:

$$(S_k f)\,(z) = z^k f\left(z^N\right), \ \forall f \in \mathbb{H}_2, \ \forall z \in \mathbb{D}, \ k = 0, 1, \ldots, N - 1. \tag{5.106}$$

Show that the operators (S_k) in (5.106) satisfy the $\mathcal{O}_N$-relations (5.104)-(5.105), i.e., that

$$S_j^* S_k = \delta_{jk} I_{\mathbb{H}_2}, \text{ and}$$

$$\sum_{j=0}^{N-1} S_j S_j^* = I_{\mathbb{H}_2};$$

hence a *representation* of $\mathcal{O}_N$ in $\mathbb{H}_2$.

Exercise 5.44 (The multivariable Toeplitz algebra). For $k \in \mathbb{N}$, set $\mathscr{H}_k = \mathbb{C}^k$ = the k-dimensional complex Hilbert space with the usual inner product:

$$\langle v, w \rangle = \sum_{j=1}^{k} \overline{v_j} w_j. \tag{5.107}$$

For $k = 1$, pick a normalized basis vector Ω. For $N > 1$, set

$$\mathscr{F}\,(\mathscr{H}_N) = \mathscr{H}_1 \oplus \sum_{n=1}^{\infty} \oplus \mathscr{H}_N^{\otimes n}. \tag{5.108}$$

(The letter $\mathscr{F}$ is for Fock-space.) For $f \in \mathscr{H}_N$, set:

$$T_f\,(\otimes_1^n h_j) \ = \ f \otimes (\otimes_1^n h_j), \text{ and} \tag{5.109}$$

$$T_f^*\,(\otimes_1^n h_j) \ = \ \langle f, h_1 \rangle \otimes_2^n h_j, \ n \in \mathbb{N}. \tag{5.110}$$

And finally, the vacuum rule:

$$T_f^* \Omega = 0. \tag{5.111}$$

1. Show that the following hold:

$$T_f^* T_g = \langle f, g \rangle_N \, I_{\mathscr{F}(\mathscr{H}_N)}, \ \forall f, g \in \mathscr{H}_N. \tag{5.112}$$

Define T_i and T_i^* from and ONB in $\mathscr{H}_N$, we get

$$\sum_{i=1}^{N} T_i T_i^* = I_{\mathscr{F}(\mathscr{H}_N)} - |\Omega \rangle \langle \Omega|. \tag{5.113}$$

The C^*-algebra generated by $\{T_f \,: \, f \in \mathscr{H}_N\}$ is called the (multivariable) *Toeplitz algebra*, and is denoted $\mathscr{T}_N$.

2. Show, with the use of (5.112)-(5.113), that there is a natural short exact sequence of C^*-algebras:

$$0 \longrightarrow \mathscr{K} \longrightarrow \mathscr{T}_N \longrightarrow \mathscr{O}_N \longrightarrow 0.$$

Compare with (5.100) in Remark 5.17.

5.10 Examples of Representations

We consider the Fourier algebra.

1. Discrete case: $l^1(\mathbb{Z})$ and the Gelfand transform

$$
\begin{aligned}
(a * b)_n &= \sum_k a_k b_{n-k} \\
(a^*)_n &= \overline{a_{-n}} \\
1_{\mathfrak{A}} &= \delta_0
\end{aligned}
$$

$$a \xrightarrow[\text{Gelfand}]{\mathcal{F}} F(z) := \sum_n a_n z^n.$$

We may specialize to $z = e^{it}$, $t \in \mathbb{R}$ mod 2π. $\{F(z)\}$ is an Abelian algebra of functions, with multiplication is given by

$$F(z)G(z) = \sum_n (a * b)_n z^n.$$

In fact, most Abelian algebras can be thought of as function algebras.

Homomorphism:

$$
\begin{aligned}
(l^1, *) &\xrightarrow{\mathcal{F}} C(\mathbb{T}^1) \\
(a_n) &\mapsto F(z).
\end{aligned}
$$

If we want to write $F(z)$ as power series, then we need to drop a_n for $n < 0$. Then $F(z)$ extends to an analytic function over the unit disk. The representation by the sequence space

$$\{a_0, a_1, \dots\}$$

was suggested by Hardy. We set

$$\|F\|_{\mathbb{H}_2}^2 = \sum_{k=0}^{\infty} |a_k|^2 \, ;$$

the natural isometric isomorphism. Rudin has two nice chapters on H^2, as a Hilbert space, a RKHS. See [Rud87, ch16].

2. Continuous case: $L^1(\mathbb{R})$

$$(f * g)(x) = \int_{-\infty}^{\infty} f(s)g(x-s)ds$$
$$f^*(x) = \overline{f(-x)}.$$

The algebra L^1 has no identity, but we may always insert one by adding δ_0. So δ_0 is the homomorphism $f \longmapsto f(0)$; and $L^1(\mathbb{R}) \cup \{\delta_0\}$ is again a Banach $*$-algebra.

The *Gelfand map* is the classical Fourier transform, i.e.,

$$f \xrightarrow[\text{Gelfand}]{\mathcal{F}} \hat{f}(\xi) = \int_{-\infty}^{\infty} f(x) e^{-i\xi x} dx \qquad (5.114)$$

where $\widehat{f * g} = \hat{f}\hat{g}$.

Remark 5.19. $C(\mathbb{T}^1)$ is called the C^*-algebra completion of l^1. $L^\infty(X, B, \mu) = L^1(\mu)^*$ is also a C^*-algebra. It is a W^*-algebra, or von Neumann algebra (see, e.g., [Sak98]). The W^* refers to the fact that its topology comes from the weak $*$-topology. Recall that $\mathcal{B}(\mathscr{H})$, for any Hilbert space, is a von Neumann algebra.

Example 5.25. Fix φ and set $uf = e^{i\theta}f(\theta)$, $vf = f(\theta - \varphi)$, restrict to $[0, 2\pi]$, i.e., 2π periodic functions.

$$vuv^{-1} = e^{i\varphi}u$$
$$vu = e^{i\varphi}uv$$

u, v generate a noncommutative C^*-algebra. See [EN12, Boc08].

Example 5.26 (Quantum Mechanics). Consider the canonical commutation relation

$$[p, q] = -iI, \quad i = \sqrt{-1},$$

where $[x, y] := xy - yx$ denotes the commutator of x and y .

The two symbols p, q generate an algebra, but they *can not* be represented by bounded operators. But we may apply bounded functions to them and get a C^*-algebra.

Exercise 5.45 (No bounded solutions to the canonical commutation relations). Show that p, q and not be represented by bounded operators. <u>Hint</u>: Take the trace.

Remark 5.20 (CAR and CCR). There are two $*$-algebras built functorially from a fixed (single) Hilbert space $\mathscr{H}$; often called the one-particle Hilbert space (in physics). The dimension $\dim \mathscr{H}$ is called *the number of degrees of freedom*. The case of interest here is when $\dim \mathscr{H} = \aleph_0$ (countably infinite). The two

*-algebras are called the CAR, and the CCR algebras, and they are extensively studied; see e.g., [BR81b]. Of the two, only $CAR(\mathscr{H})$ is a C^*-algebra. The operators arising from representations of $CCR(\mathscr{H})$ will be *unbounded*, but still having a common dense domain in the respective representation Hilbert spaces. In both cases, we have a Fock representation; and for $CAR(\mathscr{H})$ it is realized on the anti-symmetric Fock space, while for $CCR(\mathscr{H})$, it is realized in the symmetric Fock space $\Gamma_{sym}(\mathscr{H})$. There are many other representations, inequivalent to the respective Fock representations.

Let A and B be elements in a *-algebra, and introduce:

$$[A, B] = AB - BA = \text{the commutator; and} \tag{5.115}$$
$$\{A, B\} = AB + BA = \text{the anti-commutator.} \tag{5.116}$$

Let $\mathscr{H}$ be as above, then $CAR(\mathscr{H})$ is generated (as a unital C^*-algebra) by axiomatically assigned elements, $a_+(h)$, $a_+^*(h)$, $h \in \mathscr{H}$, and the relations (CAR):

$$\begin{aligned}
\{a_+(h), a_+(k)\} &= 0, \quad \forall h, h \in \mathscr{H}, \quad \text{and} \\
\{a_+(h), a_+^*(k)\} &= \langle h, k \rangle_{\mathscr{H}} \mathbb{1}.
\end{aligned} \tag{5.117}$$

By contrast, $CCR(\mathscr{H})$ is generated by a system (axiomatic), $a_-(h)$, $a_-^*(h)$, $h \in \mathscr{H}$, subject to the CCRs:

$$\begin{aligned}
[a_-(h), a_-(k)] &= 0, \quad \forall h, h \in \mathscr{H}, \quad \text{and} \\
[a_-(h), a_-^*(k)] &= \langle h, k \rangle_{\mathscr{H}} \mathbb{1}.
\end{aligned} \tag{5.118}$$

The CAR in $CAR(\mathscr{H})$ from (5.117) is for "Canonical Anti-commutation relations", while the CCR in $CCR(\mathscr{H})$ from (5.118) is for "Canonical Commutation relations"; and $CAR(\mathscr{H})$ is called the CAR-algebra, while $CCR(\mathscr{H})$ is called the CCR-algebra. Both of them are *-algebras, but only $CAR(\mathscr{H})$ is a C^*-algebra; see also [BR81b].

The respective *Fock States* $\omega_{Fock}^{\pm}$ on the two *-algebras are specified as follows:

$$\omega_{Fock}^+ \left(a_+(h) a_+^*(k) \right) = \langle h, k \rangle_{\mathscr{H}}, \quad \text{and} \tag{5.119}$$
$$\omega_{Fock}^- \left(a_-(h) a_-^*(k) \right) = \langle h, k \rangle_{\mathscr{H}}. \tag{5.120}$$

In both cases, we have

$$\omega_{Fock}^{\pm} \left(a_{\pm}^*(h) a_{\pm}(h) \right) = 0, \quad \forall h \in \mathscr{H}; \tag{5.121}$$

the vacuum property.

For the corresponding Fock representations $\pi_{\pm}$ we have:

$$\left\{ \pi_+(h), \pi_+^*(k) \right\} = \langle h, k \rangle_{\mathscr{H}} I_{\Gamma_{antisym}(\mathscr{H})}, \quad \text{and} \tag{5.122}$$
$$\left[\pi_-(h), \pi_-^*(k) \right] = \langle h, k \rangle_{\mathscr{H}} I_{\Gamma_{sym}(\mathscr{H})}, \tag{5.123}$$

where I on the RHS of the formulas refers to the respective identity operators, indicated with subscripts.

Exercise 5.46 (The CAR($\mathcal{H}$)-Fock-state). Show that the two rules (5.119) and (5.121) for the CAR-Fock state in fact determine a state on the C^*-algebra CAR($\mathcal{H}$).

Hint: Show that, for all $n, m \in \mathbb{N}$, and all system $h_1, \cdots, h_n, k_1, \cdots, k_m$ of vectors in $\mathcal{H}$, the state ω^+_{Fock} must satisfy

$$\omega^+_{Fock}\left(a\left(h_1\right) \cdots a\left(h_n\right) a^*\left(k_1\right) \cdots a^*\left(k_m\right)\right)$$
$$= \delta_{n,m} \det\left(\left[\langle h_i, k_j \rangle_{\mathcal{H}}\right]^n_{i,j=1}\right) \tag{5.124}$$

where the $n \times n$ matrix on the RHS in (5.124) is

$$\begin{bmatrix} \langle h_1, k_1 \rangle_{\mathcal{H}} & \cdots & \langle h_1, k_n \rangle_{\mathcal{H}} \\ \vdots & & \vdots \\ \langle h_n, k_1 \rangle_{\mathcal{H}} & \cdots & \langle h_n, k_n \rangle_{\mathcal{H}} \end{bmatrix},$$

called the Gramian.

Example 5.27. Let $\mathcal{H}$ be an infinite dimensional Hilbert space, then $\mathcal{H}$ is isometrically isomorphic to a proper subspace of itself. For example, let $\{e_n\}$ be an ONB. $\mathcal{H}_1 = \overline{span}\{e_{2n}\}$, $\mathcal{H}_2 = \overline{span}\{e_{2n+1}\}$. Let

$$V_1(e_n) = e_{2n}$$
$$V_2(e_n) = e_{2n+1}$$

then we get two isometries. Also,

$$V_1 V_1^* + V_2 V_2^* = I$$
$$V_i^* V_i = I$$
$$V_i V_i^* = P_i$$

where P_i is a selfadjoint *projection*, $i = 1, 2$ onto the respective $\mathcal{H}_i$. This is the Cuntz algebra $\mathcal{O}_2$. More general $\mathcal{O}_N$, $N > 2$.

Cuntz (in 1977) showed that this is a *simple* C^*-algebra, i.e., it does not have non-trivial closed two-sided ideals. For studies of its representations, see, e.g., [Gli60, Gli61, BJO04].

5.11   Beginning of Multiplicity Theory

The theme to follow below is an outline of multiplicity theory. But our discussion will be limited to a comparison of commutative and non-commutative multiplicity theory (MT). To appreciate the issues, the reader will recall that in finite dimensions, and in commutative MT, the question of MT is well understood; it simply refers to the eigenvalue spectrum of selfadjoint (or normal) matrices (see Examples 5.35-5.36). In this case, each eigenvector then occurs with finite multiplicity. In infinite dimensions, however, the situation is a lot more subtle: Nonetheless, for selfadjoint operators in infinite-dimensional Hilbert space (see Chapter 4), commutative MT deals with the analogous issue, only here "spectrum" is much more subtle, encompassing: point spectrum vs continuous spectrum, absolutely continuous spectrum vs singular continuous spectrum, and a discussion of families of associated spectral measures which might be mutually singular, see Section 4.2. In this setting, MT entails equivalence classes of Borel measures on $\mathbb{R}$. Moreover, the case of continuous spectrum entails direct integrals in the category of Hilbert space, see e.g., [Nel69].

Recall that the spectrum of a selfadjoint operator is real, and selfadjoint operators may be realized, up to unitary equivalence, as multiplication operators in Hilbert spaces $L^2(\mu)$ for suitable measures μ (the Spectral Representation Theorem, Theorem 4.1). As a consequence, even the case of commutative MT for selfadjoint operators in Hilbert space entails a classification of Borel measures μ on the real line $\mathbb{R}$; itself a highly non-trivial issue.

And passing to non-commutative MT, we are faced instead with the study of unitary representations of non-abelian Lie groups (or, more generally, of locally compact groups). In the non-commutative case, in the corresponding direct integral decompositions, the irreducible representations play the role of eigenvectors, or generalized eigenvectors, from the commutative MT counterpart. In the non-commutative case, a useful notion of multiplicity entails an identification of cyclic representations and their associated commutants. Note that we shall resume the study of unitary representations of non-abelian groups in Chapter 8 below.

Since representations of groups may be understood from associated representations of the corresponding (associative) group-algebras, our discussion below will begin with the latter generality.

The main question here is how to break up a representation into smaller ones. The smallest are the irreducible representations, and the next would be the multiplicity free representations.

Let $\mathfrak{A}$ be an algebra.

- *commutative*: e.g., function algebras

- *non-commutative*: e.g., matrix algebra, algebras generated by representation of non-Abelian groups

Smallest representation:

- *irreducible*: $\pi \in Rep_{irr}(\mathfrak{A}, \mathscr{H})$, where the commutant $\pi(\mathfrak{A})'$ is 1-dimensional This is the starting point of further analysis.

- *multiplicity free*: Let $\pi \in Rep(\mathfrak{A}, \mathscr{H})$. We may assume π is cyclic, since otherwise π can be decomposed into a direct sum of cyclic representations, i.e., $\pi = \oplus \pi_{cyc}$; see Theorem 5.2. Then,

$$\pi \text{ is multiplicity free} \iff \pi(\mathfrak{A})' \text{ is abelian.}$$

Fix a Hilbert space $\mathscr{H}$, and let $\mathfrak{C}$ be a $*$-algebra in $\mathscr{B}(\mathscr{H})$. The commutant $\mathfrak{C}'$ is given by

$$\mathfrak{C}' = \{X \in \mathscr{B}(\mathscr{H}) : XC = CX, \, \forall C \in \mathfrak{C}\}.$$

The commutant $\mathfrak{C}'$ is also a $*$-algebra, and

$$\mathfrak{C} \text{ is abelian} \iff \mathfrak{C} \subset \mathfrak{C}'.$$

Note that $\mathfrak{C} \subset \mathfrak{C}''$ (double-commutant.)

Theorem 5.16 (von Neumann's double commutant theorem). *If M is a von Neumann algebra, then $M = M''$.*

Proof. See, e.g., [BR79, KR97a].   □

Definition 5.18. Let $\pi \in Rep(\mathfrak{A}, \mathscr{H})$. We say that π has *multiplicity* n, $n \in \{0\} \cup \mathbb{N}$, if $\pi(\mathfrak{A})' \simeq M_n(\mathbb{C})$, i.e., the commutant $\pi(\mathfrak{A})'$ is $*$-isomorphic to the algebra of all $n \times n$ complex matrices. π is said to be *multiplicity-free* if $\pi(\mathfrak{A})' \simeq \mathbb{C}I_{\mathscr{H}}$.

Example 5.28. Let

$$A = \begin{bmatrix} 1 & & \\ & 1 & \\ & & 2 \end{bmatrix} = \begin{bmatrix} I_2 & 0 \\ 0 & 2 \end{bmatrix}.$$

Let $C \in M_3(\mathbb{C})$, then $AC = CA$ if and only if C has the form

$$C = \begin{bmatrix} a & b & \\ c & d & \\ & & 1 \end{bmatrix} = \begin{bmatrix} B & 0 \\ 0 & 1 \end{bmatrix}$$

where $B \in M_2(\mathbb{C})$.

Let A be a linear operator (not necessarily bounded) acting in the Hilbert space $\mathscr{H}$. By the Spectral Theorem (Chapter 4), we have $A = A^*$ if and only if

$$A = \int_{sp(A)} \lambda P_A(d\lambda);$$

where P_A is the corresponding projection-valued measure (PVM).

Example 5.29. The simplest example of a PVM is when $\mathscr{H} = L^2(X, \mu)$, for some compact Hausdorff space X, and $P(\omega) := \chi_\omega$, for all Borel subsets ω in X. Indeed, the Spectral Theorem states that *all* PVMs come this way.

Example 5.30. Let A be compact and selfadjoint. We may further assume that A is positive, $A \geq 0$, in the usual order of Hermitian operators (i.e., $\langle x, Ax \rangle \geq 0$, $\forall x \in \mathscr{H}$.) Then by Theorem 4.6, A has the decomposition

$$A = \sum_{n=1}^{\infty} \lambda_n P_n \tag{5.125}$$

where $\lambda'_n s$ are the eigenvalues of A, such that $\lambda_1 \geq \lambda_2 \geq \cdots \lambda_n \to 0$; and $P'_n s$ are the selfadjoint projections onto the (finite dimensional) eigenspace of λ_n. In this case, the projection-valued measure P_A is supported on $\mathbb{N}$, and $P_A(\{n\}) = P_n$, $\forall n \in \mathbb{N}$.

In (5.125), we may arrange the eigenvalues as follows:

$$\overbrace{\lambda_1 = \cdots = \lambda_1}^{s_1} > \overbrace{\lambda_2 = \cdots = \lambda_2}^{s_2} > \cdots > \overbrace{\lambda_n = \cdots = \lambda_n}^{s_n} > \cdots \to 0. \qquad (5.126)$$

We say that λ_i has multiplicity s_i, i.e., the dimension of the eigenspace of λ_i. Note that

$$\dim \mathscr{H} = \sum_{i=1}^{\infty} s_i.$$

Question: What does A look like if it is represented as the operator of multiplication by the independent variable?

Example 5.31. Let $s_1, s_2, \ldots$ be a sequence in $\mathbb{N}$, set

$$E_k = \left\{ x_1^{(k)}, \ldots, x_{s_k}^{(k)} \right\} \subset \mathbb{C}, \text{ and } E = \bigcup_{k=1}^{\infty} E_k.$$

Let $\mathscr{H} = l^2(E)$, and

$$f := \sum_{k=1}^{\infty} \lambda_k \chi_{E_k}, \text{ s.t. } \lambda_1 > \lambda_2 > \cdots > \lambda_n \to 0.$$

Let We represent A as the operator M_f of multiplication by f on $L^2(X, \mu)$. Let $E_k = \{x_{k,1}, \ldots, x_{k,s_k}\} \subset X$, and let $\mathscr{H}_k = span\{\chi_{\{x_{k,j}\}} : j \in \{1, 2, \ldots, s_k\}\}$. Let Notice that χ_{E_k} is a rank s_1 projection. M_f is compact if and only if it is of the given form.

Example 5.32. Follow the previous example, we represent A as the operator M_t of multiplication by the independent variable on some Hilbert space $L^2(\mu_f)$. For simplicity, let $\lambda > 0$ and

$$f = \lambda \chi_{\{x_1, x_2\}} = \lambda \chi_{\{x_1\}} + \lambda \chi_{\{x_2\}}$$

i.e. f is compact since it is λ times a rank-2 projection; f is positive since $\lambda > 0$. The eigenspace of λ has two dimension,

$$M_f \chi_{\{x_i\}} = \lambda \chi_{\{x_i\}}, \quad i = 1, 2.$$

Define $\mu_f(\cdot) = \mu \circ f^{-1}(\cdot)$, then

$$\mu_f = \mu(\{x_1\})\delta_\lambda \oplus \mu(\{x_2\})\delta_\lambda \oplus \text{cont. sp } \delta_0$$

and

$$L^2(\mu_f) = L^2(\mu(\{x_1\})\delta_\lambda) \oplus L^2(\mu(\{x_2\})\delta_\lambda) \oplus L^2(\text{cont. sp } \delta_0).$$

Define $U : L^2(\mu) \to L^2(\mu_f)$ by

$$(Ug) = g \circ f^{-1}.$$

U is unitary, and the following diagram commute:

$$
\begin{array}{ccc}
L^2(X,\mu) & \xrightarrow{\;M_f\;} & L^2(X,\mu) \\
\downarrow{\scriptstyle U} & & \downarrow{\scriptstyle U} \\
L^2(\mathbb{R},\mu_f) & \xrightarrow{\;M_t\;} & L^2(\mathbb{R},\mu_f)
\end{array}
$$

To check U preserves the L^2-norm,

$$
\begin{aligned}
\|Ug\|^2 &= \int \left\| g \circ f^{-1}(\{x\}) \right\|^2 d\mu_f \\
&= \left\| g \circ f^{-1}(\{\lambda\}) \right\|^2 + \left\| g \circ f^{-1}(\{0\}) \right\|^2 \\
&= |g(x_1)|^2 \, \mu(\{x_1\}) + |g(x_2)|^2 \, \mu(\{x_2\}) + \int_{X\setminus\{x_1,x_2\}} |g(x)|^2 \, d\mu \\
&= \int_X |g(x)|^2 \, d\mu.
\end{aligned}
$$

To see U diagonalizes M_f,

$$
\begin{aligned}
M_t Ug &= \lambda g(x_1) \oplus \lambda g(x_2) \oplus 0 g(t)\chi_{X\setminus\{x_1,x_2\}} \\
&= \lambda g(x_1) \oplus \lambda g(x_2) \oplus 0 \\
UM_f g &= U(\lambda g(x)\chi_{\{x_1,x_2\}}) \\
&= \lambda g(x_1) \oplus \lambda g(x_2) \oplus 0.
\end{aligned}
$$

Thus

$$M_t U = U M_f.$$

Remark 5.21. Notice that f should really be written as

$$f = \lambda \chi_{\{x_1,x_2\}} = \lambda \chi_{\{x_1\}} + \lambda \chi_{\{x_2\}} + 0 \chi_{X\setminus\{x_1,x_2\}}$$

since 0 is also an eigenvalue of M_f, and the corresponding eigenspace is the kernel of M_f.

Example 5.33. diagonalize M_f on $L^2(\mu)$ where $f = \chi_{[0,1]}$ and μ is the Lebesgue measure on $\mathbb{R}$.

Example 5.34. diagonalize M_f on $L^2(\mu)$ where

$$
f(x) = \begin{cases} 2x & x \in [0,1/2] \\ 2 - 2x & x \in [1/2,1] \end{cases}
$$

and μ is the Lebesgue measure on $[0,1]$.

Remark 5.22. see direct integral and disintegration of measures.

In general, let A be a selfadjoint operator acting on $\mathscr{H}$. Then there exists a second Hilbert space K, a measure ν on $\mathbb{R}$, and unitary transformation F : $\mathscr{H} \to L^2_K(\mathbb{R}, \nu)$ such that
$$M_t F = FA$$
for measurable function $\varphi : \mathbb{R} \to K$,
$$\|\varphi\|_{L^2_K(\nu)} = \int \|\varphi(t)\|^2_K \, d\nu(t) < \infty.$$

Examples that do have multiplicities in finite dimensional linear algebra:

Example 5.35. 2-d, λI, $\{\lambda I\}' = M_2(\mathbb{C})$ which is not Abelian. Hence $mult(\lambda) = 2$.

Example 5.36. 3-d,
$$\begin{bmatrix} \lambda_1 & & \\ & \lambda_1 & \\ & & \lambda_2 \end{bmatrix} = \begin{bmatrix} \lambda_1 I & \\ & \lambda_2 \end{bmatrix}$$
where $\lambda_1 \neq \lambda_2$. The commutant is
$$\begin{bmatrix} B & \\ & b \end{bmatrix}$$
where $B \in M_2(\mathbb{C})$, and $b \in \mathbb{C}$. Therefore the commutant is isomorphic to $M_2(\mathbb{C})$, and multiplicity is equal to 2.

Example 5.37. The example of M_φ with repetition.
$$M_\varphi \oplus M_\varphi : L^2(\mu) \oplus L^2(\mu) \to L^2(\mu) \oplus L^2(\mu)$$
$$\begin{bmatrix} M_\varphi & \\ & M_\varphi \end{bmatrix} \begin{bmatrix} f_1 \\ f_2 \end{bmatrix} = \begin{bmatrix} \varphi f_1 \\ \varphi f_2 \end{bmatrix}$$
the commutant is this case is isomorphic to $M_2(\mathbb{C})$. If we introduces tensor product, then representation space is also written as $L^2(\mu) \otimes V_2$, the multiplication operator is amplified to $M_\varphi \otimes I$, whose commutant is represented as $I \otimes V_2$. Hence it's clear that the commutant is isomorphic to $M_2(\mathbb{C})$. To check
$$\begin{aligned} (\varphi \otimes I)(I \otimes B) &= \varphi \otimes B \\ (I \otimes B)(\varphi \otimes I) &= \varphi \otimes B. \end{aligned}$$

A summary of relevant numbers from the Reference List

For readers wishing to follow up sources, or to go in more depth with topics above, we suggest: [Arv76, BR79, Mac85, BD91, Dou80, Cob67, BJ97b, BJ97a, GJ87, AD03, Alp01, BJ02, Cun77, Con90, Dix81, Gli61, KR97a, KR97b, Seg50, Sak98, Tay86, MJD+15, Hal13, Hal15, KS03].

5.A   The Fock-state, and Representation of CCR, Realized as Malliavin Calculus

We now resume our analysis of the representation of the canonical commutation relations (CCR)-algebra induced by the canonical Fock state; see eq (5.118) in Remark 5.20 above. In our analysis below, we shall make use of the following details from Chapters 3, 4, 5, and 7: Brownian motion, Itō-integrals, and the Malliavin derivative, see especially sections 2.4, 4.5, 4.5, and 7.1.

The representation we study results from GNS applied to the Fock state. While the conclusions regarding this CCR-representation (to be discussed below) hold more generally, we have chosen to illustrate the idea here only in the simplest case, that of the standard Brownian/Wiener process. Hence our initial Hilbert space will be $\mathscr{L} = L^2(0,\infty)$. (But the conclusions in fact hold in the generality of Gaussian processes indexed by Hilbert space.)

From the initial Hilbert space $\mathscr{L}$, we build the $*$-algebra $\mathrm{CCR}(\mathscr{L})$ as in Section 5.10 above. We will show that the Fock state on $\mathrm{CCR}(\mathscr{L})$ corresponds to the Wiener measure $\mathbb{P}$. Moreover the corresponding representation π of $\mathrm{CCR}(\mathscr{L})$ will be acting on the Hilbert space $L^2(\Omega, \mathbb{P})$ in such that for every k in $\mathscr{L}$, the operator $\pi(a(k))$ is the Malliavin derivative in the direction of k. We caution that the representations of the $*$-algebra $\mathrm{CCR}(\mathscr{L})$ are by unbounded operators, but the operators in the range of the representations will be defined on a single common dense domain.

The general setting is as follows: Let $\mathscr{L}$ be a fixed Hilbert space, and let $\mathrm{CCR}(\mathscr{L})$ be the $*$-algebra on the generators $a(k)$, $a^*(l)$, $k, l \in \mathscr{L}$, and subject to the relations for the CCR-algebra, see Section 5.10:

$$[a(k), a(l)] = 0, \quad \text{and} \tag{5.127}$$

$$[a(k), a^*(l)] = \langle k, l \rangle_{\mathscr{L}} \mathbb{1} \tag{5.128}$$

where $[\cdot, \cdot]$ is the commutator bracket.

A representation π of $\mathrm{CCR}(\mathscr{L})$ consists of a fixed Hilbert space $\mathscr{H} = \mathscr{H}_\pi$ (the representation space), a dense subspace $\mathscr{D}_\pi \subset \mathscr{H}_\pi$, and a $*$-homomorphism $\pi : \mathrm{CCR}(\mathscr{L}) \longrightarrow End(\mathscr{D}_\pi)$ such that

$$\mathscr{D}_\pi \subset dom(\pi(A)), \quad \forall A \in \mathrm{CCR}. \tag{5.129}$$

The representation axiom entails the commutator properties resulting from (5.127)-(5.128); in particular π satisfies

$$[\pi(a(k)), \pi(a(l))] F = 0, \quad \text{and} \tag{5.130}$$

$$\left[\pi(a(k)), \pi(a(l))^*\right] F = \langle k, l \rangle_{\mathscr{L}} F, \tag{5.131}$$

$\forall k, l \in \mathscr{L}$, $\forall F \in \mathscr{D}_\pi$; where $\pi(a^*(l)) = \pi(a(l))^*$.

In the application below, we take $\mathscr{L} = L^2(0,\infty)$, and $\mathscr{H}_\pi = L^2(\Omega, \mathcal{F}_\Omega, \mathbb{P})$ where $(\Omega, \mathcal{F}_\Omega, \mathbb{P})$ is the standard Wiener probability space, and

$$\Phi_t(\omega) = \omega(t), \quad \forall \omega \in \Omega, \ t \in [0,\infty). \tag{5.132}$$

For $k \in \mathscr{L}$, we set

$$\Phi(k) = \int_0^\infty k(t) \, d\Phi_t \ \text{(the Itō-Wiener integral.)}$$

The dense subspace $\mathscr{D}_\pi \subset \mathscr{H}_\pi$ is generated by the polynomial fields:
For $n \in \mathbb{N}$, $h_1, \cdots, h_n \in \mathscr{L} = L^2_\mathbb{R}(0, \infty)$, $p \in \mathbb{R}^n \longrightarrow \mathbb{R}$ a polynomial in n real variables, set

$$F = p(\Phi(h_1), \cdots, \Phi(h_n)), \quad \text{and} \tag{5.133}$$

$$\pi(a(k)) F = \sum_{j=1}^n \left(\frac{\partial}{\partial x_j} p\right) (\Phi(h_1), \cdots, \Phi(h_n)) \langle h_j, k \rangle. \tag{5.134}$$

We saw in Section 3.3.1 that $\mathscr{D}_\pi$ is an algebra under pointwise product and that

$$\pi(a(k))(FG) = (\pi(a(k)) F) G + F(\pi(a(k)) G), \tag{5.135}$$

$\forall k \in \mathscr{L}$, $\forall F, G \in \mathscr{D}_\pi$. Equivalently, $T_k := \pi(a(k))$ is a derivation in the algebra $\mathscr{D}_\pi$ (relative to pointwise product.)

Theorem 5.17. *With the operators $\pi(a(k))$, $k \in \mathscr{L}$, we get a $*$-representation $\pi : CCR(\mathscr{L}) \longrightarrow End(\mathscr{D}_\pi)$, i.e., $\pi(a(k))$ = the Malliavin derivative in the direction k,*

$$\pi(a(k)) F = \langle T(F), k \rangle_{\mathscr{L}}, \quad \forall F \in \mathscr{D}_\pi, \forall k \in \mathscr{L}. \tag{5.136}$$

Proof. We begin with the following $\qquad\qquad\qquad\qquad\qquad\qquad\qquad\qquad \square$

Lemma 5.7. *Let π, $CCR(\mathscr{L})$, and $\mathscr{H}_\pi = L^2(\Omega, \mathcal{F}_\Omega, \mathbb{P})$ be as above. For $k \in \mathscr{L}$, we shall identify $\Phi(k)$ with the unbounded multiplication operator in $\mathscr{H}_\pi$:*

$$\mathscr{D}_\pi \ni F \longmapsto \Phi(k) F \in \mathscr{H}_\pi. \tag{5.137}$$

For $F \in \mathscr{D}_\pi$, we have $\pi(a(k))^* F = -\pi(a(k)) F + \Phi(k) F$; or in abbreviated form:

$$\pi(a(k))^* = -\pi(a(k)) + \Phi(k) \tag{5.138}$$

valid on the dense domain $\mathscr{D}_\pi \subset \mathscr{H}_\pi$.

Proof. This follows from the following computation for $F, G \in \mathscr{D}_\pi$, $k \in \mathscr{L}$.
Setting $T_k := \pi(a(k))$, we have

$$\mathbb{E}(T_k(F) G) + \mathbb{E}(F T_k(G)) = \mathbb{E}(T_k(FG)) = \mathbb{E}(\Phi(k) FG).$$

Hence $\mathscr{D}_\pi \subset dom(T_k^*)$, and $T_k^*(F) = -T_k(F) + \Phi(k) F$, which is the desired conclusion (5.138). $\qquad\qquad\qquad\qquad\qquad\qquad\qquad\qquad\qquad\qquad \square$

Proof of Theorem 5.17 continued. It is clear that the operators $T_k = \pi(a(k))$ form a commuting family. Hence on $\mathscr{D}_\pi$, we have for $k, l \in \mathscr{L}$, $F \in \mathscr{D}_\pi$:

$$[T_k, T_l^*](F) = [T_k, \Phi(l)](F)$$

$$= T_k \left(\Phi \left(l \right) F \right) - \Phi \left(l \right) \left(T_k \left(F \right) \right)$$
$$= T_k \left(\Phi \left(l \right) \right) F = \langle k, l \rangle_{\mathscr{L}} F$$

which is the desired commutation relation (5.128).

The remaining check on the statements in the theorem are now immediate.

□

Corollary 5.12. *The state on $CCR\left(\mathscr{L} \right)$ which is induced by π and the constant function $\mathbb{1}$ in $L^2 \left(\Omega, \mathbb{P} \right)$ is the Fock-vacuum-state, ω_{Fock}^-.*

Proof. The assertion will follow once we verify the following two conditions:

$$\int_\Omega \left(T_k^* T_k \left(\mathbb{1} \right) \right) d\mathbb{P} = 0 \tag{5.139}$$

and

$$\int_\Omega T_k T_l^* \left(\mathbb{1} \right) d\mathbb{P} = \langle k, l \rangle_{\mathscr{L}} \tag{5.140}$$

for all $k, l \in \mathscr{L}$.

This in turn is a consequence of our discussion of eqs (5.120)-(5.121) above: The Fock state ω_{Fock}^- is determined by these two conditions. Our assertions (5.139)-(5.140) in turn follow from: $T_k \left(\mathbb{1} \right) = 0$, and $\left(T_k T_l^* \right) \left(\mathbb{1} \right) = \langle k, l \rangle_{\mathscr{L}} \mathbb{1}$. □

Corollary 5.13. *For $k \in L_{\mathbb{R}}^2 \left(0, \infty \right)$ we get a family of selfadjoint multiplication operators $T_k + T_k^* = M_{\Phi(k)}$ on $\mathscr{D}_\pi$ where $T_k = \pi \left(a \left(k \right) \right)$. Moreover, the von Neumann algebra generated by these operators is $L^\infty \left(\Omega, \mathcal{F}_\Omega, \mathbb{P} \right)$, i.e., the maximal abelian L^∞-algebra of all multiplication operators in $\mathscr{H}_\pi = L^2 \left(\Omega, \mathcal{F}_\Omega, \mathbb{P} \right)$.*

Chapter 6

Completely Positive Maps

"Completely positive maps on von Neumann algebras or between C^*-algebras have fascinated me since my days as a graduate student."
— William B. Arveson

"... the development of mathematics is not something one can predict, and it would be foolish to try. One reason we love doing mathematics is that we don't know what lies ahead that future research will uncover."
— Alain Connes

The study of completely positive maps dates back five decades, but because of a recent observation of Arveson[1] (see e.g., [Arv09a, Arv09c, Arv09b]), they have acquired a brand new set of applications; applications to *quantum information theory* (QIT). In this framework, one studies completely positive maps on matrix algebras. They turn out to be the objects that are dual to *quantum channels*. Even more: Arveson proved the converse: that the study of quantum channels reduces to the study of *unital completely positive maps* of matrix algebras. This work is part of QIT, and it is still ongoing, with view to the study of *entanglement, entropy* and *channel-capacity*.

In the last chapter we studied two questions from the use of algebras of operators in quantum physics: "Where does the Hilbert space come from?" And "What are the algebras of operators from which the selfadjoint observables must be selected?" An answer is given in "the Gelfand-Naimark-Segal (GNS) theorem;" a direct correspondence between states and cyclic representations. But *states* are scalar valued positive definite functions on ∗-algebras. For a host of applications, one must instead consider *operator valued* "states." For this a different notion of positivity is needed, "*complete positivity.*"

---

[1]Appendix C includes a short biographical sketch of W. Arveson. See also [Arv72, Arv98].

The GNS construction gives a bijection between states and cyclic representations. An extension to the GNS construction is Stinespring's completely positive maps. It appeared in an early paper by Stinespring in 1955 [Sti55]. Arveson in 1970's greatly extended Stinespring's result using tensor product [Arv72]. He showed that completely positive maps are the key in multivariable operator theory, and in noncommutative dynamics.

In the details below, the following concepts and terms will play an important role; (for reference, see also Appendix F): Tensor product, completely positive mappings, Stinespring's theorem, representations of Cuntz algebras, filter bank.

6.1   Motivation

Let $\mathfrak{A}$ be a $*$-algebra with identity. Recall that a functional $w : \mathfrak{A} \to \mathbb{C}$ is a *state* if $w(1_{\mathfrak{A}}) = 1$, $w(A^*A) \geq 0$. If $\mathfrak{A}$ was a C^*-algebra, $A \geq 0 \Leftrightarrow sp(A) \geq 0$, hence we may take $B = \sqrt{A}$ and $A = B^*B$.

Given a state w, the GNS construction gives a Hilbert space $\mathscr{K}$, a cyclic vector $\Omega \in \mathscr{K}$, and a representation $\pi : \mathfrak{A} \to \mathscr{B}(\mathscr{K})$, such that

$$w(A) = \langle \Omega, \pi(A)\Omega \rangle$$

$$\mathscr{K} = \overline{span}\{\pi(A)\Omega : A \in \mathfrak{A}\}.$$

Moreover, the Hilbert space is unique up to unitary equivalence.

Stinespring modified the GNS construction as follows: Instead of a state $w : \mathfrak{A} \to \mathbb{C}$, he considered a positive map $\varphi : \mathfrak{A} \to \mathscr{B}(\mathscr{H})$, i.e., φ maps positive elements in $\mathfrak{A}$ to positive operators in $\mathscr{B}(\mathscr{H})$. φ is a natural extension of w, since $\mathbb{C}$ can be seen as a 1-dimensional Hilbert space, and w is a positive map $w : \mathfrak{A} \to \mathscr{B}(\mathbb{C})$. He further realized that φ being a positive map is not enough to produce a Hilbert space and a representation. It turns out that the condition to put on φ is *complete positivity*:

Definition 6.1. Let $\mathfrak{A}$ be a $*$-algebra. A map $\varphi : \mathfrak{A} \to \mathscr{B}(\mathscr{H})$ is *completely positive*, if for all $n \in \mathbb{N}$,

$$\varphi \otimes I_{M_n} : \mathfrak{A} \otimes M_n \to \mathscr{B}(\mathscr{H} \otimes \mathbb{C}^n) \tag{6.1}$$

maps positive elements in $\mathfrak{A} \otimes M_n$ to positive operators in $\mathscr{B}(\mathscr{H} \otimes \mathbb{C}^n)$. φ is called a completely positive map, or a CP map. (CP maps are developed primarily for non-Abelian algebras.)

The algebra M_n of $n \times n$ matrices can be seen as an n^2-dimensional Hilbert space with an ONB given by the matrix units $\{e_{ij}\}_{i,j=1}^n$. It is also a $*$-algebra generated by $\{e_{ij}\}_{i,j=1}^n$ such that

$$e_{ij}e_{kl} = \begin{cases} e_{il} & j = k \\ 0 & j \neq k. \end{cases}$$

Members of $\mathfrak{A} \otimes M_n$ are of the form

$$\sum_{i,j} A_{ij} \otimes e_{ij}.$$

In other words, $\mathfrak{A} \otimes M_n$ consists of precisely the $\mathfrak{A}$-valued $n \times n$ matrices. Similarly, members of $\mathscr{H} \otimes \mathbb{C}^n$ are the n-tuple column vectors with $\mathscr{H}$-valued entries.

Let $I_{M_n} : M_n \to \mathscr{B}(\mathbb{C}^n)$ be the identity representation of M_n onto $\mathscr{B}(\mathbb{C}^n)$. Then,

$$\varphi \otimes I_{M_n} : \mathfrak{A} \otimes M_n \to \mathscr{B}(\mathscr{H}) \otimes \mathscr{B}(\mathbb{C}^n) \, (= \mathscr{B}(\mathscr{H} \otimes \mathbb{C}^n)) \tag{6.2}$$

$$\varphi \otimes I_{M_n} \left(\sum_{i,j} A_{ij} \otimes e_{ij} \right) = \sum_{i,j} \varphi(A_{ij}) \otimes e_{ij}. \tag{6.3}$$

Note the RHS in (6.3) is an $n \times n$ matrix with $\mathscr{B}(\mathscr{H})$-valued entries.

Remark 6.1. The algebra $\mathscr{B}(\mathbb{C}^n)$ of all bounded operators on $\mathbb{C}^n$ is generated by the rank-one operators, i.e.,

$$I_{M_n}(e_{ij}) = |e_i\rangle\langle e_j|. \tag{6.4}$$

Hence the e_{ij} on the LHS of (6.3) is seen as an element in the algebra $M_n(\mathbb{C})$, i.e., $n \times n$ complex matrices; while on the RHS of (6.3), e_{ij} is treated as the rank one operator $|e_i\rangle\langle e_j| \in \mathscr{B}(\mathbb{C}^n)$. Using Dirac's notation (Definition 2.23), when we look at e_{ij} as operators, we may write

$$e_{i,j}(e_k) = |e_i\rangle\langle e_j| \, |e_k\rangle = \begin{cases} |e_i\rangle & j = k \\ 0 & j \neq k \end{cases}$$

$$e_{i,j}e_{kl} = |e_i\rangle\langle e_j| \, |e_k\rangle\langle e_l| = \begin{cases} |e_i\rangle\langle e_l| & j = k \\ 0 & j \neq k \end{cases}$$

This also shows that I_{M_n} is in fact an algebra isomorphism.

The CP condition in (6.1) is illustrated in the following diagram.

$$\otimes \begin{cases} \mathfrak{A} \to \mathscr{B}(\mathscr{H}) : & A \mapsto \varphi(A) \\ M_n \to M_n : & x \mapsto I_{M_n}(X) = X \ \text{(identity representation of } M_n) \end{cases}$$

It is saying that if $\sum_{i,j} A_{ij} \otimes e_{ij}$ is a positive element in the algebra $\mathfrak{A} \otimes M_n$, then the $n \times n$ $\mathscr{B}(\mathscr{H})$-valued matrix $\sum_{i,j} \varphi(A_{ij}) \otimes e_{ij}$ is a positive operator acting on the Hilbert space $\mathscr{H} \otimes \mathbb{C}^n$.

Specifically, take any $v = \sum_{k=1}^{n} v_k \otimes e_k$ in $\mathscr{H} \otimes \mathbb{C}^n$, we must have

$$\left\langle \sum_l v_l \otimes e_l, \left(\sum_{i,j} \varphi(A_{ij}) \otimes e_{ij} \right) \left(\sum_k v_k \otimes e_k \right) \right\rangle$$

$$\begin{aligned}
&= \left\langle \sum_l v_l \otimes e_l, \sum_{i,j,k} \varphi(A_{ij})v_k \otimes e_{ij}(e_k) \right\rangle \\
&= \left\langle \sum_l v_l \otimes e_l, \sum_{i,j} \varphi(A_{ij})v_j \otimes e_i \right\rangle \\
&= \sum_{i,j,l} \langle v_l, \varphi(A_{ij})v_j \rangle \langle e_l, e_i \rangle \\
&= \sum_{i,j} \langle v_i, \varphi(A_{ij})v_j \rangle \geq 0.
\end{aligned} \tag{6.5}$$

Using matrix notation, the CP condition is formulated as:

For all $n \in \mathbb{N}$, and all $v \in \mathscr{H} \otimes \mathbb{C}^n$, i.e.,

$$v = \sum_{k=1}^n v_k \otimes e_k = \begin{bmatrix} v_1 \\ \vdots \\ v_n \end{bmatrix}$$

we have

$$\begin{bmatrix} v_1 & v_2 & \cdots & v_n \end{bmatrix} \begin{bmatrix} \varphi(A_{11}) & \varphi(A_{12}) & \cdots & \varphi(A_{1n}) \\ \varphi(A_{21}) & \varphi(A_{22}) & \cdots & \varphi(A_{2n}) \\ \vdots & \vdots & \ddots & \vdots \\ \varphi(A_{n1}) & \varphi(A_{n2}) & \cdots & \varphi(A_{nn}) \end{bmatrix} \begin{bmatrix} v_1 \\ v_2 \\ \vdots \\ v_n \end{bmatrix} \geq 0. \tag{6.6}$$

6.2   CP v.s. GNS

The GNS construction can be reformulated as a special case of the Stinespring's theorem [Sti55].

Let $\mathfrak{A}$ be a $*$-algebra, given a state $\varphi : \mathfrak{A} \to \mathbb{C}$, there exists a triple $(\mathscr{K}, \Omega, \pi)$, all depending on φ, such that

$$\varphi(A) = \langle \Omega, \pi(A)\Omega \rangle_{\mathscr{K}}$$

where

$$\begin{aligned}
\Omega &= \pi(1_{\mathfrak{A}}) \in \mathscr{K} \\
\mathscr{K} &= \overline{span}\{\pi(A)\Omega : A \in \mathfrak{A}\}.
\end{aligned}$$

The 1-dimensional Hilbert space $\mathbb{C}$ is thought of being embedded into $\mathscr{K}$ (possibly infinite dimensional) via

$$\mathbb{C} \ni t \xrightarrow{V} t\Omega \in \mathbb{C}\Omega \tag{6.7}$$

where $\mathbb{C}\Omega$ = the one-dimensional subspace in $\mathscr{K}$ generated by the unit cyclic vector Ω.

Lemma 6.1. *The map V in (6.7) is an isometry, such that $V^*V = I_{\mathbb{C}} : \mathbb{C} \to \mathbb{C}$, and*

$$VV^* : \mathscr{K} \to \mathbb{C}\Omega \qquad (6.8)$$

is the projection from $\mathscr{K}$ onto the 1-d subspace $\mathbb{C}\Omega$ in $\mathscr{K}$.
  Moreover,

$$\varphi(A) = V^*\pi(A)V, \ \forall A \in \mathfrak{A}. \qquad (6.9)$$

Proof. Let $t \in \mathbb{C}$, then $\|Vt\|_{\mathscr{K}} = \|t\Omega\|_{\mathscr{K}} = |t|$, and so V is an isometry.
  For all $\xi \in \mathscr{K}$, we have

$$\langle \xi, Vt \rangle_{\mathscr{K}} = \langle V^*\xi, t \rangle_{\mathbb{C}} = t\overline{V^*\xi}.$$

By setting $t = 1$, we get

$$V^*\xi = \overline{\langle \xi, V1 \rangle_{\mathscr{K}}} = \overline{\langle \xi, \Omega \rangle_{\mathscr{K}}} = \langle \Omega, \xi \rangle_{\mathscr{K}} \Longleftrightarrow V^* = \langle \Omega, \cdot \rangle_{\mathscr{K}}.$$

Therefore,

$$V^*Vt = V^*(t\Omega) = \langle \Omega, t\Omega \rangle_{\mathscr{K}} = t, \ \forall t \in \mathbb{C} \Longleftrightarrow V^*V = I_{\mathbb{C}}$$

$$VV^*\xi = V(\langle \Omega, \xi \rangle_{\mathscr{K}}) = \langle \Omega, \xi \rangle_{\mathscr{K}} \Omega, \ \forall \xi \in \mathscr{K} \Longleftrightarrow VV^* = |\Omega\rangle\langle\Omega|.$$

It follows that

$$
\begin{aligned}
\varphi(A) &= \langle \Omega, \pi(A)\Omega \rangle_{\mathscr{K}} \\
&= \langle V1, \pi(A)V1 \rangle_{\mathscr{K}} \\
&= \langle 1, V^*\pi(A)V1 \rangle_{\mathbb{C}} \\
&= V^*\pi(A)V, \ \forall A \in \mathfrak{A}
\end{aligned}
$$

which is the assertion in (6.9). $\qquad \square$

  In other words, $\Omega \longmapsto \pi(A)\Omega$ sends the unit vector Ω from the 1-dimensional subspace $\mathbb{C}\Omega$ to the vector $\pi(A)\Omega \in \mathscr{K}$, and $\langle \Omega, \pi(A)\Omega \rangle_{\mathscr{K}}$ cuts off the resulting vector $\pi(A)\Omega$ and only preserves the component corresponding to the 1-d subspace $\mathbb{C}\Omega$. Notice that the unit vector Ω is obtained from embedding the constant $1 \in \mathbb{C}$ via the map V, i.e., $\Omega = V1$. In matrix notation, if we identify $\mathbb{C}$ with its image $\mathbb{C}\Omega$ in $\mathscr{K}$, then $\varphi(A)$ is put into a matrix corner (see Remark 6.2):

$$\pi(A) = \begin{bmatrix} \varphi(A) & * \\ * & * \end{bmatrix}$$

so that when acting on vectors,

$$\varphi(A) = \begin{bmatrix} \Omega & 0 \end{bmatrix} \begin{bmatrix} \varphi(A) & * \\ * & * \end{bmatrix} \begin{bmatrix} \Omega \\ 0 \end{bmatrix}.$$

Equivalently,

$$\varphi(A) = P_1\pi(A) : P_1\mathscr{K} \to \mathbb{C};$$

where $P_1 := VV^* = |\Omega\rangle\langle\Omega| = \text{rank-1 projection on } \mathbb{C}\Omega$.

Stinespring's construction is a generalization of the above formulation: Let $\mathfrak{A}$ be a $*$-algebra, given a CP map $\varphi : \mathfrak{A} \to \mathscr{B}(\mathscr{H})$, there exists a Hilbert space $\mathscr{K} \, (= \mathscr{K}_\varphi)$, an isometry $V : \mathscr{H} \to \mathscr{K}$, and a representation $\pi \, (= \pi_\varphi) : \mathfrak{A} \to \mathscr{K}$, such that

$$\varphi(A) = V^* \pi(A) V, \; \forall A \in \mathfrak{A}.$$

Notice that this construction starts with a possibly infinite dimensional Hilbert space $\mathscr{H}$ (instead of the 1-dimensional Hilbert space $\mathbb{C}$), the map V embeds $\mathscr{H}$ into a bigger Hilbert space $\mathscr{K}$. If $\mathscr{H}$ is identified with its image in $\mathscr{K}$, then $\pi(A)$ is put into a matrix corner,

$$\begin{bmatrix} \pi(A) & * \\ * & * \end{bmatrix}$$

so that when acting on vectors,

$$\varphi(A)\xi = \begin{bmatrix} V\xi & 0 \end{bmatrix} \begin{bmatrix} \pi(A) & * \\ * & * \end{bmatrix} \begin{bmatrix} V\xi \\ 0 \end{bmatrix}.$$

This can be formulated alternatively:

For every CP map $\varphi : \mathfrak{A} \to \mathscr{B}(\mathscr{H})$, there is a dilated Hilbert space $\mathscr{K} \, (= \mathscr{K}_\varphi) \supset \mathscr{H}$, a representation $\pi \, (= \pi_\varphi) : \mathfrak{A} \to \mathscr{B}(\mathscr{K})$, such that

$$\varphi(A) = P_\mathscr{H} \pi(A)$$

i.e., $\pi(A)$ can be put into a matrix corner. $\mathscr{K}$ is chosen as minimal in the sense that

$$\mathscr{K} = \overline{span}\{\pi(A)(Vh) : A \in \mathfrak{A}, h \in \mathscr{H}\}.$$

$$\begin{array}{ccc} \mathscr{H} & \xrightarrow{\;V\;} & \mathscr{K} \\ {\scriptstyle \varphi(A)}\downarrow & & \downarrow{\scriptstyle \pi(A)} \\ \mathscr{H} & \xrightarrow{\;V\;} & \mathscr{K} \end{array}$$

Note 7. The containment $\mathscr{H} \subset \mathscr{K}$ comes after the identification of $\mathscr{H}$ with its image in $\mathscr{K}$ under the isometric embedding V. We write $\varphi(A) = P_H \pi(A)$, as opposed to $\varphi(A) = P_H \pi(A) P_H$, since $\varphi(A)$ only acts on the subspace $\mathscr{H}$.

6.3   Stinespring's Theorem

Theorem 6.1 (Stinespring [Sti55]). *Let $\mathfrak{A}$ be a $*$-algebra. The following are equivalent:*

1. *$\varphi : \mathfrak{A} \to \mathscr{B}(\mathscr{H})$ is a completely positive map, and $\varphi(1_\mathfrak{A}) = I_\mathscr{H}$.*

2. *There exists a Hilbert space $\mathscr{K}$, an isometry $V : \mathscr{H} \to \mathscr{K}$, and a representation $\pi : \mathfrak{A} \to \mathscr{B}(\mathscr{K})$ such that*

$$\varphi(A) = V^* \pi(A) V, \; \forall A \in \mathfrak{A}. \tag{6.10}$$

3. *If the dilated Hilbert space $\mathscr{K}$ is taken to be minimal, then it is unique up to unitary equivalence. Specifically, if there are two systems $(V_i, \mathscr{K}_i, \pi_i)$, $i = 1, 2$, satisfying*

$$
\begin{aligned}
\varphi(A) &= V_i^* \pi_i(A) V_i & (6.11) \\
\mathscr{K}_i &= \overline{span}\{\pi_i(A)Vh \ : \ A \in \mathfrak{A}, h \in \mathscr{H}\} & (6.12)
\end{aligned}
$$

then there exists a unitary operator $W : \mathscr{K}_1 \to \mathscr{K}_2$ so that

$$
W\pi_1 = \pi_2 W. \tag{6.13}
$$

Remark 6.2. The interpretation of part 2 of the conclusion in the theorem is as follows: Since the operator V in (6.10) is an isometry, we may therefore identify the initial Hilbert space $\mathscr{H}$ as a subspace of $\mathscr{K}$; or alternatively, think of $\mathscr{K}$ as an enlargement Hilbert space (also called a "dilation"). The interpretation of the assertion in (6.10) is therefore: For every CP mapping φ, we get the existence of an associated pair $(\mathscr{K}, \pi)$ where $\mathscr{K}$ is an enlargement Hilbert space, and π is a representation by operators in $\mathscr{K}$, realized in such a way that the given CP mapping φ becomes a "matrix-corner" of π.

Proof of Theorem 6.1.
    (Part (3), uniqueness) Let $(V_i, \mathscr{K}_i, \pi_i)$, $i = 1, 2$, be as in the statement satisfying (6.11)-(6.12). Define

$$
W\pi_1(A)h = \pi_2(A)Vh
$$

then W is an isometry, since

$$
\begin{aligned}
\|\pi_i(A)Vh\|_{\mathscr{K}}^2 &= \langle \pi_i(A)Vh, \pi_i(A)Vh \rangle_{\mathscr{K}} \\
&= \langle h, V^* \pi_i(A^*A)Vh \rangle_{\mathscr{H}} \\
&= \langle h, \varphi(A^*A)h \rangle_{\mathscr{H}}.
\end{aligned}
$$

Hence W extends uniquely to a unitary operator $W : \mathscr{K}_1 \to \mathscr{K}_2$. To see that W intertwines π_1, π_2, notice that a typical vector in $\mathscr{K}_i$ is $\pi_i(A)Vh$, and

$$
\begin{aligned}
W\pi_1(B)\pi_1(A)Vh &= W\pi_1(BA)Vh \\
&= \pi_2(BA)Vh \\
&= \pi_2(B)\pi_2(A)Vh \\
&= \pi_2(B)W\pi_1(A)Vh.
\end{aligned}
$$

Since such vectors are dense in the respective dilated space, we conclude that $W\pi_1 = \pi_2 W$, so (6.13) holds. $\qquad \square$

Note 8. $\|\pi_1(B)\pi_1(A)Vh\|^2 = \langle h, V^*\pi_1(A^*B^*BA)Vh \rangle$. Fix $A \in \mathscr{B}(\mathscr{H})$, the map $B \mapsto A^*BA$ is an automorphism on $\mathscr{B}(\mathscr{H})$.

Proof. $(2) \implies (1)$

Now suppose $\varphi(A) = V^* \pi(A) V$, and we verify it is completely positive. Since positive elements in $\mathfrak{A} \otimes M_n$ are sums of the operator matrix

$$\sum_{i,j} A_i^* A_j \otimes e_{ij} = \begin{bmatrix} A_1^* \\ A_2^* \\ \vdots \\ A_n^* \end{bmatrix} \begin{bmatrix} A_1 & A_2 & \cdots & A_n \end{bmatrix}$$

it suffices to show that

$$\varphi \otimes I_{M_n} \left(\sum_{i,j} A_i^* A_j \otimes e_{ij} \right) = \sum_{i,j} \varphi\left(A_i^* A_j\right) \otimes e_{ij}$$

is a positive operator in $\mathscr{B}(\mathscr{H} \otimes \mathbb{C}^n)$, i.e., need to show that for all $v \in \mathscr{H} \otimes \mathbb{C}^n$

$$\begin{bmatrix} v_1 & v_2 & \cdots & v_n \end{bmatrix} \begin{bmatrix} \varphi\left(A_1^* A_1\right) & \varphi\left(A_1^* A_2\right) & \cdots & \varphi\left(A_1^* A_n\right) \\ \varphi\left(A_2^* A_1\right) & \varphi\left(A_2^* A_2\right) & \cdots & \varphi\left(A_2^* A_n\right) \\ \vdots & \vdots & \ddots & \vdots \\ \varphi\left(A_n^* A_1\right) & \varphi\left(A_n^* A_2\right) & \cdots & \varphi\left(A_n^* A_n\right) \end{bmatrix} \begin{bmatrix} v_1 \\ v_2 \\ \vdots \\ v_n \end{bmatrix} \geq 0.$$
(6.14)

This is true, since

$$\begin{aligned}
\mathrm{RHS}_{(6.14)} &= \sum_{i,j} \langle v_i, \varphi\left(A_i^* A_j\right) v_j \rangle_{\mathscr{H}} \\
&= \sum_{i,j} \langle v_i, V^* \pi\left(A_i^* A_j\right) V v_j \rangle_{\mathscr{H}} \\
&= \sum_{i,j} \langle \pi\left(A_i\right) V v_i, \pi\left(A_j\right) V v_j \rangle_{\mathscr{K}} \\
&= \left\| \sum_i \pi\left(A_i\right) V v_i \right\|_{\mathscr{K}}^2 \geq 0.
\end{aligned}$$

$(1) \implies (2)$

Given a completely positive map φ, we construct $\mathscr{K} \, (= \mathscr{K}_\varphi)$, $V \, (= V_\varphi)$ and $\pi \, (= \pi_\varphi)$. Recall that $\varphi : \mathfrak{A} \to \mathscr{B}(\mathscr{H})$ is a CP map means that for all $n \in \mathbb{N}$,

$$\varphi \otimes I_{M_n} : \mathfrak{A} \otimes M_n \to \mathscr{B}(\mathscr{H} \otimes M_n)$$

is positive, and

$$\varphi \otimes I_{M_n}(1_{\mathfrak{A}} \otimes I_{M_n}) = I_{\mathscr{H}} \otimes I_{M_n}.$$

The condition on the identity element can be stated using matrix notation as

$$\begin{bmatrix} \varphi & 0 & \cdots & 0 \\ 0 & \varphi & \cdots & 0 \\ \vdots & \vdots & \ddots & \vdots \\ 0 & 0 & \cdots & \varphi \end{bmatrix} \begin{bmatrix} 1_{\mathfrak{A}} & 0 & \cdots & 0 \\ 0 & 1_{\mathfrak{A}} & \cdots & 0 \\ \vdots & \vdots & \ddots & \vdots \\ 0 & 0 & \cdots & 1_{\mathfrak{A}} \end{bmatrix} = \begin{bmatrix} I_{\mathscr{H}} & 0 & \cdots & 0 \\ 0 & I_{\mathscr{H}} & \cdots & 0 \\ \vdots & \vdots & \ddots & \vdots \\ 0 & 0 & \cdots & I_{\mathscr{H}} \end{bmatrix}.$$

Let K_0 be the algebraic tensor product $\mathfrak{A} \otimes \mathscr{H}$, i.e.,

$$K_0 = span \left\{ \sum_{finite} A_i \otimes \xi_i : A \in \mathfrak{A}, \xi \in \mathscr{H} \right\}.$$

Define a sesquilinear form $\langle \cdot, \cdot \rangle_\varphi : K_0 \times K_0 \to \mathbb{C}$, by

$$\left\langle \sum_{i=1}^n A_i \otimes \xi_i, \sum_{j=1}^n B_j \otimes \eta_j \right\rangle_\varphi := \sum_{i,j} \langle \xi_i, \varphi\left(A_i^* B_j\right) \eta_j \rangle_\mathscr{H}. \tag{6.15}$$

By the CP condition (6.1), we have

$$\left\langle \sum_{i=1}^n A_i \xi_i, \sum_{j=1}^n A_j \xi_j \right\rangle_\varphi = \sum_{i,j} \langle \xi_i, \varphi(A_i^* A_j) \xi_j \rangle_\mathscr{H} \geq 0.$$

Let $N := \left\{ v \in K_0 : \langle v, v \rangle_\varphi = 0 \right\}$. Since the Schwarz inequality holds for any sesquilinear form, it follows that

$$N = \left\{ v \in K_0 : \langle s, v \rangle_\varphi = 0, \ \forall s \in K_0 \right\}.$$

Thus N is a closed subspace in K_0. Let $\mathscr{K} (= \mathscr{K}_\varphi)$ be the Hilbert space by completing K_0/N with respect to

$$\|\cdot\|_\mathscr{K} := \langle \cdot, \cdot \rangle_\varphi^{1/2}.$$

Let $V : \mathscr{H} \to K_0$, by

$$V\xi := 1_\mathfrak{A} \otimes \xi, \ \forall \xi \in \mathscr{H}.$$

Then,

$$\begin{aligned}
\|V\xi\|_\varphi^2 &= \langle 1_\mathfrak{A} \otimes \xi, 1_\mathfrak{A} \otimes \xi \rangle_\varphi \\
&= \langle \xi, \varphi(1_\mathfrak{A}^* 1_\mathfrak{A}) \xi \rangle_\mathscr{H} \\
&= \langle \xi, \xi \rangle_\mathscr{H} = \|\xi\|_\mathscr{H}^2
\end{aligned}$$

i.e., V is isometric, and so $\mathscr{H} \xrightarrow{V} K_0$ is an isometric embedding.
Claim.
   (i) $V^*V = I_\mathscr{H}$;
   (ii) $VV^* =$ projection from K_0 on the subspace $1_\mathfrak{A} \otimes \mathscr{H}$.
   Indeed, for any $A \otimes \eta \in K_0$, we have

$$\begin{aligned}
\langle A \otimes \eta, V\xi \rangle_\varphi &= \langle A \otimes \eta, 1_\mathfrak{A} \otimes \xi \rangle_\varphi \\
&= \langle \eta, \varphi(A^*) \xi \rangle_\mathscr{H} \\
&= \langle \varphi(A^*)^* \eta, \xi \rangle_\mathscr{H}
\end{aligned}$$

which implies that
$$V^*(A \otimes \eta) = \varphi(A^*)^*\eta.$$
It follows that

$$V^*V\xi = V^*(1_{\mathfrak{A}} \otimes \xi) = \varphi(1_{\mathfrak{A}}^*)^*\xi = \xi, \ \forall \xi \in \mathcal{H}$$

i.e., $V^*V = I_{\mathcal{H}}$. Moreover, for any $A \otimes \eta \in K_0$,

$$VV^*(A \otimes \eta) = V(\varphi(A^*)^*\eta) = 1_{\mathfrak{A}} \otimes \varphi(A^*)^*\eta.$$

This proves the claim. It is clear that the properties of V pass to the dilated space $\mathcal{K} (= \mathcal{K}_\varphi) = cl_\varphi (K_0/N)$.

To finish the proof of the theorem, define $\pi (= \pi_\varphi)$ as follows: Set

$$\pi(A) \left(\sum_j B_j \otimes \eta_j \right) := \sum_j AB_j \otimes \eta_j, \ \forall A \in \mathfrak{A}$$

and extend it to $\mathcal{K}$.
   For all $\xi, \eta \in \mathcal{H}$, then,

$$
\begin{aligned}
\langle \xi, V^*\pi(A)V\eta \rangle_{\mathcal{H}} &= \langle V\xi, \pi(A)V\eta \rangle_{\mathcal{K}} \\
&= \langle 1_{\mathfrak{A}} \otimes \xi, \pi(A)1_{\mathfrak{A}} \otimes \eta \rangle_{\mathcal{K}} \\
&= \langle 1_{\mathfrak{A}} \otimes \xi, A \otimes \eta \rangle_{\mathcal{K}} \\
&= \langle \xi, \varphi(1_{\mathfrak{A}}^*A)\eta \rangle_{\mathcal{H}} \\
&= \langle \xi, \varphi(A)\eta \rangle_{\mathcal{H}}.
\end{aligned}
$$

We conclude that $\varphi(A) = V^*\pi(A)V$, for all $A \in \mathfrak{A}$. $\square$

Application of Stinespring's Theorem to Representations of $\mathcal{O}_N$

In a number of applications, only one of the two axioms of the Cuntz relations is satisfied, see (6.27) in Definition 6.2. Note that the combined system (6.27) entails a system of isometries which map a fixed Hilbert space onto a system of subspaces which are mutually orthogonal. Below we show that if the orthogonality requirement is dropped, so only (6.16) holds for a system of bounded operators, then it is still possible to realize the Cuntz relations (6.27) in an enlarged (dilated) Hilbert space in such a way that (6.27) holds in the compressed Hilbert space. A system of operators satisfying (5.22) is called a "row-isometry."

Corollary 6.1. *Let $N \in \mathbb{N}$, $N > 1$, and let $A_i \in \mathcal{B}(\mathcal{H})$, $1 \leq i \leq N$, be a system of operators in a Hilbert space $\mathcal{H}$ such that*

$$\sum_{i=1}^{N} A_i^*A_i = I_{\mathcal{H}}; \qquad (6.16)$$

then there is a second Hilbert space $\mathscr{K}$, *and an isometry* $V : \mathscr{H} \to \mathscr{K}$, *and a representation* $\pi \in Rep\,(\mathscr{O}_N, \mathscr{K})$ *such that*

$$V^* \pi(s_i) V = A_i^*, \ 1 \le i \le N, \tag{6.17}$$

where $\{s_i\}_{i=1}^N$ *are generators for* $\mathscr{O}_N$.

Proof. Given $\mathscr{O}_N$ with generators $\{s_i\}_{i=1}^N$, then set

$$\varphi(s_i s_j^*) = A_i^* A_j$$

using (6.16), it is easy to see that φ is completely positive.

Now let $(\pi, \mathscr{K})$ be the pair obtained from Theorem 6.1 (Stinespring); then as a block-matrix of operators, we have as follows

$$\pi\,(s_i)^* = \begin{bmatrix} A_i & * \\ \mathbf{0} & * \end{bmatrix} \tag{6.18}$$

relative to the splitting

$$\mathscr{K} = V\mathscr{H} \oplus (\mathscr{K} \ominus V\mathscr{H}), \tag{6.19}$$

and so $V^* \pi\,(s_i)^* V = A_i$, which is equivalent to (6.17). $\qquad\square$

6.4 Applications

In Stinespring's theorem, the dilated space comes from a general principle (using positive definite functions) when building Hilbert spaces out of the given data. We illustrate this point with a few familiar examples.

Example 6.1. In linear algebra, there is a bijection between inner product structures on $\mathbb{C}^n$ and positive-definite $n \times n$ matrices. Specifically, $\langle \cdot, \cdot \rangle : \mathbb{C}^n \times \mathbb{C}^n \to \mathbb{C}$ is an inner product if and only if there exists a positive definite matrix A such that

$$\langle v, w \rangle_A = v^* A w$$

for all $v, w \in \mathbb{C}^n$. We think of $\mathbb{C}^n$ as $\mathbb{C}$-valued functions on $\{1, 2, \ldots, n\}$, then $\langle \cdot, \cdot \rangle_A$ is an inner product built on the function space.

This is then extended to infinite dimensional space.

Example 6.2. If F is a positive definite function on $\mathbb{R}$, then on $K_0 = span\{\delta_x : x \in \mathbb{R}\}$, F defines a sesquilinear form $\langle \cdot, \cdot \rangle_F : \mathbb{R} \times \mathbb{R} \to \mathbb{C}$, where

$$\left\langle \sum_i c_i \delta_{x_i}, \sum_j d_j \delta_{x_j} \right\rangle_F := \sum_{i,j} \overline{c_i} d_j F(x_i, x_j), \text{ and}$$

$$\left\| \sum_i c_i \delta_{x_i} \right\|_F^2 := \left\langle \sum_i c_i \delta_{x_i}, \sum_j c_j \delta_{x_j} \right\rangle_F = \sum_{i,j} \overline{c_i} c_j F\,(x_i, x_j) \ge 0.$$

Let $N = \{v \in K_0 : \langle v, v \rangle = 0\}$, then N is a closed subspace in K_0. We get a Hilbert space: $\mathscr{K} := cl_F(K_0/N) =$ the completion of K_0/N with respect to $\|\cdot\|_F$.

What if the index set is not $\{1, 2, \dots, n\}$ or $\mathbb{R}$, but a $*$-algebra?

Example 6.3. $C(X)$, X compact Hausdorff. It is a C^*-algebra, where $\|f\| := \sup_x |f(x)|$. By Riesz's theorem, there is a bijection between positive states (linear functionals) on $C(X)$ and Borel probability measures on X.

Let $\mathfrak{B}(X)$ be the Borel sigma-algebra on X, which is also an Abelian algebra: The associative multiplication is defined as $AB := A \cap B$. The identity element is just X.

Let μ be a probability measure, then $\mu(A \cap B) \geq 0$, for all $A, B \in \mathfrak{B}(X)$, and $\mu(X) = 1$. Hence μ is a state. As before, we apply the GNS construction. Set

$$K_0 = span\{\delta_A : A \in \mathfrak{M}\} = span\{\chi_A : A \in \mathfrak{M}\}.$$

Note the index set here is $\mathfrak{B}(X)$, and $\sum_i c_i \delta_{A_i} = \sum_i c_i \chi_{A_i}$, i.e., these are precisely the simple functions. Define

$$\left\langle \sum_i c_i \chi_{A_i}, \sum_j d_j \chi_{B_j} \right\rangle := \sum_{i,j} \overline{c_i} d_j \mu(A_i \cap B_j)$$

which is positive definite, since

$$\left\langle \sum_i c_i \chi_{A_i}, \sum_i c_i \chi_{A_i} \right\rangle = \sum_{i,j} \overline{c_i} c_j \mu(A_i \cap A_j) = \sum_i |c_i|^2 \mu(A_i) \geq 0.$$

Here, $N = \{v \in K_0 : \langle v, v \rangle = 0\} = \mu$-measure zero sets, and

$$\mathscr{H} = cl_\mu(K_0/N) = L^2(\mu).$$

Example 6.4 (GNS). Let $\mathfrak{A}$ be a $*$-algebra. The set of $\mathbb{C}$-valued functions on $\mathfrak{A}$ is $\mathfrak{A} \otimes \mathbb{C}$, i.e., functions of the form

$$\left\{ \sum_i A_i \otimes c_i = \sum_i c_i \delta_{A_i} \right\} \tag{6.20}$$

with finite summation over i. Note that $\mathbb{C}$ is naturally embedded into $\mathfrak{A} \otimes \mathbb{C}$ as $1_\mathfrak{A} \otimes \mathbb{C}$ (i.e., $c \mapsto c \delta_{1_\mathfrak{A}}$), and the latter is a 1-dimensional subspace. In order to build a Hilbert space out of (6.20), one needs a positive definite function. A state φ on $\mathfrak{A}$ does exactly the job. The sesquilinear form is given by

$$\left\langle \sum_i c_i \delta_{A_i}, \sum_i d_j \delta_{B_j} \right\rangle_\varphi := \sum_{i,j} \overline{c_i} d_j \varphi(A_i^* B_j)$$

so that

$$\left\| \sum_i c_i \delta_{A_i} \right\|_\varphi^2 = \sum_{i,j} \overline{c_i} c_j \varphi(A_i^* A_j) \geq 0.$$

Finally, let $\mathscr{K}_\varphi$ = Hilbert completion of $\mathfrak{A} \otimes \mathbb{C}/\ker\varphi$. Define $\pi(A)\delta_B := \delta_{BA}$, so a "shift" in the index variable, and extend to $\mathscr{K}_\varphi$.

In Stinespring's construction, $\mathfrak{A} \otimes \mathbb{C}$ is replaced by $\mathfrak{A} \otimes \mathscr{H}$, i.e., in stead of working with $\mathbb{C}$-valued functions on $\mathfrak{A}$, one looks at $\mathscr{H}$-valued functions. Hence we are looking at functions of the form

$$\left\{ \sum_i A_i \otimes \xi_i = \sum_i \xi_i \delta_{A_i} \right\}$$

with finite summation over i. $\mathscr{H}$ is embedded into $\mathfrak{A} \otimes \mathscr{H}$ as $1_{\mathfrak{A}} \otimes \mathscr{H}$, by $\mathscr{H} \ni 1_{\mathfrak{A}} \otimes \xi = \xi \delta_{1_{\mathfrak{A}}}$. $1_{\mathfrak{A}} \otimes \mathscr{H}$ is in general infinite dimensional, or we say that the function $\xi \delta_{1_{\mathfrak{A}}}$ at $1_{\mathfrak{A}}$ has infinite multiplicity. If $\mathscr{H}$ is separable, we are actually attaching an l^2 sequence at every point $A \in \mathfrak{A}$.

How to build a Hilbert space out of these $\mathscr{H}$-valued functions? The question depends on the choice of a quadratic form. If $\varphi : \mathfrak{A} \to \mathscr{B}(\mathscr{H})$ is positive, i.e., φ maps positive elements in $\mathfrak{A}$ to positive operators on $\mathscr{H}$, then quadratic form

$$\langle A \otimes \xi, B \otimes \eta \rangle_\varphi := \langle \xi, \varphi(A^*B)\eta \rangle_{\mathscr{H}}$$

is indeed positive definite. But when extend linearly, one is in trouble. For

$$\left\langle \sum_i A_i \otimes \xi_i, \sum_j B_j \otimes \eta_j \right\rangle_\varphi$$

$$= \sum_{i,j} \langle \xi_i, \varphi(A_i^* B_j)\eta_j \rangle_{\mathscr{H}}$$

$$= \begin{bmatrix} \xi_1 & \xi_2 & \cdots & \xi_n \end{bmatrix} \begin{bmatrix} \varphi(A_1^*B_1) & \varphi(A_1^*B_2) & \cdots & \varphi(A_1^*B_n) \\ \varphi(A_2^*B_1) & \varphi(A_2^*B_1) & \cdots & \varphi(A_2^*B_1) \\ \vdots & \vdots & \vdots & \vdots \\ \varphi(A_n^*B_1) & \varphi(A_n^*B_2) & \cdots & \varphi(A_n^*B_n) \end{bmatrix} \begin{bmatrix} \xi_1 \\ \xi_2 \\ \vdots \\ \xi_n \end{bmatrix}$$

and it is not clear why the matrix $(\varphi(A_i^* B_j))_{i,j=1}^n$ should be a positive operator acting in $\mathscr{H} \otimes \mathbb{C}^n$. But we could very well put this extra requirement into an axiom, so the CP condition (6.1).

1. We only assume $\mathfrak{A}$ is a $*$-algebra, not necessarily a C^*-algebra. $\varphi : \mathfrak{A} \to \mathscr{B}(\mathscr{H})$ is positive does not necessarily imply φ is completely positive. A counterexample for $\mathfrak{A} = M_2(\mathbb{C})$, and $\varphi : \mathfrak{A} \to B(\mathbb{C}^2) \simeq M_2(\mathbb{C})$ given by taking transpose, i.e., $A \mapsto \varphi(A) = A^{tr}$. Then φ is positive, but $\varphi \otimes I_{M_2}$ is not.

2. The operator matrix $(A_i^* A_j)$, which is also written as $\sum_{i,j} A_i^* A_j \otimes e_{ij}$ is a positive element in $\mathfrak{A} \otimes M_n$. All positive elements in $\mathfrak{A} \otimes M_n$ are in such form. This notation goes back again to Dirac, for the rank-1 projections $|v\rangle\langle v|$ are positive, and all positive operators are sums of these rank-1 operators.

3. Given a CP map $\varphi : \mathfrak{A} \to \mathscr{B}(\mathscr{H})$, we get a Hilbert space $\mathscr{K}_\varphi$, a representation $\pi : \mathfrak{A} \to B(\mathscr{K}_\varphi)$ and an isometry $V : \mathscr{H} \to \mathscr{K}_\varphi$, such that

$$\varphi(A) = V^*\pi(A)V$$

for all $A \in \mathfrak{A}$. $P = VV^*$ is a selfadjoint projection from $\mathscr{K}_\varphi$ to the image of $\mathscr{H}$ under the embedding. To see P is a projection, note that

$$P^2 = VV^*VV^* = V(V^*V)V^* = VV^*.$$

Summary

Positive maps have been a recursive theme in functional analysis. A classical example is $\mathfrak{A} = C_c(X)$ with a positive linear functional $\Lambda : \mathfrak{A} \to \mathbb{C}$, mapping $\mathfrak{A}$ into a 1-d Hilbert space $\mathbb{C}$.

In Stinespring's formulation, $\varphi : \mathfrak{A} \to \mathscr{H}$ is a CP map, then we may write $\varphi(A) = V^*\pi(A)V$ where $\pi : \mathfrak{A} \to \mathscr{K}$ is a representation on a bigger Hilbert space $\mathscr{K}$ containing $\mathscr{H}$. The containment is in the sense that $V : \mathscr{H} \hookrightarrow \mathscr{K}$ embeds $\mathscr{H}$ into $\mathscr{K}$. Notice that

$$V\varphi(A) = \pi(A)V \implies \varphi(A) = V^*\pi(A)V$$

but not the other way around. In Nelson's notes [Nel69], we use the notation $\varphi \subset \pi$ for one representation being the subrepresentation of another. To imitate the situation in linear algebra, we may want to split an operator T acting on $\mathscr{K}$ into operators action on $\mathscr{H}$ and its complement in $\mathscr{K}$. Let $P : \mathscr{K} \to \mathscr{H}$ be the orthogonal projection. In matrix language,

$$\begin{bmatrix} PTP & PTP^\perp \\ P^\perp TP & P^\perp TP^\perp \end{bmatrix}. \tag{6.21}$$

A better looking would be

$$\begin{bmatrix} PTP & 0 \\ 0 & P^\perp TP^\perp \end{bmatrix} = \begin{bmatrix} \varphi_1 & 0 \\ 0 & \varphi_2 \end{bmatrix}$$

hence

$$\pi = \varphi_1 \oplus \varphi_2.$$

Stinespring's theorem is more general, where the off-diagonal entries may not be zero.

Exercise 6.1 (Tensor with M_n). Let $\mathfrak{A} = \mathscr{B}(\mathscr{H})$, $\xi_1, \ldots, \xi_n \in \mathscr{H}$. The map

$$(\xi_1, \ldots, \xi_n) \mapsto (A\xi_1, \ldots, A\xi_n) \in \oplus^n \mathscr{H}$$

is a representation of $\mathfrak{A}$ if and only if

$$\overbrace{id_{\mathscr{H}} \oplus \cdots \oplus id_{\mathscr{H}}}^{n \text{ times}} \in Rep(\mathfrak{A}, \overbrace{\mathscr{H} \oplus \cdots \oplus \mathscr{H}}^{n \text{ times}})$$

where in matrix notation, we have

$$
\begin{bmatrix}
id_{\mathfrak{A}}(A) & 0 & \cdots & 0 \\
0 & id_{\mathfrak{A}}(A) & \cdots & 0 \\
\vdots & \vdots & \vdots & \vdots \\
0 & 0 & \cdots & id_{\mathfrak{A}}(A)
\end{bmatrix}
\begin{bmatrix}
\xi_1 \\
\xi_2 \\
\vdots \\
\xi_n
\end{bmatrix}
$$
$$
=
\begin{bmatrix}
A & 0 & \cdots & 0 \\
0 & A & \cdots & 0 \\
\vdots & \vdots & \vdots & \vdots \\
0 & 0 & \cdots & A
\end{bmatrix}
\begin{bmatrix}
\xi_1 \\
\xi_2 \\
\vdots \\
\xi_n
\end{bmatrix}
=
\begin{bmatrix}
A\xi_1 \\
A\xi_2 \\
\vdots \\
A\xi_n
\end{bmatrix}.
$$

In this case, the identity representation $id_{\mathfrak{A}} : \mathfrak{A} \to \mathscr{H}$ has multiplicity n.

Exercise 6.2 (Column operators). Let $V_i : \mathscr{H} \to \mathscr{H}$, and

$$
V := \begin{bmatrix}
V_1 \\
V_2 \\
\vdots \\
V_n
\end{bmatrix} : \mathscr{H} \to \oplus_1^n \mathscr{H}. \tag{6.22}
$$

Let $V^* : \oplus_1^n \mathscr{H} \to \mathscr{H}$ be the adjoint of V. Prove that $V^* = \begin{bmatrix} V_1^* & V_2^* & \cdots & V_n^* \end{bmatrix}$.

Proof. Let $\xi \in \mathscr{H}$, then $V\xi = \begin{bmatrix} V_1\xi \\ V_2\xi \\ \vdots \\ V_n\xi \end{bmatrix}$, and

$$
\left\langle \begin{bmatrix} \eta_1 \\ \eta_2 \\ \vdots \\ \eta_n \end{bmatrix}, \begin{bmatrix} V_1\xi \\ V_2\xi \\ \vdots \\ V_n\xi \end{bmatrix} \right\rangle = \sum_i \langle \eta_i, V_i\xi \rangle
$$

$$
= \sum_i \langle V_i^*\eta_i, \xi \rangle = \left\langle \begin{bmatrix} V_1^* & V_2^* & \cdots & V_n^* \end{bmatrix} \begin{bmatrix} \eta_1 \\ \eta_2 \\ \vdots \\ \eta_n \end{bmatrix}, \xi \right\rangle
$$

This shows that $V^* = \begin{bmatrix} V_1^* & V_2^* & \cdots & V_n^* \end{bmatrix}$. □

Exercise 6.3 (Row-isometry). Let V be as in (6.22). The following are equivalent:

1. V is an isometry, i.e., $\|V\xi\|^2 = \|\xi\|^2$, for all $\xi \in \mathscr{H}$;

2. $\sum V_i^* V_i = I_{\mathscr{H}}$;

3. $V^*V = I_{\mathcal{H}}$.

Proof. Notice that

$$\|V\xi\|^2 = \sum_i \|V_i\xi\|^2 = \sum_i \langle \xi, V_i^*V_i\xi \rangle = \left\langle \xi, \sum_i V_i^*V_i\xi \right\rangle.$$

Hence $\|V\xi\|^2 = \|\xi\|^2$ if and only if

$$\left\langle \xi, \sum_i V_i^*V_i\xi \right\rangle = \langle \xi, \xi \rangle$$

for all $\xi \in \mathcal{H}$. Equivalently, $\sum_i V_i^*V_i = I_H = V^*V$.   □

Corollary 6.2 (Krauss). *Let* $\dim\mathcal{H} = n$. *Then all the CP maps are of the form*

$$\varphi(A) = \sum_i V_i^*AV_i.$$

(The essential part here is that for any CP mapping φ, we get a system $\{V_i\}$.)

This was discovered in the physics literature by Kraus [Kra83]; see also [Cho75], and Theorem 6.2 below. The original proof was very intricate, but it is a corollary of Stinespring's theorem. When $\dim\mathcal{H} = n$, let $\{e_1, \ldots e_n\}$ be an ONB. Fix a CP map φ, and get $(V, \mathcal{K}, \pi)$. Set

$$V_i : e_i \mapsto Ve_i \in \mathcal{K}, \ i = 1, \ldots n;$$

then V_i is an isometry. So we get a system of isometries, and

$$\varphi(A) = \begin{bmatrix} V_1^* & V_2^* & \cdots & V_n^* \end{bmatrix} \begin{bmatrix} A & & & \\ & A & & \\ & & \ddots & \\ & & & A \end{bmatrix} \begin{bmatrix} V_1 \\ V_2 \\ \vdots \\ V_n \end{bmatrix}.$$

Notice that $\varphi(1) = 1$ if and only if $\sum_i V_i^*V_i = 1$.

Exercise 6.4 (Tensor products). Prove the following.

1. $\oplus_1^n \mathcal{H} \simeq \mathcal{H} \otimes \mathbb{C}^n$

2. $\sum_1^{\oplus\infty} \mathcal{H} \simeq \mathcal{H} \otimes l^2$

3. Given $L^2(X, \mathfrak{M}, \mu)$, then $L^2(X, \mathcal{H}) \simeq \mathcal{H} \otimes L^2(\mu)$; where $L^2(X, \mathcal{H})$ consists of all measurable functions $f : X \to \mathcal{H}$ such that

$$\int_X \|f(x)\|_{\mathcal{H}}^2 \, d\mu(x) < \infty$$

and

$$\langle f, g \rangle = \int_X \langle f(x), g(x) \rangle_{\mathcal{H}} \, d\mu(x).$$

4. All the normed spaces above are Hilbert spaces.

Exercise 6.5 (Using tensor product in representations). Let $(X_i, \mathfrak{M}_i, \mu_i)$, $i = 1, 2$, be measure spaces. Let $\pi_i : L^\infty(\mu_i) \to L^2(\mu_i)$ be the representation such that $\pi_i(f)$ is the operator of multiplication by f on $L^2(\mu_i)$. Hence $\pi_i \in Rep(L^\infty(X_i), L^2(\mu_i))$, and

$$\pi_1 \otimes \pi_2 \in Rep(L^\infty(X_1 \times X_2), L^2(\mu_1 \times \mu_2)),$$
$$\pi_1 \otimes \pi_2(\tilde{\varphi})\tilde{f} = \tilde{\varphi}\tilde{f}$$

for all $\tilde{\varphi} \in L^\infty(X_1 \times X_2)$, and all $\tilde{f} \in L^2(\mu_1 \times \mu_2)$.

Notation. Elementary tensors: Special form

$$\begin{aligned}
\tilde{\varphi}(x_1, x_2) &= \varphi_1(x_1)\,\varphi_2(x_2)\,, \\
\tilde{f}(x_1, x_2) &= f_1(x_1)\,f_2(x_2)\,, \\
(\pi_1 \otimes \pi_2)(\tilde{\varphi})\,f &= \pi_1(\varphi_1)\,f_1 \otimes \pi_2(\varphi_2)\,f_2.
\end{aligned}$$

Exercise 6.6 ("Transpose" is not completely positive).

1. Let $\mathfrak{A}$ be an Abelian C^*-algebra; and let $\varphi : \mathfrak{A} \to \mathscr{B}(\mathscr{H})$ be a positive mapping; then show that φ is in fact automatically completely positive.

2. Show that there are positive mappings which are <u>not</u> completely positive. <u>Hint:</u> Let M_n be the $n \times n$ complex matrices, and set

$$\varphi(A) = A^T, \ A \in M_n$$

where A^T is the transpose matrix. If $n > 1$, show that $M_n \xrightarrow{\varphi} M_n$ is positive but <u>not</u> completely positive.

6.5 Factorization

The setting below is as in the statement of Theorem 6.1.

Given $\mathfrak{A}$, φ, $\mathscr{H}$ where

$\mathfrak{A}$ is a $*$-algebra,

$\mathscr{H}$ is a Hilbert space; and

$\varphi : \mathfrak{A} \longrightarrow \mathscr{B}(\mathscr{H})$ is given and completely positive.

Terminology. When $\mathfrak{A}$, φ, $\mathscr{H}$ are as above, we say that $\varphi \in \mathrm{CP}(\mathfrak{A}, \mathscr{H})$.

Observation. Let $\mathscr{K}$ be a second Hilbert space. Let $N \in \mathbb{N}$, and let $\{V_i\}_{i=1}^N$ be a subset of $\mathscr{L}(\mathscr{H}, \mathscr{K})$ (= the bounded linear operators from $\mathscr{H}$ into $\mathscr{K}$.) Set

$$\varphi_V(A) := \sum_{i=1}^N V_i A V_i^*, \quad A \in \mathscr{B}(\mathscr{H}). \tag{6.23}$$

Then $\varphi_V \in CP(\mathscr{B}(\mathscr{H}), \mathscr{K})$.

The following theorem of Choi follows from the discussion above, and a little linear algebra. For details, see [Cho75, Kra83].

Theorem 6.2 (M. D. Choi, K. Kraus). *Let $\mathscr{H}$ and $\mathscr{K}$ be finite-dimensional Hilbert spaces; set $\dim \mathscr{H} = d_{\mathscr{H}}$, and $\dim \mathscr{K} = d_{\mathscr{K}}$. Let $\varphi : \mathscr{B}(\mathscr{H}) \longrightarrow \mathscr{B}(\mathscr{K})$ be given and completely positive, i.e., $\varphi \in CP(\mathscr{B}(\mathscr{H}), \mathscr{K})$; then φ has the form (6.23) for some system of operators*

$$V_i : \mathscr{H} \longrightarrow \mathscr{K}, \quad 1 \le i \le d_{\mathscr{H}} d_{\mathscr{K}}(:= N).$$

Proof. (Hint!) Let φ be as in the statement of the theorem, and let $\{e_{ij}\}_{i,j=1}^{d_{\mathscr{H}}}$ be the matrix-unit-system (see Remark 6.1), i.e., e_{ij} is the $d_{\mathscr{H}} \times d_{\mathscr{H}}$ matrix with entries

$$(e_{ij})_{k,l} = \delta_{i,k} \delta_{j,l}, \tag{6.24}$$

then

$$[\varphi(e_{ij})]_{i,j=1}^{d_{\mathscr{H}}} \tag{6.25}$$

is positive in $M_{d_{\mathscr{H}}} \otimes B(\mathscr{K})$ where $M_{d_{\mathscr{H}}} = B(\mathscr{H})$ (= all $d_{\mathscr{H}} \times d_{\mathscr{H}}$ complex matrices.) The matrix in (6.25) is called the Choi-matrix.

To appreciate the argument involved in (6.25), consider $\xi = (\xi_i)_{i=1}^{d_{\mathscr{H}}} \subset \mathbb{C}^{d_{\mathscr{H}}}$ then

$$\sum_i \sum_j \overline{\xi}_i \xi_j e_{ij} = |\xi\rangle\langle\xi| \tag{6.26}$$

(referring to Dirac's notation for rank-one operators), or equivalently

$$\text{LHS}_{(6.26)} = \begin{bmatrix} |\xi_1|^2 & \overline{\xi}_1\xi_2 & \cdots & \overline{\xi}_1\xi_{d_{\mathscr{H}}} \\ \xi_1\overline{\xi}_2 & |\xi_2|^2 & \cdots & \xi_1\overline{\xi}_{d_{\mathscr{H}}} \\ \vdots & \vdots & & \vdots \\ \xi_1\overline{\xi}_{d_{\mathscr{H}}} & \overline{\xi}_1\xi_{d_{\mathscr{H}}} & \cdots & |\xi_{d_{\mathscr{H}}}|^2 \end{bmatrix}.$$

$\square$

Remark 6.3. Note that the last conclusion about (6.25) holds even if φ is only assumed $d_{\mathscr{H}}$-positive (as opposed to CP.) Hence the "complete" part is automatic from the assumption that φ be positive; in this particular case.

6.6   Endomorphisms, Representations of $\mathscr{O}_N$, and Numerical Range

Let $\mathscr{H}$ be a Hilbert space, and consider endomorphisms in $\mathscr{B}(\mathscr{H})$, i.e., $\sigma : \mathscr{B}(\mathscr{H}) \longrightarrow \mathscr{B}(\mathscr{H})$, linear, and satisfy

$$
\begin{aligned}
\sigma(AB) &= \sigma(A)\sigma(B) \\
\sigma(A^*) &= \sigma(A)^*, \ \forall A, B \in \mathscr{B}(\mathscr{H}), \text{ and} \\
\sigma(I) &= I.
\end{aligned}
$$

Notation. Given $\sigma \in End(\mathscr{B}(\mathscr{H}))$, the N in the corresponding representation (6.32) below is called *Powers-index* of σ. It holds that for every $\sigma \in End(\mathscr{B}(\mathscr{H}))$, the relative commutant

$$
\mathscr{B}(\mathscr{H}) \cap \sigma(\mathscr{B}(\mathscr{H}))'
$$

is a type I_N, and this N coincides with the Powers-index.

Definition 6.2. By a *representation* π of $\mathscr{O}_N$ in $\mathscr{H}$, $\pi \in Rep(\mathscr{O}_N, \mathscr{H})$, we mean a system of isometries $(S_i)_{i=1}^N$ in $\mathscr{H}$ such that

$$
\left\{ \begin{array}{l} S_i^* S_j = \delta_{ij} I \\ \sum_i S_i S_i^* = I \end{array} \right\} \quad \text{(Cuntz relations)}. \tag{6.27}
$$

See Figure 6.1.

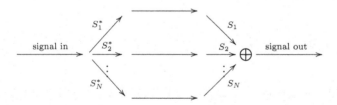

Figure 6.1: Orthogonal bands in filter bank, "in" = "out". An application of representations of the Cuntz relations (6.27).

Remark 6.4. While the relations in (6.27), called the Cuntz relations, of Definition 6.2 are axioms, they have implications for a host of applications, and Figure 6.1 is a graphic representations of (6.27) stated in a form popular in *applications to signal processing*. Effective transmission of signals (speech, or images), is possible because the transmitted signals can be divided into *frequency sub-bands*; this is done with filters. A *low-pass filter* picks out the band corresponding to frequencies in a "band" around zero, and similarly with intermediate, and high bands. The horizontal lines in Figure 6.1 represent prescribed bands. The orthogonality part of (6.27) represents *non-interference* from one band to the next. Adding the projections on the LHS in (6.27) to recover the identity operator reflects perfect reconstruction, i.e, signal out equals signal in. The projections on the LHS in (6.27) are projections onto subspaces of a total Hilbert space (of signals to be transmitted), the subspaces thus representing frequency bands. (See Figures 6.1-6.2.)

Thus Figure 6.1 represents such a filter design; there are many such, some good some not. Each one is called a "filter bank." And each one corresponds to a representation of (6.27), or equivalently a representation of the Cuntz algebra $\mathscr{O}_N$ where N is the number of band for the particular filter design.

In signal and image processing, a low-pass filter (LPF) is a filter that passes signals with a low frequency range, say, lower than a chosen cutoff, and the LPF attenuates the other signals having their frequency range higher than a cutoff. There is an analogous definition for high-pass filters (HPF). Together they constitute a system of band-pass filters. In case of more than two chosen bands for the frequency range, there will be an associated system of band-pass filters. Examples of their uses includes electronic circuits (such as a hiss filter used in audio), anti-aliasing filters for conditioning signals prior to analog-to-digital conversion, digital filters, acoustic barriers, blurring of images, and so on. Filters based on various kinds of moving average operations, are used in multi-resolutions, including MRAs for wavelets. They are also used in finance is a particular kind of low-pass filter used for removing the short-term fluctuations. The literature dealing with band-pass filters of one kind or the other is vast, see e.g., [AJL13, Alp01, BJ02, dBR66, JS07, Mal01, Mey06, MM09, KMRS05].

Exercise 6.7 (*Rep*($\mathscr{O}_N, \mathscr{H}$)). Fix $N \geq 2$, and let $\mathscr{O}_N$ denote the Cuntz C^*-algebra (see Example 5.3). By

$$\pi \in Rep\left(\mathscr{O}_N, \mathscr{H}\right) \tag{6.28}$$

we mean a homomorphism

$$\pi : \mathscr{O}_N \longrightarrow \mathscr{B}\left(\mathscr{H}\right),$$

(in particular, satisfying: $\pi\left(AB\right) = \pi\left(A\right)\pi\left(B\right)$, $\pi(A^*) = \pi\left(A\right)^*$, $\forall A, B \in \mathscr{O}_N$, and $\pi\left(1\right) = I_{\mathscr{H}}$.)

Let $\{S_i\}_{i=1}^N$ be a system of isometries in a Hilbert space $\mathscr{H}$ satisfying (6.27), called Cuntz-isometries. For all multi-indices $J = (j_1, j_2, \cdots, j_m)$, $j_i \in \{1, 2, \cdots, N\}$, set

$$\begin{aligned} s_J &:= s_{j_1} s_{j_2} \cdots s_{j_m}, \text{ and} \\ S_J &:= S_{j_1} S_{j_2} \cdots S_{j_m}. \end{aligned}$$

Show that, given a (6.27)-system $\{S_i\}_{i=1}^N$ of isometries, there is then a unique $\pi \in Rep(\mathscr{O}_N, \mathscr{H})$ such that

$$\pi\left(s_J s_K^*\right) = S_J S_K^* \tag{6.29}$$

holds for all multi-indices J, K.

Example 6.5 (Representation of the Cuntz algebra $\mathcal{O}_2$). Let $\mathcal{H} = L^2(\mathbb{T})$. In signal processing language $\mathcal{H}$ is the L^2-space of frequency functions. Set

$$(S_0 f)(x) := \cos(x) f(2x) \tag{6.30}$$

$$(S_1 f)(x) := \sin(x) f(2x) \tag{6.31}$$

as the two Cuntz operators, where $f \in \mathcal{H}$ (see Figure 6.2), and

$$2x := 2x \quad \text{mod} \ 2\pi\mathbb{Z} \ (= \text{multiples of } 2\pi).$$

(The generators S_i, $i = 1, 2$ with up-sampling, and S_i^* with down-sampling.)

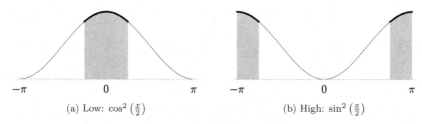

(a) Low: $\cos^2\left(\frac{x}{2}\right)$ (b) High: $\sin^2\left(\frac{x}{2}\right)$

Figure 6.2: Low / high pass filters for the Haar wavelet (frequency mod 2π).

Exercise 6.8 (Simplest Low/High Pass filter bank). Show that (6.30)-(6.31) satisfy the $\mathcal{O}_2$-Cuntz relations, i.e.,

1. S_i, $i = 0, 1$ are isometries in $\mathcal{H}$;

2. $S_0^* S_1 = 0$ (orthogonality);

3. $S_0 S_0^* + S_1 S_1^* = I_{\mathcal{H}}$.

Hint: First show that

$$(S_0^* f)(x) = \frac{1}{2}\left(\cos\left(\frac{x}{2}\right) f\left(\frac{x}{2}\right) + \cos\left(\frac{x+\pi}{2}\right) f\left(\frac{x+\pi}{2}\right)\right)$$

are similarly for $S_1^* f$. Then compute directly that

$$\|S_0^* f\|_{\mathcal{H}}^2 + \|S_1^* f\|_{\mathcal{H}}^2 = \|f\|_{\mathcal{H}}^2 .$$

Example 6.6. [2]Consider the Haar wavelet as in Example 2.7, with ϕ_0 (scaling function), φ_1 and $\psi_{j,k}$, $j, k \in \mathbb{Z}$ be as in (2.61)-(2.62). Set

$$h(n) = \begin{cases} -\frac{1}{2} & n = -1 \\ \frac{1}{2} & n = 0 \\ 0 & \text{otherwise} \end{cases}, \quad g(n) = \begin{cases} \frac{1}{2} & n = -1 \\ \frac{1}{2} & n = 0 \\ 0 & \text{otherwise} \end{cases};$$

---

[2]A summary of some highpoints from wavelets is contained in Appendix B. See also Sections 2.4.2 (pg. 46), 5.8 (pg. 198), 6.4 (pg. 233), and Chapter 6 (pg. 223).

so that $h, g \in l^2$; where g is the low-pass filter (averaging data), and h is the high-pass filter (capturing high-frequency oscillations). Let m_0 and m_1 be Fourier transform of g and h respectively, i.e.,

$$m_0(x) = \sum_{n \in \mathbb{Z}} g(n) e^{-ixn}$$

$$m_1(x) = \sum_{n \in \mathbb{Z}} h(n) e^{-ixn}$$

and $m_0, m_1 \in L^2(\mathbb{T})$. Finally set S_0^* and S_1^* as in Figure 6.3.

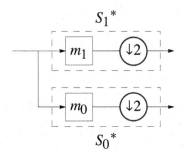

Figure 6.3: The Cuntz operators S_0 and S_1 in the Haar wavelet.

An input signal goes through the analysis filter bank (Figure 6.4) and splits into layers (frequency bands) of fine details. The original signal can be rebuilt through the synthesis filter bank (Figure 6.5), i.e., a perfect reconstruction.

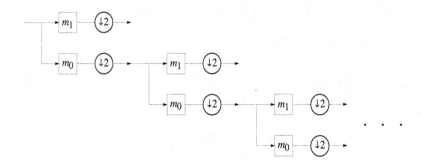

Figure 6.4: The two-channel analysis filter bank.

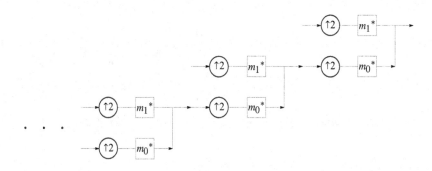

Figure 6.5: The two-channel synthesis filter bank.

Depending on the applications, the output of the analysis filter bank will go through other DSP device. For example, in data compression, insignificant coefficients are dropped; or if the task is to remove noise in the input signal, the coefficients corresponding to high frequency components (noise) are removed, and the remaining coefficients go through the synthesis filter bank. See Figures 6.6-6.7 for an illustration, and Figure 6.8 for an application in imaging processing.

We return to a much more detailed discussion of down-sampling and up-sampling in Chapter 7.

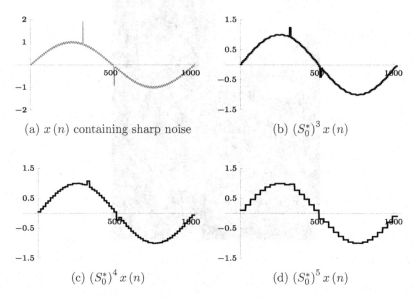

(a) $x(n)$ containing sharp noise

(b) $(S_0^*)^3 x(n)$

(c) $(S_0^*)^4 x(n)$

(d) $(S_0^*)^5 x(n)$

Figure 6.6: The outputs of $(S_0^*)^n$, $n = 3, 4, 5$.

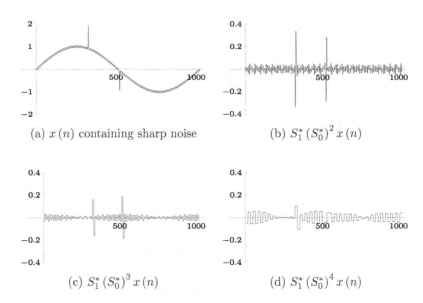

(a) $x(n)$ containing sharp noise

(b) $S_1^* (S_0^*)^2 x(n)$

(c) $S_1^* (S_0^*)^3 x(n)$

(d) $S_1^* (S_0^*)^4 x(n)$

Figure 6.7: The outputs of high-pass filters.

Figure 6.8: A coarser resolution in three directions in the plane, filtering in directions, x, y, and diagonal; — corresponding dyadic scaling in each coordinate direction. (Image cited from M.-S. Song, "*Wavelet Image Compression*" in [HJL06].)

Exercise 6.9 (Endomorphism vs representation). Let $\mathcal{H}$ be a general separable Hilbert space. The purpose below is to point out that the study of $Rep(\mathcal{O}_N, \mathcal{H})$ is essentially equivalent to that of the endomorphisms of $\mathcal{B}(\mathcal{H})$.

1. Let σ be an endomorphism in $\mathcal{B}(\mathcal{H})$ of finite index N. Show that there is a representation (S_i) of $\mathcal{O}_N$ in $\mathcal{H}$ such that

$$\sigma(A) = \sum_{i=1}^{N} S_i A S_i^*. \tag{6.32}$$

2. Let σ, $\{S_i\}$ be as in (1), and let $A \in \mathcal{B}(\mathcal{H})$; then show that

$$NR_{\sigma(A)} \subseteq NR_A. \tag{6.33}$$

In other words, endomorphisms in $\mathcal{B}(\mathcal{H})$ contract the *numerical range*.

<u>Hint</u>: Use the following three facts:

(i) The numerical range NR_A is convex; and

(ii) if $x \in \mathcal{H}$, $\|x\| = 1$, then (see Figure 6.9)

$$w_x(\sigma(A)) = \sum_{i=1}^{N} \|S_i^* x\|^2 \, w_{\frac{S_i^* x}{\|S_i^* x\|}}(A); \tag{6.34}$$

(iii) and lastly,

$$\sum_{i=1}^{N} \|S_i^* x\|^2 = 1. \tag{6.35}$$

Exercise 6.10 (Convex sets that are not numerical ranges). Give an example of a bounded convex subset of the complex plane which is not NR_A for any $A \in \mathcal{B}(\mathcal{H})$, where $\mathcal{H}$ is some Hilbert space.

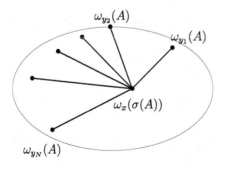

Figure 6.9: Illustration of eq. (6.34), with $y_i := \dfrac{S_i^* x}{\|S_i^* x\|}$, $i = 1, 2, \ldots, N$.

A summary of relevant numbers from the Reference List

For readers wishing to follow up sources, or to go in more depth with topics above, we suggest:

The paper [Sti55] is pioneering, starting the study of completely positive mappings in operator algebra theory. A more comprehensive list is: [Arv98, BR81a, Sti59, Tak79, BJ02, Jor06, Arv76, Fan10, BJKR84, Cun77, KR97b, Pow75, Sti55, MJD^{+}15].

Chapter 7

Brownian Motion

From its shady beginnings devising gambling strategies and counting corpses in medieval London, probability theory and statistical inference now emerge as better foundations for scientific models, especially those of the process of thinking and as essential ingredients of theoretical mathematics, even the foundations of mathematics itself.
— David Mumford

"The glory of science is to imagine more than we can prove."
— Freeman Dyson

"Not only does God play dice, but... he sometimes throws them where they cannot be seen."
— Stephen Hawking

It is intriguing that the mathematics of Brownian motion was discovered almost simultaneously more than 100 years ago by Bachelier[1] and by Einstein: In physics (Albert Einstein, 1005, "Über die von der molekularkinetischen Theorie der Wärme geforderte Bewegung von in ruhenden Flüssigkeiten suspendierten Teilchen;" On the Motion of Small Particles Suspended in a Stationary Liquid, as Required by the Molecular Kinetic Theory of Heat). And in finance (Louis Bachelier, 1900, "The Theory of Speculation").

In Einstein's paper, Brownian motion offered one of the first experimental justification for the atomic theory. The Brownian motion model for financial markets is a continuous extension of the "one-period market model" of H. Markowitz, (in fact much later than 1900).

---

[1]Appendix C includes a short biographical sketch of L. Bachelier.

Bachelier: Continuous prices of financial asset-markets evolve in time according to a geometric Brownian motion.

In the early days of the subjects of Brownian motion, and related stochastic processes (Wiener, Lévy, Itō, Malliavin), the new advances have always had an underlying core of functional analysis, and more recently non-commutative analysis. As the more specialized sub-areas of the subject developed, in some instances, this connection has perhaps been swept to the background. (This often happens in interdisciplinary themes.) Fast forward to free probability: With the fast growing research directions within "free probability," and quantum information, an underlying core of non-commutative analysis has now been added.

Disclaimer: It is the purpose of the present section to highlight these interconnections, even though we are not able to go in depth with these exciting neighboring areas. Such an in-depth treatment could easily justify a separate book; or books. Here we have limited our scope to some main ideas we hope will serve as invitations for beginning students who start out in modern analysis. A list of relevant references with details includes [AJ12, ARR13, BØSW04, CW14, Fel68, Fel71, Gro67, Gro70, Hid80, Itô06, Jør14, Loè63, Nel59b, Nel67, Par09, Seg58, Voi85, Voi06]. Even though there is a large and diverse literature, we have aimed here at a self-contained presentation. With the preliminaries from sections 1.2.4 and 3.3.1 and from chapter 5, the reader should be able to start the present chapter directly and from scratch, without first consulting the books. For readers who want to go in depth, the cited references above we believe will help.

7.1 Introduction, Applications, and Context for Path-space Analysis

In our choice of both topics and presentation of the results in the present chapter, dealing with Brownian motion and more general Gaussian processes, we have stressed a unified theme; use of Hilbert space methods. For example we realize Itō-integration as an isometry between Hilbert spaces. In fact, this is a recurrent theme, see also Chapters 2 and 12.

In the details below, the following concepts and terms will play an important role; (for reference, see also Appendix F): Probability space, mutually independent random variables, Brownian motion, Geometric Brownian motion, quadratic variation.

The concept of Brownian motion is not traditionally included in Functional Analysis. Below we offer a presentation which relies on almost all the big theorems from functional analysis, and especially on L^2-Hilbert spaces, built from probability measures on function spaces.

We have also included a brief discussion of Brownian motion in order to illustrate infinite Cartesian products (sect. 1.2.4) and unitary one-parameter group $\{U(t)\}_{t \in \mathbb{R}}$ acting in Hilbert space. (See [Jør14, Nel67, Nel59b, Hid80, Loè63].)

While the mathematical theory of Brownian motion started with L. Bachelier, A. Einstein, N. Wiener, and A. Kolmogorov, the subject has since expanded in a number of exciting directions, right up to the present. As of current, the theory of Brownian motion is at the cutting edge in a number of areas of both pure and applied mathematics. Of the next generation of researchers who influenced our present choice of topics, we mention J. Doob (applications in harmonic analysis), S. Kakutani[2] (two-dimensional Brownian motion, random ergodic theorems, spectrum of the flow of Brownian motion, K. Itō (Itō-calculus, multiple Itō-integrals), T. Hida (functionals of Brownian motion, white noise processes, and infinite dimensional unitary groups), I. E. Segal, E. Nelson, and L. Gross (applications in mathematical physics)[3].

Definition 7.1. Let $(\Omega, \mathcal{F}, \mathbb{P})$ be a *probability space*, i.e.,

- Ω = sample space

- $\mathcal{F}$ = sigma algebra of events

- $\mathbb{P}$ = probability measure defined on $\mathcal{F}$.

A function $X : \Omega \to \mathbb{R}$ is called a *random variable* if it is a measurable function, i.e., if for all intervals $(a, b) \subset \mathbb{R}$ the inverse image

$$X^{-1}((a,b)) = \{\omega \in \Omega \mid X(\omega) \in (a,b)\} \tag{7.1}$$

is in $\mathcal{F}$; and we write $\{a < X(\omega) < b\} \in \mathcal{F}$ in short-hand notation.

Definition 7.2. Events $A, B \in \mathcal{F}$ are said to be *independent* if

$$\mathbb{P}(A \cap B) = \mathbb{P}(A)\mathbb{P}(B).$$

Random variables X and Y are said to be independent iff (Def.) $X^{-1}(I)$ and $Y^{-1}(J)$ are independent for all intervals I and J.

Definition 7.3. Let $(\Omega, \mathcal{F}, \mathbb{P})$ be a fixed probability space, and let X_i be a family of random variables, where i ranges over an arbitrary index set; we say that they are *mutually independent* if, for all Borel sets $B_i \subset \mathbb{R}$, $i = 1, \cdots, n$, the events $X_i^{-1}(B_i)$ are independent.

Definition 7.4. Let μ and ν be two Borel probability measures on $\mathbb{R}$. The *convolution* of the two measures is denoted $\mu * \nu$. It is determined uniquely by the following condition:

$$\int_{\mathbb{R}} f \, d\mu(\mu * \nu) = \int_{\mathbb{R}} \int_{\mathbb{R}} f(x+y) \, d\mu(x) \, d\nu(y), \quad \forall f \in C_c(\mathbb{R}).$$

---

[2]Appendix C includes a short biographical sketch of S. Kakutani. See also [Kak48, Hid80].

[3]A summary of some highpoints from infinite dimensional analysis is contained in Appendix B. See also Sections 1.2.3 (pg. 8), 2.1 (pg. 25), 2.4 (pg. 39), and Chapter 7 (pg. 249).

If "$\widehat{}$" denotes the Fourier transform, one easily shows that $\widehat{\mu * \nu} = \widehat{\mu} \cdot \widehat{\nu}$.

Exercise 7.1 (Independent random variables). Let $(\Omega, \mathcal{F}, \mathbb{P})$ be a probability space (Definition 7.1), and let X_j, $j = 1, 2$, be two random variables. Let μ_{X_j} be the corresponding distributions with respective generating functions

$$\mathbb{E}\left(e^{i\xi X_j}\right) = \widehat{\mu}_{X_j}(\xi) = \int_{\mathbb{R}} e^{i\xi x} d\mu_{X_j}(x), \quad \xi \in \mathbb{R}.$$

Show that X_1 and X_2 are independent iff $\mu_{X_1} * \mu_{X_2} = \mu_{X_1 + X_2}$.

Hint: Use the known fact

$$\widehat{\mu_{X_1} * \mu_{X_2}}(\xi) = \widehat{\mu}_{X_1}(\xi)\,\widehat{\mu}_{X_2}(\xi), \quad \xi \in \mathbb{R}.$$

Definition 7.5. We say that X is *Gaussian* if $\exists\, m, \sigma$ (written $N\left(m, \sigma^2\right)$) such that

$$\mathbb{P}\left(\{a < X(\omega) < b\}\right) = \int_a^b \frac{1}{\sigma\sqrt{2\pi}} e^{-(x-m)^2/2\sigma^2} dx. \tag{7.2}$$

The function under the integral in (7.2) is called the Gaussian (or normal) distribution.

The normal (or Gaussian) distribution is a continuous probability distribution. A random variable with normal distribution is called Gaussian, and a stochastic process with jointly Gaussian distributions is called a Gaussian process. The central limit theorem states that, under suitable conditions on an i.i.d. system of random variables, averages converge in distribution to a Gaussian. (See Exercise 7.12.)

Definition 7.6. A family $\{X_t\}_{t \in \mathbb{R}}$ of random variables on $(\Omega, \mathcal{F}, \mathbb{P})$ is said to be a *Brownian motion* iff (Def.) for every $n \in \mathbb{N}$,

1. the random variables $X_{t_1}, X_{t_2}, \ldots, X_{t_n}$ are jointly Gaussian with

$$\mathbb{E}(X_t) = \int_\Omega X_t(\omega)\,d\mathbb{P}(\omega) = 0, \ \forall t \in \mathbb{R};$$

2. (independent increments) if $t_1 < t_2 < \cdots < t_n$ then $X_{t_{i+1}} - X_{t_i}$ and $X_{t_j} - X_{t_{j-1}}$ are independent for all $j \le i$;

3. for all $s, t \in \mathbb{R}$,

$$\mathbb{E}\left(|X_t - X_s|^2\right) = |t - s|.$$

In particular, if $0 < s < t$, then the joint distribution of $(B_s, B_t) : \Omega \longrightarrow \mathbb{R}^2$ is the 2D Gaussian density:

$$G_{\begin{pmatrix} s & s \\ s & t \end{pmatrix}}(x, y) = \frac{1}{2\pi\sqrt{s(t-s)}} \exp\left(-\frac{tx^2 - 2sxy + sy^2}{2s(t-s)}\right)$$

i.e., the 2D-Gaussian $N\left(\underbrace{(0,0)}_{\text{mean}}, \underbrace{\begin{pmatrix} s & s \\ s & t \end{pmatrix}}_{\text{covariance matrix}}\right)$.

Remark 7.1. It follows from the properties 1-3 in Definition 7.6 that Brownian motion $(X_t)_{t \in \mathbb{R}}$ satisfies the covariance formula:

$$\mathbb{E}(X_s X_t) = \frac{|s| + |t| - |s - t|}{2}$$

$$= \begin{cases} |s| \wedge |t| & \text{if } st > 0 \ (\text{"same sign"}) \\ 0 & \text{if } st \leq 0 \ (\text{"opposite sign"}). \end{cases}$$

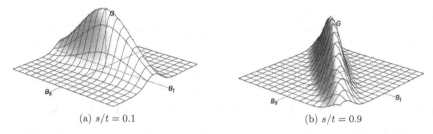

(a) $s/t = 0.1$ (b) $s/t = 0.9$

Figure 7.1: The joint distribution of (B_s, B_t) where B is standard Brownian motion and $0 < s < t$.

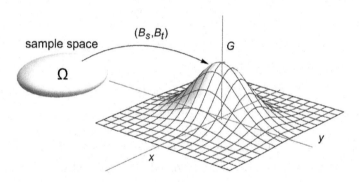

Figure 7.2: The joint distribution of $(B_s, B_t) : \Omega \longrightarrow \mathbb{R}^2$, $0 < s < t$.

Remark 7.2. Condition 2 in Definition 7.6 is called "independent increments." It states that, at every point in time t, the increments of the process up to the next future point in time are independent of all the earlier increments before t. This property is called "independent increments," and it is an important part of the definition of Brownian motion. There are many Gaussian processes that do not satisfy the "independent increments"-property; for example fractional Brownian motion with Hurst parameter H different from $1/2$.

Definition 7.7. The log-normal distribution is a continuous probability distribution. It is the distribution of a random variable X whose logarithm is normally distributed [KM04]. In other words, a random variable X is log-normally distributed iff $Y = \ln X$ has a normal distribution. See Table 7.1, and Figure 7.5.

Prices of securities under the condition that the trend and the volatility are known, may be represented by log-normally distributed random variables.

Exercise 7.2 (Quadratic variation). Let $\{X_t\}$ be the Brownian motion (Definition 7.6). Fix $T \in \mathbb{R}_+$, and consider partitions $\pi : (t_i)_{i=0}^n$ of $[0, T]$, i.e.,

$$\pi : 0 = t_0 < t_1 < t_2 < \cdots < t_n = T. \tag{7.3}$$

Set

$$\text{mesh}\,(\pi)\,\big(:= |\pi|\,\big) = \max_i \{t_i - t_{i-1}\}. \tag{7.4}$$

Then show that the limit,

$$\lim_{\text{mesh}(\pi) \to 0} \sum_i \left(X_{t_i} - X_{t_{i-1}}\right)^2 = T \tag{7.5}$$

holds a.e. on $(\Omega, \mathcal{F}, \mathbb{P})$, where $\Omega = C(\mathbb{R})$, $\mathcal{F} = \text{Cyl}$, and $\mathbb{P} = $ the Wiener measure.

Hint: Establish that

$$\mathbb{E}\left(|\triangle X_i|^2\right) = \triangle t_i, \tag{7.6}$$

$$\mathbb{E}\left(|\triangle X_i|^4\right) = 3\left(\triangle t_i\right)^2, \text{ and} \tag{7.7}$$

$$\mathbb{E}\left((\triangle X_i)^{2n-1}\right) = 0, \ n \in \mathbb{N}, \tag{7.8}$$

i.e., all the odd moments vanish; where

$$\begin{aligned}
\triangle X_i &= X_{t_i} - X_{t_{i-1}}, \text{ and} \\
\triangle t_i &= t_i - t_{i-1}, \text{ for } i = 1, 2, \ldots, n.
\end{aligned}$$

Note that $\sum_i ()^2$ on the LHS in (7.5) is a measurable function on $(\Omega, \mathcal{F}, \mathbb{P})$, while the RHS is deterministic, i.e., it is the constant function T.

Remark 7.3. Spectral Theorem and *functional calculus* are about the substitutions (see (7.9)).

$$\boxed{\begin{array}{c} A \\ \text{selfadjoint operator} \end{array}} \longrightarrow \boxed{\begin{array}{c} f : \mathbb{R} \longrightarrow \mathbb{R} \\ \text{scalar function} \\ \hline \hookrightarrow f(A) \end{array}} \tag{7.9}$$

By contrast, Itō-calculus is about substitutions of Brownian motion (at least in a special case); as follows:

$$\boxed{\begin{array}{c} B_t \\ \text{Brownian motion} \end{array}} \longrightarrow \boxed{\begin{array}{c} f : \mathbb{R} \longrightarrow \mathbb{R} \\ \text{scalar function} \\ \hline \hookrightarrow f(B_t) \end{array}} \tag{7.10}$$

See [Sto90, Yos95, Nel69, RS75, DS88b].

An application of (7.10) and Exercise 7.2, now yields the following version of *Itō's lemma*: If the function f in (7.10) is assumed C^2, then, for all $T > 0$, we have:

$$f\left(B\left(T\right)\right) - f\left(0\right) = \int_0^T f'\left(B\left(t\right)\right) dB_t + \frac{1}{2} \int_0^T f''\left(B\left(t\right)\right) dt. \qquad (7.11)$$

Exercise 7.3 (Geometric Brownian motion).

1. Apply (7.10) and (7.11) to $f\left(x\right) = \ln x$, $x \in \mathbb{R}_+$, together with (7.5) in Exercise 7.2, to show that, for $T \in \mathbb{R}_+$, the process,

$$X_T = X_0 \exp\left(\left(\mu - \frac{1}{2}\sigma^2\right) T + \sigma B_T\right) \qquad (7.12)$$

   solves the SDE for geometric Brownian motion:

$$dX_t = X_t\left(\mu dt + \sigma dB_t\right). \qquad (7.13)$$

   See Figure 7.3. The distribution of each of the random variables X_T in (7.12) is lognormal; see Table 7.1.

2. Apply (7.10) and (7.11) to $f\left(x\right) = x^2$, together with (7.5) in Exercise 7.2 to establish the following:

$$\int_0^T B_t \, dB_t = \frac{1}{2}\left(B_T^2 - T\right). \qquad (7.14)$$

   See Figure 7.4.

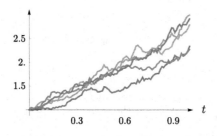

Figure 7.3: Geometric Brownian motion: 5 sample paths, with $\mu = 1$, $\sigma = 0.02$, and $X_0 = 1$.

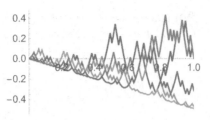

Figure 7.4: The process $\frac{1}{2}\left(B_T^2 - T\right)$ in (7.14): 5 sample paths, with $T = 1$.

In the previous exercise we explored stochastic processes derived from standard Brownian motion. We now return to explore some additional properties of Brownian motion itself:

Exercise 7.4 (A unitary one-parameter group). Using Definition 7.6, Corollary 2.2, Remark 2.9 and Example 2.3, show that there is a unique strongly continuous unitary one-parameter group $\{U(t)\}_{t \in \mathbb{R}}$ acting in $L^2\left(C\left(\mathbb{R}\right), Cyl, \mathbb{P}\right)$, determined by

$$U(t)\, X_s = X_{s+t}, \ \forall s, t \in \mathbb{R}. \tag{7.15}$$

(Here, Cyl = the sigma algebra generated by the cylinder sets; see (7.17).)

Hint: By (3) in Definition 7.6, we have

$$\mathbb{E}\left(\left|X_{t_2} - X_{t_1}\right|^2\right) = \mathbb{E}\left(\left|X_{t_2+s} - X_{t_1+s}\right|^2\right), \ \forall s, t_1, t_2 \in \mathbb{R}. \tag{7.16}$$

Hence, if $U(t)$ is defined on the generator $\{X_s \ : \ s \in \mathbb{R}\} \subset L^2(\mathbb{P})$ as in (7.15), it follows by (7.16) that it preserves the $L^2(\mathbb{P})$-norm. The remaining steps are left to the reader.

Exercise 7.5 (Infinitesimal generator). Discuss the infinitesimal generator of $\{U(t)\}_{t \in \mathbb{R}}$.

Remark 7.4. There is a list of popular kernels in probability theory, see Table 7.1. (In the generating function of the semicircle law, $J_1(z)$ is the Bessel function.)

Definition 7.8. Let $(\Omega, \mathcal{F}, \mathbb{P})$ be a probability space, and let $T_t : \Omega \longrightarrow \Omega$ be a measurable transformation group which preserves $\mathbb{P}$, i.e.,

$$(U(t)\, \psi)(\omega) = \psi(T_t(\omega)), \ \omega \in \Omega, \ t \in \mathbb{R},$$

is a unitary one-parameter group acting on $L^2(\Omega, \mathcal{F}, \mathbb{P})$.

We say that $\{T_t\}_{t \in \mathbb{R}}$ is *ergodic* if the following implication holds:

$$[A \in \mathcal{F}, \ T_t(A) = A, \ \forall t \in \mathbb{R}] \Longrightarrow \mathbb{P}(A)(1 - \mathbb{P}(A)) = 0;$$

i.e., the T_t-invariant sets in $\mathcal{F}$ must have measure 0, or measure 1.

Exercise 7.6 (An ergodic action). Returning to Brownian motion, now let $U(t)$, $t \in \mathbb{R}$, be the corresponding unitary one parameter group acting in $L^2(\Omega, \mathbb{P})$; see (7.15). Show that $\{U(t)\}_{t \in \mathbb{R}}$ is induced by an *ergodic action*.

	values	distributions	generating func $\xi \in \mathbb{R}$
uniform	$a \le x \le b$	$\dfrac{1}{b-a}$	$\dfrac{e^{ib\xi} - e^{ia\xi}}{i\xi(b-a)}$
exponential (λ)	$x \ge 0$	$\lambda e^{-\lambda x}$	$\left(1 - \dfrac{i\xi}{\lambda}\right)^{-1}$
Gaussian $N(m, \sigma^2)$	$x \in \mathbb{R}$	$\dfrac{1}{\sigma\sqrt{2\pi}} e^{-\frac{1}{2}\left(\frac{x-m}{\sigma}\right)^2}$	$e^{im\xi - \frac{1}{2}\sigma^2\xi^2}$
Cauchy	$x \in \mathbb{R}$	$\dfrac{1}{\pi(1+x^2)}$	$e^{-|\xi|}$
χ^2 (chi-square)	$x \ge 0$	$\dfrac{e^{-\frac{x}{2}} x^{\frac{\nu}{2}-1}}{2^{\frac{\nu}{2}} \Gamma\left(\frac{\nu}{2}\right)}$	$(1 - 2i\xi)^{-\frac{\nu}{2}}$
Gamma	$x \ge 0$	$\dfrac{x^{\gamma-1} e^{-x}}{\Gamma(\gamma)}$, $\gamma > 0$	$(1 - i\xi)^{-\gamma}$
Lévy	$x \ge \mu_0$	$\sqrt{\dfrac{c_0}{2\pi}} e^{-\frac{c_0}{2(x-\mu_0)}} (x-\mu_0)^{-\frac{3}{2}}$	$e^{i\mu_0\xi - \sqrt{-2i\,c_0\xi}}$
semicircle	$x \in [-r, r]$	$\dfrac{2}{\pi r^2} \sqrt{r^2 - x^2}$	$2\dfrac{J_1(r\xi)}{r\xi}$
log-normal	$x > 0$	$\dfrac{1}{\sigma\sqrt{2\pi}} \dfrac{1}{x} e^{-\frac{1}{2}\left(\frac{\ln x - \mu}{\sigma}\right)^2}$	undefined

Table 7.1: Some commonly occurring probability measures on $\mathbb{R}$, or on $\mathbb{R}_+$, and their corresponding *generating functions*. P.S.: Note that the Lévy distribution has both its mean and its variance $= \infty$. See Figure 7.5.

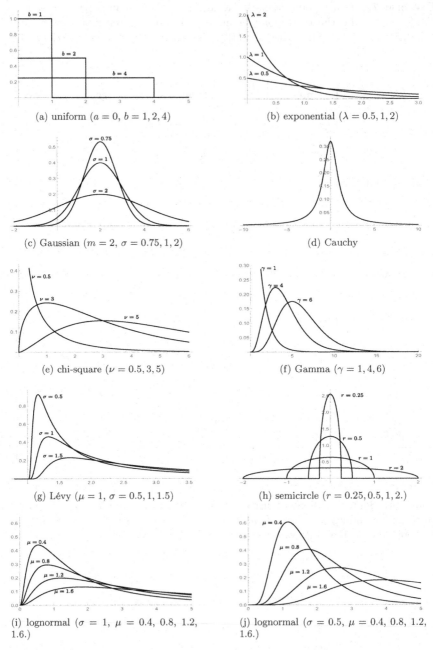

Figure 7.5: Some commonly occurring probability distributions. See Table 7.1 for explicit formulas for the respective distributions, and their parameters.

7.2   The Path Space

We now turn to the details regarding path-space. The object is to realize a probability space $(\Omega, \mathcal{F}, \mathbb{P})$ where the sample space Ω is a set of paths (functions on $\mathbb{R}$), $\mathcal{F}$ is a sigma-algebra of events (cylinder-subsets of Ω corresponding to choices of finite samples). There is in fact a host of seemingly different (but equivalent) representations of the underlying probability space. In our presentation of Brownian motion B given below, for every value of t, the random variable $B(t)$ simply samples functions ω from Ω at t, i.e., $B_t(\omega) = \omega(t)$. We have specialized here to the version of Brownian motion indexed by "one-dimensional" time, but there are generalizations based on the same idea, *mutatis mutandis* for processes with different and more subtle index sets.

Brownian motion is in fact characterized by a set of axioms; see Definition 7.6 above. Hence, with our present choice, the probability measure $\mathbb{P}$ is a measure on path space, the Wiener measure.

We now turn to the existence of this measure, and its properties, Theorem 7.1 below. The measure $\mathbb{P}$ goes by the name Wiener measure, path-space measure, or white noise-measure (see the cited references at the end of the present ch 7.) Below we make a few remarks about $\mathbb{P}$, and our choice of probability space Ω to be "the" sample space of paths. It is possible to use for Ω the set of all the continuous functions; and this is a good choice. If by contrast we take for Ω instead all functions, we still have $\mathbb{P}$, and it can now be shown that the subset of the total Ω consisting of the continuous functions will have outer measure 1 with respect to $\mathbb{P}$. Even smaller subsets (e.g., spaces of Lipschitz functions with Lipschitz constant $< 1/2$) also have outer measure 1 with respect to $\mathbb{P}$. By contrast, Wiener measure $\mathbb{P}$ assigns outer measure 0 to the subset of Ω consisting of all differentiable functions. The latter fact, due to N. Wiener and A. Kolmogorov, can be visualized intuitively by simulations. Note that, from a glance at Figures 7.3 and 7.9, the "wiggly" appearance of Monte Carlo simulation[4] of Brownian motion, and of processes derived from Brownian motion. Translation: Sample-paths are continuous, but non-differentiable, with probability 1.

The sample-space as a path-space

Theorem 7.1 (see e.g., [Nel67]). *Set* $\Omega = C(\mathbb{R}) = $ *(all continuous real valued function on $\mathbb{R}$), $\mathcal{F} = $ the sigma algebra generated by cylinder sets, i.e., determined by finite systems* $t_1, \ldots, t_n$, *and intervals* $J_1, \ldots, J_n$;

$$Cyl(t_1, \ldots, t_n, J_1, \ldots, J_n) = \left\{ \omega \in C(\mathbb{R}) \mid \omega(t_i) \in J_i, \ i = 1, 2 \ldots, n \right\}. \quad (7.17)$$

The measure $\mathbb{P}$ is determined by its value on cylinder sets, and an integral over Gaussians; it is called the Wiener-measure. Set

$$X_t(\omega) = \omega(t), \ \forall t \in \mathbb{R}, \omega \in \Omega (= C(\mathbb{R})). \quad (7.18)$$

---

[4]A summary of some highpoints from Wiener measure and Monte Carlo simulation is contained in Appendix B. See also Sections 2.4, 7.2, and Chapter 8 (pg. 273).

If $0 < t_1 < t_2 < \cdots < t_n$, and the cylinder set is as equation (7.17), then

$$\mathbb{P}\left(Cyl\left(t_1, \ldots, t_n, J_1, \ldots, J_n\right)\right)$$
$$= \int_{J_1} \cdots \int_{J_n} g_{t_1}\left(x_1\right) g_{t_2-t_1}\left(x_2 - x_1\right) \cdots g_{t_n-t_{n-1}}\left(x_n - x_{n-1}\right) dx_1 \cdots dx_n$$

where

$$g_t\left(x\right) = \frac{1}{\sqrt{2\pi t}} e^{-x^2/2t}, \ \forall t > 0,$$

i.e., the $N(0,t)$-Gaussian. See Figure 7.6.

Exercise 7.7 (Reflection of Brownian motion). Let $\{B\left(t\right)\}_{t\in[0,\infty)}$ be standard Brownian motion. For $t \in \mathbb{R}_+$, set

$$X\left(t\right) := t B\left(\frac{1}{t}\right). \tag{7.19}$$

1. Show that a.e. with respect to $(\Omega, \mathcal{F}, \mathbb{P})$, we have

$$\lim_{t\to 0+} X\left(t\right) = 0. \tag{7.20}$$

2. When extended by the limit (7.20), setting $X\left(0\right) = 0$, show that X is then again a "copy" of Brownian motion on $[0, \infty)$.

3. For $s, t \in [0, \infty)$, such that $st \leq 1$, show that $\mathbb{E}\left(B\left(s\right) X\left(t\right)\right) = st$.

4. Let $dB\left(s\right)$ denote the element of integration for the Brownian motion $B\left(s\right)$; and similarly, $dX\left(t\right)$ denotes the Itō-differential for the reflected Brownian motion $X\left(t\right)$. Consider $s, t \in [0, 1]$. Prove that

$$\mathbb{E}\left(dB\left(s\right) dX\left(t\right)\right) = ds\,dt, \text{ i.e.,}$$
$$\mathbb{E}\left(\left(\int_0^1 f dB\right)\left(\int_0^1 g dX\right)\right) = \left(\int_0^1 f\left(s\right) ds\right)\left(\int_0^1 g\left(t\right) dt\right)$$

for all $f, g \in L^2\left(0, 1\right)$.

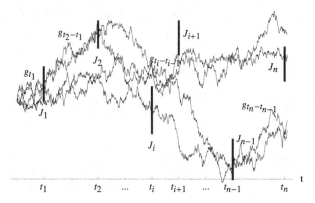

Figure 7.6: Stochastic processes indexed by time: A cylinder set C is a special subset of the space of all paths, i.e., functions of a time variable. A fixed cylinder set C is specified by a finite set of sample point on the time-axis (horizontal), and a corresponding set of "windows" (intervals on the vertical axis). When sample points and intervals are given, we define the corresponding cylinder set C to be the set of all paths that pass through the respective windows at the sampled times. In the figure we illustrate sample points (say future relative to $t = 0$). Imagine the set C of all outcomes with specification at the points $t_1, t_2, \ldots$ etc.

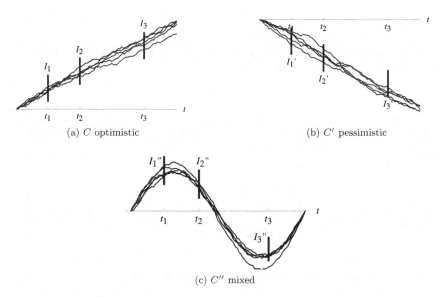

(a) C optimistic   (b) C' pessimistic

(c) C'' mixed

Figure 7.7: The cylinder sets C, C', and C''.

Exercise 7.8 (Iterated Itō-integrals). Using Theorem 7.1 and Wiener path space measure $\mathbb{P}$ in (7.18), and let $\{B(t) \; ; \; t \in [0, \infty)\}$ be the standard Brownian motion, see (7.18).

1. Show that, for all $n \in \mathbb{N}$, the iterated Itō-integrals

$$J_n(t) := \int_0^t \left(\int_0^{s_n} \int_0^{s_{n-1}} \cdots \left(\int_0^{s_3} \left(\int_0^{s_2} dB(s_1) \right) dB(s_2) \right) \cdots dB(s_{n-1}) \right) dB(s_n)$$

$$(0 \le s_1 \le s_2 \le \cdots \le s_n \le t)$$

exists; and that $J_n(t) \in L^2(\Omega, \mathcal{F}, \mathbb{P})$.

2. For $n = 0, 1, 2, \cdots$, let H_n denote the sequence of Hermite polynomials,

$$H_n(x) := (-1)^n e^{\frac{x^2}{2}} \left(\frac{d}{dx} \right)^n \left(e^{-\frac{x^2}{2}} \right), \quad x \in \mathbb{R};$$

see also Table 2.2. (In particular, $H_0(x) = 1$, $H_1(x) = x$, $H_2(x) = x^2 - 1, \cdots$.) Show that then, for all $t \in \mathbb{R}_+$, we have:

$$J_n(t) = \frac{t^{n/2}}{n!} H_n \left(\frac{B(t)}{\sqrt{t}} \right), \quad n \in \mathbb{N}.$$

Exercise 7.9 (Symmetric Fock space as a random Gaussian space). Let $\mathscr{H}$ be a Hilbert space over $\mathbb{R}$, and let $(\Omega, \mathcal{F}, \mathbb{P})$ be a probability space such that

$$\mathscr{H} \ni h \longrightarrow \widetilde{h} \in L^2(\Omega, \mathcal{F}, \mathbb{P}) \tag{7.21}$$

offers a Gaussian representation.

We assume that $(\Omega, \mathcal{F}, \mathbb{P})$ may be realized as a Gelfand-triple with $\Omega = \mathcal{S}'$ (Schwartz' tempered distributions), and

$$\mathcal{S} \hookrightarrow \mathscr{H} \hookrightarrow \mathcal{S}'$$

satisfying the conditions of Gelfand-triples (see Sections 2.24, 10.5.) In particular, if $h \in \mathscr{H}$, then the random variable $\widetilde{h}$ is such that $\widetilde{h}(\xi) = \langle h, \xi \rangle$ for all $\xi \in \mathscr{H}$.

Note that $\mathbb{E}(\widetilde{h}) = 0$, and

$$\mathbb{E}\left(e^{i\widetilde{h}} \right) = e^{-\frac{1}{2}\|h\|_{\mathscr{H}}^2}, \quad \forall h \in \mathscr{H}. \tag{7.22}$$

It follows that:

$$\mathbb{E}\left(\widetilde{h}_1 \widetilde{h}_2 \right) = \langle h_1, h_2 \rangle_{\mathscr{H}} \tag{7.23}$$

holds for $\forall h_1, h_2 \in \mathscr{H}$.

Let $\Gamma_s(\mathscr{H})$ be the symmetric Fock space over $\mathscr{H}$, with

$$\langle \Gamma_s(e^{h_1}), \Gamma_s(e^{h_2}) \rangle_{\Gamma_s(e^{\mathscr{H}})} = \sum_{n=0}^{\infty} \frac{\langle h_1, h_2 \rangle_{\mathscr{H}}^n}{n!} = e^{\langle h_1, h_2 \rangle_{\mathscr{H}}}; \text{ and} \tag{7.24}$$

$$\Gamma_s\left(e^h\right) = \left[\frac{1}{\sqrt{n!}} h \otimes h \otimes \cdots \otimes h \atop \underbrace{\qquad\qquad}_{n \text{ times}}\right]_0^\infty = \left[\frac{1}{\sqrt{n!}} h^{\otimes n}\right]_0^\infty . \qquad (7.25)$$

1. Show that the vectors $\Gamma_s\left(e^h\right)$ in (7.25) span a dense subspace in the Hilbert space $\Gamma_s\left(e^{\mathscr{H}}\right)$.

2. Show that the assignment:

$$\Gamma_s\left(e^h\right) \longrightarrow \exp\left(\widetilde{h}\left(\cdot\right) - \frac{1}{2}\|h\|_{\mathscr{H}}^2\right) \qquad (7.26)$$

extends by linearity and closure to yield a unitary isometric isomorphism of $\Gamma_s\left(e^{\mathscr{H}}\right)$ onto $L^2\left(\Omega, \mathcal{F}, \mathbb{P}\right)$.

Infinite-product measure

Let $\Omega = \prod_{k=1}^\infty \{1, -1\}$ be the infinite Cartesian product of $\{1, -1\}$ with the product topology. Ω is compact and Hausdorff by Tychonoff's theorem.

For each $k \in \mathbb{N}$, let $X_k : \Omega \to \{1, -1\}$ be the kth coordinate projection, and assign probability measures μ_k on Ω so that $\mu_k \circ X_k^{-1}\{1\} = a$ and $\mu_k \circ X_k^{-1}\{-1\} = 1-a$, where $a \in (0, 1)$. The collection of measures $\{\mu_k\}$ satisfies the consistency condition, i.e., μ_k is the restriction of μ_{k+1} onto the kth coordinate space. By Kolmogorov's extension theorem, there exists a unique probability measure $\mathbb{P}$ on Ω so that the restriction of $\mathbb{P}$ to the kth coordinate is equal to μ_k. (See, e.g., [Hid80].)

It follows that $\{X_k\}$ is a sequence of *independent identically distributed* (i.i.d.) random variables in $L^2\left(\Omega, \mathbb{P}\right)$ with $\mathbb{E}\left(X_k\right) = 0$ and $Var[X_k^2] = 1$; and $L^2\left(\Omega, \mathbb{P}\right) = \overline{span}\{X_k\}$.

Remark 7.5. Let $\mathscr{H}$ be a separable Hilbert space with an orthonormal basis $\{u_k\}$. The map $\varphi : u_k \longmapsto X_k$ extends linearly to an isometric embedding of $\mathscr{H}$ into $L^2\left(\Omega, \mathbb{P}\right)$. Moreover, let $\mathcal{F}_+(\mathscr{H})$ be the *symmetric Fock space*. $\mathcal{F}_+(\mathscr{H})$ is the closed span of the algebraic tensors $u_{k_1} \otimes \cdots \otimes u_{k_n}$, thus φ extends to an isomorphism from $\mathcal{F}_+(\mathscr{H})$ to $L^2\left(\Omega, \mathbb{P}\right)$.

Exercise 7.10 (The "fair-coin" measure). Let $\Omega = \prod_1^\infty \{-1, 1\}$, and let μ be the "fair-coin" measure $\left(\frac{1}{2}, \frac{1}{2}\right)$ on $\{\pm 1\}$ (e.g., "Head v.s. Tail"), let $\mathcal{F}$ be the cylinder sigma-algebra of subsets of Ω. Let $\mathbb{P} = \prod_1^\infty \mu$ be the infinite-product measure on Ω. Set $Z_k\left(\omega\right) = \omega_k$, $\omega = \left(\omega_i\right) \in \Omega$. Finally, let $\{\psi_j\}_{j \in \mathbb{N}}$ be an ONB in $L^2\left(0, 1\right)$, and set

$$X_t\left(\omega\right) := \sum_{j=1}^\infty \left(\int_0^t \psi_j\left(s\right) ds\right) Z_j\left(\omega\right), \ t \in [0, 1], \omega \in \Omega.$$

Show that the $\{X_t\}_{t \in [0,1]}$ is the Brownian motion, where time "t" is restricted to $[0, 1]$.

Hint: Let $t_1, t_2 \in [0, 1]$, then

$$
\begin{aligned}
\mathbb{E}\left(X_{t_1} X_{t_2}\right) &= \mathbb{E}\left(\sum_j \left(\int_0^{t_1} \psi_j(s)\,ds\right) Z_j \cdot \sum_k \left(\int_0^{t_2} \psi_k(s)\,ds\right) Z_k\right) \\
&= \sum_j \sum_k \int_0^{t_1} \psi_j(s)\,ds \int_0^{t_2} \psi_k(s)\,ds \underbrace{\mathbb{E}\left(Z_j Z_k\right)}_{=\delta_{jk}} \\
&= \sum_j \left\langle \chi_{[0,t_1]}, \psi_j \right\rangle_{L^2} \left\langle \chi_{[0,t_2]}, \psi_j \right\rangle_{L^2} \\
&= \left\langle \chi_{[0,t_1]}, \chi_{[0,t_2]} \right\rangle_{L^2(0,1)} = t_1 \wedge t_2 \; (:= \min(t_1, t_2)).
\end{aligned}
$$

Semigroups of operators

The theory of semigroups of operators is a central theme in Functional Analysis and its applications. While we shall refer here to books regarding the general theory (see e.g., [BR79, JM84, DS88b, GJ87, Hid80, HØUZ10, Kat95, Lax02, Nel64, Nel69, Yos95]), we shall include below in detail only a selection of examples, most notable the Ornstein-Uhlenbeck semigroup. The latter plays a role both in the general theory and in a host of applications to quantum physics and to study of stochastic processes, especially to Itō-calculus.

As for the general theory, it centers around a theorem named after Hille[5], Yosida and Phillips (the HYP-theorem). They proved a fundamental theorem giving a precise correspondence between on the one hand, a given strongly continuous semigroup, and on the other, its associated infinitesimal generator. It is a necessary and sufficient condition for a given linear operator with dense domain to be the infinitesimal generator for a semigroup.

The theorem has a host of applications to partial differential equations and to Markov processes. (Other names are Feller and Miyadera.) Hille and Yosida independently discovered the result in 1948. In all non-trivial cases, the infinitesimal generator is an unbounded closed linear operator with dense domain. Via an introduction of bounded resolvent operators, the HYP-theorem characterizes the particular unbounded operators which are generators of strongly continuous one-parameter semigroups acting on Banach space. The special case of contraction semigroup includes the Ornstein-Uhlenbeck semigroup of wide use in the theory of Markov processes.

Exercise 7.11 (The Ornstein-Uhlenbeck semigroup).

1. Let $\{B_t\}$ denote standard Brownian motion. Show that the solution to the stochastic differential equation (SDE)

$$dX_t = \sqrt{2}\,(dB_t) - X_t dt \qquad (7.27)$$

---

[5] Appendix C includes a short biographical sketch of E. Hille. See also [HP57].

is

$$X_t = e^{-t}X_0 + \sqrt{2}\int_0^t e^{s-t}dB_s \tag{7.28}$$

where the RHS in (7.28) is an Itō-integral. See Remark 7.6 and Figure 7.8.

2. For $x \in \mathbb{R}$, and f a continuous bounded function $f : \mathbb{R} \longrightarrow \mathbb{R}$ ($f \in C_b(\mathbb{R})$), set

$$S_t f(x) := \mathbb{E}_x(f(X_t)) = \mathbb{E}(f(X_t) \mid X_0 = x), \tag{7.29}$$

and let $d\gamma(y) = \frac{1}{\sqrt{2\pi}}e^{-\frac{y^2}{2}}dy$ denote the standard $N(0,1)$-Gaussian. Show that

$$(S_t f)(x) = \int_{\mathbb{R}} f\left(e^{-t}x + \sqrt{1 - e^{-2t}}\, y\right)d\gamma(y). \tag{7.30}$$

3. Also show that $\{S_t\}_{t \geq 0}$ is a strongly continuous Markov semigroup, i.e., that the following hold:

   (a) $S_0 = id$;

   (b) For $\forall f \in C_b(\mathbb{R})$, the mapping $t \longrightarrow S_t f \in L^2(\mathbb{R}, \gamma)$ is continuous;

   (c) For $\forall t_1, t_2 \in \mathbb{R}_+$, we have $S_{t_1} S_{t_2} = S_{t_1 + t_2}$ (the semigroup law);

   (d) $S_t \mathbb{1} = \mathbb{1}$, $t \in \mathbb{R}_+$, where $\mathbb{1}$ denotes the constant function "one."

   (e) $\|S_t f\|_\infty \leq \|f\|_\infty$, $\forall f \in C_b(\mathbb{R})$, $\forall t \in \mathbb{R}_+$;

   (f) $\int_{\mathbb{R}}(S_t f_1) f_2 d\gamma = \int_{\mathbb{R}} f_1 (S_t f_2) d\gamma$ holds for all $f_i \in L^2(\gamma)$, $i = 1, 2$, and all $t \in \mathbb{R}_+$; i.e., each S_t is selfadjoint in $L^2(\gamma)$.

   (g) Show that the infinitesimal generator of S_t is as follows: If f is differentiable, then

$$\frac{S_t f(x) - f(x)}{t} \xrightarrow[t \longrightarrow 0]{} \left(\frac{d}{dx}\right)^2 f(x) - x\frac{d}{dx}f(x).$$

Remark 7.6. Motivated by Newton's second law of motion, the OU-process is proposed to model a random external driving force. In 1D, the process is the solution to the following stochastic differential equation

$$dv_t = -\gamma v_t dt + \beta dB_t, \ \gamma, \beta > 0. \tag{7.31}$$

Here, $-\gamma v_t$ is the dissipation, βdB_t denotes a random fluctuation, and B_t is the standard Brownian motion.

Assuming the particle starts at $t = 0$. The solution to (7.31) is a Gaussian stochastic process such that

$$\begin{aligned}
\mathbb{E}[v_t] &= v_0 e^{-\gamma t}\\
var[v_t] &= \frac{\beta^2}{2\gamma}\left(1 - e^{-2\gamma t}\right);
\end{aligned}$$

with v_0 being the initial velocity. See Figure 7.8. Moreover, the process has the following covariance function

$$c(s,t) = \frac{\beta^2}{2\gamma} \left(e^{-\gamma|t-s|} - e^{-\gamma|s+t|} \right).$$

If we wait long enough, it turns to a stationary process such that

$$c(s,t) \sim \frac{\beta^2}{2\gamma} e^{-\gamma|t-s|}.$$

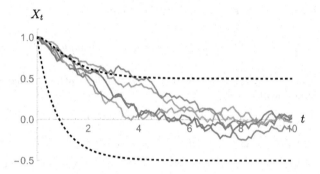

Figure 7.8: Simulation of the solution X_t to the SDE (7.31), i.e., the Ornstein-Uhlenbeck process. It governs a massive Brownian particle under the influence of friction. It is stationary, and Markovian. Over time, the process tends to drift towards its long-term mean: such a process is called mean-reverting. (Monte-Carlo simulation of OU-process with 5 sample paths. $\beta = \gamma = 1$, $v_0 = 1$, $t \in [0, 10]$)

7.3   Decomposition of Brownian Motion

Let $\{X_t\}_{t \in \mathbb{R}}$ be a Brownian motion process, so that it is a mean zero Gaussian process in $L^2(\Omega, \mathcal{F}, \mathbb{P})$, with

$$\mathbb{E}(X_s X_t) = \int_\Omega X_s X_t d\mathbb{P} = s \wedge t \, (= \min(s,t)); \qquad (7.32)$$

where $\mathbb{P}$ is the Wiener measure on the path-space Ω. It follows that the corresponding increment process satisfies

$$X_t - X_s \sim N(0, |t-s|), \quad \forall s, t \in \mathbb{R}. \qquad (7.33)$$

See also Definition 7.6.

Building the measure space $(\Omega, \mathcal{F}, \mathbb{P})$ is a fancy version of Riesz's representation theorem [Rud87, Theorem 2.14]. One may choose

$$\Omega = \prod_t \overline{\mathbb{R}}.$$

Equipped with the Tychonoff's product topology, Ω is then a compact Hausdorff space. Here, $\overline{\mathbb{R}} = (\mathbb{R} \cup \{\infty\})^{\sim}$ denotes the one-point compactification of $\mathbb{R}$.

Now introduce the random variables $X_t : \Omega \to \mathbb{R}$, by

$$X_t(\omega) = \omega(t), \ t \in \mathbb{R};$$

i.e., X_t is the continuous linear functional of evaluation at t on Ω. (See Theorem 7.1, and [Nel59b] for detail.)

Remark 7.7. For Brownian motion, the increment of the process ΔX_t, in statistical sense, is proportional to $\sqrt{\Delta t}$ (see (7.33)), i.e.,

$$\Delta X_t \sim \sqrt{\Delta t}.$$

This implies that the subset of differentiable functions in Ω has measure zero. In this sense, the trajectory of Brownian motion is nowhere differentiable.

Let K be the covariance functions of Brownian motion process as in (7.32). Consider the integral kernel $K : [0,1] \times [0,1] \to \mathbb{R}$, $K(s,t) = s \wedge t$, by

$$Kf(x) = \int_0^1 (x \wedge y)f(y)dy.$$

Then K is a compact operator in $L^2[0,1]$. Moreover, Kf is a solution to the differential equation

$$-\frac{d^2}{dx^2}u = f$$

satisfying the boundary conditions, $u(0) = u'(1) = 0$.

A very important application of the Spectral Theorem of compact operators is to decompose the Brownian motion process:

$$B_t(\omega) = \sum_{n=1}^{\infty} \frac{2}{(2n+1)\pi} \sqrt{2} \sin\left(\frac{(2n+1)\pi}{2}t\right) Z_n(\omega) \qquad (7.34)$$

where $Z_n \sim N(0,1)$; and

$$s \wedge t = \sum_{n=0}^{\infty} \frac{4}{((2n+1)\pi)^2} \left[\sqrt{2}\sin\left(\frac{(2n+1)\pi}{2}s\right)\sqrt{2}\sin\left(\frac{(2n+1)\pi}{2}t\right)\right].$$
$$\qquad (7.35)$$

The reader will be able to verify (7.34) directly from the spectral theorem applied to the integral operator K. But (7.34) is also an immediate consequence of the conclusion in Exercise 7.10 above.

Eq. (7.34) is the Karhunen-Loève expansion of Brownian motion. See Theorem 7.2 in Section 7.4, and Appendix C. For more details, we refer to [JS09, JS07, Jor06].

Remark 7.8. Consider the Hardy space $\mathbb{H}_2$, and the operator S from Exercise 5.33. Writing $f(z) = \sum_{n=0}^{\infty} x_n z^n$, we get

$$(Sf)(z) = f(z^N) = x_0 + x_1 z^N + x_2 z^{2N} + \cdots; \qquad (7.36)$$

and

$$(S^* f)(z) = x_0 + x_N z + x_{2N} z^2 + x_{3N} z^3 + \cdots; \qquad (7.37)$$

so in symbol-space, S^* acts as follows:

$$\begin{array}{c} (x_0, x_1, \cdots, x_{N-1}, x_N, x_{N+1}, \cdots, x_{2N}, x_{2N+1}, \cdots) \\ S^* \quad \downarrow \qquad\qquad\qquad\qquad\qquad\qquad\qquad\qquad\qquad (7.38) \\ (x_0, x_N, x_{2N}, x_{3N}, \cdots); \end{array}$$

so down-sampling $\simeq$ "decimation" $\simeq$ killing time-signals x_k when $N + k$, i.e., k is not divisible by N.

The projection SS^* is:

$$\begin{array}{c} \left(x_0, x_1, \cdots, x_{N-1}, x_N, x_{N+1}, \cdots, x_{2N-1}, x_{2N}, x_{2N+1}, \cdots, x_{3N-1}, x_{3N}, x_{3N+1}, \cdots \right) \\ SS^* \quad \downarrow \qquad\qquad\qquad\qquad\qquad\qquad\qquad\qquad\qquad (7.39) \\ \left(x_0, 0, \cdots, 0, x_N, 0, \cdots, 0, x_{2N}, 0 \cdots, 0, x_{3N}, 0, \cdots \right) \end{array}$$

If the coordinates in $\mathbb{H}_2$ label the i.i.d. random variables $Z_k(\cdot)$ in the expansion (7.34) for Brownian motion, then downsampling corresponds to *conditional expectation*; conditioning on "less information", i.e., leaving out the "decimated coordinates" in the expansion (7.34) for Brownian motion.

Exercise 7.12 (The Central Limit Theorem). Look up the *Central Limit Theorem* (CLT), and prove the following approximation formula for Brownian motion:

Let π be the "fair-coin-measure" on the two outcomes $\{\pm 1\}$, i.e., winning or loosing one unit, and let $\Omega = \times_{\mathbb{N}} \{\pm 1\}$, $\mathbb{P} = \times_{\mathbb{N}} \pi$ be the corresponding infinite product measure. On Ω, set

$$W_k(\omega) = \omega_k, \quad \omega = (\omega_1, \omega_2, \ldots) \in \Omega, \ k = 1, 2, \ldots; \text{ and}$$

$$S_n(\cdot) = \frac{1}{\sqrt{n}} \sum_{k=1}^{n} W_k(\cdot). \qquad (7.40)$$

Let X_t denote Brownian motion. Then show that

$$X_t(\cdot) = \lim_{n \to \infty} \frac{1}{\sqrt{n}} \sum_{k=1}^{\lfloor n t \rfloor} W_k(\cdot) \qquad (7.41)$$

where $\lfloor n t \rfloor$ denotes the largest integer $\leq n t$.

Hint: A good reference to the CLT is [CW14]. First apply the CLT to the sequence S_n in (7.40). Figure 7.9 illustrates the approximation formula in (7.41). ($X_1 = S$.)

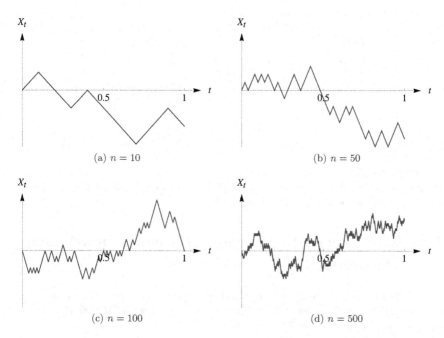

(a) $n = 10$

(b) $n = 50$

(c) $n = 100$

(d) $n = 500$

Figure 7.9: Monte-Carlo simulation of the standard Brownian motion process $\{X_t \ : \ 0 \le t \le 1\}$, where $\mathbb{E}\,(X_t) = 0$, and $\mathbb{E}\,(X_s X_t) = s \wedge t = \min\,(s, t)$. For $n = 10, 50, 100, 500$, set $X_0 = 0$ and $X_{j/n}^{(n)} = n^{-1/2} \sum_{k=1}^{j} W_k$. Applying linear interpolation between sample points $\{j/n : j = 0, \ldots, n\}$ yields the n-point approximation $X_t^{(n)}$, which converges in measure to X_t (standard BM restricted to the unit interval), as $n \to \infty$.

The Central Limit Theorem (CLT) states that the limit, $n \to \infty$, of the sequence S_n in (7.40) is a copy of $N\,(0, 1)$-random variable; i.e., $\lim_{n \to \infty} S_n\,(\cdot) = S\,(\cdot)$ exists; and

$$\mathbb{P}\,(\{\omega \mid a \le S\,(\omega) \le b\}) = \int_a^b \frac{1}{\sqrt{2\pi}} e^{-\frac{1}{2}x^2}\,dx,$$

for all intervals $(a, b) \subset \mathbb{R}$. Applying this to (7.41), we get existence of X_t as a limit, and $X_t \sim N\,(0, t)$, i.e.,

$$\mathbb{P}\,(\{\omega \mid a \le X_t\,(\omega) \le b\}) = \int_a^b \frac{1}{\sqrt{2\pi t}} e^{-\frac{1}{2t}x^2}\,dx.$$

We claim that

$$\mathbb{E}_{\mathbb{P}}\,(X_s X_t) = s \wedge t. \tag{7.42}$$

Below we sketch the argument for the assertion in (7.42).

Fix $s, t \in \mathbb{R}_+$, say $s < t$, and $n \in \mathbb{N}$; then

$$
\mathbb{E}\left(\frac{1}{\sqrt{n}}\left(\sum_{j=1}^{\lfloor n s \rfloor} W_j\right)\frac{1}{\sqrt{n}}\left(\sum_{k=1}^{\lfloor n t \rfloor} W_k\right)\right) = \frac{1}{n}\sum_{j=1}^{\lfloor n s \rfloor}\sum_{k=1}^{\lfloor n t \rfloor}\mathbb{E}\left(W_j W_k\right)
$$

$$
= \frac{1}{n}\sum_{j=1}^{\lfloor n s \rfloor}\sum_{k=1}^{\lfloor n t \rfloor}\delta_{j,k}
$$

$$
= \frac{\lfloor n s \rfloor}{n} \to s, \text{ as } n \to \infty.
$$

Hence by the CLT, the desired conclusion in (7.42) follows.

This is the key step in proving that the limit X_t in (7.41) is Brownian motion. The remaining steps are left to the readers.

7.4   The Spectral Theorem, and Karhunen-Loève Decomposition

The Karhunen-Loève theorem (the KL theorem) is a kind of converse to the decomposition from Exercise 7.10 and Remark 7.7. See, e.g., [LMCL09, GDV06], and (7.34)-(7.35) in Section 7.3.

Theorem 7.2 (Karhunen-Loève). *Let $(T, \mathcal{F}_T, \mu)$ be a measure space, and let, $T \ni t \longmapsto X_t$, be a Gaussian process relative to a fixed probability space $(\Omega, \mathcal{F}_\Omega, \mathbb{P})$. We assume the following conditions hold:*

1. *X_t is real valued Gaussian for all $t \in T$,*

2. *$\mathbb{E}(X_t) = 0$, $\forall t \in T$,*

3. *$K_X(s,t) = \mathbb{E}(X_s X_t)$ is well defined on $T \times T$; in particular, $X_t \in L^2(\Omega, \mathcal{F}_\Omega, \mathbb{P})$, $\forall t \in T$;*

4. *$(T_X f)(t) = \int_T K_X(t,s) f(s) \, d\mu(s)$, $\forall f \in L^2(T, \mu)$, defines a non-singular compact selfadjoint integral operator in $L^2(T, \mu)$.*

Then, if $(h_n)_{n \in \mathbb{N}}$ is an orthonormal basis (ONB) in $L^2(T, \mu)$ such that the spectral representation for T_X is

$$
T_X = \sum_{n \in \mathbb{N}} \lambda_n |h_n\rangle\langle h_n|, \quad \lambda_n > 0; \tag{7.43}
$$

then

$$
Z_n := \int_T X_t(\cdot) h_n(t) \, d\mu(t), \quad n \in \mathbb{N} \tag{7.44}
$$

is an i.i.d. system, Gaussian with

$$
Z_n \sim N(0, \lambda_n), \tag{7.45}
$$

mean zero and variance λ_n. (The variance λ_n changes with n.)
  Moreover,

$$X_t(\cdot) = \sum_{n \in \mathbb{N}} h_n(t) Z_n(\cdot) \quad on \quad \Omega, \tag{7.46}$$

is an orthogonal $L^2(\Omega, \mathcal{F}_\Omega, \mathbb{P})$-expansion for $(X_t)_{t \in T}$, called the Karhunen-Loève expansion.

Exercise 7.13 (Karhunen-Loève). Prove the K-L theorem from the Spectral Theorem, Theorem 4.6.

Remark 7.9. Alternatively, with

$$W_n := \frac{1}{\sqrt{\lambda_n}} \int_T X_t(\cdot) h_n(t) \, d\mu(t),$$

we get i.i.d. with $W_n \sim N(0, 1)$.

7.5 Large Matrices Revisited

The pioneering work by Wigner (on the semicircle-law for the distribution of the spectrum of "large" Hermitian matrices with independent entries, see e.g. [LF01]), and Dyson on pair-correlations, have spun off a number of very exciting research area. In the 1980ties, Freeman Dyson and Hugh Montgomery discovered together a fascinating connection between quantum physics and what is now known as the Montgomery pair correlation conjecture; the latter referring to the zeros of the Zeta function. Since these developments have a core of non-commutative functional analysis, we have included a very brief sketch of a few highpoints. And we have concluded with a literature guide.

Since large matrices and limit distributions have played a role in several topics in the present chapter, we mention here yet a different one; but now without proofs. Readers will find a complete treatment, for example in [Wig58, SS98].

Setting. For all $N \in \mathbb{N}$, consider a symmetric (random) matrix

$$X = \begin{bmatrix} X_{1,1}^{(N)} & \cdots & X_{1,N}^{(N)} \\ \vdots & & \vdots \\ X_{N,1}^{(N)} & \cdots & X_{N,N}^{(N)} \end{bmatrix} \tag{7.47}$$

$$X_{i,j}^{(N)} = X_{j,i}^{(N)} \quad \text{(real valued)} \tag{7.48}$$

where the entries are i.i.d. random variables ("i.i.d" is short for independent identically distributed), mean 0, and variance m^2. We further assume that all the moments are finite, and with at most exponential bounds.

Let $a, b \in \mathbb{R}$, $a < b$ be fixed, and set

$$V_N^{(a,b)} := \text{ number of eigenvalues of } X^{(N)} \text{ that} \tag{7.49}$$

fall in the interval $\left(a\sqrt{N}, b\sqrt{N}\right)$.

Then the following limit exists, i.e., the semicircle law holds for the limit distribution of the eigenvalues:

$$\lim_{N\to\infty} \frac{\mathbb{E}\left(V_N^{(a,b)}\right)}{N} = \frac{1}{2\pi m^2} \int_a^b \sqrt{4m^2 - x^2}dx. \tag{7.50}$$

A summary of relevant numbers from the Reference List

For readers wishing to follow up sources, or to go in more depth with topics above, we suggest: [Hid80, HØUZ10, Itô04, Itô06, Itô07, Loè63, Nel67, Par09, Gro70, Gro64, Nel64, AJ12, AJS14, BM13, Sch58, SS09, Jør14, Seg54, Seg53, Seg58, Gro67, LMCL09, GDV06].

Chapter 8

Lie Groups, and their Unitary Representations

"Every axiomatic (abstract) theory admits, as is well known, an unlimited number of concrete interpretations besides those from which it was derived. Thus we find applications in fields of science which have no relation to the concepts of random event and of probability in the precise meaning of these words."
— A.N. Kolmogorov

"The miracle of the appropriateness of the language of mathematics for the formulation of the laws of physics is a wonderful gift which we neither understand nor deserve."
— Eugene Paul Wigner

"Nowadays group theoretical methods—especially those involving characters and representations, pervade all branches of quantum mechanics."
— George Whitelaw Mackey

"The universe is an enormous direct product of representations of symmetry groups."
— Hermann Weyl

In the present chapter we shall treat a selection of themes from the theory of unitary representations of Lie groups. Our selection (see also Chapters 2-4) is dictated in turn by a number of issues from non-commutative analysis, all directly related to quantum physics. One such theme is a correspondence (dating back to Dirac and to Wigner) linking elementary particles to irreducible (unitary) representations of the Poincaré group, say G. By this interpretation,

composite quantum systems will entail a notion of direct integral of irreducible
unitary representations; the latter in turn indexed by "pure states." Quantum
mechanical observables, in this correspondence, become selfadjoint operators in
Hilbert space $\mathcal{H}$. States become norm-one vectors in $\mathcal{H}$, quantum states.

In some cases, rather than the unitary representation of G acting on $\mathcal{H}$,
there are associated representations of the Lie algebra $\mathfrak{g}$ of G, and of its associated universal enveloping algebra (see inside for definitions), and the relevant
observables are to be found in the latter representations. Starting with a unitary
representation $\mathcal{U}$ of G it is customary to think of the associated representations
of the Lie algebra $\mathfrak{g}$ of G, as the differential $d\mathcal{U}$ calculated as an infinitesimal
representation variant of $\mathcal{U}$ itself. Since quantum observables are typically unbounded, the relevant representations $d\mathcal{U}$ will be by unbounded operators in
$\mathcal{H}$.

Hence the topics from non-commutative analysis that must be brought to
bear on this, and related themes from quantum theory, entail explicit realizations of the spectral resolution for unbounded selfadjoint operators, and a discussion of non-commuting unbounded operators with dense domain in Hilbert
space.

While the topic of unitary representations and quantum physics is big, we
have limited our present chapter to the parts of it that make direct connections
to the tools we already developed in chapters 3–5 above. Using this, the reader
will be able to follow the details below without additional background. The
chapter concludes with a guide to the literature. Of special relevance are the
books: [Mac92, JM84, Dix81, Ørs79, Tay86].

Outline   As part of our discussion of spectral theory and harmonic analysis,
we had occasion to study unitary one-parameter groups $\mathcal{U}(t)$, $t \in \mathbb{R}$. Stated
differently (see Chapters 3-5), a unitary one-parameter group acting on a Hilbert
space $\mathcal{H}$, is a *strongly continuous unitary representation* of the group $\mathbb{R}$ with
addition; — so it is an element in $Rep(\mathbb{R}, \mathcal{H})$. Because of applications to
physics, to non-commutative harmonic analysis, to stochastic processes, and to
geometry, it is of interest to generalize to $Rep(G, \mathcal{H})$ where G is some more
general group, other than $(\mathbb{R}, +)$, for example, G may be a matrix group, a *Lie
group*, both compact and non-compact, or more generally, G may be a *locally
compact group*.

In Chapters 3-4 we studied the canonical commutation-relations for the
quantum mechanical momentum and position operators P, respectively Q. Below we outline how this problem can be restated as a result about a unitary
representation of the matrix group G_3 of all upper triangular 3×3 matrices over
$\mathbb{R}$. This is a special unitary irreducible representation $\mathcal{U}$ in $Rep(G_3, L^2(\mathbb{R}))$. It is
called the Schrödinger representation, and the group G_3 is called the Heisenberg
group. We shall need the Stone-von Neumann uniqueness theorem, outlined in
the appendix below. (Also see [vN32b, vN31].) Its proof will follow from a more
general result which is included inside the present chapter. The Stone-von Neumann uniqueness theorem states that every unitary irreducible representation

of G_3 is unitarily equivalent to the Schrödinger representation.

We studied operators in Hilbert space of relevance to quantum physics. A source of examples is relativistic physics. The symmetry group of Einstein's theory is a particular Lie group, called the *Poincaré group*. The study of its unitary representations is central to relativistic physics. But it turns out that there is a host of diverse applications (including harmonic analysis) where other groups arise. Below we offer a glimpse of the theory of unitary representations, and its many connections to operators in Hilbert space.

Two pedantic points regarding unbounded operators. The first is the distinction between "selfadjoint" vs "essentially selfadjoint." An operator is said to be essentially selfadjoint if its closure is selfadjoint. The second is the distinction between selfadjoint and skewadjoint. The difference there is just a multiple of $i\left(=\sqrt{-1}\right)$.

These distinctions play a role in the study of unitary representations of Lie groups, but are often swept under the rug, especially in the physics literature. Every *unitary representation* of a Lie group has a *derived representation* of the corresponding Lie algebra. The individual operators in a derived representation are skewadjoint; — but to get a common dense domain for all these operators, we must resort to essentially skewadjointness. Nonetheless, indeed there are choices of common dense domains (e.g., C^∞-vectors), but the individual operators in the derived representation will then only be *essentially skewadjoint* there. For more details on this, see e.g., [Pou72].

The chapter concludes with an overview of the five duality operations which have been used up to now, including their basic properties.

Particle physics and representation theory, as per Eugene Wigner
Particles in quantum mechanics (QM) are governed by particle-wave duality. A given QM particle has an associated system of non-commuting observables, momentum, position, spin, etc.; each with its distribution. QM states are written as vectors (in "ket" form), all referring to a Hilbert space $\mathscr{H}$. Hence, feasible types of QM particles will be classified by the possibilities for $\mathscr{H}$; more precisely the associated projective Hilbert space $P\mathscr{H}$. Reason: Any two vectors that differ by a scalar factor (or in physics terminology, two "kets" that differ by a "phase factor") correspond to the same physical quantum state.

Let G be a QM *symmetry group* (for example the Poincaré group in relativistic physics), i.e., a group of symmetries (leaving invariant the laws of physics.) Wigner outlined a correspondence between QM particles and representations, beginning with projective group representations of G, i.e., referring to $P\mathscr{H}$. (For example, the projective condition guarantees that applying a symmetry transformation, then applying its inverse transformation, will restore the original quantum state.)

Conclusion: Any QM particle is associated with a unique representation of G on a projective vector space $P\mathscr{H}$. But now, Wigner's Theorem states that these projective representations may in fact be realized as unitary representations, or possibly anti-unitary. The *irreducible representations* in the correspondence are the *elementary particles*. Properties of the unitary representations therefore yield observable features of elementary particles. Hence, every QM particle corresponds to a representation of a suitable symmetry group G, and if we can classify the group representations of G, we will have explicit information about the possibilities for the Hilbert space $\mathscr{H}$ of quantum mechanical states, and therefore what types of particles can exist.

In the details below, the following concepts and terms will play an important role; (for reference, see also Appendix F): Positive definite functions on groups, central extensions, the exponential mapping of Lie theory, intertwining operators, induced representations, imprimitivity theorem, The Stone-von Neumann uniqueness theorem.

8.1   Motivation

The following non-commutative Lie groups will be of special interest to us because of their applications to physics, and to a host of areas within mathematics; they are: the Heisenberg group $G = H_3$, the $ax + b$ group $G = S_2$; and $SL_2(\mathbb{R})$. In outline:

Definition 8.1.

- $G = H_3$, in real form $\simeq \mathbb{R}^3$, with multiplication

$$(a, b, c)(a', b', c') = (a + a', b + b', c + c' + ab'), \tag{8.1}$$

$\forall (a, b, c)$, and $(a', b', c') \in \mathbb{R}^3$. This is also matrix-multiplication when (a, b, c) has the form

$$\begin{pmatrix} 1 & a & c \\ 0 & 1 & b \\ 0 & 0 & 1 \end{pmatrix}.$$

In complex form, $z \in \mathbb{C}$, $c \in \mathbb{R}$, we have

$$(z, c)(z', c') = (z + z', c + c' + \Im(\bar{z}z')). \tag{8.2}$$

- $G = S_2$, the group of transformation $x \mapsto ax + b$ where $a \in \mathbb{R}_+$, and $b \in \mathbb{R}$, with multiplication

$$(a, b)(a', b') = (aa', b + ab'). \tag{8.3}$$

This is also the matrix-multiplication when (a, b) has the form

$$\begin{pmatrix} a & b \\ 0 & 1 \end{pmatrix}.$$

- $G = SL_2(\mathbb{R}) = 2 \times 2$ matrices

$$\begin{pmatrix} a & b \\ c & d \end{pmatrix}$$

over $\mathbb{R}$, with $ad - bc = 1$. Note that $SL_2(\mathbb{R})$ is locally isomorphic to $SU(1,1) = 2 \times 2$ matrices over $\mathbb{C}$,

$$\begin{pmatrix} \alpha & \beta \\ \overline{\beta} & \overline{\alpha} \end{pmatrix}$$

such that $|\alpha|^2 - |\beta|^2 = 1$. In both cases, the multiplication in G is matrix-multiplication for 2×2 matrices.

The four groups, and their harmonic analysis will be studied in detail inside this chapter.

An important question in the theory of unitary representations is the following: For a given Lie group G, *what are its irreducible unitary representations, up to unitary equivalence?* One aims for lists of these irreducibles. The question is an important part of non-commutative harmonic analysis. When answers are available, they have important implications for physics and for a host of other applications, but complete lists are hard to come by; and the literature on the subject is vast. We refer to the book [Tay86], and its references for an overview.

To begin with, the tools going into obtaining lists of the equivalence classes of irreducible representations, differ from one class of Lie groups to the other. Cases in point are the following classes, nilpotent, solvable, and semisimple. The Heisenberg group is in the first class, the $ax + b$ group in the second, and $SL_2(\mathbb{R})$ in the third. By a theorem of Stone and von Neumann, the classes of irreducibles for the Heisenberg group are indexed by a real parameter h; they are infinite dimensional for non-zero values of h, and one dimensional for $h = 0$.

For the $ax + b$ group, there are just two classes of unitary irreducibles. The verification of this can be made with the use of Mackey's theory of induced representations [Mac88]. But the story is much more subtle in the semisimple cases, even for $SL_2(\mathbb{R})$. The full list is divided up in series of representations (principal, continuous, discrete, and complementary series representations), and the paper [JÓ00] outlines some of their properties. But the details of this are far beyond the scope of the present book.

We now turn to some:

Exercise 8.1 (Semidirect product $G \circledS V$). Let V be a finite-dimensional vector space, and let $G \subset GL(V)$ be a subgroup of the corresponding general linear group. Set

$$(g, v)(g', v') := (gg', g(v') + v) \tag{8.4}$$

for all $g, g' \in G$, and $v, v' \in V$.

1. Show that with (8.4) we get a new group; called the semidirect product.

2. In the group $G \circledS V$, show that

$$(g, v)^{-1} = \left(g^{-1}, -g^{-1}(v) \right), \ g \in G, \ v \in V;$$

and conclude from this that V identifies as a normal subgroup in $G \circledS V$.

3. Show that, if G is a Lie group, then so is $G \circledS V$.

4. Show that, within the Lie algebra of $G \circledS V$, the vector space V identifies as an ideal.

General Considerations Every group G is also a $*$-semigroup, with the $*$ operation

$$g^* := g^{-1}.$$

But G is not a complex $*$ algebra yet, in particular, multiplication by a complex number is not defined. As a general principal, G can be embedded into the $*$-algebra

$$\mathfrak{A}_G = G \otimes \mathbb{C} = \mathbb{C}\text{-valued functions on } G.$$

Note that $\mathfrak{A}_G$ carries a natural pointwise multiplication and a scalar multiplication, given by

$$(g \otimes c_g)(h \otimes c_h) = gh \otimes c_g c_h$$
$$t(g \otimes c_g) = g \otimes t c_g$$

for all $g, h \in G$ and $c_g, c_h, t \in \mathbb{C}$. The $*$ operation extends from G to $\mathfrak{A}_G$ as

$$(g \otimes c_g)^* = g^{-1} \otimes \overline{c_{g^{-1}}}. \tag{8.5}$$

Remark 8.1. The $*$ operation so defined (as in (8.5)) is the only way to make it a period-2, conjugate linear, anti-automorphism. That is, $(tA)^* = \bar{t}A^*$, $A^{**} = A$, and $(AB)^* = B^*A^*$, for all $A, B \in \mathfrak{A}_G$, and all $t \in \mathbb{C}$.

There is a bijective correspondence between representations of G and representations of $\mathfrak{A}_G$. For if $\pi \in Rep(G, \mathscr{H})$, it then extends to $\tilde{\pi} = \pi \otimes Id_{\mathbb{C}} \in Rep(\mathfrak{A}_G, \mathscr{H})$, by

$$\tilde{\pi}(g \otimes c_g) = \pi_g \otimes c_g. \tag{8.6}$$

$id_{\mathbb{C}}$ denotes the identity representation $\mathbb{C} \longrightarrow \mathbb{C}$. Conversely, if $\rho \in Rep(\mathfrak{A}_G, \mathscr{H})$ then

$$\pi = \rho\big|_{G \otimes 1}$$

is a representation of the group $G \simeq G \otimes 1$.

Remark 8.2. The notation $g \otimes c_g$ is usually abbreviated as c_g. Thus pointwise multiplication takes the form

$$c_g d_h = l_{gh}.$$

Equivalently, we write

$$l_g = \sum_h c_h d_{h^{-1}g},$$

which is the usual convolution. In more details:

$$\left(\sum_g c_g \pi(g) \right) \left(\sum_h d_h \pi(h) \right) = \sum_g \underbrace{\left(\sum_h c_h d_{h^{-1}g} \right)}_{=l_g} \pi(g).$$

The $*$ operation now becomes

$$c_g^* = \overline{c_{g^{-1}}}.$$

More generally, every locally compact group has a left (and right) Haar measure. Thus the above construction has a "continuous" version. The above construction of $\mathfrak{A}_G$ then yields the Banach $*$-algebra $L^1(G)$. Again, there is a bijection between $Rep(G, \mathscr{H})$ and $Rep(L^1(G), \mathscr{H})$. In particular, the "discrete" version is recovered if the measure μ is discrete, in which case $\mathfrak{A}_G = l^1(G)$.

Definition 8.2. Let G be a group, and $\psi : G \to \mathbb{C}$ be a function. We say that ψ is *positive definite* iff (Def.) for all $n \in \mathbb{N}$, all $g_1, \ldots, g_n \in G$, and all $c_1, \ldots, c_n \in \mathbb{C}$, we have

$$\sum_{j=1}^{n} \sum_{k=1}^{n} \overline{c_j} c_k \psi \left(g_j^{-1} g_k \right) \geq 0.$$

Often we also assume that $\psi(e) = 1$.

Recall (Chapter 5) that a state on a $*$-algebra $\mathfrak{A}$ is a positive linear functional on $\mathfrak{A}$, normalized when appropriate. The link between states and representations is given by the GNS construction, Theorem 5.3. Below we point out that there is also a direct link between the GNS construction and a canonical correspondence between: (i) positive definite functions on a group G, on one side, and (ii) cyclic unitary representations of G on the other. In detail: Given a group G, we consider the associated group algebra $\mathfrak{A}_G$ as a $*$-algebra (Remark 8.2). In the next Exercise, show that if $\mathfrak{A}_G$ is the group algebra of some group G, then every positive definite function on G (Definition 8.2) extends canonically to a state on $\mathfrak{A}_G$. Moreover, all states on $\mathfrak{A}_G$ arise this way.

Exercise 8.2 (Positive definite functions and states). Show that every positive definite function ψ on G extends by linearity to a state $\widetilde{\psi}$ on the group algebra $\mathbb{C}[G]$, by setting

$$\widetilde{\psi} \left(\sum_g c_g g \right) := \sum_g c_g \psi(g),$$

for all (finite) linear expressions $\sum_g c_g g$.

Exercise 8.3 (Contractions). Let $\mathscr{H}$ be a Hilbert space, and let T be a *contraction*, i.e., $T \in \mathscr{B}(\mathscr{H})$, satisfying one of the two equivalent conditions:

$$\|T\| \leq 1 \Longleftrightarrow I - T^*T \geq 0 \text{ in the order on Hermitian operators.}$$

1. For $G = \mathbb{Z}$, define $\psi : \mathbb{Z} \longrightarrow \mathbb{C}$ as follows:

$$\psi(n) = \begin{cases} T^n = \underbrace{T \circ \cdots \circ T}_{n \text{ times}} & \text{if } n \geq 0 \\ (T^*)^{|n|} & \text{if } n < 0. \end{cases} \tag{8.7}$$

Show that ψ is positive definite.

2. Conclude that $\widetilde{\psi}$ is completely positive (see Chapter 6).

3. Apply (1) & (2) to conclude the existence of a triple $(V, \mathscr{K}, \mathcal{U})$, where $\mathscr{K}$ is a Hilbert space, $V : \mathscr{H} \longrightarrow \mathscr{K}$ is isometric, $\mathcal{U} : \mathscr{K} \longrightarrow \mathscr{K}$ is a unitary operator; and we have

$$T^n = V^* \mathcal{U}^n V, \ \forall n \in \mathbb{N}. \tag{8.8}$$

Historical Note. The system $(V, \mathscr{K}, \mathcal{U})$ above, satisfying (8.8), is called a *unitary dilation* [Sch55] (details in Chapter 6.)

Exercise 8.4 (Central Extension). Let V be a vector space over $\mathbb{R}$, and let $B : V \times V \to \mathbb{R}$ be a function.

1. Then show that $V \times \mathbb{R}$ turns into a group G when the operation in G is defined as follows:

$$(u, \alpha)(v, \beta) = (u + v, \alpha + \beta + B(u, v)), \ \forall \alpha, \beta \in \mathbb{R}, \ \forall u, v \in V, \tag{8.9}$$

if and only if B satisfies

$$\begin{aligned} B(u, 0) &= B(0, u) = 0, \ \forall u \in V; \text{ and} \\ B(u, v) + B(u + v, w) &= B(u, v + w) + B(v, w), \ \forall u, v, w \in V. \end{aligned} \tag{8.10}$$

2. Assuming that (8.10) is satisfied; show that the group inverse under the operation (8.9) is

$$(u, \alpha)^{-1} = (-u, -\alpha - B(u, -u)), \ \forall \alpha \in \mathbb{R}, u \in V. \tag{8.11}$$

8.2   Unitary One-Parameter Groups

"The mathematical landscape is full if groups of unitary operators. The ..., strongly continuous one-parameter groups $U(t)$, $-\infty < t < \infty$, come mostly from three sources: processes where energy is conserved, such as those governed by wave equations of all sorts; process where probability is preserved, for instance, ones governed by Schrödinger equations; and Hamiltonian and other measure-preserving flows."
— Peter Lax, from [Lax02]

Let $\mathscr{H}$ be a Hilbert space, and let $P(\cdot)$ be a *projection valued measure* (PVM),

$$P : \mathcal{B}(\mathbb{R}) \longrightarrow Proj(\mathscr{H}) \tag{8.12}$$

i.e., defined on the sigma-algebra of all Borel subsets in $\mathbb{R}$. (See Definition 3.10.)

Then, as we saw, the integral (operator-valued)

$$U(t) = \int_{\mathbb{R}} e^{i\lambda t} P(d\lambda), \ t \in \mathbb{R} \tag{8.13}$$

is well defined, and yields a strongly continuous one-parameter group acting on $\mathcal{H}$; equivalently,

$$U \in Rep_{uni}(\mathbb{R}, \mathcal{H}) \tag{8.14}$$

where U is defined by (8.13).

The theorem of M.H. Stone states that the converse holds as well, i.e., every U, as in (8.14), corresponds to a unique P, a PVM, such that (8.13) holds.

Exercise 8.5 (von Neumann's ergodic theorem). Let $\{U(t)\}_{t \in \mathbb{R}}$ be a strongly continuous one-parameter group with PVM, $P(\cdot)$. Denote by $P(\{0\})$ the value of P on the singleton $\{0\}$.

1. Show that

$$P(\{0\})\mathcal{H} = \{h \in \mathcal{H} \ : \ U(t)h = h, \ \forall t \in \mathbb{R}\}. \tag{8.15}$$

(We set $\mathcal{H}_0 := P(\{0\})\mathcal{H}$.)

2. Establish the following limit conclusion:

$$\lim_{T \to \infty} \frac{1}{2T} \int_{-T}^{T} U(t)\, dt = P(\{0\}). \tag{8.16}$$

Hint:

$$\frac{1}{2T} \int_{-T}^{T} U(t)\, dt = \int_{\mathbb{R}} \frac{\sin(\lambda T)}{\lambda T} P(d\lambda)$$

holds for all $T \in \mathbb{R}_+$.

8.3 Group - Algebra - Representations

In physics, we are interested in representation of symmetry groups, which preserve inner product or energy, and naturally leads to *unitary representations*. Every unitary representation can be decomposed into *irreducible representations*; the latter amounts to *elementary particles* which can not be broken up further. In practice, quite a lot work goes into finding irreducible representations of symmetry groups. Everything we learned about algebras is also true for groups. The idea is to go from groups to algebras and then to representations.

We summarize the basic definitions:

- $\pi_G \in Rep(G, \mathcal{H})$
$$\begin{cases} \pi(g_1 g_2) & = \pi(g_1)\pi(g_2) \\ \pi(e_G) & = I_{\mathcal{H}} \\ \pi(g^{-1}) & = \pi(g)^* \end{cases}$$

- $\pi_{\mathfrak{A}} \in Rep(\mathfrak{A}, \mathscr{H})$

$$\begin{cases} \pi(A_1 A_2) & = \pi(A_1)\pi(A_2) \\ \pi(1_{\mathfrak{A}}) & = I_{\mathscr{H}} \\ \pi(A^*) & = \pi(A)^* \end{cases}$$

Case 1. G is discrete $\longrightarrow \mathfrak{A} = G \otimes l^1$

$$\left(\sum_g a(g)g\right)\left(\sum_g b(h)h\right) \;=\; \sum_{g,h} a(g)b(h)gh$$

$$= \sum_{g'}\sum_h a(g'h^{-1})b(h)g'$$

$$\left(\sum_g c(g)g\right)^* = \left(\sum_g \overline{c(g^{-1})}g\right)$$

where $c^*(g) = \overline{c(g^{-1})}$. The multiplication of functions in $\mathfrak{A}$ is a generalization of convolutions.

Case 2. G is locally compact $\longrightarrow \mathfrak{A} = G \otimes L^1(\mu) \simeq L^1(G)$.

Definition 8.3. Let $\mathcal{B}(G)$ be the Borel sigma-algebra of G. A regular Borel measure λ is said to be *left (resp. right) invariant*, if $\lambda(gE) = \lambda(E)$ (resp. $\lambda(Eg) = \lambda(Eg)$), for all $g \in G$, and $E \in \mathcal{B}(G)$. λ is called a *left (resp. right) Haar measure* accordingly.

Note that λ is left invariant iff

$$\lambda'(E) := \lambda\left(E^{-1}\right)$$

is right invariant, where $E^{-1} = \left\{g \in G \,:\, g^{-1} \in E\right\}$, for all $E \in \mathcal{B}(G)$. Hence one may choose to work with either a left or right invariant measure.

Theorem 8.1. *Every locally compact group G has a left Haar measure, unique up to a multiplicative constant.*

For the existence *of Haar measures*, one first proves the easy case when G is compact, and then extends to locally compact cases. For non compact groups, the left / right Haar measures could be different. Many non compact groups have no Haar measure. In applications, the Haar measures are usually constructed explicitly.

Remark 8.3. While we shall omit proofs of the existence of Haar measure in the most general case of locally compact groups G, we mention here that, in the case when the group G is further assumed compact, there is an elegant proof of the existence of Haar measure, based on the Markov–Kakutani fixed-point theorem. Further note that compact groups are unimodular.

The M-K fixed-point theorem states that any commuting family $\mathscr{F}$ of continuous affine self-mappings in a given compact convex subset K in a locally convex topological vector space automatically will have a common fixed point; i.e., a point x in K which is fixed by all the maps from $\mathscr{F}$. (An elegant approach to Haar measure for the case of compact G is given in [Zim90].)

Given a *left Haar measure* λ_L, and $g \in G$, then

$$E \longmapsto \lambda_L(Eg), \ E \in \mathcal{B}(G)$$

is also left invariant. Hence, by Theorem 8.1,

$$\lambda_L(Eg) = \triangle_G(g)\lambda_L(E) \tag{8.17}$$

for some constant $\triangle_G(g) \in \mathbb{R}\backslash\{0\}$. Note that $\triangle_G$ is well-defined, independent of the choice of λ_L. Moreover,

$$\begin{aligned}
\lambda_L(Egh) &= \triangle_G(h)\lambda_L(Eg) = \triangle_G(h)\triangle_G(g)\lambda_L(E) \\
\lambda_L(Egh) &= \triangle_G(gh)\lambda_L(E)
\end{aligned}$$

and it follows that $\triangle_G : G \longrightarrow \mathbb{R}_\times$ is a homomorphism, i.e.,

$$\triangle_G(gh) = \triangle_G(h)\triangle_G(g), \ \forall g, h \in G. \tag{8.18}$$

Definition 8.4. $\triangle_G$ is called the *modular function* of G. G is said to be *unimodular* if $\triangle_G \equiv 1$.

Corollary 8.1. *Every compact group G is unimodular.*

Proof. Since $\triangle_G$ is a homomorphism, $\triangle_G(G)$ is compact in $\mathbb{R}_\times$, so $\triangle_G \equiv 1$. $\square$

Corollary 8.2. *For all $f \in C_c(G)$, and $g \in G$,*

$$\int_G f(\cdot g)\,d\lambda_L = \triangle_G(g^{-1})\int_G f\,d\lambda_L. \tag{8.19}$$

Equivalently, we get the substitution formula:

$$d\lambda_L(\cdot g) = \triangle_G(g)\,d\lambda_L(\cdot). \tag{8.20}$$

Similarly,

$$\int_G f(g^{-1}\cdot)\,d\lambda_R = \triangle_G(g^{-1})\int_G f\,d\lambda_R; \tag{8.21}$$

i.e.,

$$d\lambda_R(g\cdot) = \triangle_G(g^{-1})\,d\lambda_R(\cdot). \tag{8.22}$$

Proof. It suffices to check for characteristic functions. Fix $E \in \mathcal{B}(G)$, then

$$\begin{aligned}
\int_G \chi_E(\cdot g)\,d\lambda_L &= \int_G \chi_{Eg^{-1}}\,d\lambda_L \\
&= \lambda_L(Eg^{-1}) \\
&\underset{(8.17)}{=} \triangle_G(g^{-1})\lambda_L(E)
\end{aligned}$$

$$= \Delta_G \left(g^{-1}\right) \int_G \chi_E d\lambda_L$$

hence (8.19)-(8.20) follow from this and a standard approximation.

For the right Haar measure, recall that $E \mapsto \lambda_L \left(E^{-1}\right)$ is right invariant, and so $\lambda_R (E) = c\lambda_L \left(E^{-1}\right)$, for some constant $c \in \mathbb{R}\backslash \{0\}$. (In fact, more is true; see Theorem 8.2 below.) Therefore,

$$
\begin{aligned}
\lambda_R (gE) &= c\lambda_L \left(E^{-1}g^{-1}\right) \\
&= c\Delta_G \left(g^{-1}\right) \lambda_L \left(E^{-1}\right) \\
&= \Delta_G \left(g^{-1}\right) \lambda_R (E).
\end{aligned}
$$

This yields (8.21)-(8.22). $\square$

Theorem 8.2. *Let G be a locally compact group, then the two Haar measures are mutually absolutely continuous, i.e., $\lambda_L \ll \lambda_R \ll \lambda_L$.*

Specifically, fix λ_L, and set

$$\lambda_R (E) := \lambda_L \left(E^{-1}\right), \ E \in \mathcal{B}(G);$$

then

$$\frac{d\lambda_R}{d\lambda_L} (g) = \Delta_G \left(g^{-1}\right) = \text{Radon-Nikodym derivative.}$$

Proof. Note that $\Delta_G d\lambda_R$ is left invariant. Indeed,

$$
\begin{aligned}
\Delta_G (g\cdot) \, d\lambda_R (g\cdot) &= \underbrace{\left(\Delta_G (g) \Delta_G (\cdot)\right)}_{(8.18)} \underbrace{\left(\Delta_G \left(g^{-1}\right) d\lambda_R (\cdot)\right)}_{(8.22)} \\
&= \Delta_G (\cdot) \, d\lambda_R (\cdot).
\end{aligned}
$$

Hence, by the uniqueness of the Haar measure, we have

$$\Delta_G d\lambda_R = c \, d\lambda_L$$

for some constant $c \in \mathbb{R}\backslash \{0\}$. One then checks that $c \equiv 1$. $\square$

Corollary 8.3. *If λ_L is a left Haar measure on G, then*

$$d\lambda_L \left(g^{-1}\right) = \Delta_G \left(g^{-1}\right) d\lambda_L (g). \tag{8.23}$$

Similarly, if λ_R is a right Haar measure, then

$$d\lambda_R \left(g^{-1}\right) = \Delta_G (g) \, d\lambda_R (g). \tag{8.24}$$

Remark 8.4. In the case $l^1(G)$, $\lambda_L = \lambda_R =$ the counting measure, which is unimodular, hence Δ_G does not appear.

In $L^1(G)$, we define

$$
\begin{aligned}
(\varphi \star \psi)(g) &:= \int_G \varphi\left(gh^{-1}\right)\psi(h)\,d\lambda_R(h) \qquad (8.25)\\
&= \int_G \varphi\left(h^{-1}\right)\psi(hg)\,d\lambda_R(h)\\
&= \int_G \varphi(h)\psi\left(h^{-1}g\right)\underbrace{\triangle_G(h)\,d\lambda_R(h)}_{d\lambda_L(h)}
\end{aligned}
$$

and

$$
\varphi^*(g) := \overline{\varphi\left(g^{-1}\right)}\triangle_G(g). \qquad (8.26)
$$

The choice of (8.26) preserves the L^1-norm. Indeed,

$$
\begin{aligned}
\int_G |\varphi^*|\,d\lambda_R &= \int_G \left|\varphi\left(g^{-1}\right)\right|\triangle_G(g)\,d\lambda_R(g)\\
&= \int_G |\varphi(g)|\,\triangle_G\left(g^{-1}\right)\,d\lambda_R\left(g^{-1}\right)\\
&= \int_G |\varphi|\,d\lambda_R
\end{aligned}
$$

where $\triangle_G\left(g^{-1}\right)d\lambda_R\left(g^{-1}\right) = d\lambda_R(g)$ by (8.24).

$L^1(G)$ is a Banach *-algebra, and $L^1(G) = L^1$-completion of $C_c(G)$. (Fubini's theorem shows that $f \star g \in L^1(G)$, for all $f, g \in L^1(G)$.)

We may also use left Haar measure in (8.25). Then, we set

$$
\begin{aligned}
(\varphi * \psi)(g) &:= \int_G \varphi(h)\psi\left(h^{-1}g\right)\,d\lambda_L(h) \qquad (8.27)\\
&= \int_G \varphi(gh)\psi\left(h^{-1}\right)\,d\lambda_L(h)\\
&= \int_G \varphi\left(gh^{-1}\right)\psi(h)\underbrace{\triangle_G\left(h^{-1}\right)\,d\lambda_L(h)}_{d\lambda_R(h)}
\end{aligned}
$$

and set

$$
\varphi^*(g) := \overline{\varphi\left(g^{-1}\right)}\triangle_G\left(g^{-1}\right). \qquad (8.28)
$$

There is a bijection between representations of groups and representations of algebras.

Given a unitary representation $\pi \in Rep(G, \mathscr{H})$, let dg denote the Haar measure in $L^1(G)$, then we get the group algebra representation $\pi_{L^1(G)} \in Rep\left(L^1(G), \mathscr{H}\right)$, where

$$
\begin{aligned}
\pi_{L^1(G)}(\varphi) &= \int_G \varphi(g)\pi(g)\,dg\\
\pi_{L^1(G)}(\varphi^*) &= \pi_{L^1(G)}(\varphi)^*.
\end{aligned}
$$

Indeed, one checks that

$$\pi_{L^1(G)}\left(\varphi_1 \star \varphi_2\right) = \pi_{L^1(G)}\left(\varphi_1\right)\pi_{L^1(G)}\left(\varphi_2\right).$$

Conversely, given a representation of $L^1(G)$, let (φ_i) be a sequence in L^1 such that $\varphi_i \to \delta_g$. Then

$$\int \varphi_i(h)\pi(h)g\,dh \to \pi(g),$$

i.e., the limit is a representation of G.

Remark 8.5. Let G be a matrix group, then $x^{-1}dx$, $x \in G$, is left translation invariant. For if $y \in G$, then

$$(yx)^{-1} d\,(yx) = x^{-1}\left(y^{-1}y\right)dx = x^{-1}dx.$$

Now assume $\dim G = n$, and so $x^{-1}dx$ contains n linearly independent differential forms, $\sigma_1, \ldots, \sigma_n$; and each σ_j is left translation invariant. Thus $\sigma_1 \wedge \cdots \wedge \sigma_n$ is a left invariant volume form, i.e., the left Haar measure. Similarly, the right Haar measure can be constructed from $dx \cdot x^{-1}$, $x \in G$.

.

8.3.1   Example − $ax + b$ group

Let $G = \left\{\begin{bmatrix} a & b \\ 0 & 1 \end{bmatrix} : a \in \mathbb{R}_+,\, b \in \mathbb{R}\right\}$.

- Multiplication

$$\begin{bmatrix} a' & b' \\ 0 & 1 \end{bmatrix}\begin{bmatrix} a & b \\ 0 & 1 \end{bmatrix} = \begin{bmatrix} a'a & a'b + b' \\ 0 & 1 \end{bmatrix}$$

- Inverse

$$\begin{bmatrix} a & b \\ 0 & 1 \end{bmatrix}^{-1} = \begin{bmatrix} \frac{1}{a} & -\frac{b}{a} \\ 0 & 1 \end{bmatrix}.$$

G is isomorphic to the transformation group $x \mapsto ax + b$; where composition gives

$$x \mapsto ax + b \mapsto a'\,(ax + b) + b' = aa'x + (a'b + b').$$

Remark 8.6. Setting $a = e^t$, $a' = e^{t'}$, $aa' = e^t e^{t'} = e^{t+t'}$, i.e., multiplication aa' can be made into addition.

The left Haar measure is given as follows:

Let $g = \begin{bmatrix} a & b \\ 0 & 1 \end{bmatrix} \in G$, so that

$$g^{-1}dg = \frac{1}{a}\begin{bmatrix} 1 & -b \\ 0 & a \end{bmatrix}\begin{bmatrix} da & db \\ 0 & 0 \end{bmatrix}$$

$$= \frac{1}{a} \begin{bmatrix} da & db \\ 0 & 0 \end{bmatrix}.$$

Hence we get two left invariant (linear independent) differential forms:

$$\frac{da}{a} \quad \text{and} \quad \frac{db}{a}.$$

Set

$$d\lambda_L(g) = d\lambda_L(x, y) := \frac{1}{x^2} dx \wedge dy; \quad \left(g = \begin{bmatrix} x & y \\ 0 & 1 \end{bmatrix}, \ x \in \mathbb{R}_+ \right).$$

Indeed, λ_L is left invariant. To check this, consider

$$g = \begin{bmatrix} a & b \\ 0 & 1 \end{bmatrix}, \ h = \begin{bmatrix} a' & b' \\ 0 & 1 \end{bmatrix}, \text{ and}$$

$$h^{-1}g = \begin{bmatrix} \frac{1}{a'} & -\frac{b'}{a'} \\ 0 & 1 \end{bmatrix} \begin{bmatrix} a & b \\ 0 & 1 \end{bmatrix} = \begin{bmatrix} \frac{a}{a'} & \frac{b-b'}{a'} \\ 0 & 1 \end{bmatrix};$$

then

$$\begin{aligned}
\int_G f\left(h^{-1}g\right) d\lambda_L(g) &= \int_0^\infty \int_{-\infty}^\infty f\left(\frac{a}{a'}, \frac{b-b'}{a'}\right) \frac{da \wedge db}{a^2} \\
&= \int_0^\infty \int_{-\infty}^\infty f(s, t) \frac{d(a's) \wedge d(a't + b')}{(a's)(a's)} \\
&= \int_0^\infty \int_{-\infty}^\infty f(s, t) \frac{ds \wedge dt}{s^2}
\end{aligned}$$

where we set

$$s = \frac{a}{a'}, \ da = a'ds$$

$$t = \frac{b-b'}{a'}, \ db = a'dt$$

so that

$$\frac{da \wedge db}{a^2} = \frac{a'^2 ds \wedge dt}{a'^2 s^2} = \frac{ds \wedge dt}{s^2}.$$

For the right Haar measure, note that

$$\begin{aligned}
\int f\left(gh^{-1}\right)(dg)\, g^{-1} &= \int f(g')\, d(g'h)(g'h)^{-1} \\
&= \int f(g')(dg')\left(hh^{-1}\right)g'^{-1} \\
&= \int f(g')(dg')\, g'^{-1}.
\end{aligned}$$

Since

$$(dg)\, g^{-1} = \begin{bmatrix} da & db \\ 0 & 0 \end{bmatrix} \begin{bmatrix} \frac{1}{a} & -\frac{b}{a} \\ 0 & 1 \end{bmatrix} = \begin{bmatrix} \frac{da}{a} & -\frac{b\, da}{a} + db \\ 0 & 0 \end{bmatrix}$$

we then set

$$d\lambda_R(g) := d\lambda_R(a,b) = \frac{da \wedge db}{a}.$$

Check:

$$g = \begin{bmatrix} a & b \\ 0 & 1 \end{bmatrix}, \quad h = \begin{bmatrix} a' & b' \\ 0 & 1 \end{bmatrix},$$

$$gh^{-1} = \begin{bmatrix} a & b \\ 0 & 1 \end{bmatrix} \begin{bmatrix} \frac{1}{a'} & -\frac{b'}{a'} \\ 0 & 1 \end{bmatrix} = \begin{bmatrix} \frac{a}{a'} & -\frac{ab'}{a'} + b \\ 0 & 1 \end{bmatrix}$$

and so

$$\begin{aligned}
\int_G f\left(gh^{-1}\right) d\lambda_R(g) &= \int_0^\infty \int_{-\infty}^\infty f\left(\frac{a}{a'}, -\frac{ab'}{a'} + b\right) \frac{da \wedge db}{a} \\
&= \int_0^\infty \int_{-\infty}^\infty f\left(s,t\right) \frac{a' ds \wedge dt}{a' s} \\
&= \int_0^\infty \int_{-\infty}^\infty f\left(s,t\right) \frac{ds \wedge dt}{s}
\end{aligned}$$

with a change of variable:

$$s = \frac{a}{a'}, \quad da = a' ds$$

$$t = -\frac{ab'}{a'} + b, \quad db = dt$$

$$\frac{da \wedge db}{a} = \frac{(a' ds) \wedge dt}{a' s} = \frac{ds \wedge dt}{s}.$$

8.4 Induced Representations

Most of the Lie groups considered here fall in the following class:

Let V be a finite-dimensional vector space over $\mathbb{R}$ (or $\mathbb{C}$). We will consider the real case here, but the modifications needed for the complex case are straightforward.

Let $q : V \times V \to \mathbb{R}$ be a non-degenerate bilinear form, such that $v \to q(v, \cdot) \in V^*$ is 1-1. Let $GL(V)$ be the general linear group for V, i.e., all invertible linear maps $V \to V$. (If a basis in V is chosen, this will be a matrix-group.)

Lemma 8.1. *Set*

$$G(q) = \{g \in GL(V) \mid q(gu, gv) = q(u,v), \ \forall u, v \in V\}. \tag{8.29}$$

Then $G(q)$ is a Lie group, and its Lie algebra consists of all linear mappings $X : V \to V$ such that

$$q(Xu, v) + q(u, Xv) = 0, \ \forall u, v \in V. \tag{8.30}$$

Proof. Fix a linear mapping $X : V \to V$, and set

$$g_X(t) = \sum_{n=0}^{\infty} \frac{t^n}{n!} X^n = \exp(tX)$$

i.e., the matrix-exponential. Note that $g_X(t)$ satisfies (8.29) for all $t \in \mathbb{R}$ iff X satisfies (8.30). To see this, differentiate: i.e., compute

$$\frac{d}{dt} q\left(\exp(tX), \exp(tX) v\right)$$

using that q is assumed bilinear. $\qquad\square$

We will address two questions:

1. How to induce a representation of a group G from a representation of the a subgroup $\Gamma \subset G$?

2. Given a representation of a group G, how to test whether it is induced from a representation of a subgroup $\Gamma \subset G$? (See [Mac88].)

The main examples we will study are the Lie groups of

- $ax + b$

- Heisenberg

- $SL_2(\mathbb{R})$

- Lorentz

- Poincaré

Among these, the $ax + b$, Heisenberg and Poincaré groups are semidirect product groups. Their representations are induced from normal subgroups.

In more detail, the five groups in the list above are as follows:

The $ax + b$ group is the group of 2×2 matrices $\begin{pmatrix} a & b \\ 0 & 1 \end{pmatrix}$ where $a \in \mathbb{R}_+$, and $b \in \mathbb{R}$.

The Heisenberg group is the group of upper triangular 3×3 real matrices $\begin{pmatrix} 1 & x & z \\ 0 & 1 & y \\ 0 & 0 & 1 \end{pmatrix}$, $x, y, z \in \mathbb{R}$.

The group $SL_2(\mathbb{R})$ is the group of all 2×2 matrices $\begin{pmatrix} a & b \\ c & d \end{pmatrix}$ satisfying $a, b, c, d \in \mathbb{R}$, and $ad - bc = 1$.

The Lorentz group is the group $L = G(q)$ defined by (8.29) where $V = \mathbb{R}^4$, (space-times in physics) and

$$q(x_0, x_1, x_2, x_3) = -x_0^2 + x_1^2 + x_2^2 + x_3^2;$$

so q is the non-degenerate quadratic form with one minus sign, and three plus signs.

The Poincaré group P is the semidirect product $P = L \circledS \mathbb{R}^4$, where the group-product in P is as follows:

$$(g, v)(g', v') = (gg', v + gv')$$

for all $g, g' \in L$, and all $v, v' \in \mathbb{R}^4$.

Exercise 8.6 $(SL_2(\mathbb{R})$ and Lemma 8.1). Consider $V = \mathbb{R}^2$ in the setting of Lemma 8.1. Set

$$q(x) = -x_0^2 + x_1^2, \quad \forall x = (x_0, x_1) \in \mathbb{R}^2.$$

Let $G := G(q)$.

1. Show that
$$G = \left\{ g \ : \ 2 \times 2 \text{ invertible}, \ g^T J g = J \right\}$$
   where g^T is the transposed matrix, and $J = \begin{bmatrix} 1 & 0 \\ 0 & -1 \end{bmatrix}$.

2. Show that every $g \in G$ satisfies $\det(g) \in \{\pm 1\}$.

Exercise 8.7 (The Heisenberg group as a semidirect product). Consider the following two subgroups A and B in the Heisenberg group H:

$$A = \begin{bmatrix} 1 & x & 0 \\ 0 & 1 & 0 \\ 0 & 0 & 1 \end{bmatrix}, \quad \text{and} \quad B = \begin{bmatrix} 1 & 0 & z \\ 0 & 1 & y \\ 0 & 0 & 1 \end{bmatrix}.$$

1. Verify that A and B are both Abelian subgroups under the matrix-multiplication of H.

2. Show that H becomes a semidirect product
$$H = A \circledS B$$
   where the action α_x of A, as a group of automorphisms in B, is as follows:
$$\alpha_x(y, z) = (y, z + xy), \quad \forall x, y, z \in \mathbb{R}.$$

Exercise 8.8 (The invariant complex vector fields from the Heisenberg group). In its complex form, the Heisenberg group takes the form $\mathbb{C} \times \mathbb{R}$, $z \in \mathbb{C}$, $c \in \mathbb{R}$; and with group multiplication:

$$(z, c)(z', c') = (z + z', c + c' + 2\Im(\bar{z}z')) \tag{8.31}$$

Set $\frac{\partial}{\partial z} = \frac{1}{2} \left(\frac{\partial}{\partial x} - i \frac{\partial}{\partial y} \right)$, and $\frac{\partial}{\partial \bar{z}} = \frac{1}{2} \left(\frac{\partial}{\partial x} + i \frac{\partial}{\partial y} \right)$, or in abbreviated form

$$\partial = \partial_z = \frac{1}{2} (\partial_x - i\partial_y), \quad \text{and} \quad \bar{\partial} = \bar{\partial}_z = \frac{1}{2} (\partial_x + i\partial_y), \tag{8.32}$$

so

$$4\left(\partial_x^2 + \partial_y^2\right) = \partial\bar{\partial} = \bar{\partial}\partial.$$ (8.33)

Show that a basis for the left-invariant vector fields on H is as follows:

$$\partial_z - i\bar{z}\partial_c, \quad \bar{\partial}_z + iz\partial_c, \quad \text{and} \quad i\partial_c,$$ (8.34)

with commutator

$$\left[\partial_z - i\bar{z}\partial_c, \bar{\partial}_z + iz\partial_c\right] = 2i\partial_c.$$ (8.35)

Representation Theory.

It is extremely easy to find representations of Abelian subgroups. Unitary irreducible representation of Abelian subgroups are one-dimensional, but the induced representation [Mac88] on an enlarged Hilbert space is infinite dimensional.

Exercise 8.9 (The Campbell-Baker-Hausdorff formula). The exponential function is arguably the most important function in analysis. The Campbell-Baker-Hausdorff formula (below) illustrates the role of non-commutativity in this.

Let G and $\mathfrak{g}$ be as above, and let $\mathfrak{g} \xrightarrow{\exp} G$ be the exponential mapping. See Figure 8.1.

1. Show that there is a convergent series with terms of degree > 1 being iterated commutators $Z(X, Y)$ with

$$\exp X \exp Y = \exp Z(X, Y), \text{ and}$$ (8.36)

$$Z(X, Y) = X + Y + \frac{1}{2}[X, Y] + \frac{1}{12}\left([X, [X, Y]] + [Y, [Y, X]]\right) + \cdots$$ (8.37)

2. Use combinatorics and algebra in order to derive an algorithm for the terms "$+ \cdots$" in (8.37). This is the Baker–Campbell–Hausdorff formula; see, e.g., [HS68].

3. Show that

$$Z(X, Y) + Z(-X, -Y) = 0.$$

Exercise 8.10 (The Lie algebra of a central extension). Let V be a vector space over $\mathbb{R}$, and $B : V \times V \to \mathbb{R}$ a *cocycle*, i.e.,

$$B(u, v) + B(u + v, w) = B(u, v + w) + B(v, w), \text{ for } \forall u, v, w, \in V.$$ (8.38)

Let $G_B = V \times \mathbb{R}$ be the corresponding Lie group:

$$(u, \alpha)(v, \beta) = (u + v, \alpha + \beta + B(u, v)), \text{ for } \forall \alpha, \beta \in \mathbb{R}, \forall u, v \in V.$$ (8.39)

1. Show that the Lie algebra $La(G_B)$ of G_B is $V \times \mathbb{R}$ itself with $0 \times \mathbb{R} \subseteq$ the center; and with Lie bracket $[\cdot, \cdot]$ given by

$$[u, v] = B(u, v) - B(v, u), \text{ for } \forall u, v \in V.$$ (8.40)

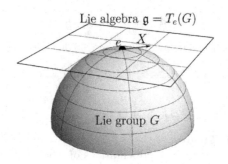

Figure 8.1: G and $\mathfrak{g}$ (Lie algebra, Lie group, and exponential mapping).

2. Show that the exponential mapping $\exp_{G_B}$ is the trivial mapping

$$\exp_{G_B} \underbrace{(u,\alpha)}_{\in La(G_B)} = \underbrace{(u,\alpha)}_{\in G_B}, \text{ for } \forall u \in V, \alpha \in \mathbb{R}.$$

Example 8.1 (The $ax + b$ group $(a > 0)$). $G = \{(a,b)\}$, where $(a,b) = \begin{bmatrix} a & b \\ 0 & 1 \end{bmatrix}$. The multiplication rule is given by

$$
\begin{aligned}
(a,b)(a',b') &= (aa', ab' + b) \\
(a,b)^{-1} &= (\tfrac{1}{a}, -\tfrac{b}{a}).
\end{aligned}
$$

$\Gamma = \{(1,b)\}$ is a one-dimensional Abelian, normal subgroup of G. We check that

- Abelian: $(1,b)(1,c) = (1, c+b)$

- normal: $(x,y)(1,b)(x,y)^{-1} = (1, xb)$, note that this is also Ad_g acting on the normal subgroup Γ

- The other subgroup $\{(a,0)\}$ is isomorphic to the multiplicative group $(\mathbb{R}_+, \times)$. Because we have

$$(a,0)(a',0) = (aa', 0)$$

  by the group multiplication rule above.

- Notice that $(\mathbb{R}_+, \times)$ is not a normal subgroup, since

$$(a,b)(x,0)(\tfrac{1}{a}, -\tfrac{b}{a}) = (ax,b)(\tfrac{1}{a}, -\tfrac{b}{a}) = (x - bx + b).$$

Γ is unimodular, hence it is just a copy of $\mathbb{R}$. Its invariant measure is the Lebesgue measure on $\mathbb{R}$.

The multiplicative group $(\mathbb{R}_+, \times)$ acts on the additive group $(\mathbb{R}, +)$ by

$$\varphi : (\mathbb{R}_+, \times) \;\mapsto\; Aut((\mathbb{R}, +))$$
$$\varphi_a(b) \;=\; ab$$

check:

$$(a, b)(a', b') = (aa', b + \varphi_a(b')) = (aa', b + ab')$$

$$(a, b)^{-1} = (a^{-1}, \varphi_{a^{-1}}(b^{-1})) = (a^{-1}, a^{-1}(-b)) = (\frac{1}{a}, -\frac{b}{a})$$

$$(a, b)(1, x)(a, b^{-1}) \;=\; (a, b + \varphi_a(x))(a, b^{-1})$$
$$=\; (a, b + ax)(\frac{1}{a}, -\frac{b}{a})$$
$$=\; (1, ax) = \varphi_a(x).$$

Example 8.2. The Lie algebra of G is given by $X = \begin{bmatrix} 1 & 0 \\ 0 & 0 \end{bmatrix}, = \begin{bmatrix} 0 & 1 \\ 0 & 0 \end{bmatrix}$.
We check that

$$e^{tX} = \begin{bmatrix} e^t & 0 \\ 0 & 1 \end{bmatrix}$$

which is subgroup $(\mathbb{R}_+, \times)$; and

$$e^{sY} = I + sY + 0 + \cdots + 0 = \begin{bmatrix} 1 & s \\ 0 & 1 \end{bmatrix}$$

which is subgroup $(\mathbb{R}, +)$. We also have $[X, Y] = Y$.

Example 8.3. Form $L^2(\mu_L)$ where μ_L is the left Haar measure. Then $\pi : g \to \pi(g)f(x) = f(g^{-1}x)$ is a unitary representation in $L^2(\mu_L)$. Specifically, if $g = (a, b)$ then

$$f(g^{-1}x) = f(\frac{x}{a}, \frac{y - b}{a}).$$

Differentiate along the a direction we get

$$\tilde{X}f \;=\; \frac{d}{da}\Big|_{a=1, b=0} f(\frac{x}{a}, \frac{y - b}{a}) = (-x\frac{\partial}{\partial x} - y\frac{\partial}{\partial y})f(x, y)$$
$$\tilde{Y}f \;=\; \frac{d}{db}\Big|_{a=1, b=0} f(\frac{x}{a}, \frac{y - b}{a}) = -\frac{\partial}{\partial y}f(x, y)$$

therefore we have the vector field

$$\tilde{X} \;=\; -x\frac{\partial}{\partial x} - y\frac{\partial}{\partial y}$$
$$\tilde{Y} \;=\; -\frac{\partial}{\partial y}$$

or equivalently we get the Lie algebra representation $d\pi$ on $L^2(\mu_L)$. Notice that

$$
\begin{aligned}
[\tilde{X}, \tilde{Y}] &= \tilde{X}\tilde{Y} - \tilde{Y}\tilde{X} \\
&= (-x\frac{\partial}{\partial x} - y\frac{\partial}{\partial y})(-\frac{\partial}{\partial y}) - (-\frac{\partial}{\partial y})(-x\frac{\partial}{\partial x} - y\frac{\partial}{\partial y}) \\
&= x\frac{\partial^2}{\partial x \partial y} + y\frac{\partial^2}{\partial y^2} - (x\frac{\partial^2}{\partial x \partial y} + \frac{\partial}{\partial y} + y\frac{\partial^2}{\partial y^2}) \\
&= -\frac{\partial}{\partial y} \\
&= \tilde{Y}.
\end{aligned}
$$

Notice that $\tilde{X}$ and $\tilde{Y}$ can be obtained by the exponential map as well.

$$
\begin{aligned}
\tilde{X}f &= \frac{d}{dt}\Big|_{t=0} f(e^{-tX}x) \\
&= \frac{d}{dt}\Big|_{t=0} f((e^{-t}, 1)(x, y)) \\
&= \frac{d}{dt}\Big|_{t=0} f(e^{-t}x, e^{-t}y + 1) \\
&= (-x\frac{\partial}{\partial x} - y\frac{\partial}{\partial y})f(x, y)
\end{aligned}
$$

$$
\begin{aligned}
\tilde{Y}f &= \frac{d}{dt}\Big|_{t=0} f(e^{-tY}x) \\
&= \frac{d}{dt}\Big|_{t=0} f((1, -t)(x, y)) \\
&= \frac{d}{dt}\Big|_{t=0} f(x, y - t) \\
&= -\frac{\partial}{\partial y} f(x, y)
\end{aligned}
$$

Example 8.4. We may parametrize the Lie algebra of the $ax + b$ group using (x, y) variables. Build the Hilbert space $L^2(\mu_L)$. The unitary representation $\pi(g)f(\sigma) = f(g^{-1}\sigma)$ induces the follows representations of the Lie algebra

$$
\begin{aligned}
d\pi(\tilde{s})f(\sigma) &= \frac{d}{dx}\Big|_{s=0} f(e^{-sX}\sigma) = \tilde{X}f(\sigma) \\
d\pi(\tilde{t})f(\sigma) &= \frac{d}{dy}\Big|_{t=0} f(e^{-tY}\sigma) = \tilde{Y}f(\sigma).
\end{aligned}
$$

Hence in the parameter space $(s, t) \in \mathbb{R}^2$ we have two usual derivative operators $\partial/\partial s$ and $\partial/\partial t$, where on the manifold we have

$$
\begin{aligned}
\frac{\partial}{\partial s} &= -x\frac{\partial}{\partial x} - y\frac{\partial}{\partial y} \\
\frac{\partial}{\partial t} &= -\frac{\partial}{\partial y}.
\end{aligned}
$$

The usual positive Laplacian on $\mathbb{R}^2$ translates to

$$
\begin{aligned}
-\Delta &= \left(\frac{\partial}{\partial s}\right)^2 + \left(\frac{\partial}{\partial t}\right)^2 \\
&= (\tilde{X})^2 + (\tilde{Y})^2 \\
&= \left(-x\frac{\partial}{\partial x} - y\frac{\partial}{\partial y}\right)\left(-x\frac{\partial}{\partial x} - y\frac{\partial}{\partial y}\right) + \left(-\frac{\partial}{\partial y}\right)^2 \\
&= x^2\frac{\partial^2}{\partial x^2} + 2xy\frac{\partial^2}{\partial x\partial y} + (y^2+1)\frac{\partial^2}{\partial y^2} + x\frac{\partial}{\partial x} + y\frac{\partial}{\partial y},
\end{aligned}
$$

where we used $\left(x\frac{\partial}{\partial x}\right)^2 = x^2\left(\frac{\partial}{\partial x}\right)^2 + x\frac{\partial}{\partial x}$. This is in fact an elliptic operator, since the matrix

$$
\begin{bmatrix} x^2 & xy \\ xy & y^2+1 \end{bmatrix}
$$

has trace $trace = x^2 + y^2 + 1 \geq 1$, and det $= x^2 \geq 0$. If instead we have "y^2" then the determinant is the constant zero.

The term "$y^2 + 1$" is essential for Δ being elliptic. Also note that all the coefficients are analytic functions in the (x, y) variables.

Example 8.5. Heisenberg group $G = \{a, b, c\}$ where

$$
(a, b, c) = \begin{bmatrix} 1 & a & c \\ 0 & 1 & b \\ 0 & 0 & 1 \end{bmatrix}.
$$

The multiplication rule is given by

$$
\begin{aligned}
(a, b, c)(a', b', c') &= (a + a', b + b', c + ab' + c') \\
(a, b, c)^{-1} &= (-a, -b, -c + ab).
\end{aligned}
$$

The subgroup $\Gamma = \{(0, b, c)\}$ where

$$
(0, b, c) = \begin{bmatrix} 1 & 0 & c \\ 0 & 1 & b \\ 0 & 0 & 1 \end{bmatrix}
$$

is two dimensional, Abelian and normal.

- Abelian: $(0, b, c)(0, b', c') = (0, b + b', c + c')$

- normal:

$$
\begin{aligned}
(a, b, c)(0, x, y)(a, b, c)^{-1} &= (a, b, c)(0, x, y)(-a, -b, -c + ab) \\
&= (a, b + x, c + y + ax)(-a, -b, -c + ab) \\
&= (0, x, y + ax + ab - ab) \\
&= (0, x, ax + y)
\end{aligned}
$$

Note that this is also Ad_g acting on the Lie algebra of Γ.

The additive group $(\mathbb{R}, +)$ acts on $\Gamma = \{(0, b, c)\} \simeq (\mathbb{R}^2, +)$ by

$$\varphi : (\mathbb{R}, +) \quad \rightarrow \quad Aut(\Gamma)$$

$$\varphi(a) \begin{bmatrix} c \\ b \end{bmatrix} = \begin{bmatrix} 1 & a \\ 0 & 1 \end{bmatrix} \begin{bmatrix} c \\ b \end{bmatrix}$$

$$= \begin{bmatrix} c + ab \\ b \end{bmatrix}$$

check:

$$
\begin{aligned}
(a, (b, c))(a', (b', c')) &= (a + a', (b, c) + \varphi(a)(b', c')) \\
&= (a + a', (b, c) + (b', c' + ab')) \\
&= (a + a', b + b', c + c' + ab) \\
(a, (b, c))^{-1} &= (-a, \varphi_{a^{-1}}(-b, -c)) \\
&= (-a, (-b, -c + ab)) \\
&= (-a, -b, -c + ab) \\
(a, b, c)(0, b', c')(a, b, c)^{-1} &= (a, b + b', c + c' + ab')(-a, -b, -c + ab) \\
&= (0, b', c' + ab') \\
&= \varphi_a \begin{bmatrix} c' \\ b' \end{bmatrix}
\end{aligned}
$$

Exercise 8.11 (Weyl's commutation relation). Let $\mathscr{H}$ be a Hilbert space, and let P, Q be a pair of selfadjoint operators such that

$$e^{isP} e^{itQ} = e^{ist} e^{itQ} e^{isP} \tag{8.41}$$

holds for all $s, t \in \mathbb{R}$; then prove that P and Q have a common dense invariant domain $\mathscr{D}$ such that $P = \overline{P|_{\mathscr{D}}}$, $Q = \overline{Q|_{\mathscr{D}}}$; and

$$[P, Q] = -i\, I \tag{8.42}$$

holds on $\mathscr{D}$; more precisely,

$$PQ\varphi - QP\varphi = -i\, \varphi, \tag{8.43}$$

holds for all $\varphi \in \mathscr{D}$.

Hint: One way of proving this is to show that when P and Q satisfy (8.41), then we automatically get a unitary representation of the Heisenberg group (see Sections 8.1, 8.4, and Example 8.5), and for $\mathscr{D}$ we can take the corresponding Gårding Space (Section 8.7). However there is also a direct proof from first principles.

Caution. The converse implication does not hold: There are selfadjoint solutions to (8.43) which do not have a counterpart relation (8.41); see [JM84, KL14a, JM80]. (Eq (8.41) is called the Weyl relation.)

8.4.1   Integral operators and induced representations

This also goes under the name of "Mackey machine" [Mac88]. Its modern formulation is in the context of completely positive map.

Let G be a locally compact group, and $\Gamma \subset G$ be a closed subgroup. Let dx (resp. $d\xi$) be the right Haar measure on G (resp. Γ), and $\triangle$ (resp. δ) be the corresponding modular function. Recall the modular function comes in when the translation is put on the wrong side, i.e.,

$$\int_G f\left(g^{-1}x\right) dx = \triangle\left(g^{-1}\right) \int_G f\left(x\right) dx$$

or equivalently,

$$\triangle\left(g\right) \int_G f\left(g^{-1}x\right) dx = \int_G f\left(x\right) dx.$$

Form the quotient $M = \Gamma\backslash G$ space, and let $\pi : G \to \Gamma\backslash G$ be the quotient map (the covering map). M carries a transitive G action.

group	right Haar measure	modular function
G	dg	$\triangle$
Γ	$d\xi$	δ

Note 9. M is called *a fundamental domain* or a *homogeneous* space. M is a group if and only if Γ is a normal subgroup in G. In general, M may not be a group, but it is still a very important manifold.

Note 10. μ is an *invariant measure* on M, if $\mu\left(Eg\right) = \mu\left(E\right)$, $\forall g \in G$. μ is *quasi-invariant*, if $\mu(E) = 0 \Leftrightarrow \mu(Eg) = 0$, $\forall g$. In general there is no invariant measures on M, but only quasi-invariant measures.

G has an invariant measure if and only if G is unimodular (e.g. Heisenberg group.) Not all groups are unimodular. A typical example is the $ax + b$ group.

Define $\tau : C_c\left(G\right) \to C_c\left(M\right)$ by

$$(\tau\varphi)\left(\pi\left(x\right)\right) = \int_\Gamma \varphi\left(\xi x\right) d\xi. \tag{8.44}$$

Note 11. Since φ has compact support, the integral in (8.44) is well-defined. τ is called *conditional expectation*. It is the summation of φ over the orbit Γx. Indeed, for fixed x, if ξ runs over Γ then ξx runs over Γx. We may also say $\tau\varphi$ is a Γ-periodic extension, by looking at it as a function defined on G. For if $\xi_1 \in \Gamma$, we have

$$(\tau\varphi)\left(\xi_1 x\right) = \int_\Gamma \varphi\left(\xi\xi_1 x\right) d\xi = \tau\varphi\left(x\right)$$

using the fact that $d\xi$ right-invariant. Thus $\tau\varphi$, viewed as a function on G, is Γ-periodic, i.e., $(\tau\varphi)\left(\xi x\right) = (\tau\varphi)\left(x\right)$, $\forall \xi \in \Gamma$.

Lemma 8.2. τ *is surjective.*

Proof. Suppose f is Γ-periodic, choose $\psi \in C_c(G)$ such that $\tau\psi \equiv 1$. Then $\psi f \in C_c(G)$, and

$$
\begin{aligned}
(\tau(\psi f))(x) &= \int_\Gamma \psi(\xi x) f(\xi x)\, d\xi \\
&= \int_\Gamma \psi(\xi x) f(x)\, d\xi \\
&= f(x) \int_\Gamma \psi(\xi x)\, d\xi \\
&= f(x)(\tau\psi)(x) = f(x).
\end{aligned}
$$

$\square$

Example 8.6. For $G = \mathbb{R}$, $\Gamma = \mathbb{Z}$, $d\xi =$ counting measure on $\mathbb{Z}$, we have

$$
(\tau\varphi)(\pi(x)) = \int_\Gamma \varphi(\xi x)\, d\xi = \sum_{n \in \mathbb{Z}} \varphi(n + x), \quad \forall \varphi \in C_c(\mathbb{R}).
$$

Since φ has compact support, $\varphi(n + x)$ vanishes for all but a finite number of n. Hence $\tau\varphi$ contains a finite summation, and so it is well-defined. Moreover, for all $n_0 \in \mathbb{Z}$, it follows that

$$
(\tau\varphi)(n_0 + x) = \sum_{n \in \mathbb{Z}} \varphi(n_0 + n + x) = \sum_{n \in \mathbb{Z}} \varphi(n + x) = (\tau\varphi)(x).
$$

Hence $\tau\varphi$ is translation invariant by integers, i.e., $\tau\varphi$ (as a function on $\mathbb{R}$) is $\mathbb{Z}$-periodic.

Let $L : \Gamma \to V$ be a unitary representation of Γ on a Hilbert space V. We now construct a unitary representation $ind_\Gamma^G : G \to \mathcal{H}$ of G on an enlarged Hilbert space $\mathcal{H}$.

Let F_* be the set of function $f : G \to V$ so that

$$
f(\xi g) = \rho(\xi)^{1/2} L_\xi f(g), \quad \forall \xi \in \Gamma \tag{8.45}
$$

where $\rho = \frac{\delta}{\Delta}$, i.e., $\rho(\xi) = \frac{\delta(\xi)}{\Delta(\xi)}$, $\forall \xi \in \Gamma$. For all $f \in F_*$, let

$$
(R_g f)(\cdot) := f(\cdot g)
$$

be the right-translation of f by $g \in G$.

Lemma 8.3. $R_g f \in F_*$. *That is, F_* is invariant under right-translation by $g \in G$.*

Proof. To see this, let $f \in F_*$, $\xi \in \Gamma$, then

$$
(R_g f)(\xi x) = f(\xi x g) = \rho(\xi)^{1/2} L_\xi f(xg) = \rho(\xi)^{1/2} L_\xi (R_g f)(x)
$$

so that $R_g f \in F_*$.

$\square$

Note 12. We will defined an inner product on F_* so that $\|f(\xi \cdot)\|_{new} = \|f(\cdot)\|_{new}$, $\forall \xi \in \Gamma$. Eventually, we will define the induced representation $U^{ind} := ind_\Gamma^G (L)$ by

$$\left(U_g^{ind} f \right)(\cdot) := (R_g f)(\cdot)$$

not on F_*, but pass to a quotient space. The factor $\rho(\xi)^{1/2}$ comes in as we are going to construct a quasi-invariant measure on $\Gamma \backslash G$.

To construct F_*, let $\varphi \in C_c(G)$, and set

$$f(g) := \int_\Gamma \rho^{1/2} \left(\xi^{-1} \right) L \left(\xi^{-1} \right) f(\xi g) \, d\xi.$$

Now if $\xi_1 \in \Gamma$ then

$$
\begin{aligned}
f(\xi_1 g) &= \int_\Gamma \rho^{1/2} \left(\xi^{-1} \right) L \left(\xi^{-1} \right) f(\xi \xi_1 g) \, d\xi \\
&= \int_\Gamma \rho^{1/2} \left(\left(\xi \xi_1^{-1} \right)^{-1} \right) L \left(\left(\xi \xi_1^{-1} \right)^{-1} \right) f(\xi g) \, d\xi \\
&= \rho^{1/2} (\xi_1) L (\xi_1) \int_\Gamma \rho^{1/2} \left(\xi^{-1} \right) L \left(\xi^{-1} \right) f(\xi g) \, d\xi \\
&= \rho^{1/2} (\xi_1) L (\xi_1) f(g).
\end{aligned}
$$

The proof of Lemma 8.2 shows that all functions in F_* are obtained this way.

Note 13. Let's ignore the factor $\rho(\xi)^{1/2}$ for a moment. L_ξ is unitary implies that for all $f \in F_*$,

$$\|f(\xi \cdot)\|_V = \|L_\xi f(\cdot)\|_V = \|f(\cdot)\|_V, \quad \forall \xi \in \Gamma.$$

Since Hilbert spaces exist up to unitary equivalence, $L_\xi f(g)$ and $f(g)$ really are the same function. As ξ running through Γ, ξg running through Γg. Thus $\|f(\xi g)\|$ is a constant on the orbit Γg. It follows that $f(\xi g)$ is in fact a V-valued function defined on the quotient $M = \Gamma \backslash G$ (i.e., quasi-Γ-periodic). We will later use these functions as multiplication operators.

Example 8.7. The Heisenberg group is unimodular, so $\rho \equiv 1$.

Example 8.8. For the $ax + b$ group,

$$
\begin{aligned}
d\lambda_R &= \frac{da\,db}{a} \\
d\lambda_L &= \frac{da\,db}{a^2} \\
\triangle &= \frac{d\lambda_L}{d\lambda_R} = \frac{1}{a}
\end{aligned}
$$

On the Abelian normal subgroup $\Gamma = \{(1, b)\}$, we have $a = 1$ and $\triangle(\xi) = 1$. Γ is unimodular, $\delta(\xi) = 1$. Therefore, $\rho(\xi) = 1$, $\forall \xi \in \Gamma$.

For all $f \in F_*$, the map $\mu_{f,f} : C_c(M) \to \mathbb{C}$ given by

$$\mu_{f,f} : \tau\varphi \longmapsto \int_G \|f(g)\|_V^2 \, \varphi(g) \, dg, \quad \varphi \in C_c(G) \tag{8.46}$$

is a positive linear functional. By Riesz's theorem, there exists a unique Radon measure $\mu_{f,f}$ on M, such that

$$\int_G \|f(g)\|_V^2 \, \varphi(g) dg = \int_M (\tau\varphi) d\mu_{f,f}.$$

Lemma 8.4. *(8.46) is a well-defined positive linear functional.*

Proof. Suppose $\varphi \in C_c(G)$ such that $\tau\varphi \equiv 0$ on M. It remains to verify that $\mu_{f,f}(\tau\varphi) = 0$, see (8.46). For this, we choose $\psi \in C_c(G)$ such that $\tau\psi \equiv 1$ on M, and so

$$
\begin{aligned}
\int_G \|f(g)\|_V^2 \, \varphi(g) \, dg \; &= \; \int_G \|f(g)\|_V^2 \, (\tau\psi)(\pi(g)) \, \varphi(g) \, dg \\
&= \; \int_G \|f(g)\|_V^2 \, \varphi(g) \left(\int_\Gamma \psi(\xi g) \, d\xi \right) dg \\
&\underset{\text{Fubini}}{=} \; \int_\Gamma \left(\int_G \|f(g)\|_V^2 \, \varphi(g) \, \psi(\xi g) \, dg \right) d\xi \\
&= \; \int_\Gamma \left(\int_G \|f(\xi^{-1}g)\|_V^2 \, \varphi(\xi^{-1}g) \, \psi(g) \, \triangle(\xi) \, dg \right) d\xi \\
&\underset{\text{Fubini}}{=} \; \int_G \psi(g) \left(\int_\Gamma \|f(\xi^{-1}g)\|_V^2 \, \varphi(\xi^{-1}g) \, \triangle(\xi) \, d\xi \right) dg \\
&= \; \int_G \psi(g) \left(\int_\Gamma \|f(\xi g)\|_V^2 \, \varphi(\xi g) \, \triangle(\xi^{-1}) \, \delta(\xi) \, d\xi \right) dg \\
&\underset{(8.45)}{=} \; \int_G \psi(g) \, \|f(g)\|_V^2 \left(\int_\Gamma \varphi(\xi g) \, d\xi \right) dg \\
&= \; \int_G \psi(g) \, \|f(g)\|_V^2 \, \underbrace{(\tau\varphi)(\pi(g))}_{\equiv 0} dg = 0.
\end{aligned}
$$

$\square$

Note 14. Recall that given a measure space $(X, \mathfrak{M}, \mu)$, let $f : X \to Y$. Define a linear functional $\Lambda : C_c(Y) \to \mathbb{C}$ by

$$\Lambda\varphi := \int \varphi(f(x)) d\mu(x).$$

Λ is positive, hence by Riesz's theorem, there exists a unique regular Borel measure μ_f on Y so that

$$\Lambda\varphi = \int_Y \varphi d\mu_f = \int_X \varphi(f(x)) d\mu(x).$$

It follows that $\mu_f = \mu \circ f^{-1}$.

Note 15. Under current setting, we have a covering map $\pi : G \to \Gamma\backslash G =: M$, and the right Haar measure μ on G. Thus we may define a measure $\mu \circ \pi^{-1}$. However, given $\varphi \in C_c(M)$, $\varphi(\pi(x))$ may not have compact support, or equivalently, $\pi^{-1}(E)$ is Γ periodic. For example, take $G = \mathbb{R}$, $\Gamma = \mathbb{Z}$, $M = \mathbb{Z}\backslash\mathbb{R}$. Then $\pi^{-1}([0, 1/2))$ is $\mathbb{Z}$-periodic, which has infinite Lebesgue measure. What we really need is some map so that the inverse of a subset of M is restricted to a single Γ period. This is essentially what τ does: from $\tau\varphi \in C_c(M)$, get the inverse image $\varphi \in C_c(G)$. Even if φ is not restricted to a single Γ period, φ always has compact support.

Hence we get a family of measures indexed by elements in F_*. If choosing $f, g \in F_*$ then we get complex measures $\mu_{f,g}$ (using polarization identity.)

- Define $\|f\|^2 := \mu_{f,f}(M)$, $\langle f, g \rangle := \mu_{f,g}(M)$.

- Complete F_* with respect to this norm to get an enlarged Hilbert space $\mathscr{H}$.

- Define the induced representation $U^{ind} := ind_\Gamma^G (L)$ on $\mathscr{H}$ as

$$\left(U_g^{ind} f\right)(x) = f(xg)$$

U^{ind} is unitary, in particular,

$$\|U_g^{ind} f\|_{\mathscr{H}} = \|f\|_{\mathscr{H}}, \quad \forall g \in G.$$

Note 16. $\mu_{f,g}(M) = \int_M (\tau\varphi)(\xi)\, d\xi$ with $\tau\varphi \equiv 1$. What is φ then? It turns out that φ could be constant 1 over a single Γ-period, or equivalently, φ could spread out to a finite number of Γ-periods. In the former case,

$$
\begin{aligned}
\|f\|^2 &= \int_G \|f(g)\|_V^2\, \varphi(g)\, dg \\
&= \int_{1\text{-period}} \|f(g)\|_V^2\, (\tau\varphi)(g)\, dg \\
&= \int_{1\text{-period}} \|f(g)\|_V^2\, dg \\
&= \int_M \|f(g)\|_V^2\, dg.
\end{aligned}
$$

Define $P(\psi)f(x) := \psi(\pi(x))f(x)$, for $\psi \in C_c(M)$, $f \in \mathscr{H}$, $x \in G$. Note $\{P(\psi) \mid \psi \in C_c(M)\}$ is the Abelian algebra of multiplication operators.

Lemma 8.5. *We have*

$$U_g^{ind} P(\psi) U_{g^{-1}}^{ind} = P(\psi(\cdot g)).$$

Proof. One checks that

$$U_g^{ind} P(\psi) f(x) \quad = \quad U_g^{ind} \psi(\pi(x)) f(x)$$

$$\begin{aligned}
&= \psi(\pi(xg))f(xg) \\
P(\psi(\cdot g))U_g^{ind}f(x) &= P(\psi(\cdot g))f(xg) \\
&= \psi(\pi(xg))f(xg).
\end{aligned}$$

$\square$

Conversely, how to recognize induced representations? Answer:

Theorem 8.3 (Imprimitivity Theorem [Ørs79]). *Let G be a locally compact group with a closed subgroup Γ. Let $M = \Gamma\backslash G$. Suppose the system (U, P) satisfies the covariance relation*

$$U_g P(\psi)U_{g^{-1}} = P(\psi(\cdot g)),$$

and $P(\cdot)$ is non-degenerate. Then, there exists a unitary representation $L \in Rep(\Gamma, V)$ such that $U \cong ind_\Gamma^G(L)$.

Remark 8.7. $P(\cdot)$ is non-degenerate if

$$P(C_c(M))\mathscr{H} = \{P(\psi)a : \psi \in C_c(M), a \in \mathscr{H}\}$$

is dense in $\mathscr{H}$.

8.5   Example - Heisenberg group

Let $G = \{(a, b, c)\}$ be the *Heisenberg group*, where

$$(a, b, c) = \begin{bmatrix} 1 & a & c \\ 0 & 1 & b \\ 0 & 0 & 1 \end{bmatrix}.$$

The multiplication rule is given by

$$\begin{aligned}
(a, b, c)(a', b', c') &= (a + a', b + b', c + c' + ab') \\
(a, b, c)^{-1} &= (-a, -b, -c + ab).
\end{aligned}$$

The subgroup $\Gamma = \{(0, b, c)\}$ where

$$(1, b, c) = \begin{bmatrix} 1 & 0 & c \\ 0 & 1 & b \\ 0 & 0 & 1 \end{bmatrix}$$

is two dimensional, Abelian and normal.

- Abelian: $(0, b, c)(0, b', c') = (0, b + b', c + c')$

- normal:

$$
\begin{aligned}
(a,b,c)(0,x,y)(a,b,c)^{-1} &= (a,b,c)(0,x,y)(-a,-b,-c+ab) \\
&= (a,b+x,c+y+ax)(-a,-b,-c+ab) \\
&= (0,x,y+ax+ab-ab) \\
&= (0,x,ax+y)
\end{aligned}
$$

i.e., $Ad : G \to GL(\mathfrak{n})$, as

$$
\begin{aligned}
Ad_g(n) &:= gng^{-1} \\
(x,y) &\mapsto (x,ax+y)
\end{aligned}
$$

the orbit is a 2-d transformation.

Fix $h \in \mathbb{R}\backslash\{0\}$. Recall the *Schrödinger representation* of G on $L^2(\mathbb{R})$

$$
U_g f(x) = e^{ih(c+bx)}f(x+a). \tag{8.47}
$$

See Definition 3.3 and Example 3.1.

Theorem 8.4. *The Schrödinger representation is induced.*

Proof. We show that the Schrödinger representation is induced from a unitary representation L on the subgroup Γ.

Note the Heisenberg group is a non Abelian unimodular Lie group ($\triangle = 1$, $\delta = 1$, and so $\rho \equiv 1$.) The Haar measure on G is just the product measure $dxdydz$ on $\mathbb{R}^3$. Conditional expectation becomes integrating out the variables correspond to the subgroup.

1. Let $L \in Rep(\Gamma, V)$ where $\Gamma = \{(0,b,c)\}$, $V = \mathbb{C}$,

$$
L_{\xi(b,c)} = e^{ihc}.
$$

The complex exponential comes in since we want a unitary representation. The subgroup $\{(0,0,c)\}$ is the center of G. (What is the induced representation? Is it unitarily equivalent to the Schrödinger representation?)

2. Look for the family F_* of functions $f : G \to \mathbb{C}$ (V is the 1-d Hilbert space $\mathbb{C}$), such that

$$
f(\xi(b,c)g) = L_\xi f(g).
$$

Since

$$
f(\xi(b,c)g) = f((0,b,c)(x,y,z)) = f(x,b+y,c+z), \quad \text{and}
$$

$$
L_{\xi(b,c)}f(g) = e^{ihc}f(x,y,z)
$$

so f satisfies

$$
f(x,b+y,c+z) = e^{ihc}f(x,y,z).
$$

That is, we may translate the y, z variables by arbitrary amount, and the only price to pay is the multiplicative factor e^{ihc}. Therefore f is really a function defined on the quotient

$$M = \Gamma \backslash G \simeq \mathbb{R}.$$

The homogeneous space $M = \{(x, 0, 0)\}$ is identified with $\mathbb{R}$, and the invariant measure on M is simply the Lebesgue measure. It is almost clear at this point why the induced representation is unitarily equivalent to the Schrödinger representation on $L^2(\mathbb{R})$.

3. The positive linear functional $\tau\varphi \mapsto \int_G \|f(g)\|_V^2 \, \varphi(g) \, dg$ induces a measure $\mu_{f,f}$ on M. This can be seen as follows:

$$
\begin{aligned}
\int_G \|f(g)\|_V^2 \, \varphi(g) \, dg &= \int_{G \simeq \mathbb{R}^3} |f(x, y, z)|^2 \, \varphi(x, y, z) \, dxdydz \\
&= \int_{M \simeq \mathbb{R}} \left(\int_{\Gamma \simeq \mathbb{R}^2} |f(x, y, z)|^2 \, \varphi(x, y, z) \, dydz \right) dx \\
&= \int_{\mathbb{R}} |f(x, y, z)|^2 \left(\int_{\mathbb{R}^2} \varphi(x, y, z) \, dydz \right) dx \\
&= \int_{\mathbb{R}} |f(x, y, z)|^2 \, (\tau\varphi)(\pi(g)) \, dx \\
&= \int_{\mathbb{R}} |f(x, 0, 0)|^2 \, (\tau\varphi)(x) \, dx.
\end{aligned}
$$

Note that

$$
\begin{aligned}
(\tau\varphi)(\pi(g)) &= \int_{\Gamma} \varphi(\xi g) \, d\xi \\
&= \int_{\mathbb{R}^2} \varphi((0, b, c)(x, y, z)) \, dbdc \\
&= \int_{\mathbb{R}^2} \varphi(x, b + y, c + z) \, dbdc \\
&= \int_{\mathbb{R}^2} \varphi(x, b, c) \, dbdc \\
&= (\tau\varphi)(x), \quad M = \Gamma \backslash G \simeq \mathbb{R}.
\end{aligned}
$$

Hence $\Lambda : C_c(M) \to \mathbb{C}$ given by

$$\Lambda : \tau\varphi \longmapsto \int_G \|f(g)\|_V^2 \, \varphi(g) dg$$

is a positive linear functional, therefore $\Lambda = \mu_{f,f}$ and

$$\int_{\mathbb{R}^3} |f(x, y, z)|^2 \, \varphi(x, y, z) \, dxdydz = \int_{\mathbb{R}} (\tau\varphi)(x) \, d\mu_{f,f}(x).$$

4. Define

$$\|f\|_{ind}^2 := \mu_{f,f}(M) = \int_M |f|^2 \, d\xi = \int_{\mathbb{R}} |f(x, y, z)|^2 \, dx = \int_{\mathbb{R}} |f(x, 0, 0)|^2 \, dx$$

$$U_g^{ind} f\left(g'\right) := f\left(g'g\right).$$

By definition, if $g = g(a,b,c)$ and $g' = g'(x,y,z)$ then

$$
\begin{aligned}
U_g^{ind} f(g') &= f(g'g) \\
&= f((x,y,z)(a,b,c)) \\
&= f(x+a, y+b, z+c+xb)
\end{aligned}
$$

and U^{ind} is a unitary representation.

5. To see that U^{ind} is unitarily equivalent to the Schrödinger representation on $L^2(\mathbb{R})$, we set

$$W : \mathcal{H}^{ind} \to L^2(\mathbb{R}), \quad (Wf)(x) = f(x,0,0).$$

(If put other numbers into f, as $f(x,y,z)$, the result is the same, since $f \in \mathcal{H}^{ind}$ is really defined on the quotient $M = \Gamma \backslash G \simeq \mathbb{R}$.)

W is unitary:

$$\|Wf\|_{L^2}^2 = \int_{\mathbb{R}} |Wf|^2 \, dx = \int_{\mathbb{R}} |f(x,0,0)|^2 \, dx = \int_{\Gamma \backslash G} |f|^2 \, d\xi = \|f\|_{ind}^2.$$

The intertwining property: Let U_g be the Schrödinger representation, then

$$
\begin{aligned}
U_g(Wf) &= e^{ih(c+bx)} f(x+a, 0, 0) \\
WU_g^{ind} f &= W(f((x,y,z)(a,b,c))) \\
&= W(f(x+a, y+b, z+c+xb)) \\
&= W\left(e^{ih(c+bx)} f(x+a, y, z)\right) \\
&= e^{ih(c+bx)} f(x+a, 0, 0).
\end{aligned}
$$

6. Since $\{U,L\}' \subset \{L\}'$, the system $\{U,L\}$ is reducible implies L is reducible. Equivalent, $\{L\}$ is irreducible implies $\{U,L\}$ is irreducible. Since L is 1-dimensional, it is irreducible. Consequently, U^{ind} is irreducible. $\square$

Exercise 8.12 (The Schrödinger representation). Prove that for $h \neq 0$ fixed, the Schrödinger representation U^h (8.47) is irreducible.

Hint: Show that if $A \in \mathscr{B}(L^2(\mathbb{R}))$ commutes with $\{U_g^h : g \in G_{\text{Heis}}\}$, then there exists $\lambda \in \mathbb{C}$ such that $A = \lambda I_{L^2(\mathbb{R})}$, i.e., that the commutant of the representation U^h is one-dimensional.

8.5.1 $ax + b$ group

$a \in \mathbb{R}_+$, $b \in \mathbb{R}$, $g = (a,b) = \begin{bmatrix} a & b \\ 0 & 1 \end{bmatrix}$.

$$U_g f(x) = e^{iax} f(x+b)$$

could also write $a = e^t$, then

$$U_g f(x) = e^{ie^t x} f(x + b)$$

$$U_{g(a,b)} f(x) = e^{iae^x} f(x+b)$$

$$[\frac{d}{dx}, ie^x] = ie^x$$
$$[A, B] = B$$

or

$$U_{(e^t, b)} f = e^{ite^x} f(x + b)$$

$$\begin{bmatrix} 0 & b \\ 0 & 1 \end{bmatrix}$$

1-d representation. $L_b = e^{ib}$. Induce $ind_L^G \simeq$ the Schrödinger representation.

8.6 Co-adjoint Orbits

It turns out that only a small family of representations are induced. The question is how to detect whether a representation is induced. The whole theory is also under the name of "Mackey machine" [Mac52, Mac88]. The notion of "machine" refers to something that one can actually compute in practice. Two main examples are the Heisenberg group and the $ax + b$ group.

Below we address two queries. What is the mysterious parameter h that goes into the Schrödinger representation? It is a physical constant, but how to explain it in the mathematical theory?

8.6.1 Review of some Lie theory

Theorem 8.5 (Ado). *Every Lie group is diffeomorphic to a matrix group.*

Proof. While Ado's theorem is handy to have, its proof is perhaps not so enlightening. It can be found in most books on Lie groups, for example in [Jac79, pp. 202–203]. In fact it is often stated in its equivalent Lie algebra variant: Every finite-dimensional Lie algebra over a field of characteristic zero can be viewed as a Lie algebra of square matrices under the commutator bracket. □

The exponential function exp maps a neighborhood of 0 into a connected component of G containing the identity element. For example, the Lie algebra of the Heisenberg group is

$$\begin{bmatrix} 0 & * & * \\ 0 & 0 & * \\ 0 & 0 & 0 \end{bmatrix}.$$

All the Lie groups that we will ever encounter come from a quadratic form. Given a quadratic form

$$\varphi : V \times V \to \mathbb{C}$$

there is an associated group that fixes φ, i.e., we consider elements g such that

$$\varphi(gx, gy) = \varphi(x, y)$$

and define $G(\varphi)$ as the collection of these elements. $G(\varphi)$ is clearly a group. Apply the exponential map and the product rule,

$$\frac{d}{dt}\Big|_{t=0}\varphi(e^{tX}x, e^{tX}y) = 0 \iff \varphi(Xx, y) + \varphi(x, Xy) = 0$$

hence

$$X + X^{tr} = 0.$$

The determinant and trace are related by

$$\det(e^{tX}) = e^{t \cdot trace(X)} \tag{8.48}$$

thus $\det = 1$ if and only if $trace = 0$. It is often stated in differential geometry that the derivative of the determinant is equal to the trace.

Example 8.9. $\mathbb{R}^n$, $\varphi(x, y) = \sum x_i y_i$. The associated group is the orthogonal group O_n.

There is a famous cute little trick to make O_{n-1} into a subgroup of O_n. O_{n-1} is not normal in O_n. We may split the quadratic form into

$$\sum_{i=1}^{n-1} x_i^2 + 1$$

where 1 corresponds to the last coordinate in O_n. Then we may identity O_{n-1} as a subgroup of O_n

$$g \mapsto \begin{bmatrix} g & 0 \\ 0 & I \end{bmatrix}$$

where I is the identity operator.

Claim: $O_n/O_{n-1} \simeq S^{n-1}$. How to see this? Let u be the unit vector corresponding to the last dimension, look for g that fixes u, i.e., $gu = u$. Such elements form a subgroup of O_n, and it is called the *isotropy* group.

$$I_n = \{g \mid gu = u\} \simeq O_{n-1}.$$

Notice that for all $v \in S^{n-1}$, there exists $g \in O_n$ such that $gu = v$. Hence

$$g \mapsto gu$$

in onto S^{n-1}. The kernel of this map is $I_n \simeq O_{n-1}$, thus

$$O_n/O_{n-1} \simeq S^{n-1}.$$

Such spaces are called homogeneous spaces.

Example 8.10. Visualize this with O_3 and O_2.

Other examples of homogeneous spaces show up in number theory all the time. For example, the Poincaré group G/discrete subgroup.

Fix a group G, and let $N \subset G$ be a normal subgroup. The map $g \cdot g^{-1} : G \to G$ is an automorphism which fixes the identity element, hence if we differentiate it, we get a transformation in $GL(\mathfrak{g})$; i.e., we get a family of maps $Ad_g \in GL(\mathfrak{g})$ indexed by elements in G. $g \mapsto Ad_g \in GL(\mathfrak{g})$ is a representation of G, hence if it is differentiated, we get a representation of $\mathfrak{g}$, $ad_g : \mathfrak{g} \mapsto End(\mathfrak{g})$, acting on the vector space $\mathfrak{g}$.

Note that $gng^{-1} \in N$, for all $g \in G$, and all $n \in N$. Thus, for all g, the map $g\,(\cdot)\,g^{-1}$ is a transformation from N to N. Set $Ad_g(n) = gng^{-1}$. Differentiate to get $ad : \mathfrak{n} \to \mathfrak{n}$. $\mathfrak{n}$ is a vector space. Recall that linear transformations on vector spaces pass to the dual spaces, where

$$\varphi^*(v^*)(u) \;\; = \;\; v^*(\varphi(u))$$
$$\Updownarrow$$
$$\langle \Lambda^* v^*, u \rangle \;\; = \;\; \langle v^*, \Lambda u \rangle \,.$$

In order to get the transformation rules work out, one has to pass to the adjoint or the dual space, with

$$Ad_g^* : \mathfrak{n}^* \to \mathfrak{n}^*,$$

called the coadjoint representation of $\mathfrak{n}$.

Orbits of co-adjoint representation amounts precisely to equivalence classes of irreducible representations.

Example 8.11. Heisenberg group $G = \{(a,b,c)\}$ with

$$(a,b,c) = \begin{bmatrix} 1 & a & c \\ 0 & 1 & b \\ 0 & 0 & 1 \end{bmatrix}$$

normal subgroup $N = \{(0,b,c)\}$

$$(0,b,c) = \begin{bmatrix} 1 & 0 & c \\ 0 & 1 & b \\ 0 & 0 & 1 \end{bmatrix}$$

with Lie algebra $\mathfrak{n} = \{(\xi, \eta)\}$, or written:

$$(0,\xi,\eta) = \begin{bmatrix} 0 & 0 & \eta \\ 0 & 0 & \xi \\ 0 & 0 & 0 \end{bmatrix}$$

$Ad_g : \mathfrak{n} \to \mathfrak{n}$ given by

$$gng^{-1} \;\; = \;\; (a,b,c)(0,y,x)(-a,-b,-c+ab)$$

$$\begin{aligned}
&= (a, b+y, c+x+ay)(-a, -b, -c+ab) \\
&= (0, y, x+ay)
\end{aligned}$$

hence $Ad_g : \mathbb{R}^2 \to \mathbb{R}^2$

$$Ad_g : \begin{bmatrix} x \\ y \end{bmatrix} \mapsto \begin{bmatrix} x+ay \\ y \end{bmatrix}.$$

The matrix of Ad_g is (before taking adjoint) is

$$Ad_g = \begin{bmatrix} 1 & a \\ 0 & 1 \end{bmatrix}.$$

The matrix for Ad_g^* is

$$Ad_g^* = \begin{bmatrix} 1 & 0 \\ a & 1 \end{bmatrix}.$$

We use $[\xi, \eta]^T$ for the dual $\mathfrak{n}^*$; and use $[x, y]^T$ for $\mathfrak{n}$. Then

$$Ad_g^* : \begin{bmatrix} \xi \\ \eta \end{bmatrix} \mapsto \begin{bmatrix} \xi \\ a\xi + \eta \end{bmatrix}.$$

What about the orbit? In the example of O_n/O_{n-1}, the orbit is S^{n-1}.

For $\xi \in \mathbb{R}\backslash\{0\}$, the orbit of Ad_g^* is

$$\begin{bmatrix} \xi \\ 0 \end{bmatrix} \mapsto \begin{bmatrix} \xi \\ \mathbb{R} \end{bmatrix}$$

i.e., vertical lines with x-coordinate ξ. $\xi = 0$ amounts to fixed point, i.e. the orbit is a fixed point.

The simplest orbit is when the orbit is a fixed point. i.e.

$$Ad_g^* : \begin{bmatrix} \xi \\ \eta \end{bmatrix} \mapsto \begin{bmatrix} \xi \\ \eta \end{bmatrix} \in V^*$$

where if we choose

$$\begin{bmatrix} \xi \\ \eta \end{bmatrix} = \begin{bmatrix} 0 \\ 1 \end{bmatrix}$$

it is a fixed point.

The other extreme is to take any $\xi \neq 0$, then

$$Ad_g^* : \begin{bmatrix} \xi \\ 0 \end{bmatrix} \mapsto \begin{bmatrix} \xi \\ \mathbb{R} \end{bmatrix}$$

i.e., get vertical lines indexed by the x-coordinate ξ. In this example, a cross section is a subset of $\mathbb{R}^2$ that intersects each orbit at precisely one point. Every cross section in this example is a Borel set in $\mathbb{R}^2$.

We don't always get measurable cross sections. An example is the construction of non-measurable set as was given in Rudin's book. Cross section is a Borel set that intersects each coset at precisely one point.

Why does it give all the equivalent classes of irreducible representations? Since we have a unitary representation $L_n \in Rep(N, V)$, $L_n : V \to V$ and by construction of the induced representation $U_g \in Rep(G, \mathscr{H})$, $N \subset G$ normal such that

$$U_g L_n U_{g^{-1}} = L_{gng^{-1}}$$

i.e.

$$L_g \simeq L_{gng^{-1}}$$

now pass to the Lie algebra and its dual

$$L_n \to LA \to LA^*.$$

8.7   Gårding Space

For a given strongly continuous unitary representation of a Lie group G, acting on a Hilbert space $\mathscr{H}$, there is an associated representation of the Lie algebra $La(G)$, but for the unitary representations of interest, one is faced with the difficulty that the operators in the derived Lie algebra representation are unbounded, and so different operators will have different domains. To even make sense of the derived Lie algebra representation as a Lie algebra of operators in $\mathscr{H}$, one must first display a common dense domain in $\mathscr{H}$ on which all the operators in the Lie algebra are represented. (In fact the same issue arises for more general representations, even when realized in Banach space, or in other topological linear spaces.) The remedy in all these cases was found by Gårding. Conclusion: *For a given representation there is always a common dense and invariant domain*, called the Gårding domain, on which all the Lie algebra operators act. Gårding's construction entails a generalized convolution operation (details below); a "smearing" in the language of physicists. We turn here to the details in the construction of the Gårding domain for a given representation.

Definition 8.5. Let $\mathcal{U}$ be a strongly continuous representation of a Lie group G, with Lie algebra $\mathfrak{g}$, and let exp : $\mathfrak{g} \to G$ denote the exponential mapping from Lie theory. For every $\varphi \in C_c^{\infty}(G)$, set

$$\mathcal{U}(\varphi) = \int_G \varphi(g) \mathcal{U}_g \, dg$$

where dg is a left-invariant Haar measure on G; and set

$$\mathscr{H}_{G\text{ård}ing} = \left\{ \mathcal{U}(\varphi) v \mid \varphi \in C_c^{\infty}(G), v \in \mathscr{H} \right\}.$$

Lemma 8.6. *Fix $X \in \mathfrak{g}$, set*

$$d\mathcal{U}(X)v = \lim_{t \to 0} \frac{\mathcal{U}(\exp(tX))v - v}{t}$$

then

$$\mathscr{H}_{G\mathring{a}rding} \subset \bigcap_{X \in \mathfrak{g}} dom\,(d\mathcal{U}(X))$$

and

$$d\mathcal{U}(X)\mathcal{U}(\varphi)v = \mathcal{U}\left(\widetilde{X}\varphi\right)v,$$

for all $\varphi \in C_c^\infty(G)$, $v \in \mathscr{H}$, where

$$\left(\widetilde{X}\varphi\right)(g) = \frac{d}{dt}\Big|_{t=0}\varphi(\exp(-tX)g),\ \forall g \in G.$$

Proof. (Hint)

$$\int_G \varphi(g)\mathcal{U}(exp(tX))\mathcal{U}(g)\,dg = \int_G \varphi(\exp(-tX)g)\mathcal{U}(g)\,dg.$$

$\square$

The Imprimitivity Theorem (Theorem 8.3) from Section 8.4 shows how to detect whether a representation is induced. Given a group G with a subgroup Γ, let $M := \Gamma \backslash G$. The map $\pi : G \to M$ is called a covering map, which sends g to its equivalent class or the coset Γg. M is given its projective topology, so π is continuous. When G is compact, many things simplify. For example, if G is compact, any irreducible representation is finite dimensional. But many groups are not compact, only locally compact. For example, the groups $ax + b$, H_3, SL_n.

Specialize to Lie groups. G and subgroup H have Lie algebras $\mathfrak{g}$ and $\mathfrak{h}$ respectively.

$$\mathfrak{g} = \left\{X\ :\ e^{tX} \in G, \forall t \in \mathbb{R}\right\}$$

Almost all Lie algebras we will encounter come from specifying a quadratic form $\varphi : G \times G \to \mathbb{C}$. φ is then uniquely determined by a Hermitian matrix A so that

$$\varphi(x,y) = x^{tr} \cdot Ay.$$

Let $G = G(\varphi) = \{g\ :\ \varphi(gx, gy) = \varphi(x,y)\}$, then

$$\frac{d}{dt}\Big|_{t=0}\varphi(e^{tX}x, e^{tX}y) = 0$$

and with an application of the product rule,

$$\begin{aligned}
\varphi(Xx, y) + \varphi(x, Xy) &= 0 \\
(Xx)^{tr} \cdot Ay + x^{tr} \cdot AXy &= 0
\end{aligned}$$

$$X^{tr}A + AX = 0$$

hence
$$\mathfrak{g} = \{X : X^{tr}A + AX = 0\}. \tag{8.49}$$

Let $U \in Rep(G, \mathscr{H})$, for $X \in \mathfrak{g}$, $U(e^{tX})$ is a strongly continuous one-parameter group of unitary operator, hence by Stone's theorem (see [vN32b, Nel69]), it must have the form

$$U(e^{tX}) = e^{itH_X} \tag{8.50}$$

for some selfadjoint operator H_X (possibly unbounded). The RHS in (8.50) is given by the Spectral Theorem (see [Sto90, Yos95, Nel69, RS75, DS88b]). We often write

$$dU(X) := iH_X$$

to indicate that $dU(X)$ is the directional derivative along the direction X. Notice that $H_X^* = H_X$ but

$$(iH_X)^* = -(iH_X)$$

i.e. $dU(X)$ is skewadjoint.

Example 8.12. $G = \{(a, b, c)\}$ Heisenberg group. $\mathfrak{g} = \{X_1 \sim a, X_2 \sim b, X_3 \sim c\}$. Take the Schrödinger representation $U_g f(x) = e^{ih(c+bx)}f(x+a)$, $f \in L^2(\mathbb{R})$.

- $U(e^{tX_1})f(x) = f(x+t)$

$$\frac{d}{dt}\Big|_{t=0}U(e^{tX_1})f(x) = \frac{d}{dx}f(x)$$
$$dU(X_1) = \frac{d}{dx}$$

- $U(e^{tX_2})f(x) = e^{ih(tx)}f(x)$

$$\frac{d}{dt}\Big|_{t=0}U(e^{tX_2})f(x) = ihxf(x)$$
$$dU(X_2) = ihx$$

- $U(e^{tX_3})f(x) = e^{iht}f(x)$

$$\frac{d}{dt}\Big|_{t=0}U(e^{tX_3})f(x) = ihf(x)$$
$$dU(X_2) = ihI$$

Notice that $dU(X_i)$ are all skew adjoint.

$$[dU(X_1), dU(X_2)] = [\frac{d}{dx}, ihx]$$
$$= ih[\frac{d}{dx}, x]$$

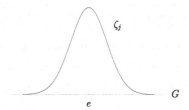

Figure 8.2: Approximation of identity.

$$= ih.$$

In case we want selfadjoint operators, replace $dU(X_i)$ by$-idU(X_i)$ and get

$$-i\,dU(X_1) = \frac{1}{i}\frac{d}{dx}$$
$$-i\,dU(X_2) = hx$$
$$-i\,dU(X_3) = hI$$

$$[\frac{1}{i}\frac{d}{dx}, hx] = \frac{h}{i}.$$

Below we answer the following question:

What is the space of functions that U_g acts on? L. Gårding /gor-ding/ (Swedish mathematician) looked for one space that always works. It's now called the Gårding space.

Start with $C_c(G)$, every $\varphi \in C_c(G)$ can be approximated by the so called Gårding functions, using the convolution argument. Define convolution as

$$\varphi \star \psi(g) = \int_G \varphi(gh)\psi(h)d_R h$$
$$\varphi \star \psi(g) = \int_G \varphi(h)\psi(g^{-1}h)d_L h$$

Take an approximation of identity ζ_j (Figure 8.2), so that

$$\varphi \star \zeta_j \to \varphi,\ j \to 0.$$

Define Gårding space as the span of the vectors in $\mathscr{H}$, given by

$$U(\varphi)v = \int \varphi(h)U(h)vd_L h$$

where $\varphi \in C_c(G)$, $v \in \mathscr{H}$, or we say

$$U(\varphi) := \int_G \varphi(h)U(h)\,d_L h.$$

Since φ vanishes outside a compact set, and since $U(h)v$ is continuous and bounded in $\|\cdot\|$, it follows that $U(\varphi)$ is well-defined.

Every representation $\mathcal{U}$ of a Lie group G induces a representation (also denote $\mathcal{U}$) of the group algebra:

Lemma 8.7. $U(\varphi_1 \star \varphi_2) = U(\varphi_1)U(\varphi_2)$ *(U is a representation of the group algebra)*

Proof. Use Fubini,

$$
\begin{aligned}
\int_G \varphi_1 \star \varphi_2(g)U(g)dg &= \iint_{G\times G} \varphi_1(h)\varphi(h^{-1}g)U(g)dhdg \\
&= \iint_{G\times G} \varphi_1(h)\varphi(g)U(hg)dhdg \ (dg \text{ is r-Haar} g \mapsto hg) \\
&= \iint_{G\times G} \varphi_1(h)\varphi(g)U(h)U(g)dhdg \\
&= \int_G \varphi_1(h)U(h)dh \int_G \varphi_2(g)U(g)dg.
\end{aligned}
$$

Choose φ to be an approximation of identity, then

$$
\int_G \varphi(g)U(g)vdg \to U(e)v = v
$$

i.e. any vector $v \in H$ can be approximated by functions in the Gårding space. It follows that

$$
\{U(\varphi)v\}
$$

is dense in $\mathscr{H}$. $\qquad\square$

Lemma 8.8. $U(\varphi)$ *can be differentiated, in the sense that*

$$
dU(X)U(\varphi)v = U(\tilde{X}\varphi)v
$$

where we use $\tilde{X}$ to denote the vector field.

Proof. Need to prove

$$
\lim_{t\to 0} \frac{1}{t}\left[(U(e^{tX}) - I)U(\varphi)v\right] = U(\tilde{X}\varphi)v.
$$

Let $v_\varphi := U(\varphi)v$, need to look at in general $U(g)v_\varphi$.

$$
\begin{aligned}
U(g)v_\varphi &= U(g)\int_G \varphi(h)U(h)vdh \\
&= \int_G \varphi(h)U(gh)dh \\
&= \int_G \triangle(g)\varphi(g^{-1}h)U(h)dh
\end{aligned}
$$

set $g = e^{tX}$. $\qquad\square$

Note 17. If assuming unimodular, $\triangle$ does not show up. Otherwise, $\triangle$ is some correction term which is also differentiable. $\tilde{X}$ acts on φ as $\tilde{X}\varphi$. $\tilde{X}$ is called the derivative of the translation operator e^{tX}.

Exercise 8.13 (The Gårding space for the Schrödinger representation). Show that Schwartz space $\mathcal{S}$ is the Gårding space for the Schrödinger representation.

Exercise 8.14 (The Lie bracket). Let U be a representation of a Lie group G, and let $dU(\cdot)$ be the derived representation, see Lemma 8.8. On the dense Gårding space, show that

$$
\begin{aligned}
dU\left([X,Y]\right) &= [dU\left(X\right), dU\left(Y\right)] \\
&= dU\left(X\right)dU\left(Y\right) - dU\left(Y\right)dU\left(X\right),
\end{aligned}
$$

where $[X, Y]$ denotes the Lie bracket of the two elements X and Y in the Lie algebra.

8.8 Decomposition of Representations

We study some examples of duality.

- $G = \mathbb{T}$, $\hat{G} = \mathbb{Z}$

$$
\begin{aligned}
\chi_n(z) &= z^n \\
\chi_n(zw) &= z^n w^n = \chi_n(z)\chi_n(w)
\end{aligned}
$$

- $G = \mathbb{R}$, $\hat{G} = \mathbb{R}$

$$
\chi_t(x) = e^{itx}
$$

- $G = \mathbb{Z}/n\mathbb{Z} \simeq \{0, 1, \cdots, n-1\}$. $\hat{G} = G$.
  This is another example where $\hat{G} = G$.
  Let $\zeta = e^{i2\pi/n}$ be the primitive n^{th}-root of unity. $k \in \mathbb{Z}_n$, $l \in \{0, 1, \ldots, n-1\}$

$$
\chi_l(k) = e^{i\frac{2\pi kl}{n}}
$$

If G is a locally compact Abelian group, $\hat{G}$ is the set of 1-dimensional representations.

$$
\hat{G} = \{\chi : g \mapsto \chi(g) \in \mathbb{T}, \chi(gh) = \chi(g)\chi(h), \text{ assumed continuous}\}.
$$

$\hat{G}$ is also a group, with group operation defined by $(\chi_1\chi_2)(g) := \chi_1(g)\chi_2(g)$. $\hat{G}$ is called the group characters.

Theorem 8.6 (Pontryagin duality). *If G is a locally compact Abelian group, then $G \simeq \hat{\hat{G}}$ (isomorphism between G and the double dual $\hat{\hat{G}}$,) where "$\simeq$" means "natural isomorphism."*

Remark 8.8. This result first appears in 1930s in the annals of math, when J. von Neumann was the editor of the journal at the time. The original paper was hand written. von Neumann rewrote it, since then the theorem became very popular, see [Rud90].

There are many groups that are not Abelian. We want to study the duality question in general. Examples:

- compact group

- finite group (Abelian, or not)

- H_3 locally compact, non-Abelian, unimodular

- $ax + b$ locally compact, non-Abelian, non-unimodular

If G is not Abelian, $\hat{G}$ is not a group. We would like to decompose $\hat{G}$ into irreducible representations. The big names in this development are Krein[1], Peter-Weyl, Weil, Segal. See [AD86, ARR13, BR79, Emc84, JÓ00, KL14b, KR97b, Rud91, Rud90, Seg50, Sto90].

Let G be a group (may not be Abelian). The right regular representation is defined as

$$R_g f(\cdot) = f(\cdot g), \text{ (translation on the right)}.$$

Then R_g is a unitary operator acting on $L^2(\mu_R)$, where μ_R is the right invariant Haar measure.

Theorem 8.7 (Krein, Weil, Segal). *Let G be locally compact unimodular (Abelian or not). Then the right regular representation decomposes into a direct integral of irreducible representations*

$$R_g = \int_{\hat{G}}^{\oplus} \text{"irrep"} \, d\mu$$

where μ is called the Plancherel measure. See [Sti59, Seg50].

Example 8.13. $G = T$, $\hat{G} = \mathbb{Z}$. Irreducible representations $\{e^{in(\cdot)}\}_n \sim \mathbb{Z}$

$$
\begin{aligned}
(U_y f)(x) &= f(x + y) \\
&= \sum_n \hat{f}(n)\chi_n(x + y) \\
&= \sum_n \hat{f}(n)e^{i2\pi n(x+y)}
\end{aligned}
$$

$$(U_y f)(0) = f(y) = \sum_n \hat{f}(n)e^{i2\pi ny}$$

The Plancherel measure in this case is the counting measure.

---

[1]Appendix C includes a short biographical sketch of M. Krein. See also [Kre46, Kre55].

Example 8.14. $G = \mathbb{R}$, $\hat{G} = \mathbb{R}$. Irreducible representations $\{e^{it(\cdot)}\}_{t\in\mathbb{R}} \sim \mathbb{R}$.

$$
\begin{aligned}
(U_y f)(x) &= f(x+y) \\
&= \int_{\mathbb{R}} \hat{f}(t)\chi_t(x+y)\,dt \\
&= \int_{\mathbb{R}} \hat{f}(t)e^{it(x+y)}\,dt
\end{aligned}
$$

$$
(U_y f)(0) = f(y) = \int_{\mathbb{R}} \hat{f}(t)e^{ity}\,dt
$$

where the Plancherel measure is the Lebesgue measure on $\mathbb{R}$.

As can be seen that Fourier series and Fourier integrals are special cases of the decomposition of the right regular representation R_g of a unimodular locally compact group. $\int^{\oplus} \implies \|f\| = \|\hat{f}\|$. This is a result that was done 30 years earlier before the non Abelian case. Classical function theory studies other types of convergence, pointwise, uniform, etc.

Example 8.15. $G = H_3$. G is unimodular, non Abelian. $\hat{G}$ is not a group.

Irreducible representations: $\mathbb{R}\backslash\{0\}$ Schrödinger representation, $\{0\}$ 1-d trivial representation

Decomposition:

$$
R_g = \int_{\mathbb{R}\backslash\{0\}}^{\oplus} U_{irrep}^h h\,dh
$$

For all $f \in L^2(G)$,

$$
(U_g f)(e) = \int^{\oplus} U^h f\, h\,dh, \quad U^h \text{ irrep.}
$$

Set

$$
\begin{aligned}
F(g) &= (R_g F)(e) \\
&= \int_{\mathbb{R}\backslash\{0\}}^{\oplus} e^{ih(c+bx)} f(x+a)\, h\,dh; \text{ then} \\
\hat{F}(h) &= \int_G (U_g^h F)\,dg
\end{aligned}
$$

Plancherel measure: $h\,dh$ and the point measure δ_0 at zero.

Example 8.16. $G = ax + b$ group, non Abelian. $\hat{G}$ not a group. 3 irreducible representations: $+, -, 0$ but G is not unimodular.

The $+$ representation is supported on $\mathbb{R}_+$, the $-$ representation on $\mathbb{R}_-$, and the 0 representation is the trivial one-dimensional representation.

The duality question may also be asked for discrete subgroups. This leads to remarkable applications in automorphic functions, automorphic forms, p-adic numbers, compact Riemann surface, hyperbolic geometry, etc.

Example 8.17. Cyclic group of order n. $G = \mathbb{Z}/n\mathbb{Z} \simeq \{0, 1, \cdots, n-1\}$. $\hat{G} = G$. This is another example where the dual group is identical to the group itself. Let $\zeta = e^{i2\pi/n}$ be the primitive n^{th}-root of unity. $k \in \mathbb{Z}_n$, $l = \{0, 1, \ldots, n-1\}$

$$\chi_l(k) = e^{i\frac{2\pi kl}{n}}.$$

In this case, Segal's theorem gives finite Fourier transform. $U : l^2(\mathbb{Z}) \to l^2(\hat{\mathbb{Z}})$ where

$$Uf(l) = \frac{1}{\sqrt{N}} \sum_k \zeta^{kl} f(k).$$

8.9 Summary of Induced Representations, the Example of d/dx

We study decomposition of group representations. Two cases: Abelian and non-Abelian. The non-Abelian case may be induced from the Abelian ones.
    non Abelian

- semi product $G = HN$ often N is normal.

- G simple. G does not have normal subgroups, i.e., the Lie algebra does not have any ideals.

Exercise 8.15 (Normal subgroups). (1) Find the normal subgroups in the Heisenberg group. (2) Find the normal subgroups in the $ax + b$ group.

Example 8.18. $SL_2(\mathbb{R})$ (non-compact)

$$\begin{pmatrix} a & b \\ c & d \end{pmatrix}, \ ad - bc = 1$$

with Lie algebra

$$sl_2(\mathbb{R}) = \{X : tr(X) = 0\}.$$

Note that sl_2 is generated by

$$\begin{pmatrix} 0 & 1 \\ 1 & 0 \end{pmatrix}, \begin{pmatrix} 0 & -1 \\ 1 & 0 \end{pmatrix}, \begin{pmatrix} 1 & 0 \\ 0 & -1 \end{pmatrix}.$$

In particular, $\begin{pmatrix} 0 & -1 \\ 1 & 0 \end{pmatrix}$ generates the one-parameter group $\begin{pmatrix} \cos t & -\sin t \\ \sin t & \cos t \end{pmatrix} \simeq$ $\mathbb{T}$ whose dual group is $\mathbb{Z}$, where

$$\chi_n(g(t)) = g(t)^n = e^{itn}.$$

May use this to induce a representation of G. This is called principle series. Need to do something else to get all irreducible representations.

A theorem by Iwasawa states that simple matrix group (Lie group) can be decomposed into

$$G = KAN$$

where K is compact, A is Abelian and N is nilpotent. For example, in the SL_2 case,

$$SL_2(\mathbb{R}) = \begin{pmatrix} \cos t & -\sin t \\ \sin t & \cos t \end{pmatrix} \begin{pmatrix} e^s & 0 \\ 0 & e^{-s} \end{pmatrix} \begin{pmatrix} 1 & u \\ 0 & 1 \end{pmatrix}.$$

The simple groups do not have normal subgroups. The representations are much more difficult.

8.9.1 Equivalence and imprimitivity for induced representations

Suppose from now on that G has a normal Abelian subgroup $N \lhd G$, and $G = H \ltimes N$. The $N \simeq \mathbb{R}^d$ and $N^* \simeq (\mathbb{R}^d)^* = \mathbb{R}^d$. In this case

$$\chi_t(\nu) = e^{it\nu}$$

for $\nu \in N$ and $t \in \hat{N} = N^*$. Notice that χ_t is a 1-d irreducible representation on $\mathbb{C}$.

Let $\mathscr{H}_t$ be the space of functions $f : G \to \mathbb{C}$ so that

$$f(\nu g) = \chi_t(\nu) f(g).$$

On $\mathscr{H}_t$, define inner product so that

$$\|f\|_{\mathscr{H}_t}^2 := \int_G |f(g)|^2 = \int_{G/N} \|f(g)\|^2 \, dm$$

where dm is the invariant measure on $N \backslash G \simeq H$.

Define $U_t = ind_N^G(\chi_t) \in Rep(G, \mathscr{H}_t)$. Define $U_t(g)f(x) = f(xg)$, for $f \in \mathscr{H}_t$. Notice that the representation space of χ_t is $\mathbb{C}$, 1-d Hilbert space; however, the representation space of U_t is $\mathscr{H}_t$ which is infinite dimensional. U_t is a family of irreducible representations indexed by $t \in N \simeq \hat{N} \simeq \mathbb{R}^d$.

Note 18. Another way to recognize induced representations is to see these functions are defined on H, not really on G.

Define the unitary transformation $W : \mathscr{H}_t \to L^2(H)$. Notice that $H \simeq N \backslash G$ is a group, and it has an invariant Haar measure. By uniqueness on the Haar measure, this has to be dm. It would be nice to cook up the same space $L^2(H)$ so that all induced representations indexed by t act on it. In other words, this Hilbert space $L^2(H)$ does not depend on t. W_t is defined as

$$W F_t(h) = F_t(h).$$

So what does the induced representation look like in $L^2(H)$ then? Recall by definition that

$$U_t(g) := W\left(ind^G_{\chi_t}(g)\right) W^*$$

and the following diagram commutes.

$$
\begin{array}{ccc}
\mathcal{H}_t & \xrightarrow{\;ind^G_{\chi_t}\;} & \mathcal{H}_t \\
\;\Big\downarrow{\scriptstyle W} & & \;\Big\downarrow{\scriptstyle W} \\
L^2(H) & \xrightarrow{\;U_t\;} & L^2(H)
\end{array}
$$

Let $f \in L^2(H)$.

$$
\begin{aligned}
U_t(g)\,f(h) &= W\left(ind^G_{\chi_t}(g)\right) W^* f(h) \\
&= \left(ind^G_{\chi_t}(g) W^* f\right)(h) \\
&= (W^* f)(hg).
\end{aligned}
$$

Since $G = H \ltimes N$, g is uniquely decomposed into $g = g_N g_H$. Hence $hg = hg_N g_H = g_N g_N^{-1} h g_N g_H = g_N \tilde{h} g_H$ and

$$
\begin{aligned}
U_t(g)f(h) &= (W^* f)(hg) \\
&= (W^* f)(g_N \tilde{h} g_H) \\
&= \chi_t(g_N)(W^* f)(\tilde{h} g_H) \\
&= \chi_t(g_N)(W^* f)(g_N^{-1} h g_N g_H).
\end{aligned}
$$

This last formula is called the Mackey machine [Mac52, Mac88].

The Mackey machine does not cover many important symmetry groups in physics. Actually most of these are simple groups. However it can still be applied. For example, in special relativity theory, we have the Poincaré group $\mathcal{L} \ltimes \mathbb{R}^4$ where $\mathbb{R}^4$ is the normal subgroup. The baby version of this is when $\mathcal{L} = SL_2(\mathbb{R})$. It was V. Bargmann[2] who first worked out the complete list, with spectral parameters, for all the equivalence classes of the unitary irreducible representations ("irreps" for short) in the special case of the "small" Lorentz group.

The Lorentz group is the matrix group which preserves the indefinite form $(-, +, +, +)$ on 4-dimensional space time. The Poincaré group is the semi-direct product of the Lorentz group with Minkowski space (i.e., with the Abelian 4-dimensional space time). It is the symmetry group of Einstein's relativity. Continuing Bargmann's work on the Lorentz group, Wigner found the unitary irreducible representations of the Poincaré group. And because the semi-direct product aspect of the Poincaré group, one could argue that Wigner discovered the Mackey machine (in an important special case), before Mackey developed his theory. Or, conversely, that what we call "the Mackey machine" was motivated by the early work by Bargmann and Wigner.

---

[2] Appendix C includes a short biographical sketch of V. Bargmann. See also [Bar47].

V. Bargmann formulated this "baby version." What Bargmann did amounted to an explicit classification of the dual of the "small Lorentz group," i.e., $SL_2 (\mathbb{R})$: Bargmann [Bar47] found all the equivalence classes of the irreducible representations of $SL_2 (\mathbb{R})$. Specifically, he wrote out the "irreps" both by explicit formulas, by spectral theory, and by labeling of the distinct series of irreps by parameters: the principal series, discrete series, holomorphic series, and the complementary series. This became the model for later work by Wigner and by Harish-Chandra (the latter for the case of semisimple Lie groups [HC57].) Wigner in turn pioneered the use of the semidirect product method, what later became the Mackey machine, (but Wigner's work was before Mackey entered the scene.)

Once we get unitary representations, differentiate it and get selfadjoint algebra of operators (possibly unbounded). These are the observables in quantum mechanics.

Example 8.19. $\mathbb{Z} \subset \mathbb{R}$, $\hat{\mathbb{Z}} = T$. $\chi_t \in T$, $\chi_t(n) = e^{itn}$. Let $\mathscr{H}_t$ be the space of functions $f : \mathbb{R} \to \mathbb{C}$ so that

$$f(n + x) = \chi_t(n)f(x) = e^{int}f(x).$$

Define inner product on $\mathscr{H}_t$ so that

$$\|f\|^2_{\mathscr{H}_t} := \int_0^1 |f(x)|^2 \, dx.$$

Define $ind^{\mathbb{R}}_{\chi_t}(y)f(x) = f(x + y)$. Claim that $\mathscr{H}_t \simeq L^2[0, 1]$. The unitary transformation is given by $W : \mathscr{H}_t \to L^2[0, 1]$

$$(WF_t)(x) = F_t(x).$$

Let's see what $ind^{\mathbb{R}}_{\chi_t}(y)$ looks like on $L^2[0, 1]$. For any $f \in L^2[0, 1]$,

$$
\begin{aligned}
\left(W \left(ind^G_{\chi_t}(y) \right) W^* f \right)(x) &= \left(ind^G_{\chi_t}(y) W^* f \right)(x) \\
&= (W^* f)(x + y).
\end{aligned}
$$

Since $y \in \mathbb{R}$ is uniquely decomposed as $y = n + x'$ for some $x' \in [0, 1)$, therefore

$$
\begin{aligned}
\left(W \left(ind^G_{\chi_t}(y) \right) W^* f \right)(x) &= (W^* f)(x + y) \\
&= (W^* f)(x + n + x') \\
&= (W^* f)(n + (-n + x + n) + x') \\
&= \chi_t(n)(W^* f)((-n + x + n) + x') \\
&= \chi_t(n)(W^* f)(x + x') \\
&= e^{itn}(W^* f)(x + x')
\end{aligned}
$$

Note 19. Are there any functions in $\mathscr{H}_t$? Yes, for example, $f(x) = e^{itx}$. If $f \in \mathscr{H}_t$, $|f|$ is 1-periodic. Therefore f is really a function defined on $\mathbb{Z}\backslash\mathbb{R} \simeq [0, 1]$. Such a function has the form

$$f(x) = \left(\sum c_n e^{i2\pi nx} \right) e^{itx} = \sum c_n e^{i(2\pi n + t)x}.$$

Any 1-periodic function g satisfies the boundary condition $g(0) = g(1)$. $f \in \mathscr{H}_t$ has a modified boundary condition where $f(1) = e^{it}f(0)$.

8.10   Connections to Nelson's Spectral Theory

In Nelson's notes [Nel69], a normal representation has the form (counting multiplicity)

$$\rho = \sum^{\oplus} n\pi\big|_{\mathscr{H}_n}, \quad \mathscr{H}_n \perp \mathscr{H}_m$$

where

$$n\pi = \pi \oplus \cdots \oplus \pi \text{ (n times)}$$

is a representation acting on the Hilbert space

$$\sum^{\oplus} K = l^2_{\mathbb{Z}_n} \otimes K.$$

In matrix form, this is a diagonal matrix with π repeated on the diagonal n times. n could be $1, 2, \ldots, \infty$. We apply this to group representations.

Locally compact group can be divided into the following types.

- Abelian

- non-Abelian: unimodular, non-unimodular

- non-Abelian: Mackey machine, semidirect product e.g. H_3, $ax + b$; simple group $SL_2(\mathbb{R})$. Even it's called simple, ironically its representation is much more difficult than the semidirect product case.

We want to apply these to group representations.

Spectral theorem says that given a normal operator A, we may define $f(A)$ for quite a large class of functions, actually all measurable functions (see [Sto90, Yos95, Nel69, RS75, DS88b]). One way to define $f(A)$ is to use the multiplication version of the spectral theorem, and let

$$f(A) = \mathcal{F}f(\hat{A})\mathcal{F}^{-1}.$$

The other way is to use the projection-valued measure version of the spectral theorem, write

$$A = \int \lambda P(d\lambda)$$

$$f(A) = \int f(\lambda) P(d\lambda).$$

The effect is ρ is a representation of the Abelian algebra of measurable functions onto operators action on some Hilbert space.

$$\rho : f \mapsto \rho(f) = f(A)$$
$$\rho(fg) = \rho(f)\rho(g).$$

To imitate Fourier transform, let's call $\hat{f} := \rho(f)$. Notice that $\hat{f}$ is the multiplication operator.

Example 8.20. $G = (\mathbb{R}, +)$, group algebra $L^1(\mathbb{R})$. Define Fourier transform

$$\hat{f}(t) = \int f(x)e^{-itx}dx.$$

$\{e^{itx}\}_t$ is a family of 1-dimensional irreducible representation of $(\mathbb{R}, +)$.

Example 8.21. Fix t, $\mathscr{H} = \mathbb{C}$, $\rho(\cdot) = e^{it(\cdot)} \in Rep(G, \mathscr{H})$. From the group representation ρ, we get a group algebra representation $\tilde{\rho} \in Rep(L^1(\mathbb{R}), \mathscr{H})$ defined by

$$\tilde{\rho}(f) = \int f(x)\rho(x)dx = \int f(x)e^{itx}dx.$$

It follows that

$$\hat{f}(\rho) \quad := \quad \tilde{\rho}(f)$$
$$\widehat{f \star g} \quad = \quad \widehat{f \star g} = \hat{f}\hat{g}$$

i.e. Fourier transform of $f \in L^1(\mathbb{R})$ is a representation of the group algebra $L^1(\mathbb{R})$ on to the 1-dimensional Hilbert space $\mathbb{C}$. The range of Fourier transform in this case is 1-d Abelian algebra of multiplication operators, multiplication by complex numbers.

Example 8.22. $\mathscr{H} = L^2(\mathbb{R})$, $\rho \in Rep(G, \mathscr{H})$ so that

$$\rho(y)f(x) := f(x + y)$$

i.e., ρ is the right regular representation. The representation space $\mathscr{H}$ in this case is infinite dimensional. From ρ, we get a group algebra representation $\tilde{\rho} \in Rep(L^1(\mathbb{R}), \mathscr{H})$ where

$$\tilde{\rho}(f) = \int f(y)\rho(y)dy.$$

Define

$$\hat{f}(\rho) := \hat{\rho}(f)$$

then $\hat{f}(\rho)$ is an operator acting on $\mathscr{H}$.

$$\hat{f}(\rho)g = \tilde{\rho}(f)g = \int f(y)\rho(y)g(\cdot)dy$$
$$= \int f(y)(R_y g)(\cdot)dy$$
$$= \int f(y)g(\cdot + y)dy.$$

If we have used the left regular representation, instead of the right, then

$$\hat{f}(\rho)g = \tilde{\rho}(f)g = \int f(y)\rho(y)g(\cdot)dy$$

$$= \int f(y)(L_y g)(\cdot) dy$$

$$= \int f(y)g(\cdot - y) dy.$$

Hence $\hat{f}(\rho)$ is the left or right convolution operator.

Back to the general case. Given a locally compact group G, form the group algebra $L^1(G)$, and define the left and right convolutions as

$$(\varphi \star \psi)(x) = \int \varphi(g)\psi(g^{-1}x) d_L g = \int \varphi(g)(L_g \psi) d_L g$$

$$(\varphi \star \psi)(x) = \int \varphi(xg)\psi(g) d_R g = \int (R_g \varphi)\psi(g) d_R g$$

Let $\rho(g) \in Rep(G, \mathscr{H})$, define $\tilde{\rho} \in Rep(L^1(G), \mathscr{H})$ given by

$$\tilde{\rho}(\psi) := \int_G \psi(g)\rho(g) dg$$

and write

$$\hat{\psi}(\rho) := \tilde{\rho}(\psi).$$

$\hat{\psi}$ is an analog of Fourier transform. If ρ is irreducible, the operators $\hat{\psi}$ forms an Abelian algebra. In general, the range of this generalized Fourier transform gives rise to a non Abelian algebra of operators.

For example, if $\rho(g) = R_g$ and $\mathscr{H} = L^2(G, d_R)$, then

$$\tilde{\rho}(\psi) = \int_G \psi(g)\rho(g) dg = \int_G \psi(g) R_g dg$$

and

$$\tilde{\rho}(\psi)\varphi = \int_G \psi(g)\rho(g)\varphi dg = \int_G \psi(g)(R_g \varphi) dg$$

$$= \int_G \psi(g)\varphi(xg) dg$$

$$= (\varphi \star \psi)(x)$$

Example 8.23. $G = H_3 \sim \mathbb{R}^3$. $\hat{G} = \{\mathbb{R}\backslash\{0\}\} \cup \{0\}$. $0 \in \hat{G}$ corresponds to the trivial representation, i.e. $g \mapsto Id$ for all $g \in G$.

$$\rho_h : G \to L^2(\mathbb{R})$$

$$\rho_h(g)f(x) = e^{ih(c+bx)}f(x+a) \simeq ind_H^G(\chi_h)$$

where H is the normal subgroup $\{b, c\}$. It is not so nice to work with $ind_H^G(\chi_h)$ directly, so instead, we work with the equivalent representations, i.e. Schrödinger representation.

$$\hat{\psi}(h) = \int_G \psi(g)\rho_h(g) dg$$

Notice that $\hat{\psi}(h)$ is an operator acting on $L^2(\mathbb{R})$. Specifically,

$$
\begin{aligned}
\hat{\psi}(h) &= \int_G \psi(g)\rho_h(g)dg \\
&= \iiint \psi(a,b,c)e^{ih(c+bx)}f(x+a)dadbdc \\
&= \iint \left(\int \psi(a,b,c)e^{ihc}dc \right) f(x+a)e^{ihbx}dadb \\
&= \iint \hat{\psi}(a,b,h)f(x+a)e^{ihbx}dadb \\
&= \int \left(\int \hat{\psi}(a,b,h)e^{ihbx}db \right) f(x+a)da \\
&= \int \hat{\psi}(a,hx,h)f(x+a)da \\
&= \left(\hat{\psi}(\cdot,h\cdot,h) \star f \right)(x)
\end{aligned}
$$

Here the $\hat{\psi}$ on the right hand side in the Fourier transform of ψ in the usual sense. Therefore the operator $\hat{\psi}(h)$ is the one so that

$$
L^2(\mathbb{R}) \ni f \mapsto \left(\hat{\psi}(\cdot,h\cdot,h) \star f \right)(x).
$$

If $\psi \in L^1(G)$, $\hat{\psi}$ is not of trace class. But if $\psi \in L^1 \cap L^2$, then $\hat{\psi}$ is of trace class.

$$
\int_{\mathbb{R}\setminus\{0\}}^{\oplus} tr\left(\hat{\psi}^*(h)\hat{\psi}(h) \right) d\mu = \int |\psi|^2 \, dg = \int \bar{\psi}\psi dg
$$

where μ is the Plancherel measure.

If the group G is non unimodular, the direct integral is lost (not orthogonal). These are related to coherent states from physics, which is about decomposing Hilbert into non orthogonal pieces.

Important observables in QM come in pairs (dual pairs). For example, position - momentum; energy - time etc. The Schwartz space $S(\mathbb{R})$ has the property that $\widehat{S(\mathbb{R})} = S(\mathbb{R})$. We look at the analog of the Schwartz space. $h \mapsto \hat{\psi}(h)$ should decrease faster than any polynomials.

Take $\psi \in L^1(G)$, X_i in the Lie algebra, form $\triangle = \sum X_i^2$. Require that

$$
\triangle^n\psi \in L^1(G), \ \psi \in C^\infty(G).
$$

For $\triangle^n$, see what happens in the transformed domain. Notice that

$$
\frac{d}{dt}\Big|_{t=0} (R_{e^{tX}}\psi) = \tilde{X}\psi
$$

where $X \mapsto \tilde{X}$ represents the direction vector X as a vector field.

Let G be any Lie group. $\varphi \in C_c^\infty(G)$, $\rho \in Rep(G, \mathcal{H})$.

$$d\rho(X)v = \int (\tilde{X}\varphi)(g)\rho(g)vdg$$

where
$$v = \int \varphi(g)\rho(g)wdg = \rho(\varphi)w. \text{ generalized convolution}$$

If $\rho = R$, the n
$$v = \int \varphi(g)R(g)$$

$$\tilde{X}(\varphi \star w) = (X\varphi) \star w.$$

Example 8.24. H_3

$$a \ \to \ \frac{\partial}{\partial a}$$

$$b \ \mapsto \ \frac{\partial}{\partial b}$$

$$c \ \mapsto \ \frac{\partial}{\partial c}$$

get standard Laplace operator. $\{\rho_h(\varphi)w\} \subset L^2(\mathbb{R})$. " $=$ " due to Dixmier. $\{\rho_h(\varphi)w\}$ is the Schwartz space.

$$\left(\frac{d}{dx}\right)^2 + (ihx)^2 + (ih)^2 = \left(\frac{d}{dx}\right)^2 - (hx)^2 - h^2.$$

Notice that
$$-\left(\frac{d}{dx}\right)^2 + (hx)^2 + h^2$$

is the Harmonic oscillator. Spectrum $= h\mathbb{Z}_+$.

8.11   Multiplicity Revisited

Let $\mathfrak{A}$ be a $*$-algebra, and let π and ρ be representations of $\mathfrak{A}$. To indicate the Hilbert space, we write $\pi \in Rep(\mathfrak{A}, \mathcal{H}_\pi)$, and $\rho \in Rep(\mathfrak{A}, \mathcal{H}_\rho)$.

Definition 8.6. Consider the following space of bounded linear operators s, t : $\mathcal{H}_\pi \longrightarrow \mathcal{H}_\rho$ which intertwine the respective representations, i.e., we have:

$$s\pi(a) = \rho(a)s, \ \forall a \in \mathfrak{A}. \tag{8.51}$$

The set of solutions s to (8.51) forms a vector space, and it is denoted $Int(\pi, \rho)$, the intertwining operators. We check the following:

$$s \in Int(\pi, \rho) \Longleftrightarrow s^* \in Int(\rho, \pi); \tag{8.52}$$

and therefore, if $s, t \in Int(\pi, \rho)$, we have:

$$s^* t \in Int(\pi, \pi).\qquad(8.53)$$

Note

$$
\begin{aligned}
Int(\pi, \pi) &= \pi(\mathfrak{A})' \text{ (commutant)}\\
&= \{A \in \mathscr{B}(\mathscr{H}_\pi) : \pi(a)A = A\pi(a), \ \forall a \in \mathfrak{A}\}.\qquad(8.54)
\end{aligned}
$$

If π is irreducible, therefore $Int(\pi, \pi)$ is one-dimensional; hence, for $\forall s, t \in Int(\pi, \rho)$,

$$s^* t = \langle s, t \rangle I_{\mathscr{H}_\pi},\qquad(8.55)$$

where $\langle s, t \rangle \in \mathbb{C}$ is uniquely determined. This form $\langle \cdot, \cdot \rangle$ is sesquilinear, and positive definite. We therefore get a Hilbert-completion of $Int(\pi, \rho)$. Let $\mathscr{H}(\pi, \rho)$ be the corresponding Hilbert space.

Definition 8.7. Let π and ρ be as above, assume that π is irreducible, and let $\mathscr{H}(\pi, \rho)$ be the corresponding Hilbert space; see (8.55). We say that π occurs in ρ m times if

$$m = \dim \mathscr{H}(\pi, \rho).\qquad(8.56)$$

Exercise 8.16 (Multiplicity). Show that the definition of multiplicity (Definition 8.7) agrees with the one used inside Chapter 8.

Exercise 8.17 (A Hilbert space of intertwiners). Let π and ρ be as above, π irreducible. Show that with the inner product defined in (8.55), $Int(\pi, \rho)$ is a Hilbert space.

Exercise 8.18 (An ONB in $Int(\pi, \rho)$). Let (s_i) be an ONB in $Int(\pi, \rho)$.

1. Show that this is a system of isometries, satisfying:

$$s_i^* s_j = \delta_{i,j} I_{\mathscr{H}_\pi}.\qquad(8.57)$$

2. For $A \in \mathscr{B}(\mathscr{H}_\pi)$, set

$$\alpha(A) := \sum_i s_i A s_i^*.\qquad(8.58)$$

Show that

$$
\begin{aligned}
\alpha(AB) &= \alpha(A)\alpha(B), \text{ and}\\
\alpha(A^*) &= \alpha(A)^*, \ \forall A, B \in \mathscr{B}(\mathscr{H}_\pi).
\end{aligned}
$$

3. What can be said about

$$\alpha(I_{\mathscr{H}_\pi}) = \sum_i s_i s_i^* \ ?$$

Exercise 8.19 (A Hilbert space of intertwining operators). Verify that the results above about $Int(\pi, \rho)$ apply to *unitary representations* π, and ρ of some given *group* G; i.e., with

$$Int(\pi, \rho) = \{s : \mathscr{H}_\pi \longrightarrow \mathscr{H}_\rho : s\pi(g) = \rho(g)s, \ \forall g \in G\}.$$

Hint: Use the above on the group algebra $\mathfrak{A}_G := \mathbb{C}[G]$.

Now consider the Heisenberg group G of all 3×3 matrices

$$g = \begin{bmatrix} 1 & a & c \\ 0 & 1 & b \\ 0 & 0 & 1 \end{bmatrix}, \ (a, b, c) \in \mathbb{R}^3.$$

Recall its Haar measure is $dg = da\, db\, dc = 3$-dimensional Lebesgue measure.

Consider the following two representations ρ and π of G (the regular representations RR, and the Schrödinger representation SR):

- (RR) $\mathscr{H}_\rho = L^2(G, dg)$, Haar measure, and

$$(\rho(g)f)(h) = f(hg), \ \forall f \in \mathscr{H}_\rho, \ \forall g, h \in G.$$

And the Schrödinger representation ($\hbar = 1$):

- (SR) $\mathscr{H}_\pi = L^2(\mathbb{R})$, Lebesgue measure, and

$$(\pi(g)F)(x) = e^{i(c+bx)}F(x+a), \ \forall F \in L^2(\mathbb{R}) = \mathscr{H}_\pi, \ \forall g \in G, \ x \in \mathbb{R}.$$

Exercise 8.20 (Specify the operators in $Int(\pi, \rho)$). Let G, π, and ρ be as above. What is the Hilbert space $Int(\pi, \rho)$?

Exercise 8.21 (A formula from Peter-Weyl [Mac92]). In case G is a compact group, look up and explain that the Peter-Weyl theorem states the following: If ρ is the regular representation, and if π is irreducible unitary, then

$$\dim(Int(\pi, \rho)) = \dim(\pi).$$

Remark 8.9. An important class of non-compact, non-commutative, locally compact groups G, and unitary representations π, for which the intertwining Hilbert spaces $Int(\pi, \rho)$ are non-zero is the class of square-integrable representations: Suppose the representation π is irreducible and square-integrable, then $Int(\pi, \rho)$ is non-zero. Here ρ denotes the regular representation of G. A representation π is square-integrable if its matrix coefficients are in $L^2(G/Z)$, where Z denotes the center of G.

initial object	first duality step	twice dual
E Banach space	E^* dual Banach space	E^{**} double-dual, $E \hookrightarrow E^{**}$ natural isometry
$S \subset \mathcal{H}$ subset of a Hilbert space	$S^\perp$ perpendicular (see Section 2.6)	$S^{\perp\perp} = \overline{span}\,(S)$, i.e., $\mathcal{H}$-closed space
A a given closable operator in Hilbert space	A^* the adjoint operator (see Section 3.1)	$A^{**} = \overline{A}$, the closure of A
$\mathcal{T} \subset \mathcal{B}(\mathcal{H})$ subset	$\mathcal{T}'$ commutant (see Section 5.7)	$\mathcal{T}''$ double commutant $=$ the von Neumann algebra generated by $\mathcal{T}$
G locally compact Abelian group	$\widehat{G}$ dual group (continuous characters of G, see Section 8.8)	$\widehat{\widehat{G}} \simeq G$ Pontryagin duality, see Theorem 8.6

Table 8.2: Overview of five duality operations.

A summary of relevant numbers from the Reference List

For readers wishing to follow up sources, or to go in more depth with topics above, we suggest: [JÓ00, Mac52, Mac85, Mac92, JM84, JPS01, JPS05, Jor11, Jor94, Jor88, Sza15, DJ08, Dix81, Jor02, Dud14, Nel59a, JM80, Seg50, Ørs79, Pou72, HJL06, DHL09, Tay86, Hal13, Hal15, Sch12].

8.A   The Stone-von Neumann Uniqueness Theorem

The "uniqueness" in the title above refers to "uniqueness up to *unitary equivalence*."

Definition 8.8. Let $\mathcal{H}_i$, $i = 1, 2$ be two Hilbert spaces, and let $S_1 = \{A_\alpha\} \subset \mathcal{B}(\mathcal{H}_1)$, and $S_2 = \{B_\alpha\} \subset \mathcal{B}(\mathcal{H}_2)$ be systems of bounded operators, where the index set $J = \{\alpha\}$ is the same for the two operator systems.

We say that S_1 and S_2 are *unitarily equivalent* iff (Def) $\exists W : \mathcal{H}_1 \to \mathcal{H}_2$, W a unitary isomorphism of $\mathcal{H}_1$ onto $\mathcal{H}_2$ such that

$$W A_\alpha = B_\alpha W, \ \forall \alpha \in J; \text{ see Fig. 8.3.} \tag{8.59}$$

We say that the system $S_1 = \{A_\alpha\}$ is *irreducible* iff (Def) the following implication holds

$$\boxed{T \in \mathscr{B}(\mathscr{H}_1),\ TA_\alpha = A_\alpha T,\ \alpha \in J} \implies T = \lambda I_1,\ \text{for some } \lambda \in \mathbb{C}; \quad (8.60)$$

i.e., the commutant is one-dimensional.

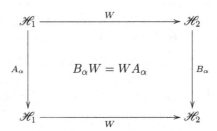

Figure 8.3: W intertwines S_1 and S_2.

Definition 8.9. The Heisenberg group G_3 is the matrix group

$$g = \begin{bmatrix} 1 & a & c \\ 0 & 1 & b \\ 0 & 0 & 1 \end{bmatrix}, \quad (a, b, c) \in \mathbb{R}^3,$$

of all upper triangular 3×3 matrices.
    Fix $h \in \mathbb{R} \backslash \{0\}$, and set

$$(\mathcal{U}_h(g) f)(x) = e^{ih(c+bx)} f(x+a) \quad (8.61)$$

$\forall g = (a, b, c) \in G_3,\ \forall f \in L^2(\mathbb{R}),\ \forall x \in \mathbb{R}.$

It is easy to see that $\mathcal{U}_h$ is a unitary irreducible representation of G_3 acting on $L^2(\mathbb{R})$, i.e., $\mathcal{U}_h \in Rep_{uni}(G_3, L^2(\mathbb{R}))$ for all $h \in \mathbb{R} \backslash \{0\}$. It is called the *Schrödinger representation.*

Theorem 8.8 (Stone-von Neumann). *Every unitary irreducible representation of G_3 in a Hilbert space (other than the trivial one-dimensional representation) is unitarily equivalent to the Schrödinger representation $\mathcal{U}_h$ for some $h \in \mathbb{R} \backslash \{0\}$.*

Proof. The proof follows from the more general result, Theorem 8.3 above; the Imprimitivity Theorem. Also see [vN32b, vN31].   □

Remark 8.10. The center of G_3 is the one-dimensional subgroup $g = (0, 0, c)$, $c \in \mathbb{R}$, and so if $\mathcal{U}_h \in Rep_{uni}(G_3, \mathscr{H})$, $\dim \mathscr{H} > 1$, then it follows that $\exists! h \in \mathbb{R} \backslash \{0\}$ such that

$$\mathcal{U}(0, 0, c) = e^{ich} I_{\mathscr{H}}.$$

Hence $\mathcal{U}$ is determined by two one-parameter groups

$$\begin{cases} \mathcal{U}_1\left(a\right) = \mathcal{U}\left(a,0,0\right), & a \in \mathbb{R}; \text{ and} \\ \mathcal{U}_2\left(b\right) = \mathcal{U}\left(0,b,0\right), & b \in \mathbb{R} \end{cases} \tag{8.62}$$

such that

$$\mathcal{U}_1\left(a\right)\mathcal{U}_2\left(b\right)\mathcal{U}_1\left(-a\right) = e^{i\,h\,a\,b}\mathcal{U}_2\left(b\right), \ \forall a,b \in \mathbb{R}. \tag{8.63}$$

The system (8.63) is called the *Weyl commutation relation*. It is the integrated form of the corresponding Heisenberg relation (for unbounded essentially selfadjoint operators). (We omit a systematic discussion of the interrelationships between the two commutation relations.)

Under the unitary equivalence $W : \mathscr{H} \to L^2\left(\mathbb{R}\right)$ from the Stone-von Neumann theorem, we get

$$\begin{cases} \left(W\mathcal{U}_1\left(a\right)W^*f\right)\left(x\right) = f\left(x+a\right), & \text{and} \\ \left(W\mathcal{U}_2\left(b\right)W^*f\right)\left(x\right) = e^{i\,h\,b}f\left(x\right), & \forall a,b,x \in \mathbb{R}, \forall f \in L^2\left(\mathbb{R}\right). \end{cases} \tag{8.64}$$

We shall use the following

Lemma 8.9. *Let $\mathcal{U}_1\left(\cdot\right)$ and $\mathcal{U}_2\left(\cdot\right)$ be the two one-parameter groups from the Weyl relation (8.63), and let P_2 be the projection valued measure corresponding to $\{\mathcal{U}_2\left(b\right)\}_{b\in\mathbb{R}}$, i.e.,*

$$\mathcal{U}_2\left(b\right) = \int_{\mathbb{R}} e^{i\,b\,\lambda}P_2\left(d\lambda\right), \ \forall b \in \mathbb{R}. \tag{8.65}$$

Then the Weyl relation (8.63) is equivalent to

$$\mathcal{U}_1\left(a\right)P_2\left(\triangle\right)\mathcal{U}_1\left(-a\right) = P_2\left(\triangle - h\,a\right), \tag{8.66}$$

$\forall a \in \mathbb{R}, \forall \triangle \in \mathcal{B}\left(\mathbb{R}\right)$, *where*

$$\triangle - h\,a = \left\{s - h\,a \mid s \in \triangle\right\}. \tag{8.67}$$

Proof. The proof is an easy application of Stone's theorem ([vN32b, Nel69]); see Appendix 3.A. □

Chapter 9

The Kadison-Singer Problem

"Born wanted a theory which would generalize these matrices or grids of numbers into something with a continuity comparable to that of the continuous part of the spectrum. The job was a highly technical one, and he counted on me for aid.... I had the generalization of matrices already at hand in the form of what is known as operators. Born had a good many qualms about the soundness of my method and kept wondering if Hilbert would approve of my mathematics. Hilbert did, in fact, approve of it, and operators have since remained an essential part of quantum theory."

— Norbert Wiener

"In science one tries to tell people, in such a way as to be understood by everyone, something that no one ever knew before. But in the case of poetry, it's the exact opposite!"

— Paul Adrien Maurice Dirac.

"It seems to be one of the fundamental features of nature that fundamental physical laws are described in terms of a mathematical equations of great beauty and power."

— Paul Adrien Maurice Dirac

In the details below, the following concepts and terms will play an important role; (for reference, see also Appendix F): The Stone-Čech compactification, states, normal, non-normal, extensions, pure states, uniqueness of pure-state extensions.

9.1   Statement of the Problem

By l^2, we shall mean the Hilbert space $l^2(\mathbb{N})$ of all square summable sequences, i.e., $x = (x_n)_{n=1}^\infty \in l^2$ iff (Def.) $\sum_{n=1}^\infty |x_n|^2 < \infty$; and we set

$$\langle x, y \rangle_2 := \sum_{n=1}^\infty \overline{x_n} y_n, \quad \forall x, y \in l^2. \tag{9.1}$$

By $\mathscr{B}(l^2)$, we mean the von Neumann algebra of all bounded linear operators in l^2, with $A \longrightarrow A^*$ defined by

$$\langle A^* x, y \rangle_2 = \langle x, Ay \rangle_2, \quad x, y \in l^2. \tag{9.2}$$

The norm on $\mathscr{B}(l^2)$ is the uniform norm; hence

$$\|A^* A\| = \|A\|^2, \quad \forall A \in \mathscr{B}(l^2); \tag{9.3}$$

see Section 2.4.3. We consider the subalgebra $\mathbb{D}$ of all diagonal operators in $\mathscr{B}(l^2)$, hence $D \in \mathbb{D}$ iff its representation relative to (9.1) has the following form:

$$D = \begin{bmatrix} d_1 & & & & & & \\ & d_2 & & & & \Large{0} & \\ & & d_3 & & & & \\ & & & \ddots & & & \\ & & & & d_n & & \\ & \Large{0} & & & & \ddots & \\ & & & & & & \ddots \end{bmatrix}$$

where $(d_n) \in l^\infty \; (= l^2(\mathbb{N}))$, and the norm $\|\cdot\|$ from (9.3) is

$$\|D\| = \|(d_n)_1^\infty\|_{l^\infty} = \sup_{n \in \mathbb{N}} |d_n|;$$

see Section 2.4.

Since $\mathbb{D} \subset \mathscr{B}(l^2)$ is a $*$-subalgebra; norm closed in fact, we may consider $\mathbb{D}$ as a C^*-subalgebra of $\mathscr{B}(l^2)$. Both are von Neumann algebras as follows: The pre-dual of $\mathscr{B}(l^2)$ is the Banach space of all trace class operators (see Section 5.3), and the pre-dual of $\mathbb{D}$ is the Banach space $l^1(\mathbb{N})$.

One easily checks that $\mathbb{D}$ is a maximal abelian subalgebra, and moreover that the spectral measure for $\mathbb{D}$ is counting measure, so purely atomic with atoms equal the points in $\mathbb{N}$. One says that the maximal abelian subalgebra $\mathbb{D}$ is discrete/purely atomic.

The Kadison-Singer Problem: Does every pure state on the (Abelian) von Neumann algebra $\mathbb{D}$ of all bounded diagonal operators on l^2 have a unique extension to a pure state on all $\mathscr{B}(l^2)$, the von Neumann algebra of all bounded operators on l^2?[1]

---

[1] Appendix C includes a short biographical sketch of R. Kadison. See also [KS59, KR97a].

We shall begin by explaining the meaning, and the significance, of the terms used in the statement of the Kadison-Singer (abbreviated KS) problem. But on the whole, our discussion of the KS problem (or conjecture) in the present book will be modest in scope. The first to say is that it was just solved by Adam Marcus, Dan Spielman, and N. Srivastava, see [MSS15]. There are several reasons for why we cannot go into proof-details for the solution: While the formulation of KS by Kadison and Singer in 1959 was in the language of operator algebras, as that subject back then was inspired by Dirac's quantum theory, it turned out that the eventual solution to KS, five decades later [MSS15], by Adam Marcus, Dan Spielman, and N. Srivastava, surprisingly, involves themes that draw on new topics, quite outside the scope of the present book. And making the connection between tools from the 2015 solution, back to the original 1959 formulation of KS, is not at all trivial. The new tools employed by Marcus, Spielman, and Srivastava are from diverse mathematical areas, and with a heavy combinatorial component, and also involving mathematical notions which we have not even defined here; for example: interlacing families, random vectors, paving, probabilistic frames, discrepancy analysis, sparsification, ... We hope readers will find it interesting to see how apparently disparate areas can meet at the crossroads in the solution of a famous problem in mathematics[2].

In view of this, we stress that even a modest attempt on our part at going into a detailed discussion of the Marcus-Spielman-Srivastava solution to KS would take us far afield; and done properly it could easily become a separate book volume. Since the Marcus-Spielman-Srivastava paper has just now appeared in the Annals 2015 [MSS15], it may also be too early for a proper book presentation. Nonetheless, we feel that a discussion here, in the present chapter, *of the original formulation* of KS is in fact appropriate. Indeed, the initial motivation for KS derives from precisely the topics which are central themes of our present book: Operator theory/algebra, positivity, states, spectral theory; and with how these mathematical themes intersect with quantum theory.

The authors of [MSS13, MSS15] have proved the Kadison-Singer conjecture in an indirect way, by proving instead Weaver's conjecture [Wea04, Sri13].

Conjecture 9.1 (KS_2). *There exist universal constants $\eta \geq 2$ and $\theta > 0$ so that the following holds. Let $w_1, \ldots, w_m \in \mathbb{C}^d$ satisfy $\|w_i\| \leq 1$ for all i, and suppose*

$$\sum_{i=1}^{m} |\langle w_i, u \rangle|^2 = \eta \tag{9.4}$$

for every unit vector $u \in \mathbb{C}^d$. Then there exists a partition S_1, S_2 of $\{1, \ldots, m\}$

---

[2]The many interconnections between the disparate areas of mathematics coming together in the proof of KS are just emerging in the literature as of this point; in particular, there is a forthcoming paper by P. Casazza, M. Bownik, A. Marcus and D. Speegle; which promises to be an authoritative source. We are grateful to P. Casazza for updates on KS. On the same theme, see also the paper "Consequences of the Marcus/Spielman/Srivastava solution of the Kadison-Singer problem," By Peter G. Casazza and Janet C. Tremain. arXiv:1407.4768v2.

so that

$$\sum_{i \in S_j} |\langle w_i, u \rangle|^2 \leq \eta - \theta$$

for every unit vector $u \in \mathbb{C}^d$ and each $j \in \{1, 2\}$.

Akemann and Anderson's projection paving conjecture [AA91, Conj. 7.1.3] follows directly from KS_2 (see [Wea04, p. 229]) .

Anderson's original paving conjecture says:

Conjecture 9.2 (Anderson Paving). *For every $\varepsilon > 0$, there is an $r \in \mathbb{N}$ such that for every $n \times n$ Hermitian matrix T with zero diagonal, there are diagonal projections $P_1, \cdots, P_r$ with $\sum_{i=1}^{r} P_i = I$ such that*

$$\|P_i T P_i\| \leq \varepsilon \|T\|, \quad \text{for } i = 1, \ldots, r.$$

Since the solution of KS by Nikhil Srivastava, Adam Marcus, and Daniel Spielman (see [MSS13, MSS15, Wea04, Sri13]), a number of equivalent formulations have been found. Now it would take us much too far afield to discuss them here. We shall only mention the following here because of its connection to the study of electric networks (see Chapter 11 below.) we give a very brief sketch, but emphasize that interested readers must be referred to the cited papers for proofs.

In summary, the electrical network problem is as follows: The Kadison-Singer problem is "nearly identical" to the following question about networks: "Given a network G, when is it possible to divide up the edges of G into two subsets — say, red edges and blue edges — such that the associated red and blue networks G_r and G_b have similar electrical properties to those of the whole network G?" It turns out that it is not always possible to do this. For instance, if the original network consists of two highly connected clusters that are linked to each other by a single edge, then that edge has an outsize importance in the network. So if that critical edge is colored red, then the blue network G_b cannot have similar electrical properties to the whole network G.

Now Weaver's problem asks whether, in the general case, this is the only obstacle to subdividing a network into similar but smaller ones (i.e., with the smaller ones having similar electrical properties to the whole): Weaver's question: If there are "enough ways" to get around in a network, can the network then be subdivided into two subnetworks with similar electrical properties? Spielman, Marcus and Srivastava answered this in the affirmative.

The Kadison-Singer problem (KS) lies at the root of how questions from quantum physics take shape in the language of functional analysis, and algebras of operators.

A brief sketch is included below, summarizing some recent advances. It is known that the solution to KS at the same time answers a host of other questions; all with applications to engineering, especially to signal processing. The notion from functional analysis here is "frame" (see [Cas13, Chr96].) A frame of vectors in Hilbert space generalizes the notion of orthonormal basis in Hilbert space.

The Kadison-Singer problem (KS) comes from functional analysis, but it was resolved (only recently) with tools from areas of mathematics quite disparate from functional analysis. More importantly, the solution to KS turned out to have important implications for a host of applied fields from engineering[3], see [CE07].

This reversal of the usual roles seem intriguing for a number of reasons:

While the applications considered so far involve problems which in one way or the other, derive from outside functional analysis itself, e.g., from physics, from signal processing, or from anyone of a number of areas of analysis, PDE, probability, statistics, dynamics, ergodic theory, prediction theory etc.; the Kadison-Singer problem is different. It comes directly from the foundational framework of functional analysis; more specifically from the axiomatic formulation of C^*-algebras. Then C^*-algebras are a byproduct of a rigorous formulation of quantum theory, as proposed by P.A.M. Dirac.[4]

From quantum theory, we have such notions as *state*, *observable*, and *measurement*. See Figure 9.1. But within the framework of C^*-algebras, each of these same terms, "state", "observable", and "measurement" also has a purely mathematical definition, see Section 4.1 in Chapter 4. Indeed C^*-algebra theory was motivated in part by the desire to make precise fundamental and conceptual questions in quantum theory, e.g., the uncertainty principle, measurement, determinacy, hidden variables, to mention a few (see for example [Emc84]). The interplay between the two sides has been extraordinarily fruitful since the birth of quantum mechanics in the 1920ties.

Cited from [KS59]:

> "The main concern of this paper is the problem of uniqueness of extensions of pure states from maximal Abelian self-adjoint algebras of operators on a Hilbert space to the algebra of all bounded operators on that space. The answer, as many of us have suspected for several years, is in negative." ... "We heard of it first from I.E. Segal and I. Kaplansky, though it is difficult to credit a problem which stems naturally from the physical interpretation and the inherent structure of a subject. This problem has arisen, in one form or another, in our work on several different occasions;..."

Now consider the following: (i) the Hilbert space $\mathscr{H} = l^2(= l^2(\mathbb{N}))$, all square summable sequences, (ii) the C^*-algebra $\mathscr{B}(l^2)$ of all bounded operators on l^2, and finally (iii) the sub-algebra $\mathfrak{A}$ of $\mathscr{B}(l^2)$ consisting of all diagonal operators, so an isomorphic copy of l^∞.

---

[3] Atiyah and Singer shared the Abel prize of 2004.

[4] P.A.M. Dirac gave a lecture at Columbia University in the late 1950's, in which he claimed without proof that pure states on the algebra of diagonal operators ($\simeq l^\infty$) extends uniquely on $\mathscr{B}(l^2)$. Kadison and Singer sitting in the audience were skeptical about whether Dirac knew what it meant to be an extension. They later formulated the conjecture in a joint paper, made precise the difference between MASAs that are continuous vs discrete. They showed that non-uniqueness holds in the continuous case.

The Kadison-Singer problem (KS), in the discrete version, is simply this: *Does every pure state of $\mathfrak{A}$ have a unique pure-state extension to $\mathscr{B}(l^2)$?*

We remark that existence (of a pure-state extension) follows from the main theorems from functional analysis of Krein and Krein-Milman, but the uniqueness is difficult. The difficulty lies in the fact that it's hard to find all states on l^∞, i.e., a dual of l^∞. The pure states of $\mathfrak{A}$ are points in the Stone-Čech compactification $\beta(\mathbb{N})$. The problem was settled in the affirmatively (uniqueness in the discrete case) only a year ago, after being open for 50 years.

Lemma 9.1. *Pure normal states on $\mathscr{B}(\mathscr{H})$ are unit vectors (in fact, the equivalent class of unit vectors.[5]) Specifically, let $u \in \mathscr{H}$, $\|u\| = 1$, then*

$$\mathscr{B}(\mathscr{H}) \ni A \longmapsto \omega_u(A) = \langle u, Au \rangle$$

is a pure state. All normal pure states on $\mathscr{B}(\mathscr{H})$ are of this form.

The pure states on $\mathscr{B}(\mathscr{H})$ not of the form ω_u, for $u \in \mathscr{H}$, $\|u\| = 1$, are called *singular pure states*.

Remark 9.1. Since l^∞ is an Abelian algebra Banach $*$-algebra, by Gelfand's theorem, $l^\infty \simeq C(X)$ where X is a compact Hausdorff space. Indeed, $X = \beta\mathbb{N}$, — the Stone-Čech compactification of $\mathbb{N}$. Points in $\beta\mathbb{N}$ are called *ultra-filters*. Pure states on l^∞ correspond to pure states on $C(\beta\mathbb{N})$, i.e., Dirac-point measures on $\beta\mathbb{N}$.

Let s be a pure state on l^∞. Using Hahn-Banach theorem one may extend s, as a linear functional, from l^∞ to $\tilde{s}$ on the Banach space $\mathscr{B}(\mathscr{H})$. However, Hahn-Banach theorem doesn't guarantee the extension remains a *pure* state. Let $E(s)$ be the set of all states on $\mathscr{B}(\mathscr{H})$ which extend s. $E(s)$ is non-empty, compact and convex in the weak $*$-topology. By Krein-Milman's theorem, $E(s) = $ closure(Extreme Points). Any extreme point will then be a pure state extension of s; but which one to choose? It's the uniqueness part that is the famous KS problem.

Exercise 9.1 (Non-normal pure states on $\mathscr{B}(l^2)$). Show that there are pure states on $\mathscr{B}(l^2)$ which do not have the form given in Lemma 9.1.

Hint:

1. The states listed in Lemma 9.1 have cardinality $c = 2^{\aleph_0}$.

2. The pure states of $C(\beta(\mathbb{N}))$ are given by points in $\beta(\mathbb{N})$, and the cardinality of $\beta(\mathbb{N})$ is

$$2^{2^{\aleph_0}} > c. \tag{9.5}$$

3. By Krien-Milman, every pure state on $\mathscr{D}$ $(\simeq l^2(\mathbb{N}))$ has a pure state extension to $\mathscr{B}(l^2)$.

4. Use (9.5) in step 2 to conclude that some of these pure state extensions to $\mathscr{B}(l^2)$ are *not* of the form given in Lemma 9.1.

Physics	Mathematics
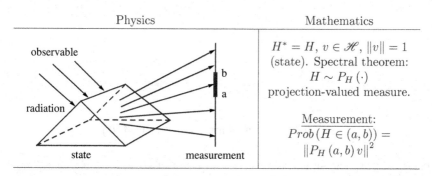	$H^* = H$, $v \in \mathscr{H}$, $\|v\| = 1$ (state). Spectral theorem: $H \sim P_H\left(\cdot\right)$ projection-valued measure. Measurement: $Prob\left(H \in (a,b)\right) = \|P_H\left(a,b\right)v\|^2$

Figure 9.1: Observable, state, measurement. Left column: An idealized physics experiment. Right: the mathematical counterpart, a selfadjoint operator H, its associated projection-valued measure P_H, and a norm-one vector v in Hilbert space.

Exercise 9.2 (The Calkin algebra and Non-normal states on $\mathscr{B}\left(l^2\right)$). Let $\mathscr{K} \subset \mathscr{B}\left(l^2\right)$ be the ideal of all compact operators in l^2; then the quotient

$$\mathscr{C} := \mathscr{B}\left(l^2\right)/\mathscr{K}$$

is called the *Calkin algebra*. Show that the quotient is a C^*-algebra.
Hint: Be careful in defining its C^*-norm.

Exercise 9.3 (The pure state $\varphi = s \circ \pi$ on $\mathscr{B}\left(l^2\right)$). Let $\pi : \mathscr{B}\left(l^2\right) \longrightarrow \mathscr{C}$ be the natural quotient mapping, and let s be a pure state on $\mathscr{C}$. Show that the composition

$$\varphi := s \circ \pi \quad \text{(see Fig 9.2.)}$$

is a pure state on $\mathscr{B}\left(l^2\right)$, and that φ does not have the form in Lemma 9.1.
Hint: Suppose to the contrary, i.e., suppose $\exists x \in l^2$, $\|x\| = 1$ such that

$$\varphi\left(A\right) = \langle x, Ax \rangle = \omega_x\left(A\right), \ \forall A \in \mathscr{B}\left(l^2\right). \quad (9.6)$$

We have $\omega_x\left(|x\rangle\langle x|\right) = 1$, but $|x\rangle\langle x| \in \mathscr{K}$, so $\varphi\left(|x\rangle\langle x|\right) = s\left(0\right) = 0$; a contradiction. Hence (9.6) cannot hold for any state-vector $x \in l^2$.

---

[5]Equivalently, pure states sit inside the projective vector space. If $\mathscr{H} = \mathbb{C}^{n+1}$, pure states is $\mathbb{C}P^n$.

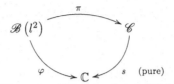

Figure 9.2: The pure state $\varphi = s \circ \pi$ on $\mathscr{B}\left(l^2\right)$

Exercise 9.4 (The Stone-Čech compactification). Extend $+$ on $\mathbb{N}$ to a "$+$" on $\beta\mathbb{N}$ (the Stone-Čech compactification).

Hint:

1. For subsets $A \subset \mathbb{N}$, and $n \in \mathbb{N}$, set $A - n := \left\{k \in \mathbb{N} \mid k + n \in A\right\}$.

2. Let F and G be ultra-filters on $\mathbb{N}$, and set

$$F + G := \{A \subset \mathbb{N} \mid \{n \in \mathbb{N}; A - n \in F\} \in G\}. \tag{9.7}$$

3. Show that $F + G$ is an ultra-filter.

4. Show that the "addition" operation "$+$" in (9.7) is an operation on $\beta\mathbb{N}$, i.e., $\beta\mathbb{N} \times \beta\mathbb{N} \longrightarrow \beta\mathbb{N}$ which is associative, but not commutative, i.e., $F + G \neq G + F$ may happen.

5. Fix $F \in \beta\mathbb{N}$, and show that

$$\beta\mathbb{N} \ni G \longmapsto F + G \in \beta\mathbb{N}$$

is continuous, where $F + G$ is defined in (9.7).

Ultra-filters define pure states of l^∞ as follows: If $(x_n)_{n\in\mathbb{N}} \in l^\infty$, and if $F \in \beta\mathbb{N}$, i.e., is an ultra-filter, then there is a well-defined limit

$$\lim_F x_n = \varphi_F(x);$$

and this defines φ_F as a state on l^∞.

9.2   The Dixmier Trace

A related use of ultra-filters yield the famous *Dixmier-trace*. For this we need ultra-filters ω on $\mathbb{N}$ with the following properties:

(i) $x_n \geq 0 \implies \lim_\omega x_n \geq 0$.

(ii) If x_n is convergent with limit x, then $\lim_\omega x_n = x$.

(iii) For $n \in \mathbb{N}$, set

$$\sigma_N(x) = \Big(\underbrace{x_1 \cdots x_1}_{N \text{ times}}, \underbrace{x_2 \cdots x_2}_{N \text{ times}}, \underbrace{x_3 \cdots x_3}_{N \text{ times}}, \cdots \Big)$$

then $\lim_\omega (x_n) = \lim_\omega (\sigma_N(x))$.

Let A be a compact operator, and assume the eigenvalues λ_k of $|A| = \sqrt{A^*A}$ as

$$\lambda_1 \geq \lambda_2 \geq \cdots, \lambda_k = \lambda_k(A)$$

and set

$$tr_{Dix,\omega}(A) = \lim_\omega \frac{1}{\log(n+1)} \sum_{k=1}^{n} \lambda_k(A). \tag{9.8}$$

Definition 9.1. We say that A has finite Dixmier trace if the limit in (9.8) is finite.

Exercise 9.5 (The Dixmier trace). Show that (9.8) is well-defined and that:

1. $A \longmapsto tr_{Dix,\omega}(A)$ is linear, and positive.

2. $tr_{Dix,\omega}(AB) = tr_{Dix,\omega}(BA)$ holds if B is bounded, and A has finite Dixmier trace.

3. If $\sum_k \lambda_k(A) < \infty$, then $tr_{Dix,\omega}(A) = 0$.

9.3 Frames in Hilbert Space

The proof of the KS-problem involves systems of vectors in Hilbert space called *frames*. For details we refer to [Cas13].

Below we include a sketch with some basic fact about frames; also called "generalized bases", see Definition 9.2 below. The general idea is that a "frame expansion" inherits some (but not all) attractive properties and features of expansions in ONBs. In frame-analysis, this then offers the desirable feature of more flexibility in a host of applications; see e.g., [CFMT11] and [Chr96, HKLW07]. But we also give up something. For example, by contrast to what holds for an ONB, non-uniqueness is a fact of life for frame expansions.

Let $\mathscr{H}$ be a separable Hilbert space, and let $\{u_k\}_{k \in \mathbb{N}}$ be an ONB, then we have the following unique representation

$$w = \sum_{k \in \mathbb{N}} \langle u_k, w \rangle_{\mathscr{H}} u_k \tag{9.9}$$

valid for all $w \in \mathscr{H}$. Moreover,

$$\|w\|_{\mathscr{H}}^2 = \sum_{k \in \mathbb{N}} |\langle u_k, w \rangle_{\mathscr{H}}|^2, \tag{9.10}$$

the Parseval-formula.

Definition 9.2. A system $\{v_k\}_{k \in \mathbb{N}}$ in $\mathscr{H}$ is called a *frame* if there are constants A, B such that $0 < A \leq B < \infty$, and

$$A \left\| w \right\|_{\mathscr{H}}^2 \leq \sum_{k \in \mathbb{N}} \left| \langle v_k, w \rangle_{\mathscr{H}} \right|^2 \leq B \left\| w \right\|_{\mathscr{H}}^2 \tag{9.11}$$

holds for all $w \in \mathscr{H}$.

Note that (9.11) generalizes (9.10). Below we show that, for frames, there is also a natural extension of (9.9).

Exercise 9.6. Let $\mathscr{H}$ be a complex Hilbert space, and for $u, v \in \mathscr{H}$, denote the rank-one operator $|u\rangle\langle v|$, adopting Dirac's notation; see Section 2.5.

Show that the following are equivalent for any system of vectors $\{u_i\}_{i \in I}$ in $\mathscr{H}$:

(i) $\sum_{i \in I} |u_i\rangle\langle v_i| = I_{\mathscr{H}}$,

(ii) $\sum_{i \in I} |\langle u_i, w \rangle|^2 = \|w\|_{\mathscr{H}}^2$, $\forall w \in \mathscr{H}$.

and

(iii) For $w \in \mathscr{H}$, set $Tw = (\langle u_i, w \rangle_{\mathscr{H}})_{i \in I}$; then $T : \mathscr{H} \longrightarrow l^2(I)$ is an isometric isomorphism of $\mathscr{H}$ into $l^2(I)$, i.e., $T^*T = I_{\mathscr{H}}$.

If (i)-(ii) hold, we say that $\{u_i\}_{i \in I}$ is a *Parseval frame* for $\mathscr{H}$; and then the expansion:

(iv) $w = \sum_{i \in I} \langle u_i, w \rangle_{\mathscr{H}} u_i$, $w \in \mathscr{H}$, holds; a reformulation of (iii).

The following is due to Marcus, Spielman, and Strivastava:

Theorem 9.1 ([MSS15]). *Let $\mathscr{H}$ be a d-dimensional real Hilbert space, $d < \infty$, and assume that $\{u_i\}_{i=1}^m$ is a finite Parseval frame. Assume some $\alpha > 0$ satisfies $\|u_i\|_{\mathscr{H}}^2 \leq \alpha$, $\forall i \in \{1, \cdots, m\}$. Then there is a partition B, R of $\{1, \cdots, m\}$, i.e., $\{1, \cdots, m\} = B \cup R$, such that*

$$\max\left\{ \left(\sum_{i \in B} |\langle u_i, w \rangle_{\mathscr{H}}|^2 - \frac{1}{2} \right), \left(\sum_{j \in R} |\langle u_j, w \rangle_{\mathscr{H}}|^2 - \frac{1}{2} \right) \right\} \leq 5\sqrt{\alpha},$$

for all $w \in \mathscr{H}_d$, s.t. $\|w\| = 1$.

Proposition 9.1. *Let $\{v_k\}_{k \in \mathbb{N}}$ be a frame in $\mathscr{H}$; then there is a dual system $\{v_k^*\}_{k \in \mathbb{N}} \subset \mathscr{H}$ such that the following representation holds:*

$$w = \sum_{k \in \mathbb{N}} \langle v_k^*, w \rangle_{\mathscr{H}} v_k \tag{9.12}$$

for all $w \in \mathscr{H}$; absolute convergence.

Proof. Define the following operator $T : \mathscr{H} \to l^2(\mathbb{N})$ by

$$Tw = (\langle v_k, w \rangle_{\mathscr{H}})_{k \in \mathbb{N}},$$

and show that the adjoint $T^* : l^2(\mathbb{N}) \to \mathscr{H}$ satisfies

$$T^*((x_k)) = \sum_{k \in \mathbb{N}} x_k v_k.$$

Hence

$$T^*Tw = \sum_{k \in \mathbb{N}} \langle v_k, w \rangle_{\mathscr{H}} v_k. \tag{9.13}$$

It follows that T^*T has a bounded inverse, in fact, $A\, I_{\mathscr{H}} \leq T^*T \leq B\, I_{\mathscr{H}}$ in the order of selfadjoint operators. As a result, $(T^*T)^{-1}$ and $(T^*T)^{-\frac{1}{2}}$ are well-defined bounded operators.

Substitute $(T^*T)^{-1}$ into (9.13) yields:

$$
\begin{aligned}
w &= \sum_{k \in \mathbb{N}} \left\langle v_k, (T^*T)^{-1} w \right\rangle_{\mathscr{H}} v_k \\
&= \sum_{k \in \mathbb{N}} \left\langle (T^*T)^{-1} v_k, w \right\rangle_{\mathscr{H}} v_k
\end{aligned}
$$

which is the desired (9.12) with $v_k^* := (T^*T)^{-1} v_k$. $\qquad\square$

Exercise 9.7 (Frames from Lax-Milgram). Show that the conclusion in Proposition 9.1 may also be obtained from an application of the Lax-Milgram lemma (see Exercise 2.18.)

Specifically, starting with a frame and given frame constants (see (9.11)), write down the corresponding sesquilinear form B in Lax-Milgram, and verify that it satisfies the premise in Lax-Milgram. Relate the frame bounds to the constants b, and c in Lax-Milgram.

Corollary 9.1. *Let $\{v_k\}$ be as in Proposition 9.1, and set $v_k^{**} := (T^*T)^{-\frac{1}{2}} v_k$, $k \in \mathbb{N}$; then*

$$w = \sum_{k \in \mathbb{N}} \langle v_k^{**}, w \rangle_{\mathscr{H}} v_k^{**}$$

holds for all $w \in \mathscr{H}$; absolute convergence.

Remark 9.2. We saw in Chapter 4 (Section 4.3) that, if $\{v_k\}_{k \in \mathbb{N}}$ is an ONB in some fixed Hilbert space $\mathscr{H}$, then

$$P(\triangle) = \sum_{k \in \triangle} |v_k\rangle\langle v_k|, \quad \triangle \in \mathcal{B}(\mathbb{R}), \tag{9.14}$$

is a projection valued measure (PVM) on $\mathbb{R}$. (See Definition 3.10.)

Suppose now that some $\{v_k\}_{k \in \mathbb{N}}$ in the expression (9.14) is only assumed to be a *frame*, see Definition 9.2.

Exercise 9.8 (Positive operator valued measures from frames). Write down the modified list of properties for $P(\cdot)$ in (9.14) which generalize the axioms of Definition 4.5 for PVMs.

Remark 9.3. It is possible to have *uniqueness* for *non-orthogonal* expansions in Hilbert space. The following theorem of Rota et al. is a case in point.

Theorem 9.2 (Rota et al. [BR60, SN53]). *Let $\mathscr{H}$ be a separable Hilbert space; let $\{e_k\}_{k\in\mathbb{N}}$ be an ONB in $\mathscr{H}$; and let $\{v_k\}_{k\in\mathbb{N}}$ be a linearly independent system of vectors in $\mathscr{H}$ such that*

$$\sum_{k=1}^{\infty} \|e_k - v_k\|^2 < \infty; \tag{9.15}$$

then every vector $u \in \mathscr{H}$ has a unique representation

$$u = \sum_{k=1}^{\infty} x_k v_k, \quad x_k \in \mathbb{C}. \tag{9.16}$$

Moreover defining

$$B\left(\sum_k x_k e_k\right) := \sum_k x_k v_k, \quad (x_k) \in l^2; \tag{9.17}$$

we get the following conclusions:

(i) *$B - I$ is compact; and*

(ii) *$ran(B) = \mathscr{H}$.*

Exercise 9.9 (The operator B). Fill in the missing details in the proof of Theorem 9.2.

The primary source on KS is the paper by R.V. Kadison and I.M. Singer [KS59]. An important early paper is [And79a] by Joel Anderson.

Since the 1970ties, the KS problem has been studied with the use of "pavings;" see e.g., [AAT14, SWZ11, CFMT11, Wea03, BT91]. While this ["pavings" and their equivalents] is an extremely interesting area, it is beyond the scope of the present book.

A summary of relevant numbers from the Reference List

For readers wishing to follow up sources, or to go in more depth with topics above, we suggest:

The pioneering paper [KS59] started the subject, and in the intervening decades there have been advances, and a discovery of the relevance of the KS-problem to a host of applied areas, especially harmonic analysis, frame theory, and signal processing. The problem was solved two years ago.

The most current paper concerning the solution to KS appears to be [MSS15] by Marcus, Spielman, and Strivastava. Paper [Cas14] by P. Casazza explains the problem and its implications. A more comprehensive citation list is: [Arv76, BR81a, Cas13, Cas14, AW14, MSS15, AAT14, And79a, BT91, Chr96, KS59, Wea03, CFMT11, Dix81, BP44, MJD⁺15, Kla15].

Within the limited scope of an Introduction, we have not been able to get in any depth regarding the proof of Kadison-Singer, and a discussion of its many equivalent. But readers will be able to explore equivalent conjectures (i.e., equivalent to KS) and a host of implications, and applications in the references cited here. The paper [CE07] is especially enlightening regarding the list of the known equivalents, e.g., what is called the relative Dixmier Property, the Paving Conjecture, the Feichtinger Conjecture, the Bourgain-Tzafriri Conjecture, and more. Also discussed there is paving-projections and open sets in the Grassmannians. In addition, there are blogs devoted to KS and its solution, and Casazza et al have papers posted on the arXiv.

Part IV

Extension of Operators

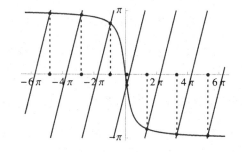

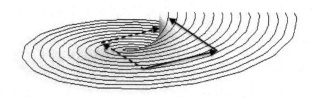

Chapter 10

Selfadjoint Extensions

"Le plus court chemin entre deux vérités dans le domaine réel
passe par le domaine complexe."
— Jacques Hadamard

"It will interest mathematical circles that the mathematical in-
struments created by the higher algebra play an essential part in
the rational formulation of the new quantum mechanics. Thus the
general proofs of the conservation theorems in Heisenberg's theory
carried out by Born and Jordan are based on the use of the theory of
matrices, which go back to Cayley and were developed by Hermite.
It is to be hoped that a new era of mutual stimulation of mechanics
and mathematics has commenced. To the physicist it will seem first
deplorable that in atomic problems we have apparently met with
such a limitation of our usual means of visualisation. This regret
will, however, have to give way to thankfulness that mathematics,
in this field too, presents us with the tools to prepare the way for
further progress."
— Niels Bohr

"Science is spectral analysis. Art is light synthesis."
— Karl Kraus

Because of dictates from applications (especially quantum physics), below we
stress questions directly related to key-issues for unbounded linear operators:
Some operator from physics may only be "formally selfadjoint" also called Hermitian; and in such cases, one ask for selfadjoint extensions (if any).

The axioms of quantum physics (see e.g., [BM13, OH13, KS02, CRKS79,
ARR13, Fan10, Maa10, Par09] for relevant recent papers), are based on Hilbert
space, and selfadjoint operators.

A quantum mechanical observable is a Hermitian (selfadjoint) linear operator mapping a Hilbert space, the space of states, into itself. The values obtained in a physical measurement are in general described by a probability distribution; and the distribution represents a suitable "average" (or "expectation") in a measurement of values of some quantum observable in a state of some prepared system. The states are (up to phase) unit vectors in the Hilbert space, and a measurement corresponds to a probability distribution (derived from a projection-valued spectral measure). The particular probability distribution used depends on both the state and the selfadjoint operator. The associated spectral type may be continuous (such as position and momentum; both unbounded) or discrete (such as spin); this depends on the physical quantity being measured.

Since the spectral theorem serves as the central tool in quantum measurements (see [Sto90, Yos95, Nel69, RS75, DS88b]), we must be precise about the distinction between linear operators with dense domain which are only Hermitian (formally selfadjoint) as opposed to selfadjoint. This distinction is accounted for by von Neumann's theory of deficiency indices [AG93, DS88b, HdSS12][1].

In the details below, the following concepts and terms will play an important role; (for reference, see also Appendix F): Deficiency spaces, and deficiency indices and existence of selfadjoint extensions, boundary triple.

10.1   Extensions of Hermitian Operators

In order to apply spectral theorem, one must work with selfadjoint operators including the unbounded ones. Some examples first.

In quantum mechanics [Pol02, PK88, CP82], to understand energy levels of atoms and radiation, the energy level comes from discrete packages. The interactions are given by Coulomb's Law where

$$H = -\triangle_{\vec{r}} + \frac{c_{jk}}{\|r_j - r_k\|}$$

and Laplacian has dimension $3 \times \#$(electrons).

In Schrödinger's wave mechanics, one needs to solve for $\psi(r,t)$ from the equation

$$H\psi = \frac{1}{i}\frac{\partial}{\partial t}\psi.$$

If we apply spectral theorem, then $\psi(t) = e^{itH}\psi(r,t=0)$. This shows that motion in quantum mechanics is governed by unitary operators. The two parts

---

[1]Starting with [vN32a, vN32c, vN32b], J. von Neumann and M. Stone did pioneering work in the 1930s on spectral theory for unbounded operators in Hilbert space; much of it in private correspondence. The first named author has from conversations with M. Stone, that the notions "deficiency-index," and "deficiency space" are due to them; suggested by MS to vN as means of translating more classical notions of "boundary values" into rigorous tools in abstract Hilbert space: closed subspaces, projections, and dimension count.

in Schrödinger equation are separately selfadjoint, but justification of the sum being selfadjoint wasn't made rigorous until 1957 when Kato wrote the book on "perturbation theory" [Kat95]. It is a study of the sum of systems of selfadjoint operators.

In Heisenberg's matrix mechanics, he suggested that one should look at two states and the transition probability between them, such that

$$\langle \psi_1, A\psi_2 \rangle = \langle \psi_1(t), A\psi_2(t) \rangle, \ \forall t.$$

If $\psi(t) = e^{itH}\psi$, then it works. In *Heisenberg's picture*, one looks at evolution of the observables $e^{-itH}Ae^{itH}$. In *Schrödinger's picture*, one looks at evolution of states. The two point of views are equivalent.

Everything so far is based on application of the spectral theorem, which requires the operators being selfadjoint in the first place.

von Neumann's index theory gives a complete classification of extensions of single Hermitian unbounded operators with dense domain in a given Hilbert space. The theory may be adapted to Hermitian representations of $*$-algebras [Nel59a].

Let A be a densely defined Hermitian operator on a Hilbert space $\mathscr{H}$, i.e. $A \subset A^*$. If B is any Hermitian extension of A, then

$$A \subset B \subset B^* \subset A^*. \tag{10.1}$$

Since the adjoint operator A^* is closed, i.e., $\mathscr{G}(A^*)$ is closed in $\mathscr{H} \oplus \mathscr{H}$, it follows that $\overline{\mathscr{G}(A)} \subset \mathscr{G}(A^*)$ is a well-defined operator graph, i.e., A is closable and $\overline{\mathscr{G}(A)} = \mathscr{G}(\overline{A})$. Thus, there is no loss of generality to assume that A is closed and only consider its closed extensions.

The containment (10.1) suggests a detailed analysis in $\mathscr{D}(A^*) \setminus \mathscr{D}(A)$. Since $\mathscr{D}(A)$ is dense in $\mathscr{H}$, the usual structural analysis in $\mathscr{H}$ (orthogonal decomposition, etc.) is not applicable. However, this structure is brought out naturally when $\mathscr{D}(A^*)$ is identified with the operator graph $\mathscr{G}(A^*)$ in $\mathscr{H} \oplus \mathscr{H}$. That is, $\mathscr{D}(A^*)$ is a Hilbert space under its graph norm. With this identification, $\mathscr{D}(A)$ becomes a closed subspace in $\mathscr{D}(A^*)$, and

$$\mathscr{D}(A^*) = \mathscr{D}(A) \oplus (\mathscr{D}(A^*) \ominus \mathscr{D}(A)). \tag{10.2}$$

The question of extending A amounts to a further decomposition

$$\mathscr{D}(A^*) \ominus \mathscr{D}(A) = S \oplus K \tag{10.3}$$

in such a way that

$$\widetilde{A} = A^*|_{\mathscr{D}(\widetilde{A})}, \text{ where} \tag{10.4}$$

$$\mathscr{D}(\widetilde{A}) = \mathscr{D}(A) \oplus S \tag{10.5}$$

defines a (closed) Hermitian operator $\widetilde{A} \supset A$.

The extension $\widetilde{A}$ in (10.4)-(10.5) is Hermitian iff the closed subspace $S \subset \mathscr{D}(A^*)$ is symmetric, in the sense that

$$\langle A^*y, x \rangle - \langle y, A^*x \rangle = 0, \ \forall x, y \in S. \tag{10.6}$$

Lemma 10.1. *Let S be a closed subspace in $\mathscr{D}(A^*)$, where $\mathscr{D}(A^*)$ is a Hilbert space under the A^*-norm. The following are equivalent.*

1. $\langle A^*y, x \rangle = \langle y, A^*x \rangle$, *for all* $x, y \in S$.

2. $\langle x, A^*x \rangle \in \mathbb{R}$, *for all* $x \in S$.

Proof. If (1) holds, setting $x = y$, we get $\langle x, A^*x \rangle = \langle A^*x, x \rangle = \overline{\langle x, A^*x \rangle}$, which implies that $\langle x, A^*x \rangle$ is real-valued.

Conversely, assume (2) is true. Since the mappings

$$(x, y) \;\mapsto\; \langle y, A^*x \rangle$$
$$(x, y) \;\mapsto\; \langle A^*y, x \rangle$$

are both sesquilinear forms on $S \times S$ (linear in the second variable, and conjugate linear in the first variable), we apply the polarization identity:

$$\langle y, A^*x \rangle \;=\; \frac{1}{4} \sum_{k=0}^{3} i^k \left\langle x + i^k y, A^* \left(x + i^k y \right) \right\rangle$$

$$\langle A^*y, x \rangle \;=\; \frac{1}{4} \sum_{k=0}^{3} i^k \left\langle A^* \left(x + i^k y \right), x + i^k y \right\rangle$$

for all $x, y \in \mathscr{D}(A^*)$. Now, since A is Hermitian, the RHSs of the above equations are equal; therefore, $\langle y, A^*x \rangle = \langle A^*y, x \rangle$, which is part (2). $\square$

Equations (10.4)-(10.5) and Lemma 10.1 set up a bijection between (closed) Hermitian extensions of A and (closed) symmetric subspaces in $\mathscr{D}(A^*) \ominus \mathscr{D}(A)$. Moreover, by Lemma 10.1, condition (10.6) is equivalent to

$$\langle x, A^*x \rangle \in \mathbb{R}, \ \forall x \in \mathscr{D}(A). \tag{10.7}$$

Let $\varphi \in \mathscr{D}(A^*)$, such that $A^*\varphi = \lambda\varphi$, $\Im\{\lambda\} \neq 0$; then $\langle \varphi, A^*\varphi \rangle = \lambda \|\varphi\|^2 \notin \mathbb{R}$. By Lemma 10.1 and (10.7), $\varphi \notin \mathscr{D}(\widetilde{A})$, where $\widetilde{A}$ is any possible Hermitian extension of A. This observation is in fact ruling out the "wrong" eigenvalues of $\widetilde{A}$. Indeed, Theorem 10.1 below shows that A is selfadjoint if and only if ALL the "wrong" eigenvalues of A^* are excluded. But first we need the following lemma.

Lemma 10.2. *Let A be a Hermitian operator in $\mathscr{H}$, then*

$$\|(A - \lambda)x\|^2 = \|(A - a)x\|^2 + |b|^2 \|x\|^2, \ \forall \lambda = a + ib \in \mathbb{C}. \tag{10.8}$$

In particular,

$$\|(A - \lambda)x\|^2 \geq |\Im\{\lambda\}|^2 \|x\|^2, \ \forall \lambda \in \mathbb{C}. \tag{10.9}$$

Proof. Write $\lambda = a + ib$, $a, b \in \mathbb{R}$; then

$$
\begin{aligned}
&\|(A - \lambda)x\|^2 \\
=\; & \langle (A - a)x - ibx, (A - a)x - ibx \rangle \\
=\; & \|(A - a)x\|^2 + |b|^2 \|x\|^2 - i(\langle(A - a)x, x\rangle - \langle x, (A - a)x\rangle) \\
=\; & \|(A - a)x\|^2 + |b|^2 \|x\|^2 \\
\geq\; & |b|^2 \|x\|^2 ;
\end{aligned}
$$

where $\langle (A - a)x, x \rangle - \langle x, (A - a)x \rangle = 0$, since $A - a$ is Hermitian. $\square$

Corollary 10.1. *Let A be a closed Hermitian operator acting in $\mathscr{H}$. Fix $\lambda \in \mathbb{C}$ with $\Im\{\lambda\} \neq 0$, then $ran(A - \lambda)$ is a closed subspace in $\mathscr{H}$. Consequently, we get the following decomposition*

$$\mathscr{H} = ran(A - \lambda) \oplus ker\left(A^* - \overline{\lambda}\right). \tag{10.10}$$

Proof. Set $B = A - \lambda$; then B is closed, and so is B^{-1}, i.e., the operator graphs $\mathscr{G}(B)$ and $\mathscr{G}\left(B^{-1}\right)$ are closed in $\mathscr{H} \oplus \mathscr{H}$. Therefore, $ran(B) \left(= dom(B^{-1})\right)$ is closed in $\|\cdot\|_{B^{-1}}$-norm. But by (10.9), B^{-1} is bounded on $ran(B)$, thus the two norms $\|\cdot\|$ and $\|\cdot\|_{B^{-1}}$ are equivalent on $ran(B)$. It follows that $ran(B)$ is also closed in $\|\cdot\|$-norm, i.e., it is a closed subspace in $\mathscr{H}$. The decomposition (10.10) follows from this. $\square$

Theorem 10.1. *Let A be a densely defined, closed, Hermitian operator in a Hilbert space $\mathscr{H}$; then the following are equivalent:*

$$\left\{\exists \lambda,\; \Im\{\lambda\} \neq 0,\; ker(A^* - \lambda) = ker\left(A^* - \overline{\lambda}\right) = 0\right\} \Longleftrightarrow \left\{A = A^*\right\}.$$

Proof. $\Longrightarrow$ By Corollary 10.1, the hypothesis in the theorem implies that

$$ran(A - \lambda) = ran\left(A - \overline{\lambda}\right) = \mathscr{H}.$$

Let $y \in \mathscr{D}(A^*)$, then

$$\langle y, (A - \lambda)x \rangle = \left\langle \left(A^* - \overline{\lambda}\right)y, x \right\rangle, \; \forall x \in \mathscr{D}(A). \tag{10.11}$$

Since $ran(A - \overline{z}) = \mathscr{H}$, $\exists y_0 \in \mathscr{D}(A)$ such that

$$\left(A^* - \overline{\lambda}\right)y = \left(A - \overline{\lambda}\right)y_0.$$

Hence, RHS of (10.11) is

$$\left\langle \left(A - \overline{\lambda}\right)y_0, x \right\rangle = \langle y_0, (A - \lambda)x \rangle. \tag{10.12}$$

Combining (10.11)-(10.12), we then get

$$\langle y - y_0, (A - \lambda)x \rangle = 0, \; \forall x \in \mathscr{D}(A).$$

Again, since $ran\,(A - \lambda) = \mathscr{H}$, the last equation above shows that $y - y_0 \perp \mathscr{H}$. In particular, $y - y_0 \perp y - y_0$, i.e.,

$$\|y - y_0\|^2 = \langle y - y_0, y - y_0 \rangle = 0.$$

Therefore, $y = y_0$, and so $y \in \mathscr{D}(A)$. This shows that $A^* \subset A$.

The other containment $A \subset A^*$ holds since A is assumed to be Hermitian. Thus, we conclude that $A = A^*$. $\qquad\qquad\square$

To capture all the "wrong" eigenvalues, we consider a family of closed subspace in $\mathscr{H}$, $ker\,(A^* - \lambda)$, where $\Im\,\{\lambda\} \neq 0$.

Theorem 10.2. *If A is a closed Hermitian operator in $\mathscr{H}$, then*

$$dim\,(ker\,(A^* - \lambda))$$

is a constant function on $\Im\,\{\lambda\} > 0$, and $\Im\,\{\lambda\} < 0$.

Proof. Fix λ with $\Im\,\{\lambda\} > 0$. For $\Im\,\{\lambda\} < 0$, the argument is similar. We proceed to verify that if $\eta \in \mathbb{C}$, close enough to λ, then $dim\,(ker\,(A^* - \eta)) = dim\,(ker\,(A^* - \lambda))$. The desired result then follows immediately.

Since A is closed, we have the following decomposition (by Corollary 10.1),

$$\mathscr{H} = ran\,(A - \overline{\lambda}) \oplus ker\,(A^* - \lambda). \tag{10.13}$$

Now, pick $x \in ker\,(A^* - \eta)$, and suppose $x \perp ker\,(A^* - \lambda)$; assuming $\|x\| = 1$. By (10.13), $\exists x_0 \in \mathscr{D}(A)$ such that

$$x = (A - \overline{\lambda})\,x_0. \tag{10.14}$$

Then,

$$
\begin{aligned}
0 = \langle (A^* - \eta)\,x, x_0 \rangle &= \langle x, (A - \overline{\eta})\,x_0 \rangle \\
&= \langle x, (A - \overline{\lambda})\,x_0 - (\overline{\eta} - \overline{\lambda})\,x_0 \rangle \\
&= \|x\|^2 - (\overline{\eta} - \overline{\lambda})\,\langle x, x_0 \rangle \\
&\geq \|x\|^2 - |\overline{\eta} - \overline{\lambda}|\,\|x\|^2 \|x_0\|^2 \quad \text{(Cauchy-Schwarz)} \\
&= 1 - |\eta - \lambda|\,\|x_0\|^2. \tag{10.15}
\end{aligned}
$$

Applying Lemma 10.2 to (10.14), we also have

$$1 = \|x\|^2 = \left\| (A - \overline{\lambda})\,x_0 \right\|^2 \geq |\Im\,\{\lambda\}|^2\,\|x_0\|^2\,;$$

substitute this into (10.15), we see that

$$0 \geq 1 - |\eta - \lambda|\,\|x_0\|^2 \geq 1 - |\eta - \lambda|\,|\Im\,\{\lambda\}|^{-2}$$

which would be a contradiction if η was close to λ.

$$\mathscr{H} \left\{ \begin{array}{c} \mathscr{D}_+ \\ \oplus \\ (A+i)\,\mathscr{D} \end{array} \quad \xrightarrow{\text{Partial Isometry}} \quad \begin{array}{c} \mathscr{D}_- \\ \oplus \\ (A-i)\,\mathscr{D} \end{array} \right\} \mathscr{H}$$
$$\xrightarrow{C_A=(A-i)(A+i)^{-1}}$$

Figure 10.1: $\mathscr{D}_\pm = \mathrm{Ker}\,(A^* \mp i)$, $\mathscr{D} = dom\,(A)$

It follows that the projection from $ker\,(A^* - \eta)$ to $ker\,(A^* - \lambda)$ is injective. For otherwise, $\exists x \in ker\,(A^* - \eta)$, $x \neq 0$, and $x \perp ker\,(A^* - \lambda)$. This is impossible as shown above. Thus,

$$dim\,(ker\,(A^* - \eta)) \leq dim\,(ker\,(A^* - \lambda)).$$

Similarly, we get the reversed inequality, and so

$$dim\,(ker\,(A^* - \eta)) = dim\,(ker\,(A^* - \lambda)).$$

$\square$

A complete characterization of Hermitian extensions of a given Hermitian operator is due to von Neumann. Theorem 10.2 suggests the following definition:

Definition 10.1. Let A be a densely defined, closed, Hermitian operator in $\mathscr{H}$. The closed subspaces

$$\begin{aligned} \mathscr{D}_\pm\,(A) &= ker\,(A^* \mp i) &(10.16)\\ &= \{\xi \in \mathscr{D}\,(A^*) : A^*\xi = \pm i\,\xi\} \end{aligned}$$

are called the *deficiency spaces* of A, and $dim\mathscr{D}_\pm\,(A)$ are called the *deficiency indices*.

For illustration, see Figure 10.1.

The role of the *Cayley-transform* $C_A := (A - i)\,(A + i)^{-1}$, and its extensions by partial isometries $\mathscr{D}_+ \longrightarrow \mathscr{D}_-$, is illustrated in Figure 10.1. The Figure further offers a geometric account of the conclusion in Theorem 10.3.

As a result we see that the two subspaces $\mathscr{D}_\pm$, also called *defect-spaces* (or *deficiency-spaces*), are non-zero precisely when the given symmetric operator A fails to be essentially selfadjoint. The respective dimensions

$$n_\pm := \dim \mathscr{D}_\pm \qquad (10.17)$$

are called *deficiency indices*. The pair (n_+, n_-) in (10.17) is called the pair of *von Neumann indices*. We note that A has selfadjoint extensions if and only if $n_+ = n_-$.

Theorem 10.3 (von Neumann). *Let A be a densely defined closed Hermitian operator acting in $\mathscr{H}$. Then*

$$\mathscr{D}\,(A^*) = \mathscr{D}\,(A) \oplus \mathscr{D}_+\,(A) \oplus \mathscr{D}_-\,(A); \qquad (10.18)$$

where $\mathscr{D}(A^*)$ is identified with its graph $\mathscr{G}(A^*)$, thus a Hilbert space under the graph inner product; and the decomposition in (10.18) refers to this Hilbert space.

Proof. By assumption, A is closed, i.e., $\mathscr{D}(A)$, identified with $\mathscr{G}(A)$, is a closed subspace in $\mathscr{D}(A^*)$.

Note that $\mathscr{D}_{\pm}(A) = ker\,(A^* \mp i)$ are closed subspaces in $\mathscr{H}$. Moreover,

$$\|x\|_{A^*}^2 = \|x\|^2 + \|A^*x\|^2 = 2\,\|x\|^2\,,\ \forall x \in \mathscr{D}_{\pm}(A)\,;$$

and so $\mathscr{D}_{\pm}(A)$, when identified with the graph of $A^*\big|_{\mathscr{D}_{\pm}(A^*)}$, are also closed subspaces in $\mathscr{D}(A^*)$.

Next, we verify the three subspaces on RHS of (10.18) are mutually orthogonal. For all $x \in \mathscr{D}(A)$, and all $x_+ \in \mathscr{D}_+(A) = ker\,(A^* - i)$, we have

$$
\begin{aligned}
\langle x_+, x\rangle_{A^*} &= \langle x_+, x\rangle + \langle A^*x_+, A^*x\rangle \\
&= \langle x_+, x\rangle - i\,\langle x_+, Ax\rangle \\
&= -i\,(\langle x_+, ix\rangle + \langle x_+, Ax\rangle) \\
&= -i\,\langle x_+, (A+i)\,x\rangle = 0
\end{aligned}
$$

where the last step follows from $x_+ \perp ran\,(A+i)$ in $\mathscr{H}$, see (10.10). Thus, $\mathscr{D}(A) \perp \mathscr{D}_+(A)$ in $\mathscr{D}(A^*)$. Similarly, $\mathscr{D}(A) \perp \mathscr{D}_-(A)$ in $\mathscr{D}(A^*)$.

Moreover, if $x_+ \in \mathscr{D}_+(A)$ and $x_- \in \mathscr{D}_-(A)$, then

$$
\begin{aligned}
\langle x_+, x_-\rangle_{A^*} &= \langle x_+, x_-\rangle + \langle A^*x_+, A^*x_-\rangle \\
&= \langle x_+, x_-\rangle + \langle i\,x_+, -i\,x_-\rangle \\
&= \langle x_+, x_-\rangle - \langle x_+, x_-\rangle = 0.
\end{aligned}
$$

Hence $\mathscr{D}_+(A) \perp \mathscr{D}_-(A)$ in $\mathscr{D}(A^*)$.

Finally, we show RHS of (10.18) yields the entire Hilbert space $\mathscr{D}(A^*)$. For this, let $x \in \mathscr{D}(A^*)$, and suppose (10.18) holds, say, $x = x_0 + x_+ + x_-$, where $x \in \mathscr{D}(A)$, $x_{\pm} \in \mathscr{D}_{\pm}(A)$; then

$$
\begin{aligned}
(A^* + i)\,x &= (A^* + i)\,(x_0 + x_+ + x_-) \\
&= (A + i)\,x_0 + 2i\,x_+.
\end{aligned}
\tag{10.19}
$$

But, by the decomposition $\mathscr{H} = ran\,(A+i) \oplus ker\,(A^* - i)$, eq. (10.10), there exist x_0 and x_+ satisfying (10.19). It remains to set $x_- := x - x_0 - x_+$, and to check $x_- \in \mathscr{D}_-(A)$. Indeed, by (10.19), we see that

$$A^*x - Ax_0 - i\,x_+ = -i\,x + i\,x_0 + i\,x_+;\ \text{i.e.,}$$

$$A^*\,(x - x_0 - x_+) = -i\,(x - x_0 - x_+)$$

and so $x_- \in \mathscr{D}_-(A)$. Therefore, we get the desired orthogonal decomposition in (10.18).

Another argument: Let $y \in \mathscr{D}(A^*)$ such that $y \perp \mathscr{D}_{\pm}(A)$ in $\mathscr{D}(A^*)$. Then, $y \perp \mathscr{D}_{+}(A)$ in $\mathscr{D}(A^*) \Longrightarrow$

$$
\begin{aligned}
0 &= \langle y, x_+ \rangle + \langle A^* y, A^* x_+ \rangle \\
&= \langle y, x_+ \rangle + \langle A^* y, i\, x_+ \rangle \\
&= i \left(\langle i\, y, x_+ \rangle + \langle A^* y, x_+ \rangle \right) \\
&= i \langle (A^* + i)\, y, x_+ \rangle, \; \forall x_+ \in \mathscr{D}_{+}(A) = ker\,(A^* - i)
\end{aligned}
$$

and so $\exists\, x_1 \in \mathscr{D}(A)$, and

$$
(A^* + i)\, y = (A + i)\, x_1. \tag{10.20}
$$

On the other hand, $y \perp \mathscr{D}_{-}(A)$ in $\mathscr{D}(A^*) \Longrightarrow$

$$
\begin{aligned}
0 &= \langle y, x_- \rangle + \langle A^* y, A^* x_- \rangle \\
&= \langle y, x_- \rangle + \langle A^* y, -i\, x_- \rangle \\
&= -i \left(\langle -i\, y, x_- \rangle + \langle A^* y, x_- \rangle \right) \\
&= i \langle (A^* - i)\, y, x_- \rangle, \; \forall x_- \in \mathscr{D}_{-}(A) = ker\,(A^* + i);
\end{aligned}
$$

hence $\exists\, x_2 \in \mathscr{D}(A)$, and

$$
(A^* - i)\, y = (A - i)\, x_2. \tag{10.21}
$$

Subtracting (10.20)-(10.21) then gives

$$
y = \frac{x_1 + x_2}{2} \in \mathscr{D}(A).
$$

$\square$

Remark 10.1. More generally, there is a family of decompositions

$$
\mathscr{D}(A^*) = \mathscr{D}(A) + ker\,(A^* - z) + ker\,(A^* - \bar{z}), \; \forall z \in \mathbb{C}, \Im\,\{z\} \neq 0. \tag{10.22}
$$

However, in the general case, we lose orthogonality.

Proof. Give $z \in \mathbb{C}$, $\Im\,\{z\} \neq 0$, suppose $x \in \mathscr{D}(A^*)$ can be written as

$$
x = x_0 + x_+ + x_-;
$$

where $x_0 \in \mathscr{D}(A)$, $x_+ \in ker\,(A^* - z)$, and $x_- \in ker\,(A^* - \bar{z})$. Then

$$
\begin{aligned}
A^* x &= A x_0 + z x_+ + \bar{z} x_- \\
\bar{z} x &= \bar{z} x_0 + \bar{z} x_+ + \bar{z} x_-
\end{aligned}
$$

and

$$
(A^* - \bar{z})\, x = (A - \bar{z})\, x_0 + (z - \bar{z})\, x_+. \tag{10.23}
$$

Now, we start with (10.23). By the decomposition

$$\mathscr{H} = ran\,(A - \overline{z}) \oplus ker\,(A^* - z),$$

there exist unique x_0 and x_+ such that (10.23) holds. This defines x_0 and x_+. Then, set

$$x_- := x - x_0 - x_+;$$

and it remains to check $x_- \in ker\,(A^* - \overline{z})$. Indeed, by (10.23), we have

$$A^*x - Ax_0 - zx_+ = \overline{z}x - \overline{z}x_0 - \overline{z}x_+, \text{ i.e.,}$$

$$A^*\,(x - x_0 - x_+) = \overline{z}\,(x - x_0 - x_+)$$

thus, $x_- \in ker\,(A^* - \overline{z})$. $\square$

Remark 10.2. In the general decomposition (10.22), if $f = x + x_+ + x_-$, $g = y + y_+ + y_-$ where $f, g \in \mathscr{D}\,(A)$, $x_+, y_+ \in ker\,(A^* - z)$, and $x_-, y_- \in ker\,(A^* - \overline{z})$; then

$$
\begin{aligned}
& \langle g, A^* f \rangle - \langle A^* g, f \rangle \\
= & \ \langle y + y_+ + y_-, A^*\,(x + x_+ + x_-) \rangle - \langle A^*\,(y + y_+ + y_-), x + x_+ + x_- \rangle \\
= & \ \langle y + y_+ + y_-, Ax + zx_+ + \overline{z}x_- \rangle - \langle Ay + zy_+ + \overline{z}y_-, x + x_+ + x_- \rangle \\
= & \ \underbrace{\langle y, Ax + zx_+ + \overline{z}x_- \rangle - \langle Ay, x + x_+ + x_- \rangle}_{0} + \\
& \ \underbrace{\langle y_+ + y_-, Ax \rangle - \langle zy_+ + \overline{z}y_-, x \rangle}_{0} + \\
& \ \langle y_+ + y_-, zx_+ + \overline{z}x_- \rangle - \langle zy_+ + \overline{z}y_-, x_+ + x_- \rangle \\
= & \ \langle y_+, zx_+ \rangle - \langle zy_+, x_+ \rangle + \langle y_-, \overline{z}x_- \rangle - \langle \overline{z}y_-, x_- \rangle \\
& \ + \langle y_+, \overline{z}x_- \rangle + \langle y_-, zx_+ \rangle - \langle zy_+, x_- \rangle - \langle \overline{z}y_-, x_+ \rangle \\
= & \ (z - \overline{z})\langle y_+, x_+ \rangle + (\overline{z} - z)\langle y_-, x_- \rangle + \\
& \ \underbrace{\overline{z}\langle y_+, x_- \rangle + z\langle y_-, x_+ \rangle - \overline{z}\langle y_+, x_- \rangle - z\langle y_-, x_+ \rangle}_{0} \\
= & \ (z - \overline{z})\,(\langle y_+, x_+ \rangle - \langle y_-, x_- \rangle).
\end{aligned}
$$

The RHS in the formula is called a "boundary term." The reason for this is that it specializes to the more familiar notion of "boundary term" which arises in integration-by-part-formulas; or, more generally, in Gauss-Green-Stokes formulas.

Theorem 10.4 (von Neumann). *Let A be a densely defined closed Hermitian operator in $\mathscr{H}$.*

1. *The (closed) Hermitian extensions of A are indexed by partial isometries with initial space in $\mathscr{D}_+\,(A)$ and final space in $\mathscr{D}_-\,(A)$.*

2. Given a partial isometry U as above, the Hermitian extension $\widetilde{A_U} \supset A$ is determined as follows:

$$\widetilde{A_U}\left(x + (1 + U)x_+\right) = Ax + i\left(1 - U\right)x_+, \text{ where}$$

$$\mathscr{D}\left(\widetilde{A_U}\right) = \{x + x_+ + Ux_+ : x \in \mathscr{D}(A), x_+ \in \mathscr{D}_+(A)\}$$

(10.24)

Proof. By the discussion in (10.6) and (10.7), and Lemma 10.1, it remains to characterize the closed symmetric subspaces S in $\mathscr{D}_+(A) \oplus \mathscr{D}_-(A)$ ($\subset \mathscr{D}(A^*)$). For this, let $x = x_+ + x_-$, $x_\pm \in \mathscr{D}_\pm(A)$, then

$$
\begin{aligned}
\langle x, A^*x \rangle &= \langle x_+ + x_-, A(x_+ + x_-) \rangle \\
&= \langle x_+ + x_-, i(x_+ - x_-) \rangle \\
&= i\left(\|x_+\|^2 - \|x_-\|^2 - 2i\Im\{\langle x_+, x_- \rangle\}\right) \\
&= i\left(\|x_+\|^2 - \|x_-\|^2\right) + 2\Im\{\langle x_+, x_- \rangle\}.
\end{aligned}
$$

(10.25)

Thus,

$$\langle x, A^*x \rangle \in \mathbb{R}, \ \forall x \in S$$

$$\Updownarrow$$

$$S = \{(x_+, x_-) : \|x_+\| = \|x_-\|, \ x_\pm \in \mathscr{D}_\pm(A)\}$$

$$\Updownarrow$$

[the vanishing of the boundary term at x]

i.e., S is identified with the graph of a partial isometry, say U, with initial space in $\mathscr{D}_+(A)$ and final space in $\mathscr{D}_-(A)$. □

Corollary 10.2. *Let A be a densely defined, closed, Hermitian operator on $\mathscr{H}$, and set $d_\pm = \dim(\mathscr{D}_\pm(A))$; then*

1. *A is maximally Hermitian if and only if one of the deficiency indices is 0;*

2. *A has a selfadjoint extension if and only if $d_+ = d_- \neq 0$;*

3. *$\overline{A}$ is selfadjoint if and only if $d_+ = d_- = 0$.*

Proof. Immediate from Theorem 10.3 and Theorem 10.4. □

Semigroup of operators

Let $\mathscr{H}$ be a Hilbert space, and suppose $S_t : \mathscr{H} \longrightarrow \mathscr{H}$, $t \in [0, \infty)$ is a family of bounded operators such that

1. $S_{t_1}S_{t_2} = S_{t_1+t_2}$, $\forall t_1, t_2 \in [0, \infty)$,

2. $S_0 = id$,

3. $\|S_t x\| \le \|x\|$, $\forall x \in \mathscr{H}$, $\forall t \in \mathbb{R}_+$.

4. For $\forall x \in \mathscr{H}$, $t \longrightarrow S_t x \in \mathscr{H}$, is continuous.

We then say that $\{S_t\}$ is a *contraction semigroup* in $\mathscr{H}$. Examples of semigroups include the Ornstein-Uhlenbeck semigroup discussed in Sections 5.4.8 (pg. 179), and 7.2 (pg. 259).

Exercise 10.1 (Dissipative generators). Let $\{S_t\}_{t \in \mathbb{R}_+}$ be a contraction semigroup in $\mathscr{H}$, and let A be its infinitesimal generator, i.e., with dense domain

$$ dom\,(A) := \left\{ x \in \mathscr{H} \;;\; \frac{S_t x - x}{t} \xrightarrow[t \longrightarrow \infty]{} Ax;\ \text{the limit exists} \right\} \qquad (10.26) $$

We shall write $S_t = e^{tA}$, $t \in \mathbb{R}_+$. Prove that A satisfies:

$$ \Re\{\langle x, Ax\rangle\} \le 0, \quad \forall x \in dom\,(A), \qquad (10.27) $$

i.e., $\langle x, Ax\rangle + \langle Ax, x\rangle \le 0$, $\forall x \in dom\,(A)$. (We say that A is *dissipative*.)

Theorem 10.5 (Phillips). *The generators of contraction semigroups in Hilbert space are precisely the maximal dissipative operators.*

Proof. See e.g., [Lax02, LP89, RS75, Yos95].   □

Exercise 10.2 (The index values $(d_+, 0)$, and $(0, d_-)$). Let $\mathscr{H}$ be a Hermitian symmetric operator with deficiency indices (d_+, d_-), and set $A = iH$, $i = \sqrt{-1}$.

1. Show that $S(t) = e^{tA}$ is a semigroup of isometries iff H has index-values of the form $(d_+, 0)$.

2. Show that $S(t) = e^{tA}$ is a semigroup of co-isometries iff H has index-values of the form $(0, d_-)$.

3. Prove that if $\{S(t)\}$ is as in part 1, then $t \longrightarrow S(t)^*$, $t > 0$, is as in part 2.

Example 10.1. Let $\mathscr{H} = L^2(0, \infty)$, and for $t \in \mathbb{R}_+$, set

$$ (S(t)f)(x) = \begin{cases} f(x-t) & \text{if } x \ge t, \\ 0 & \text{if } x \in [0,t). \end{cases} \qquad (10.28) $$

Then $S(t)$ is a semigroup of isometries, see Figure 10.2a.
    The corresponding semigroup of co-isometries is as follows:

$$ (S(t)^* f)(x) = f(x+t), \quad \forall x, t \in [0, \infty); \qquad (10.29) $$

see Figure 10.2b, i.e., translation and truncation.

The respective infinitesimal generators (A and A^*) of the two semigroups $\{S(t)\}_{t \in \mathbb{R}_+}$ and $\{S(t)^*\}_{t \in \mathbb{R}_+}$ from (10.28) and (10.29) are as follows: The operators are $f \longrightarrow f'$ in both cases, but with domains as follows:

$$dom(A) = \left\{ f \in L^2(0, \infty) \ ; \ f' \in L^2(0, \infty), \ f(0) = 0 \right\};$$

and

$$dom(A^*) = \left\{ f \in L^2(0, \infty) \ ; \ f' \in L^2(0, \infty) \right\}.$$

Note $A \subset A^*$ is a *proper* containment.

Analytic Vectors. A theorem of E. Nelson [Nel59a] states that a Hermitian symmetric operator H is essentially selfadjoint iff it has a dense space of analytic vectors. (A vector v is analytic for H if $v \in dom(H^n)$ for all $n \in \mathbb{N}$, and $\exists$ constants $C_0, C_1 < \infty$ such that $\|H^n v\| \leq C_0 C_1^n n!$, $\forall n \in \mathbb{N}$. The constants C_i, $i = 0, 1$, depend on the vector v.)

Exercise 10.3 (Absence of analytic vectors). Show that the operator $A = iH$ in Example 10.1 (see Figure 10.2a) does not have any non-zero analytic vectors.
   Hint: Recall the Hilbert space in this example is $\mathscr{H} = L^2(0, \infty)$.

Example 10.2. $d_+ = d_- = 1$. General: Let $e_\pm$ be corresponding pair of normalized deficiency vectors. Then $e_+ \mapsto z e_-$ is the unitary operator sending one to the other eigenvalue. It is clear that $|z| = 1$. Hence the selfadjoint extension is indexed by $U_1(\mathbb{C})$.

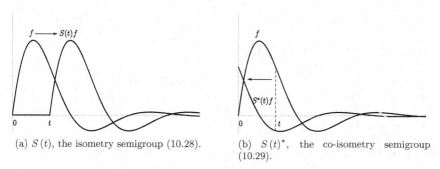

(a) $S(t)$, the isometry semigroup (10.28).   (b) $S(t)^*$, the co-isometry semigroup (10.29).

Figure 10.2: A semigroup of isometries in $L^2(0, \infty)$.

Example 10.3. $d_+ = d_- = 2$, get a family of extensions indexed by $U_2(\mathbb{C})$.

There is a simple criterion to test whether a Hermitian operator has equal deficiency indices.

Definition 10.2. An operator $J : \mathscr{H} \to \mathscr{H}$ is called a *conjugation* if

- J is conjugate linear, i.e., $J(cx) = \bar{c}x$, for all $x \in \mathscr{H}$, and all $c \in \mathbb{C}$,

- $J^2 = 1$, and

- $\langle Jx, Jy \rangle = \langle y, x \rangle$, for all $x, y \in \mathscr{H}$.

Theorem 10.6 (von Neumann). *Let A be a densely defined closed Hermitian operator in $\mathscr{H}$. Set $d_{\pm} = dim(\mathscr{D}_{\pm}(A))$. Suppose $AJ = JA$, where J is a conjugation, then $d_{+} = d_{-}$. In particular, A has selfadjoint extensions.*

Proof. Note that, by definition, we have $\langle Jx, y \rangle = \langle Jx, J^2 y \rangle = \langle Jy, x \rangle$, for all $x, y \in \mathscr{H}$.

We proceed to show that J commutes with A^*. For this, let $x \in \mathscr{D}(A)$, $y \in \mathscr{D}(A^*)$, then

$$\langle JA^*y, x \rangle = \langle Jx, A^*y \rangle = \langle AJx, y \rangle = \langle JAx, y \rangle = \langle Jy, Ax \rangle. \tag{10.30}$$

It follows that $x \mapsto \langle Jy, Ax \rangle$ is bounded, and $Jy \in \mathscr{D}(A^*)$. Thus, $J\mathscr{D}(A^*) \subset \mathscr{D}(A^*)$. Since $J^2 = 1$, $\mathscr{D}(A^*) = J^2 \mathscr{D}(A^*) \subset J\mathscr{D}(A^*)$; therefore, $J\mathscr{D}(A^*) = \mathscr{D}(A^*)$. Moreover, (10.30) shows that $JA^* = A^*J$.

Now if $x \in \mathscr{D}_{+}(A)$, then

$$A^*Jx = JA^*x = J(ix) = -iJx$$

i.e., $J\mathscr{D}_{+}(A) \subset \mathscr{D}_{-}(A)$. Similarly, $J\mathscr{D}_{-}(A) \subset \mathscr{D}_{+}(A)$.

Using $J^2 = 1$ again, $\mathscr{D}_{-}(A) = J^2 \mathscr{D}_{-}(A) \subset J\mathscr{D}_{+}(A)$; and we conclude that $J\mathscr{D}_{+}(A) = \mathscr{D}_{-}(A)$.

Since the restriction of J to $\mathscr{D}_{+}(A)$ preserves orthonormal basis, we then get $dim(\mathscr{D}_{+}(A)) = dim(\mathscr{D}_{-}(A))$. □

10.2   Cayley Transform

There is an equivalent characterization of Hermitian extensions, taking place entirely in $\mathscr{H}$ and without the identification of $\mathscr{D}(A^*) \simeq \mathscr{G}(A^*)$, where $\mathscr{G}(A^*)$ is seen as a Hilbert space under its graph inner product. This is the result of the following observation.

Lemma 10.3. *Let A be a Hermitian operator acting in $\mathscr{H}$; then*

$$\|(A \pm i)x\|^2 = \|x\|^2 + \|Ax\|^2, \ \forall x \in \mathscr{D}(A). \tag{10.31}$$

Proof. See Lemma 10.2. Or, a direct computation shows that

$$\begin{aligned}
\|(A+i)x\|^2 &= \langle (A+i)x, (A+i)x \rangle \\
&= \|x\|^2 + \|Ax\|^2 + i(\langle Ax, x \rangle - \langle x, Ax \rangle) \\
&= \|x\|^2 + \|Ax\|^2;
\end{aligned}$$

where $\langle Ax, x \rangle - \langle x, Ax \rangle = 0$ since A is Hermitian. □

Theorem 10.7 (Cayley transform). *Let A be a densely defined, closed, Hermitian operator in $\mathscr{H}$.*

1. *The following subspaces in $\mathscr{H}$ are isometrically isomorphic:*

$$ran\,(A \pm i) \simeq \mathscr{G}\,(A) \simeq \mathscr{D}\,(A).$$

In particular, $ran\,(A \pm i)$ are closed subspace in $\mathscr{H}$.

2. *The map $C_A : ran\,(A + i) \to ran\,(A - i)$ by*

$$(A + i)\,x \mapsto (A - i)\,x, \ \forall x \in \mathscr{D}\,(A) \tag{10.32}$$

is isometric. Equivalently,

$$C_A x = (A - i)\,(A + i)^{-1}\,x \tag{10.33}$$

for all $x \in ran\,(A + i)$.

3. *Moreover,*

$$A = i\,(1 + C_A)\,(1 - C_A)^{-1}. \tag{10.34}$$

Proof. By (10.3), $ran\,(A \pm i)$ are isometric to the graph of A, and the latter is closed (as a subset in $\mathscr{H} \oplus \mathscr{H}$) since A is closed (i.e., $\mathscr{G}\,(A)$ is closed). Thus, $ran\,(A \pm i)$ are closed in $\mathscr{H}$. Note this is also a result of Corollary 10.1.

The mapping (10.33) being isometric follows from (10.31).

By (10.32), we have

$$(1 - C_A)\,((A + i)\,x) = (A + i)\,x - (A - i)\,x = 2ix$$
$$(1 + C_A)\,((A + i)\,x) = (A + i)\,x + (A - i)\,x = 2Ax$$

for all $x \in \mathscr{D}\,(A)$. It follows that

$$(1 + C_A)\,(1 - C_A)^{-1}\,(2ix) = (1 + C_A)\,((A + i)\,x) = 2Ax; \text{ i.e.,}$$

$$Ax = i\,(1 + C_A)\,(1 - C_A)^{-1}\,x, \ \forall x \in \mathscr{D}\,(A)$$

which is (10.34). $\qquad\square$

Theorem 10.8. *Suppose A is densely defined, closed, and Hermitian in $\mathscr{H}$. Then the family of (closed) Hermitian extensions of A is indexed by partial isometries U with initial space in $\mathscr{D}_+\,(A)$ and final space in $\mathscr{D}_-\,(A)$. Given U, the corresponding extension $\widetilde{A}_U \supset A$ is determined by*

$$\widetilde{A}_U(x + (1 - U)\,x_+) = x + i\,(1 + U)\,x_+, \ where$$

$$dom(\widetilde{A}_U) = \{x + (1 - U)\,x_+ : x \in \mathscr{D}\,(A)\,, x_+ \in \mathscr{D}_+\,(A)\}$$

Moreover, $\widetilde{A}_U$ is selfadjoint if and only if U is unitary from $\mathscr{D}_+\,(A)$ onto $\mathscr{D}_-\,(A)$.

Proof. Since A is closed, we get the following decompositions (Corollary 10.1)

$$\mathscr{H} = ran\,(A+i) \oplus ker\,(A^* - i)$$
$$= ran\,(A-i) \oplus ker\,(A^* + i).$$

By Theorem 10.7, $C_A : ran\,(A+i) \to ran\,(A-i)$ is isometric. Consequently, getting a Hermitian extension of A amounts to choosing a partial isometry U with initial space in $ker\,(A^* - i)\,(= \mathscr{D}_+\,(A))$ and final space in $ker\,(A^* + i)\,(= \mathscr{D}_-\,(A))$, such that

$$C_{\widetilde{A}_U} := C_A \oplus U$$

is the Cayley transform of $\widetilde{A}_U \supset A$.

Given U as above, for all $x \in \mathscr{D}\,(A)$, $x_+ \in \mathscr{D}_+\,(A)$, we have

$$C_{\widetilde{A}_U}\,((A+i)\,x \oplus x_+) = (A-i)\,x \oplus Ux_+.$$

Then,

$$(1 - C_{\widetilde{A}_U})\,((A+i)\,x \oplus x_+) = ((A+i)\,x + x_+) - ((A-i)\,x + Ux_+)$$
$$= 2ix + (1-U)\,x_+$$
$$(1 + C_{\widetilde{A}_U})\,((A+i)\,x \oplus x_+) = ((A+i)\,x + x_+) + ((A-i)\,x + Ux_+)$$
$$= 2Ax + (1+U)\,x_+;$$

and so

$$i(1 + C_{\widetilde{A}_U})(1 - C_{\widetilde{A}_U})^{-1}\left(x + \frac{1}{2i}\,(1-U)\,x_+\right) = Ax + \frac{1}{2}\,(1+U)\,x_+.$$

The theorem follows by setting $x_+ := 2iy_+$. $\square$

10.3   Boundary Triple

In applications, especially differential equations, it is convenient to characterize selfadjoint extensions using boundary conditions. For recent applications, see [JPT13, JPT12, JPT14]. A slightly modified version can be found in [dO09].

Let A be a densely defined, closed, Hermitian operator acting in a Hilbert space $\mathscr{H}$. Assume A has deficiency indices (d,d), $d > 0$, and so A has non-trivial selfadjoint extensions. By von Neumann's theorem (Theorem 10.3), for all $x,y \in \mathscr{D}\,(A^*)$, we have the following decomposition,

$$\begin{aligned} x &= x_0 + x_+ + x_- \\ y &= y_0 + y_+ + y_- \end{aligned}$$

where $x_0, y_0 \in \mathscr{D}\,(A)$, $x_+, y_+ \in \mathscr{D}_+\,(A)$, and $x_-, y_- \in \mathscr{D}_-\,(A)$. Then,

$$\langle y, A^*x \rangle - \langle A^*y, x \rangle$$

$$
\begin{aligned}
&= \quad \langle y_0 + y_+ + y_-, A x_0 + i \left(x_+ - x_- \right) \rangle - \\
&\qquad \langle A y_0 + i \left(y_+ - y_- \right), x_0 + x_+ + x_- \rangle \\
&= \quad \underbrace{\langle y_0, A x_0 \rangle - \langle A y_0, x_0 \rangle}_{0} + \underbrace{\langle y_0, i \left(x_+ - x_- \right) \rangle - \langle A y_0, x_+ + x_- \rangle}_{0} + \\
&\qquad \underbrace{\langle y_+ + y_-, A x_0 \rangle - \langle i \left(y_+ - y_- \right), x_0 \rangle}_{0} + \\
&\qquad \langle y_+ + y_-, i \left(x_+ - x_- \right) \rangle - \langle i \left(y_+ - y_- \right), x_+ + x_- \rangle \\
&= \quad 2i \left\{ \langle y_+, x_+ \rangle - \langle y_-, x_- \rangle \right\}. \qquad\qquad (10.35)
\end{aligned}
$$

Therefore, we see that

$$
\left[x, y \in \mathscr{D}(\widetilde{A}), \ \widetilde{A} \supset A, \ \text{Hermitian extension} \right] \iff \left[\text{RHS of (10.35) vanishes} \right].
$$

For selfadjoint extensions, this is equivalent to choosing a partial isometry U from $\mathscr{D}_+ (A)$ onto $\mathscr{D}_- (A)$, and setting

$$
x_- = U x_+, \ y_- = U y_+; \ \text{so that}
$$

$$
\begin{aligned}
\langle y, A^* x \rangle - \langle A^* y, x \rangle &= \quad 2i \left\{ \langle y_+, x_+ \rangle - \langle U y_+, U x_+ \rangle \right\} \\
&= \quad 2i \left\{ \langle y_+, x_+ \rangle - \langle y_+, x_+ \rangle \right\} = 0.
\end{aligned}
$$

The discussion above leads to the following definition:

Definition 10.3. Let A be a densely defined, closed, Hermitian operator in $\mathscr{H}$. Suppose A has deficiency indices (d, d), $d > 0$. A boundary space for A is a triple $(\mathscr{H}_b, \rho_1, \rho_2)$ consisting of a Hilbert space $\mathscr{H}_b$ and two linear maps $\rho_1, \rho_2 : \mathscr{D}(A^*) \to \mathscr{H}_b$, such that

1. $\rho_i (\mathscr{D}(A^*))$ is dense in $\mathscr{H}_b$, $i = 1, 2$; and

2. for all $x, y \in \mathscr{D}(A^*)$, $\exists c \neq 0$, such that

$$
\langle y, A^* x \rangle - \langle A^* y, x \rangle = c \left[\langle \rho_1 (y), \rho_1 (x) \rangle_b - \langle \rho_2 (y), \rho_2 (x) \rangle_b \right]. \qquad (10.36)
$$

Remark 10.3. In (10.35), we set

$$
\mathscr{H}_b = \mathscr{D}_+ (A)
$$

$$
\rho_1 (x_0 + x_+ + x_-) = x_+
$$

$$
\rho_2 (x_0 + x_+ + x_-) = U x_+
$$

for any $x = x_0 + x_+ + x_-$ in $\mathscr{D}(A^*)$. Then $(\mathscr{H}_b, \rho_1, \rho_2)$ is a boundary space for A. In this special case, ρ_1, ρ_2 are surjective. It is clear that the choice of a boundary triple is not unique. In applications, $\mathscr{H}_b$ is usually chosen to have the same dimension as $\mathscr{D}_\pm (A)$.

Consequently, Theorem 10.4 can be restated as follows.

Theorem 10.9. *Let A be a densely defined, closed, Hermitian operator in $\mathcal{H}$. Suppose A has deficiency indices (d, d), $d > 0$. Let $(\mathcal{H}_b, \rho_1, \rho_2)$ be a boundary triple. Then the selfadjoint extensions of A are indexed by unitary operators U : $\mathcal{H}_b \to \mathcal{H}_b$, such that given U, the corresponding selfadjoint extension $\widetilde{A_U} \supset A$ is determined by*

$$\widetilde{A_U} = A^* \big|_{\mathscr{D}(\widetilde{A_U})}, \text{ where}$$

$$\mathscr{D}\left(\widetilde{A_U}\right) = \{x \in \mathscr{D}(A^*) : U\rho_1(x) = \rho_2(x)\}.$$

Certain variations of Theorem 10.9 are convenient in the boundary value problems (BVP) of differential equations. In [DM91, GG91], a boundary triple $(\mathcal{H}_b, \beta_1, \beta_2)$ is defined to satisfy

$$\langle x, A^*y \rangle = \langle A^*x, y \rangle = c' \left[\langle \beta_1(x), \beta_2(y) \rangle_b - \langle \beta_2(x), \beta_1(y) \rangle_b \right] \tag{10.37}$$

for all $x, y \in \mathscr{D}(A^*)$; and c' is some nonzero constant. Also, see [JPT13, JPT12, JPT14].

The connection between (10.36) and (10.37) is via the bijection

$$\left\{ \begin{array}{rcl} \rho_1 & = & \beta_1 + i\beta_2 \\ \rho_2 & = & \beta_1 - i\beta_2 \end{array} \right\} \Longleftrightarrow \left\{ \begin{array}{rcl} \beta_1 & = & \dfrac{\rho_1 + \rho_2}{2} \\ \beta_2 & = & \dfrac{\rho_1 - \rho_2}{2i} \end{array} \right\}. \tag{10.38}$$

Lemma 10.4. *Under the bijection (10.38), we have*

$$\langle \rho_1(x), \rho_1(y) \rangle_b - \langle \rho_2(x), \rho_2(y) \rangle_b = 2i \left(\langle \beta_1(x), \beta_2(y) \rangle_b - \langle \beta_2(x), \beta_1(y) \rangle_b \right).$$

Proof. For convenience, we suppress the variables x, y. Then a direct computation shows that,

$$\begin{aligned} & \langle \rho_1, \rho_1 \rangle_b - \langle \rho_2, \rho_2 \rangle_b \\ = \; & \langle \beta_1 + i\beta_2, \beta_1 + i\beta_2 \rangle_b - \langle \beta_1 - i\beta_2, \beta_1 - i\beta_2 \rangle_b \\ = \; & i \langle \beta_1, \beta_2 \rangle_b - i \langle \beta_2, \beta_1 \rangle_b + i \langle \beta_1, \beta_2 \rangle_b - i \langle \beta_2, \beta_1 \rangle_b \\ = \; & 2i \left(\langle \beta_1, \beta_2 \rangle_b - \langle \beta_2, \beta_1 \rangle_b \right) \end{aligned}$$

which is the desired conclusion. $\square$

Theorem 10.10. *Given a boundary triple $(\mathcal{H}_b, \beta_1, \beta_2)$ satisfying (10.37), the family of selfadjoint extensions $\widetilde{A_U} \supset A$ is indexed by unitary operators U : $\mathcal{H}_b \to \mathcal{H}_b$, such that*

$$\widetilde{A_U} = A^* \big|_{\mathscr{D}(\widetilde{A_U})}, \text{ where} \tag{10.39}$$

$$\mathscr{D}\left(\widetilde{A_U}\right) = \{x \in \mathscr{D}(A^*) : (1 - U)\beta_1(x) = i(1 + U)\beta_2(x)\}. \tag{10.40}$$

Proof. By Theorem 10.9, we need only pick a unitary operator $U : \mathscr{H}_b \to \mathscr{H}_b$, such that $\rho_2 = U\rho_1$. In view of the bijection (10.38), this yields

$$\beta_1 - i\beta_2 = U(\beta_1 + i\beta_2) \iff (1 - U)\beta_1 = i(1 + U)\beta_2$$

and the theorem follows. $\qquad\square$

Example 10.4. Let $A = -i\frac{d}{dx}\Big|_{\mathscr{D}(A)}$, and

$$\mathscr{D}(A) = \left\{ f \in L^2(0,1) : f' \in L^2(0,1), f(0) = f(1) = 0 \right\}.$$

Then $A^* = -i\frac{d}{dx}\Big|_{\mathscr{D}(A^*)}$, where

$$\mathscr{D}(A^*) = \left\{ f : f, f' \in L^2(0,1) \right\}.$$

For all $f, g \in \mathscr{D}(A^*)$, using integration by parts, we get

$$\langle g, A^* f \rangle - \langle A^* g, f \rangle = -i\overline{g(x)}f(x)\Big|_0^1 = -i\left(\overline{g(1)}f(1) - \overline{g(0)}f(0) \right).$$

Let $\mathscr{H}_b = \mathbb{C}$, i.e., one-dimensional, and set

$$\rho_1(f) = f(1), \quad \rho_2(f) = f(0); \text{ then}$$

$$\langle g, A^* f \rangle - \langle A^* g, f \rangle = -i\left(\langle \rho_1(g), \rho_1(f) \rangle_b - \langle \rho_2(g), \rho_2(f) \rangle_b \right).$$

Therefore, $(\mathscr{H}_b, \rho_1, \rho_2)$ is a boundary triple.

The family of selfadjoint extensions of A is given by the unitary operator

$$e^{i\theta} : \mathscr{H}_b \to \mathscr{H}_b, \quad s.t. \ \rho_2 = e^{i\theta}\rho_1;$$

i.e.,

$$\widetilde{A}_\theta = -i\frac{d}{dx}\Big|_{\{f \in \mathscr{D}(A^*) : f(0) = e^{i\theta}f(1)\}}.$$

Example 10.5. Let $Af = -f''$, with $\mathscr{D}(A) = C_c^\infty(0,\infty)$. Since A is Hermitian and $A \geq 0$, it follows that it has equal deficiency indices. Also, $\mathscr{D}(A^*) = \{f, f'' \in L^2(0,\infty)\}$, and $A^* f = -f''$, $\forall f \in \mathscr{D}(A^*)$.

For $f, g \in \mathscr{D}^*(A)$, we have

$$\langle g, A^* f \rangle = -\int_0^\infty \overline{g}f'' = -\left([\overline{g}f' - \overline{g}'f]_0^\infty + \int_0^\infty \overline{g}''f \right)$$

$$= (\overline{g}f')(0) - (\overline{g}'f)(0) - \int_0^\infty \overline{g}''f$$

$$= (\overline{g}f')(0) - (\overline{g}'f)(0) + \langle A^* g, f \rangle$$

and so

$$\langle g, A^* f \rangle - \langle A^* g, f \rangle = (\overline{g}f')(0) - (\overline{g}'f)(0).$$

Now, set $\mathscr{H}_b = \mathbb{C}$, i.e., one-dimensional, and

$$\beta_1 (\varphi) = \varphi (0), \quad \beta_2 (\varphi) = \varphi' (0); \text{ then}$$

$$\langle g, A^* f \rangle - \langle A^* g, f \rangle = \langle \beta_1 (g), \beta_2 (f) \rangle_b - \langle \beta_2 (g), \beta_1 (f) \rangle_b.$$

This defines the boundary triple.

The selfadjoint extensions are parameterized by $e^{i\theta}$, where

$$\left(1 - e^{i\theta}\right) \beta_1 (f) = i \left(1 + e^{i\theta}\right) \beta_2 (f); \text{ i.e.,}$$

$$f (0) = z f' (0), \ f \in \mathscr{D} (A^*)$$

where

$$z = i \frac{1 + e^{i\theta}}{1 - e^{i\theta}}.$$

We take the convention that $z = \infty \iff f' (0) = 0$, i.e., the Neumann boundary condition.

Example 10.6. $Af = -f''$, $\mathscr{D} (A) = C_c^\infty (0, 1)$; then

$$\mathscr{D} (A^*) = \left\{ f, f'' \in L^2 (0, 1) \right\}.$$

Integration by parts gives

$$\langle g, A^* f \rangle = - \int_0^1 \overline{g} f''$$

$$= - [\overline{g} f' - \overline{g}' f]_0^1 - \int_0^1 \overline{g}'' f$$

$$= - [\overline{g} f' - \overline{g}' f]_0^1 + \langle A^* g, f \rangle.$$

Thus,

$$\langle g, A^* f \rangle - \langle A^* g, f \rangle = [(\overline{g} f') (0) + (\overline{g}' f) (1)] - [(\overline{g}' f) (0) + (\overline{g} f') (1)]$$

$$= \langle \beta_1 (g), \beta_2 (f) \rangle_b - \langle \beta_2 (g), \beta_1 (f) \rangle_b$$

where

$$\beta_1 (\varphi) = \begin{bmatrix} \varphi (0) \\ \varphi' (1) \end{bmatrix}, \ \beta_2 (\varphi) = \begin{bmatrix} \varphi' (0) \\ \varphi (1) \end{bmatrix}.$$

The boundary space is $\mathscr{H}_b = \mathbb{C}^2$, i.e., 2-dimensional

The family of selfadjoint extensions is parameterized by $U \in M (2, \mathbb{C})$. Given U, the corresponding extension $\widetilde{A_U}$ is determined by

$$\widetilde{A_U} = A^* \Big|_{\mathscr{D}(\widetilde{A_U})}, \text{ where}$$

$$\mathscr{D} \left(\widetilde{A_U} \right) = \{ f \in \mathscr{D} (A^*) : (1 - U) \beta_1 (f) = i (1 + U) \beta_2 (f) \}.$$

Remark 10.4. Another choice of the boundary map:

$$\langle g, A^*f \rangle - \langle A^*g, f \rangle = [(\overline{g}f')(0) - (\overline{g}f')(1)] - [(\overline{g}'f)(0) - (\overline{g}'f)(1)]$$
$$= \langle \beta_1(g), \beta_2(f) \rangle_b - \langle \beta_2(g), \beta_1(f) \rangle_b$$

where

$$\beta_1(\varphi) = \begin{bmatrix} \varphi(0) \\ \varphi(1) \end{bmatrix}, \quad \beta_2(\varphi) = \begin{bmatrix} \varphi'(0) \\ -\varphi'(1) \end{bmatrix}.$$

The selfadjoint boundary condition leads to

$$(1 - U) \begin{bmatrix} f(0) \\ f(1) \end{bmatrix} = i(1 + U) \begin{bmatrix} f'(0) \\ -f'(1) \end{bmatrix}.$$

For $U = 1$, we get the Neumann boundary condition:

$$f'(0) = f'(1) = 0.$$

For $U = -1$, we get the Dirichlet boundary condition:

$$f(0) = f(1) = 0.$$

Exercise 10.4 (From selfadjoint extension to unitary one-parameter group).

1. For each of the selfadjoint extensions from Theorem 10.10, write down the corresponding unitary one-parameter group; and identify it as an induced representation; induction $\mathbb{Z} \longrightarrow \mathbb{R}$; see Section 8.4.

2. Same question for the selfadjoint extension operators computed in Examples 10.5 and 10.6.

Exercise 10.5 (Realize index-values). Give examples of Hermitian symmetric operators with dense domain such that the deficiency indices fall in any of the four cases, $(0,0)$, $(0,1)$, $(1,0)$ and $(1,1)$.

Exercise 10.6 (Realize the cases (d, d) for semibounded operators). (i) Show that if a Hermitian symmetric operators with dense domain is semibounded, its deficiency indices must be equal, i.e., of the form (d, d). (ii) In this family give examples when d is finite, and when it is infinite.

Exercise 10.7 (The N-covering surface of $\mathbb{R}^2 \backslash \{(0, 0)\}$). Fix $N \in \mathbb{N}$, and let M be the N-covering surface of $\mathbb{R}^2 \backslash \{(0, 0)\}$. Under polar coordinates, M is covered in a single coordinate patch as $x = r \cos \theta$, $y = r \sin \theta$, where $r \in \mathbb{R}_+$, and $\theta \in [0, 2\pi N)$; and it has the induced metric $ds^2 = dr^2 + r^2 d\theta^2$ with volume form $dV = r dr d\theta$.

1. Show that $L^2(M)$ has the following decomposition:

$$L^2(M) = \sum_{k \in \mathbb{Z}}^{\oplus} \left(L^2(\mathbb{R}_+, r dr) \otimes \text{span}\{e^{i\theta k/N}\} \right). \tag{10.41}$$

<u>Hint</u>: Apply Fourier series expansion for $f \in L^2(M)$ in the θ variable.

2. Show that the formal 2D Laplacian in $L^2(M)$ takes the form

$$\Delta = \sum_{k \in \mathbb{Z}}^{\oplus} \left(\frac{1}{r}\frac{d}{dr}\left(r\frac{d}{dr} \right) - \frac{(k/N)^2}{r^2} \right) \otimes 1. \qquad (10.42)$$

3. Moreover, setting $W : L^2(\mathbb{R}_+, rdr) \to L^2(\mathbb{R}_+, dr)$ by

$$Wf(r) := r^{1/2}f(r), \qquad (10.43)$$

show that W is unitary and such that

$$W\Delta W^* = \sum_{k \in \mathbb{Z}}^{\oplus} \left(l_{k/N} \otimes 1 \right), \text{ where} \qquad (10.44)$$

$$l_{k/N} := \frac{d^2}{dr^2} - \frac{(k/N)^2 - 1/4}{r^2} \qquad (10.45)$$

4. Recall that $l_\nu|_{C_c^\infty(\mathbb{R}_+)}$ is essentially selfadjoint iff $|\nu| \geq 1$. (See, e.g., [AG93].) Use this to deduce that the operator

$$\Delta|_{C^\infty(M)}$$

has deficiency indices $(2N - 1, 2N - 1)$ in $L^2(M)$.

Exercise 10.8 (The logarithmic Riemann surface). Let M be Riemann surface of $\log z$, i.e., the ∞-covering surface of $\mathbb{R} \setminus \{(0,0)\}$. Deduce from Exercise 10.7 that the corresponding Laplacian $\Delta|_{C^\infty(M)}$ has deficiency indices (∞, ∞). What are the defect vectors?

Exercise 10.9 (Formally commuting Hermitian symmetric operators on a common dense domain which are not commuting in the strong sense). Let M be Riemann surface of $\log z$.

1. Show that each operator $\frac{\partial}{\partial x_j}$, $j = 1, 2$, on $C_c^\infty(M)$ is essentially skew-adjoint in $L^2(M)$, i.e.,

$$-\left(\frac{\partial}{\partial x_j}\Big|_{C_c^\infty(M)} \right)^* = \text{closure}\left(\frac{\partial}{\partial x_j}\Big|_{C_c^\infty(M)} \right), \quad j = 1, 2. \qquad (10.46)$$

2. Show that the unitary one-parameter groups generated by $\overline{\frac{\partial}{\partial x_j}\Big|_{C_c^\infty(M)}}$, $j = 1, 2$, are *not* strongly commuting; see Figure 10.3. Equivalently, the two skew-adjoint operators in part (1) are *not* strongly commuting. (By a well-known theorem of Nelson, this also shows that the Laplacian

$$L := \left(\frac{\partial}{\partial x_1} \right)^2 + \left(\frac{\partial}{\partial x_2} \right)^2 \text{ on } C_c^\infty(M) \qquad (10.47)$$

is *not* essentially selfadjoint.)

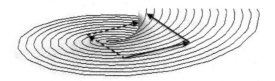

Figure 10.3: Translation of a locally defined φ on M to different sheets.

10.4   The Friedrichs Extension

Since every semibounded Hermitian symmetric operator A has equal deficiency indices (see, e.g., Exercise 10.6), it follows that it will automatically have selfadjoint extensions. From the analysis of partial isometries, we see that in general there will be selfadjoint extensions that are not semibounded. And even among the semibounded extensions, some may have a lower bound different from that of the initial operator A. Now, K. O. Friedrichs[2] showed that every semibounded Hermitian symmetric operator A has a selfadjoint extension such that its lower spectral bound is the same as the lower bound of A. The extension found by Friedrichs is called the Friedrichs extension, and as we show below, it is obtained from a certain completion procedure applied to the initial operator A.

It is known that in the general setting of semibounded Hermitian symmetric operators, there are other selfadjoint extensions having the same lower bound. They were found by M. G. Krein, but Krein's theory is beyond the scope of the present book; see however the citations after the present chapter.

Let $A : \mathscr{D} \to \mathscr{H}$ be an operator with dense domain $dom(A) := \mathscr{D}$ in $\mathscr{H}$, such that

$$\langle \varphi, A\varphi \rangle \geq \|\varphi\|^2, \quad \forall \varphi \in \mathscr{D}. \tag{10.48}$$

Set $\mathscr{H}_A :=$ Hilbert completion of $\mathscr{D}$ with respect to the

$$\|\varphi\|_A := \langle \varphi, A\varphi \rangle^{\frac{1}{2}}. \tag{10.49}$$

Then $\varphi \to \varphi$ defines a contraction $J : \mathscr{H}_A \to \mathscr{H}$, extending $J\varphi = \varphi$, for $\varphi \in \mathscr{D}$. Note that (10.48) $\Leftrightarrow$

$$\|J\varphi\| \leq \|\varphi\|_A, \quad \forall \varphi \in \mathscr{H}_A,$$

(see (10.49).)

Remark 10.5. We will make use of two inner products: $\langle \cdot, \cdot \rangle$ in $\mathscr{H}$, and $\langle \cdot, \cdot \rangle_A$ (with subscript A) in $\mathscr{H}_A$.

We have

$$\langle J\varphi, f \rangle = \langle \varphi, J^* f \rangle_A \tag{10.50}$$

see Figure 10.4: $\varphi \in \mathscr{H}_A$, $f \in \mathscr{H}$, $J^* f \in \mathscr{H}_A$.

Note both J and J^* are contractions with respect to the respective norms, so

$$\|J^* f\|_A \leq \|f\|, \quad \forall f \in \mathscr{H}. \tag{10.51}$$

---

[2]Appendix C includes a short biographical sketch of K. O. Friedrichs. See also [Fri80, FL28].

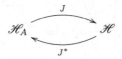

Figure 10.4: The operator J and its adjoint.

So $JJ^* : \mathscr{H} \to \mathscr{H}$ is a contractive selfadjoint operator in $\mathscr{H}$, and $(JJ^*)^{\frac{1}{2}}$ is well defined by the Spectral Theorem.

Theorem 10.11 (Friedrichs extension). *Let $A, J, \mathscr{D}, \mathscr{H}, \mathscr{H}_A$ be as above. Then there is a selfadjoint extension $\widetilde{A} \supset A$ in $\mathscr{H}$ such that*

$$\langle \widetilde{A}x, y \rangle = \langle x, y \rangle_1, \ \forall x \in dom(\widetilde{A}), \forall y \in dom(A) (= \mathscr{D}).$$

Proof. The theorem is established in three steps:

Step 1. $JJ^*A\varphi = \varphi, \ \forall \varphi \in \mathscr{D}$.

Step 2. JJ^* is invertible (easy from (10.48).)

Step 3. $A \subset (JJ^*)^{-1}$, where $(JJ^*)^{-1}$ is selfadjoint; it is the Friedrichs extension of A. Note Step 3 is immediate from step 1 by definition.

Proof of Step 1. Since $\mathscr{D}$ is dense in $\mathscr{H}$, it is enough to prove that

$$\langle \psi, JJ^*A\varphi \rangle = \langle \psi, \varphi \rangle, \quad \forall \varphi, \psi \in \mathscr{D}. \tag{10.52}$$

Let $\varphi, \psi \in \mathscr{D}$, then

$$
\begin{aligned}
\mathrm{LHS}_{(10.52)} &= \langle JJ^*\psi, A\varphi \rangle \\
&= \langle J^*\psi, A\varphi \rangle \\
&= \langle J^*\psi, \varphi \rangle_A \ \text{(by (10.48) \& (10.49))} \\
&= \langle \psi, J\varphi \rangle \ \text{(by (10.50) and use } J^{**} = J) \\
&= \langle \psi, \varphi \rangle = \mathrm{RHS}_{(10.52)}.
\end{aligned}
$$

Hence Step 1 follows.   □

Although we already outlined one approach to the Friedrichs extension in Theorem 10.11, we feel that the alternative reasoning below is of independent interest. See also the discussion in Section 10.5.

Let A be a densely defined Hermitian operator in a Hilbert space $\mathscr{H}$. A is *semi-bounded* if $A \geq c > -\infty$, in the sense that, $\langle x, Ax \rangle \geq c \langle x, x \rangle, \ \forall x \in dom(A)$. Set

$$L_A := \inf \left\{ \langle x, Ax \rangle : x \in dom(A), \|x\| = 1 \right\} \tag{10.53}$$

and L_A is called the *lower bound* of A.

In the following discussion, we first assume $A \geq 1$ and eventually drop the constraint.

Let $\mathscr{H}_1 =$ completion of $dom\,(A)$ with respect to the inner product

$$\langle x, y\rangle_1 := \langle x, Ay\rangle\,. \tag{10.54}$$

Theorem 10.12. *Let $A \geq I$, i.e., $\langle x, Ax\rangle \geq \|x\|^2$, $\forall x \in dom\,(A)$; and let $\mathscr{H}$, and $\mathscr{H}_1$ be as above. Then*

1. $\|x\| \leq \|x\|_1$, $\forall x \in dom\,(A)$.

2. $\|\cdot\|$ and $\|\cdot\|_1$ are topologically consistent, i.e., the identity map

$$\varphi : dom\,(A) \to dom\,(A)$$

   extends by continuity to
$$\widetilde{\varphi} : \mathscr{H}_1 \hookrightarrow \mathscr{H}$$

   such that
$$\|x\| \leq \|x\|_1\,, \ \forall x \in \mathscr{H}_1. \tag{10.55}$$

   Therefore, $\mathscr{H}_1$ is identified as a dense subspace in $\mathscr{H}$.

3. *Moreover,*
$$\langle y, x\rangle_1 = \langle y, Ax\rangle\,, \ \forall x \in dom\,(A)\,, \forall y \in \mathscr{H}_1. \tag{10.56}$$

4. *Define*

$$\widetilde{A} := A^*\Big|_{dom(\widetilde{A})}, \ where$$
$$dom(\widetilde{A}) := dom(\widetilde{A}^*) \cap \mathscr{H}_1.$$

   Then $\widetilde{A} = \widetilde{A}^*$, and $L_{\widetilde{A}} = L_A$.

Proof. (1)-(2) The assumption $A \geq 1$ implies that

$$\|x\|^2 = \langle x, x\rangle \leq \langle x, Ax\rangle = \|x\|_1^2\,, \ \forall x \in dom\,(A)\,.$$

Hence φ is continuous and the norm ordering passes to the completions of $dom\,(A)$ with respect to $\|\cdot\|_1$ and $\|\cdot\|$. Therefore (10.55) holds.

Next, we verify that $\widetilde{\varphi}$ is injective (i.e., $ker\widetilde{\varphi} = 0$.) Suppose $(x_n) \subset dom\,(A)$ such that $x_n \xrightarrow{\|\cdot\|_1} x \in \mathscr{H}_1$, and $x_n \xrightarrow{\|\cdot\|} 0$. We must show that $\|x\|_1 = 0$. But

$$\|x\|_1^2 = \lim_{m,n\to\infty} \langle x_m, x_n\rangle_1 \quad (\text{limit exists by assumption})$$
$$= \lim_{m,n\to\infty} \langle x_m, Ax_n\rangle$$
$$= \lim_{n\to\infty} \langle 0, Ax_n\rangle = 0.$$

In the computation, we used the fact that

$$|\langle x_m - x, Ax_n\rangle| \leq \|x_m - x\|\,\|Ax_n\|$$

$$\leq \|x_m - x\|_1 \|Ax_n\| \to 0, \text{ as } m \to \infty.$$

(3) Let $(y_n) \subset dom(A)$, and $\|y_n - y\|_1 \to 0$. For all $x \in dom(A)$, we have

$$\langle y, x \rangle_1 = \lim_{n \to \infty} \langle y_n, x \rangle_1 = \lim_{n \to \infty} \langle y_n, Ax \rangle = \langle y, Ax \rangle.$$

Equivalently,

$$\begin{aligned}
|\langle y_n, Ax \rangle - \langle y, Ax \rangle| &= |\langle y_n - y, Ax \rangle| \\
&\leq \|y_n - y\| \|Ax\| \\
&\leq \|y_n - y\|_1 \|Ax\| \to 0, \text{ as } n \to \infty.
\end{aligned}$$

(4) For all $x, y \in dom(\widetilde{A})$, $\exists (x_n), (y_n) \subset dom(A)$ such that $\|x_n - x\|_1 \to 0$ and $\|y_n - y\|_1 \to 0$. Hence the following limit exists:

$$\lim_{m,n \to \infty} \langle x_m, Ay_n \rangle \left(= \lim_{m,n \to \infty} \langle x_m, y_n \rangle_1 \right).$$

Consequently,

$$\begin{aligned}
\lim_{m \to \infty} \lim_{n \to \infty} \langle x_m, Ay_n \rangle &= \lim_{m \to \infty} \lim_{n \to \infty} \langle Ax_m, y_n \rangle \\
&= \lim_{m \to \infty} \langle Ax_m, y \rangle \\
&= \lim_{m \to \infty} \langle x_m, A^* y \rangle \\
&= \lim_{m \to \infty} \langle x_m, \widetilde{A}y \rangle = \langle x, \widetilde{A}y \rangle
\end{aligned}$$

and

$$\begin{aligned}
\lim_{n \to \infty} \lim_{m \to \infty} \langle x_m, Ay_n \rangle &= \lim_{n \to \infty} \langle x, Ay_n \rangle \\
&= \lim_{n \to \infty} \langle A^* x, y_n \rangle \\
&= \lim_{n \to \infty} \langle \widetilde{A}x, y_n \rangle = \langle \widetilde{A}x, y \rangle.
\end{aligned}$$

Thus, $\widetilde{A}$ is Hermitian.

Fix $y \in \mathscr{H}$. The map $x \longmapsto \langle y, x \rangle$, $\forall x \in dom(A) \subset \mathscr{H}_1$, is linear and satisfies

$$|\langle y, x \rangle| \leq \|y\| \|x\| \leq \|y\| \|x\|_1.$$

Hence it extends to a unique bounded linear functional on $\mathscr{H}_1$, as $dom(A)$ is dense in $\mathscr{H}_1$.

By Riesz's theorem, there exists unique $h_y \in \mathscr{H}_1$ such that

$$\langle y, x \rangle = \langle h_y, x \rangle_1, \ \forall x \in \mathscr{H}_1. \tag{10.57}$$

In particular,

$$\langle y, x \rangle = \langle h_y, x \rangle_1 = \langle h_y, Ax \rangle, \ \forall x \in dom(A).$$

Then, $h_y \in \mathscr{H}_1 \cap dom(A^*) = dom(\widetilde{A})$, and $\widetilde{A}h_y = y$. Therefore, $ran(\widetilde{A}) = \mathscr{H}$. Note we have established the identity

$$\langle \widetilde{A}y, x \rangle = \langle y, x \rangle_1, \ \forall y \in dom(\widetilde{A}), \forall x \in dom(A). \tag{10.58}$$

$\square$

Claim 10.1. $ran(\widetilde{A}) = \mathscr{H}$ implies that $\widetilde{A}$ is selfadjoint. In fact, for all $x \in dom(\widetilde{A})$ and $y \in dom(\widetilde{A}^*)$, we have

$$\langle y, \widetilde{A}x \rangle = \langle \widetilde{A}^*y, x \rangle = \langle \widetilde{A}h, x \rangle = \langle h, \widetilde{A}x \rangle$$

where $\widetilde{A}^*y = \widetilde{A}h$, for some $h \in dom(\widetilde{A})$, using the assumption $ran(\widetilde{A}) = \mathscr{H}$. Thus,

$$\langle y - h, \widetilde{A}x \rangle = 0, \ \forall x \in dom(\widetilde{A});$$

i.e., $y - h \perp ran(\widetilde{A}) = \mathscr{H}$. Therefore, $y = h$, and so $y \in dom(\widetilde{A})$.

Proof. Finally, we show $L_{\widetilde{A}} = L_A$. By the definition of lower bound, $dom(A) \subset dom(\widetilde{A})$ implies $L_{\widetilde{A}} \leq L_A$. On the other hand, let $(x_n) \subset dom(A)$ such that $x_n \xrightarrow{\|\cdot\|_1} x \in dom(\widetilde{A})$, then

$$\langle x, \widetilde{A}x \rangle = \lim_{n\to\infty} \langle x, \widetilde{A}x_n \rangle = \lim_{n\to\infty} \langle x, Ax_n \rangle$$
$$= \lim_{n\to\infty} \langle x, x_n \rangle_1 = \langle x, x \rangle_1 \geq L_A \langle x, x \rangle$$

which shows that $L_{\widetilde{A}} \geq L_A$. $\square$

Remark 10.6. In the proof of Theorem 10.12, we established an embedding $\psi : \mathscr{H} \hookrightarrow \mathscr{H}_1^*$ by $\psi : y \mapsto h_y$ with the defining equation (10.57). And we define $\widetilde{A}h_y = y$, i.e., $h_y = \widetilde{A}^{-1}y$. It follows that $dom(\widetilde{A}) = ran(\psi)$, and $ran(\widetilde{A}) = \mathscr{H}$.

Theorem 10.13. *The Friedrichs extension of A is the unique selfadjoint operator satisfying*

$$\langle \widetilde{A}x, y \rangle = \langle x, y \rangle_1, \ \forall x \in dom(\widetilde{A}), \forall y \in dom(A).$$

See (10.58).

Proof. Suppose $A \subset B, C \subset A^*$, and B, C selfadjoint, satisfying

$$\langle Bx, y \rangle = \langle x, y \rangle_1, \ \forall x \in dom(B), \forall y \in dom(A)$$
$$\langle Cx, y \rangle = \langle x, y \rangle_1, \ \forall x \in dom(C), \forall y \in dom(A).$$

Then, for all $x, y \in dom(A)$, we have

$$\langle Bx, y \rangle = \langle Cx, y \rangle = \langle x, Cy \rangle (= \langle x, Ay \rangle).$$

Fix $x \in dom(A)$, and the above identify passes to $y \in dom(C)$. Therefore, $y \in B^* = B$, and $By = Cy$. This shows $C \subset B$. Since

$$C = C^* \supset B^* = B$$

i.e., $C \supset B$, it then follows that $B = C$. $\square$

Remark 10.7. If A is only assumed to be semi-bounded, i.e., $A \geq c > -\infty$, then $B := A - c + 1 \geq 1$, and we get the Friedrichs extension $\tilde{B}$ of B; and $\tilde{B} - 1 + c$ is the Friedrichs extension of A.

10.5   Rigged Hilbert Space

In the construction of Friedrichs extensions of semi-bounded operators, we have implicitly used the idea of rigged Hilbert spaces. We now study this method systematically and recover the Friedrichs extension as a special case.

Let $\mathscr{H}_0$ be a Hilbert space with inner product $\langle \cdot, \cdot \rangle_0$, and $\mathscr{H}_1$ be a dense subspace in $\mathscr{H}_0$, which by itself, is a Hilbert space with respect to $\langle \cdot, \cdot \rangle_1$. Further, assume the ordering

$$\|x\|_0 \leq \|x\|_1, \ \forall x \in \mathscr{H}_1. \tag{10.59}$$

Hence, the identity map

$$id : \mathscr{H}_1 \hookrightarrow \mathscr{H}_0 \tag{10.60}$$

is continuous with a dense image.

Let $\mathscr{H}_{-1}$ be the space of bounded *conjugate* linear functionals on $\mathscr{H}_1$. By Riesz's theorem, $\mathscr{H}_{-1}$ is identified with $\mathscr{H}_1$ via the map

$$\mathscr{H}_{-1} \to \mathscr{H}_1, \ f \mapsto \xi_f, \ \text{s.t.} \tag{10.61}$$

$$f(x) = \langle x, \xi_f \rangle_1, \ \forall x \in \mathscr{H}_1. \tag{10.62}$$

Then $\mathscr{H}_{-1}$ is a Hilbert space with respect to the inner product

$$\langle f, g \rangle_{-1} = \langle \xi_f, \xi_g \rangle_1, \ \forall f, g \in \mathscr{H}_{-1}. \tag{10.63}$$

Remark 10.8. The map $f \mapsto \xi_f$ in (10.61) is linear. For if $c \in \mathbb{C}$, then $\langle x, \xi_{cf} \rangle_1 = cf(x) = c \langle x, \xi_f \rangle_1 = \langle x, c\xi_f \rangle_1$, for all $x \in \mathscr{H}_1$; i.e., $(\xi_{cf} - c\xi_f) \perp \mathscr{H}_1$, and so $\xi_{cf} = c\xi_f$.

Theorem 10.14. *The mapping*

$$\mathscr{H}_0 \hookrightarrow \mathscr{H}_{-1}, \ x \mapsto \langle \cdot, x \rangle_0, \ \forall x \in \mathscr{H}_0 \tag{10.64}$$

is linear, injective, continuous, and having a dense image in $\mathscr{H}_{-1}$.

Proof. Since $cx \mapsto \langle \cdot, cx \rangle = c \langle \cdot, x \rangle$, the mapping in (10.64) is linear.

For all $x \in \mathscr{H}_0$, we have 9.9

$$|\langle y, x \rangle_0| \leq \|x\|_0 \|y\|_0 \overset{(10.59)}{\leq} \|x\|_0 \|y\|_1, \ \forall y \in \mathscr{H}_1; \tag{10.65}$$

hence $\langle \cdot, x \rangle_0$ is a bounded conjugate linear functional on $\mathscr{H}_1$, i.e., $\langle \cdot, x \rangle_0 \in \mathscr{H}_{-1}$. Moreover, by (10.65),

$$\|\langle \cdot, x \rangle_0\|_{-1} \leq \|x\|_0. \tag{10.66}$$

If $\langle \cdot, x \rangle_0 \equiv 0$ in $\mathscr{H}_{-1}$, then $\langle y, x \rangle_0 = 0$, for all $y \in \mathscr{H}_1$. Since $\mathscr{H}_1$ is dense in $\mathscr{H}_0$, it follows that $x = 0$ in $\mathscr{H}_0$. Thus, (10.64) is injective.

Now, if $f \perp \{\langle \cdot, x \rangle_0 : x \in \mathscr{H}_1\}$ in $\mathscr{H}_{-1}$, then

$$\langle f, \langle \cdot, x \rangle_0 \rangle_{-1} \overset{(10.63)}{=} \langle \xi_f, \xi_{\langle \cdot, x \rangle_0} \rangle_1 \overset{(10.61)}{=} \langle \xi_f, x \rangle_0 = 0, \ \forall x \in \mathscr{H}_1. \tag{10.67}$$

Thus, $\|\xi_f\|_0 = 0$, since $\mathscr{H}_1$ is dense in $\mathscr{H}_0$. Since $id : \mathscr{H}_1 \hookrightarrow \mathscr{H}_0$ is injective, and $\xi_f \in \mathscr{H}_1$, it follows that $\|\xi_f\|_1 = 0$. This, in turn, implies $\|f\|_{-1} = 0$, and so $f = 0$ in $\mathscr{H}_{-1}$. Consequently, the image of $\mathscr{H}_1$ (resp. $\mathscr{H}_0$ as it contains $\mathscr{H}_1$) under (10.64) is dense in $\mathscr{H}_{-1}$. $\qquad\square$

Combining (10.59) and Theorem 10.14, we get the triple of Hilbert spaces

$$\mathscr{H}_1 \overset{(10.60)}{\longrightarrow} \mathscr{H}_0 \overset{(10.64)}{\longrightarrow} \mathscr{H}_{-1}. \tag{10.68}$$

The following are immediate:

1. All mappings in (10.68) are injective, continuous (in fact, contractive), having dense images.

2. The map $x \mapsto \xi_{\langle \cdot, x \rangle_0}$ is a contraction from $\mathscr{H}_1 \subset \mathscr{H}_0$ into $\mathscr{H}_0$. This follows from the estimate:

$$\begin{aligned} \|\xi_{\langle \cdot, x \rangle_0}\|_0 &\overset{(10.59)}{\leq} \|\xi_{\langle \cdot, x \rangle_0}\|_1 \\ &\overset{(10.63)}{=} \|\langle \cdot, x \rangle_0\|_{-1} \\ &\overset{(10.66)}{\leq} \|x\|_0 \overset{(10.59)}{\leq} \|x\|_1, \ \forall x \in \mathscr{H}_1. \end{aligned} \tag{10.69}$$

In particular, for $x \in \mathscr{H}_1$, $x \neq \xi_{\langle \cdot, x \rangle_0}$ in general.

3. The canonical bilinear form $\langle \cdot, \cdot \rangle : \mathscr{H}_1 \times \mathscr{H}_{-1} \to \mathbb{C}$ is given by

$$f(x) = \langle x, \xi_f \rangle_1, \ \forall x \in \mathscr{H}_1, \forall f \in \mathscr{H}_{-1}. \tag{10.70}$$

In particular, if $f = \langle \cdot, y \rangle_0$, $y \in \mathscr{H}_0$, then

$$\langle x, \xi_f \rangle_1 = \langle x, \xi_{\langle \cdot, y \rangle_0} \rangle_1 = \langle x, y \rangle_0, \ \forall x \in \mathscr{H}_1.$$

4. By Theorem 10.14, $\mathscr{H}_0$ is dense in $\mathscr{H}_{-1}$, and

$$\begin{aligned} \langle x, y \rangle_{-1} &:= \langle \langle \cdot, x \rangle_0, \langle \cdot, y \rangle_0 \rangle_{-1} \\ &= \langle \xi_{\langle \cdot, x \rangle_0}, \xi_{\langle \cdot, y \rangle_0} \rangle_1 \\ &= \langle x, \xi_{\langle \cdot, y \rangle_0} \rangle_0 = \langle \xi_{\langle \cdot, x \rangle_0}, y \rangle_0, \ \forall x, y \in \mathscr{H}_0. \end{aligned} \tag{10.71}$$

Combined with the order relation $\|x\|_{-1} \leq \|x\|_0$ for all $x \in \mathscr{H}_0$, we see that

$$|\langle x, y \rangle_{-1}| = \left| \langle x, \xi_{\langle \cdot, y \rangle_0} \rangle_0 \right|$$

$$\leq \|x\|_0 \, \|\xi_{\langle \cdot, y \rangle_0}\|_{-1}$$
$$= \|x\|_0 \, \|y\|_{-1} \, .$$
$$\leq \|x\|_0 \, \|y\|_0 \, , \, \forall x, y \in \mathscr{H}_0.$$

Thus $\langle \cdot, \cdot \rangle_{-1}$ is a continuous extension of $\langle \cdot, \cdot \rangle_0$.

Theorem 10.15. *Let* $\mathscr{H}_1 \hookrightarrow \mathscr{H}_0 \hookrightarrow \mathscr{H}_{-1}$ *be the triple in (10.68). Define* $B : \mathscr{H}_0 \to \mathscr{H}_0$ *by*

$$B : \underset{\mathscr{H}_0}{x} \xrightarrow{(10.64)} \underset{\mathscr{H}_{-1}}{\langle \cdot, x \rangle_0} \xrightarrow{(10.61)} \underset{\mathscr{H}_1}{\xi_{\langle \cdot, x \rangle_0}} \xrightarrow{(10.60)} \underset{\mathscr{H}_0}{\xi_{\langle \cdot, x \rangle_0}} . \qquad (10.72)$$

Then,

1. *For all* $x \in \mathscr{H}_1$, *and all* $y \in \mathscr{H}_0$,

$$\langle x, By \rangle_1 = \langle x, y \rangle_0 . \qquad (10.73)$$

In particular,

$$\langle x, y \rangle_{-1} = \langle Bx, By \rangle_1$$
$$= \langle x, By \rangle_0 = \langle Bx, y \rangle_0 , \, \forall x, y \in \mathscr{H}_0; \qquad (10.74)$$

where $\langle x, y \rangle_{-1} := \langle \xi_{\langle \cdot, x \rangle_0}, \xi_{\langle \cdot, y \rangle_0} \rangle_1$, *as defined in (10.71).*

2. *B is invertible.*

3. $\mathrm{ran}\,(B)$ *is dense in both* $\mathscr{H}_1$ *and* $\mathscr{H}_0$.

4. $0 \leq B \leq 1$. *In particular, B is a bounded selfadjoint operator on* $\mathscr{H}_0$.

Proof. $\square$

1. For $y \in \mathscr{H}_0$, $By = \xi_{\langle \cdot, y \rangle_0} \in \mathscr{H}_1$, where $\xi_{\langle \cdot, y \rangle_0}$ is given in (10.61)-(10.62). Thus,

$$\langle x, By \rangle_1 = \langle x, \xi_{\langle \cdot, y \rangle_0} \rangle_1 = \langle x, y \rangle_0 , \, \forall x \in \mathscr{H}_1.$$

(10.74) follows from this.

(a) If $\|Bx\|_0 = 0$, then $\|Bx\|_1 = 0$, since $Bx \in \mathscr{H}_1$ and $\mathscr{H}_1 \hookrightarrow \mathscr{H}_0$ is injective. But then

$$\|Bx\|_1 \overset{(10.74)}{=} \|x\|_{-1} = 0 \implies \|x\|_0 = 0$$

since $\mathscr{H}_0 \hookrightarrow \mathscr{H}_{-1}$ is injective. This shows that B is injective.

(b) Since $\mathscr{H}_0 \hookrightarrow \mathscr{H}_{-1}$ is dense, and $\mathscr{H}_{-1} \simeq \mathscr{H}_1$, it follows that $ran\,(B)$ is dense in $\mathscr{H}_1$. Now if $y \in \mathscr{H}_0$, and $\langle y, Bx \rangle_0 = 0$, for all $x \in \mathscr{H}_0$, then

$$\langle By, Bx \rangle_1 = 0, \ \forall x \in \mathscr{H}_0;$$

equivalently,

$$\langle y, x \rangle_{-1} = 0, \ \forall x \in \mathscr{H}_0.$$

Since $\mathscr{H}_0 \hookrightarrow \mathscr{H}_{-1}$ is dense, we have $\|y\|_{-1} = 0$. But $y \in \mathscr{H}_0$ and $\mathscr{H}_0 \hookrightarrow \mathscr{H}_{-1}$ is injective, it follows that $\|y\|_0 = 0$, i.e., $y = 0$ in $\mathscr{H}_0$. Therefore, $ran\,(B)$ is also dense in $\mathscr{H}_0$.

(c) For all $x \in \mathscr{H}_0$, we have

$$\langle x, Bx \rangle_0 \overset{(10.73)}{=} \langle Bx, Bx \rangle_1 \geq 0 \Longrightarrow B \geq 0.$$

On the other hand,

$$\langle x, Bx \rangle_0 = \langle Bx, Bx \rangle_1 \overset{(10.74)}{=} \langle x, x \rangle_{-1} \overset{(10.66)}{\leq} \langle x, x \rangle_0 ;$$

and so $B \leq 1$. Since B is positive and bounded, it is selfadjoint.

Another argument:

$$\|Bx\|_0 \leq \|Bx\|_1 = \|x\|_{-1} \leq \|x\|_0 , \ \forall x \in \mathscr{H}_0.$$

In view of applications, it is convenient to reformulate the previous theorem in terms of B^{-1}.

Theorem 10.16. *Let B, $\mathscr{H}_1$, $\mathscr{H}_0$, $\mathscr{H}_{-1}$, be as in Theorem 10.15, set $A := B^{-1}$, then*

1. $A = A^*$, $A \geq 1$.

2. $dom\,(A)$ *is dense in $\mathscr{H}_1$ and $\mathscr{H}_0$, and $ran\,(A) = \mathscr{H}_0$.*

3. *For all $y \in dom\,(A)$, $x \in \mathscr{H}_1$,*

$$\langle x, y \rangle_1 = \langle x, Ay \rangle_0 . \tag{10.75}$$

In particular,

$$\langle x, y \rangle_1 = \langle Ax, Ay \rangle_{-1}$$
$$= \langle Ax, y \rangle_0 = \langle x, Ay \rangle_0 , \ \forall x, y \in dom\,(A). \tag{10.76}$$

4. *There is a unique selfadjoint operator in $\mathscr{H}_0$ satisfying (10.75).*

Proof. Part (1)-(3) are immediate by Theorem 10.15. For (4), suppose A, B are selfadjoint in $\mathscr{H}_0$ such that
  (i) $dom\,(A)$, $dom\,(B)$ are contained in $\mathscr{H}_1$, dense in $\mathscr{H}_0$;
  (ii)

$$\langle x, y \rangle_1 = \langle x, Ay \rangle_0 \,, \, \forall x \in \mathscr{H}_1, y \in dom\,(A)$$
$$\langle x, y \rangle_1 = \langle x, By \rangle_0 \,, \, \forall x \in \mathscr{H}_1, y \in dom\,(B)\,.$$

Then, for all $x \in dom\,(A)$ and $y \in dom\,(B)$,

$$\langle x, By \rangle_0 = \langle x, y \rangle_1 = \langle Ax, y \rangle_0 \,.$$

Thus, $x \mapsto \langle Ax, y \rangle_0$ is a bounded linear functional on $dom\,(A)$, and so $y \in dom\,(A^*) = dom\,(A)$ and $A^*y = Ay = By$; i.e., $A \supset B$. Since A, B are selfadjoint, then

$$B = B^* \subset A^* = A.$$

Therefore $A = B$.   □

Theorem 10.17. *Let* $\mathscr{H}_1, \mathscr{H}_0, A$ *as in Theorem 10.16. Then*

1. $\mathscr{H}_1 = dom\,(A^{1/2})$, *and*

$$\langle x, y \rangle_1 = \left\langle A^{1/2}x, A^{1/2}y \right\rangle_0 \,, \, \forall x, y \in \mathscr{H}_1. \tag{10.77}$$

2. *For all* $x, y \in \mathscr{H}_0$,

$$\langle x, y \rangle_{-1} = \left\langle A^{-1/2}x, A^{-1/2}y \right\rangle_0 \,. \tag{10.78}$$

Since $\mathscr{H}_0$ *is dense in* $\mathscr{H}_{-1}$, *then* $\mathscr{H}_{-1} =$ *completion of* $\mathscr{H}_0$ *under the* $\left\| A^{-1/2}\cdot \right\|_0$-*norm.*

3. *For all* $x \in dom\,(A)$,

$$\|Ax\|_{-1} = \|x\|_1 \left(= \left\| A^{1/2}x \right\|_0 \right). \tag{10.79}$$

Consequently, the map $dom\,(A) \ni x \mapsto Ax \in \mathscr{H}_0$ *extends by continuity to a unitary operator from* $\mathscr{H}_1 \left(= dom\,(A^{1/2}) \right)$ *onto* $\mathscr{H}_{-1}$, *which is precisely the inverse of (10.61)-(10.62).*

Proof. (1) This is the result of the following observations:
                                                                                      □

a. $dom\,(A) \subset dom\,(A^{1/2})$. With the assumption $A \geq 1$, the containment is clear. The assertion also holds in general: By spectral theorem,

$$x \in dom\,(A) \iff \int \left(1 + |\lambda|^2 \right) \|P\,(d\lambda)\,x\|_0^2 < \infty;$$

where $P(\cdot)$ is the projection-valued measure (PVM) of A, defined on the set of all Borel sets in $\mathbb{R}$, and $d\mu_x := \|P(d\lambda)x\|_0^2$ is a finite positive Borel measure on $\mathbb{R}$. Thus,

$$\int (1 + |\lambda|) \|P(d\lambda)x\|_0^2 < \infty$$

since $L^2 \subset L^1$ when the measure is finite. But this is equivalent to $x \in dom\left(A^{1/2}\right)$.

(a) For any Hermitian operator T satisfying $T \geq c > 0$, we have the estimate:

$$\|Tx\| \leq \|x\| + \|Tx\| \leq (1+c)\|Tx\|$$

Thus, the graph norm of T is equivalent to $\|T\cdot\|$.

(b) $dom(A)$ is dense in $dom\left(A^{1/2}\right)$. Note $dom\left(A^{1/2}\right)$ is a Hilbert space with respect to the $A^{1/2}$-graph norm. By the discussion above, $\|\cdot\|_{A^{1/2}} \simeq \|A^{1/2}\cdot\|_0$.
Let $y \in dom\left(A^{1/2}\right)$, then

$$y \perp dom(A) \text{ in } dom\left(A^{1/2}\right)$$

$$\Updownarrow$$

$$\left\langle A^{1/2}y, A^{1/2}x \right\rangle_0 = 0, \ \forall x \in dom\left(A^{1/2}\right)$$

$$\Updownarrow$$

$$\langle y, Ax \rangle_0 = 0, \ \forall x \in dom\left(A^{1/2}\right)$$

$$\Updownarrow$$

$$y = 0 \text{ in } \mathcal{H}_0$$

(c) $dom(A)$ is dense in $\mathcal{H}_1$. See theorems 10.15-10.16.

(d) $\left\|A^{1/2}x\right\|_0 = \|x\|_1, \ \forall x \in dom(A)$. Indeed,

$$\langle x, x \rangle_1 \overset{(10.75)}{=} \langle x, Ax \rangle_0 = \left\langle A^{1/2}x, A^{1/2}x \right\rangle_0, \ \forall x \in dom(A).$$

Conclusion: (i) $dom(A)$ is dense in $\mathcal{H}_1$ and $dom\left(A^{1/2}\right)$; (ii) $\|\cdot\|_1$ and $\left\|A^{1/2}\cdot\right\|_0$ agree on $dom(A)$. Therefore the closures of $dom(A)$ in $\mathcal{H}_1$ and $dom\left(A^{1/2}\right)$ are identical. This shows $\mathcal{H}_1 = dom\left(A^{1/2}\right)$. (10.77) is immediate.

Proof. (2) Given $x, y \in \mathcal{H}_0$,

$$\langle x, y \rangle_{-1} \overset{(10.74)}{=} \left\langle A^{-1}x, A^{-1}y \right\rangle_1 \overset{(10.77)}{=} \left\langle A^{-1/2}x, A^{-1/2}y \right\rangle_0.$$

(3) Given $x \in dom\,(A)$, we have

$$\|Ax\|_{-1} \stackrel{(10.74)}{=} \left\|A^{-1}\,(Ax)\right\|_1 = \|x\|_1 \stackrel{(10.77)}{=} \left\|A^{1/2}x\right\|_0.$$

$\square$

Application: The Friedrichs extension revisited.

Theorem 10.18 (Friedrichs). *Let A be a densely defined Hermitian operator acting in $\mathscr{H}_0$, and assume $A \geq 1$. There exists a selfadjoint extension $S \supset A$, such that $L_S = L_A$, i.e., A and S have the same lower bound.*

Proof. Given A, construct the triple $\mathscr{H}_1 \hookrightarrow \mathscr{H}_0 \hookrightarrow \mathscr{H}_{-1}$ as in Theorem 10.12, so that $\mathscr{H}_1 = cl_{\langle \cdot, A \cdot \rangle_0}\,(dom\,(A))$, and

$$\langle y, x \rangle_1 = \langle y, Ax \rangle_0,\ \forall x \in dom\,(A), \forall y \in \mathscr{H}_1. \tag{10.80}$$

By Theorem 10.16, there is a densely defined selfadjoint operator S in $\mathscr{H}_0$, such that (i) $\mathscr{H}_1 = dom\,\left(S^{1/2}\right) \supset dom\,(S)$; (ii)

$$\langle y, x \rangle_1 = \left\langle S^{1/2}y, S^{1/2}x \right\rangle_0,\ \forall x, y \in \mathscr{H}_1. \tag{10.81}$$

Combing (10.80)-(10.81), we get

$$\langle y, Ax \rangle_0 = \left\langle S^{1/2}y, S^{1/2}x \right\rangle_0,\ \forall x \in dom\,(A), \forall y \in dom\,\left(S^{1/2}\right).$$

Therefore, $S^{1/2}x \in dom\,\left(S^{1/2}\right)$, i.e., $x \in dom\,(S)$, and

$$\langle y, Ax \rangle_0 = \langle y, Sx \rangle_0,\ \forall x \in dom\,(A), \forall y \in dom\,\left(S^{1/2}\right).$$

Since $\mathscr{H}_1 = dom\,\left(S^{1/2}\right)$ is dense in $\mathscr{H}_0$, we conclude that $Sx = Ax$, for all $x \in dom\,(A)$. Thus, $S \supset A$.

Clearly, $S \supset A$ implies $L_A \geq L_S$. On the other hand,

$$\|x\|_1^2 = \langle x, Ax \rangle_0 \geq L_A \langle x, x \rangle_0 = L \|x\|_0^2,\ \forall x \in dom\,(A)$$

and the inequality passes by continuity to all $x \in \mathscr{H}_1$; i.e.,

$$\|x\|_1^2 \geq L_A \|x\|_0^2,\ \forall x \in \mathscr{H}_1.$$

This is equivalent (by Theorem 10.16) to

$$\left\langle S^{1/2}x, S^{1/2}x \right\rangle_0 \geq L_A \langle x, x \rangle_0,\ \forall x \in dom\,\left(S^{1/2}\right) (= \mathscr{H}_1).$$

In particular,

$$\langle x, Sx \rangle_0 \geq L_A \langle x, x \rangle_0,\ \forall x \in dom\,(S)$$

and so $L_S \geq L_A$. Therefore, $L_A = L_S$. $\square$

A summary of relevant numbers from the Reference List

For readers wishing to follow up sources, or to go in more depth with topics above, we suggest: [AG93, Nel69, Dev72, DS88b, Kre46, Jor08, Rud91, Sto51, Sto90, FL28, Fug82, GJ87, JM80, vN35, JPT13, JP14, RS75, Hel13].

Chapter 11

Unbounded Graph-Laplacians

"Mathematics is an experimental science, and definitions do not come first, but later on."
— Oliver Heaviside

"It is nice to know that the computer understands the problem. But I would like to understand it too."
— Eugene Wigner

"Knowing is not enough; we must apply."
— Göthe

"Chance is a more fundamental conception than causality."
— Max Born

Below we study selfadjoint operators, and extensions in a particular case arising in the study of *infinite graphs*; the operators here are infinite discrete Laplacians.

As an application of the previous chapter, we consider the Friedrichs extension of discrete Laplacian in infinite networks [JP10, JP13a, JP13b, JT15a, JT15b].

By an *electrical network* we mean a graph G of vertices and edges satisfying suitable conditions which allow for computation of *voltage distribution* from a network of prescribed *resistors* assigned to the edges in G. The mathematical axioms are prescribed in a way that facilitates the use of the laws of *Kirchhoff* and *Ohm* in computing voltage distributions and *resistance distances* in G. It will be more convenient to work with prescribed conductance functions c on G. Indeed with a choice of *conductance function* c specified we define two crucial tools for our analysis, a *graph Laplacian* $\Delta \, (= \Delta_c,)$ a discrete version of more classical notions of Laplacians, and an *energy Hilbert space* $\mathscr{H}_E$.

Because of statistical consideration, and our use of random walk models, we focus our study on infinite electrical networks, i.e., the case when the graph G is countable infinite. In this case, for realistic models the graph Laplacian Δ_c will then be an unbounded operator with dense domain in $\mathscr{H}_E$, Hermitian and semibounded. Hence it has a unique Friedrichs extension.

Large networks arise in both pure and applied mathematics, e.g., in graph theory (the mathematical theory of networks), and more recently, they have become a current and fast developing research area; with applications including a host of problems coming from for example internet search, and social networks. Hence, of the recent applications, there is a change in outlook from finite to infinite.

More precisely, in traditional graph theoretical problems, the whole graph is given exactly, and we are then looking for relationships between its parameters, variables and functions; or for efficient algorithms for computing them. By contrast, for very large networks (like the Internet), variables are typically not known completely; — in most cases they may not even be well defined. In such applications, data about them can only be collected by indirect means; hence random variables and local sampling must be used as opposed to global processes.

Although such modern applications go far beyond the setting of large electrical networks (even the case of infinite sets of vertices and edges), it is nonetheless true that the framework of large electrical networks is helpful as a basis for the analysis we develop below; and so our results will be phrased in the setting of large electrical networks, even though the framework is much more general.

The applications of "large" or infinite graphs are extensive, counting just physics; see for example [BCD06, RAKK05, KMRS05, BC05, TD03, VZ92].

In discrete harmonic analysis, two operations play a key role, the Laplacian Δ, and the Markov operator P. An infinite network is a pair of sets, V vertices, and E, edges. In addition to this, one specifies a conductance function c . This is a function c defined on the edge set E. There are then two associated operators Δ and P are defined from, and they depend on the entire triple (V, E, c). For many problems one of the two operators is even used in the derivation of properties of the other. Both represent actions (operations) on appropriate spaces of functions, i.e., functions defined on the infinite set of vertices V. For the networks of interest to us, the vertex set V will be infinite, and we are therefore faced with a variety of choices of infinite dimensional function spaces. Because of spectral theory, the useful choices will be Hilbert spaces.

But even restricting to Hilbert spaces, there are at least three natural candidates: (i) the plain l^2 sequence space, so an l^2-space of functions on V, (ii) a suitably weighted l^2-space, and finally (iii), an energy Hilbert space $\mathscr{H}_E$. (The latter is an abstraction of more classical notions of Dirichlet spaces.) Which one of the three to use depends on the particular operator considered, and also on the questions asked.

We note that in infinite network models, both the Laplacian Δ, and the Markov operator P will have infinite by infinite matrix representations. Each of these infinite by infinite matrices is special, in that, as an infinite by infinite matrix, it will have non-zero entries localized only in finite bands containing the infinite matrix-diagonal (i.e., they are infinite banded matrices). This makes the algebraic matrix operations well defined.

Functional analytic and spectral theoretic tools enter the analysis as follows: In passing to appropriate Hilbert spaces, we arrive at classes of Hilbert space-operators, and the operators in question will be Hermitian. But the Laplacian Δ will be typically be an unbounded operator, albeit semibounded. By contrast we show that there is a weighted l^2-space such that the corresponding Markov operator P is a bounded, selfadjoint operator, and that it has its spectrum contained in the finite interval $[-1, 1]$. We caution, that in general this spectrum may be continuous, or have a mix of spectral types, continuous (singular or Lebesgue), and discrete.

In the details below, the following concepts and terms will play an important role; (for reference, see also Appendix F): Conductance function, graph Laplacian, reversible transition probabilities.

11.1 Basic Setting

Let V be a countable discrete set, and let $E \subset V \times V$ be a subset such that:

1. $(x, y) \in E \iff (y, x) \in E$; $x, y \in V$;

2. $\#\{y \in V \mid (x, y) \in E\}$ is finite, and > 0 for all $x \in V$;

3. $(x, x) \notin E$; and

4. $\exists o \in V$ such that for all $y \in V$ $\exists x_0, x_1, \ldots, x_n \in V$ with $x_0 = o$, $x_n = y$, $(x_{i-1}, x_i) \in E$, $\forall i = 1, \ldots, n$. (This property is called connectedness.)

5. If a conductance function c is given we require $c_{x_{i-1}x_i} > 0$. See Definition 11.1 below.

Definition 11.1. A function $c : E \to \mathbb{R}_+ \cup \{0\}$ is called *conductance function* if

1. $c(e) \geq 0$, $\forall e \in E$; and

2. Given $x \in V$, $c_{xy} > 0$, $c_{xy} = c_{yx}$, for all $(xy) \in E$.

If $x \in V$, we set

$$c(x) := \sum_{(xy) \in E} c_{xy}. \tag{11.1}$$

The summation in (11.1) is denoted $x \sim y$; i.e., $x \sim y$ if $(xy) \in E$.

Definition 11.2. When c is a conductance function (see also Definition 11.1) we set $\Delta = \Delta_c$ (the corresponding *graph Laplacian*

$$(\Delta u)(x) = \sum_{y \sim x} c_{xy}(u(x) - u(y)) = c(x)u(x) - \sum_{y \sim x} c_{xy}u(y). \tag{11.2}$$

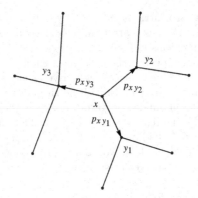

Figure 11.1: Transition probabilities p_{xy} at a vertex x (in V).

Given $G = (V, E, c)$ as above, and let $\Delta = \Delta_c$ be the corresponding graph Laplacian. With a suitable ordering on V, we obtain the following banded $\infty \times \infty$ matrix-representation for Δ (eq. (11.3)). We refer to [GLS12] for a number of applications of infinite banded matrices.

$$
\begin{bmatrix}
c(x_1) & -c_{x_1 x_2} & 0 & \cdots & \cdots & \cdots & \cdots & 0 & \cdots \\
-c_{x_2 x_1} & c(x_2) & -c_{x_2 x_3} & 0 & \cdots & \cdots & \cdots & \vdots & \cdots \\
0 & -c_{x_3 x_2} & c(x_3) & -c_{x_3 x_4} & 0 & \cdots & \cdots & 0 & \cdots \\
\vdots & 0 & \ddots & \ddots & \ddots & \ddots & \vdots & \vdots & \cdots \\
\vdots & & 0 & \ddots & \ddots & \ddots & \ddots & 0 & \vdots & \cdots \\
\vdots & 0 & \cdots & 0 & -c_{x_n x_{n-1}} & c(x_n) & -c_{x_n x_{n+1}} & 0 & \cdots \\
\vdots & \vdots & \cdots & \cdots & 0 & \ddots & \ddots & \ddots & \ddots
\end{bmatrix}
$$
(11.3)

Remark 11.1 (**Random Walk**). If (V, E, c) is given as in Definition 11.2, then for $(x, y) \in E$, set

$$p_{xy} := \frac{c_{xy}}{c(x)} \tag{11.4}$$

and note then $\{p_{xy}\}$ in (11.4) is a system of transition probabilities, i.e., $\sum_y p_{xy} = 1$, $\forall x \in V$, see Figure 11.1.

A Markov-random walk on V with transition probabilities (p_{xy}) is said to be *reversible* iff $\exists$ a positive function $\widetilde{c}$ on V such that

$$\widetilde{c}(x) \, p_{xy} = \widetilde{c}(y) \, p_{yx}, \; \forall (xy) \in E. \tag{11.5}$$

Theorem 11.1 ([Jor08, Woj09, Woj08]). *Let* $G = (V, E, c)$ *be as above,* V: *vertices,* E: *edges, and* $c : E \longrightarrow \mathbb{R}_+$ *a given conductance function; we assume finite range so that the* $\infty \times \infty$ *matrix in (11.3) is banded.*

Then the banded $\infty \times \infty$ matrix in (11.3) defines an essentially selfadjoint operator in $l^2(V)$, with dense domain equal to all finitely supported functions on V.

11.1.1 Infinite Path Space

Let $G = (V, E, c)$ be as above, and introduce the corresponding transition-probabilities, $\text{Prob}(x \to y) = p_{xy} = \frac{c_{xy}}{c(x)}$, from (11.4), indexed by the edges, i.e., $(x, y) \in E$.

Definition 11.3. An *infinite path* in G is a system of edges $(x_i x_{i+1})$, $i = 0, 1, 2, \cdots$.

With the assumption above, the space of all infinite paths Ω_x having a fixed starting vertex x, i.e., $x_0 = x$, is compact relative to the Tychonoff topology.

Lemma 11.1. *Let $G = (V, E, c, p)$ be as above, and let $x \in V$ be fixed. Let*

$$\Omega_x = \{\text{all infinite paths } \omega = (y_i y_{i+1})_{i=0}^{\infty} \text{ such that } y_0 = x.\}$$

Equip Ω_x with its cylinder sigma-algebra $\mathcal{F}_x$ (see e.g., (2.6) and (7.17).) Then there is a unique probability measure $\mathbb{P}_x$ on $(\Omega_x, \mathcal{F}_x)$ such that $\mathbb{P}_x$ is given on cylinder sets as follows:
If $(x_0 x_1 \cdots x_N)$ satisfy $x_0 = x$, $(x_i x_{i+1}) \in E$, set:

$$\mathcal{F}_x \ni Cyl^{(x_0 x_1 \cdots x_N)} = \{\omega \in \Omega_x \mid \omega = ((y_i y_{i+1}))_{i=0}^{\infty}, \ y_i = x_i, \ 0 \le i \le N\},$$

and

$$\mathbb{P}_x\left(Cyl^{(x_0 x_1 \cdots x_N)}\right) = p_{x_0 x_1} p_{x_1 x_2} \cdots p_{x_{N-1} x_N}. \tag{11.6}$$

Proof. This is a standard application of Kolmogorov consistency, a standard inductive/projective limit construction, and a use of the property $\sum_{y \sim z} p_{zy} = 1$, $\forall z \in V$. Starting with $\mathbb{P}_x$ as in (11.6), the consistency check and Kolmogorov implies that $\mathbb{P}_x$ extends from cylinders (see (11.6)) to all of $\mathcal{F}_x$. □

11.2 The Energy Hilbert Spaces $\mathscr{H}_E$

Let $G = (V, E, c)$ be an infinite connected network introduced above.

Definition 11.4. Set $\mathscr{H}_E :=$ completion of the space of all compactly supported functions $u : V \to \mathbb{C}$ with respect to

$$\langle u, v \rangle_{\mathscr{H}_E} := \frac{1}{2} \sum \sum_{(x,y) \in E} c_{xy} (\overline{u(x)} - \overline{u(y)})(v(x) - v(y)), \text{ and} \tag{11.7}$$

$$\|u\|_{\mathscr{H}_E}^2 := \frac{1}{2} \sum \sum_{(x,y) \in E} c_{xy} |u(x) - u(y)|^2; \tag{11.8}$$

then $\mathscr{H}_E$ is a Hilbert space [JP10, JT15b].

Lemma 11.2. *For all $x, y \in V$, there is a unique real-valued $\underline{dipole}$ vector $v_{xy} \in \mathscr{H}_E$ such that*

$$\langle v_{xy}, u \rangle_{\mathscr{H}_E} = u(x) - u(y), \ \forall u \in \mathscr{H}_E.$$

Proof. Apply Riesz' theorem.   □

Exercise 11.1 (Gaussian free field (GFF) [SS09]). Let $G = (V, E, c)$ be as in the setting of Section 11.2, and let $\mathscr{H}_E$ be the corresponding energy Hilbert space with inner product, and $\mathscr{H}_E$-norm as in (11.7)-(11.8).

1. Show that there is a probability space $\left(\Omega, \mathcal{F}, \mathbb{P}^{(G)}\right)$ and a Gaussian field X_φ, indexed by $\varphi \in \mathscr{H}_E$ (real valued) such that $E_{\mathbb{P}}(X_\varphi) = 0$, and

$$\mathbb{E}_{\mathbb{P}^{(G)}}\left(e^{iX_\varphi}\right) = e^{-\frac{1}{2}\|\varphi\|^2_{\mathscr{H}_E}}; \tag{11.9}$$

    in particular,

$$\mathbb{E}_{\mathbb{P}^{(G)}}(X_\varphi X_\psi) = \langle \varphi, \psi \rangle_{\mathscr{H}_E} \tag{11.10}$$

    for all $\varphi, \psi \in \mathscr{H}_E$.

2. Show that X arises from a Gaussian point process $\{X_x\}_{x \in V}$ such that

$$\mathbb{E}_{\mathbb{P}^{(G)}}(X_x X_\varphi) = \langle v_{xo}, \varphi \rangle_{\mathscr{H}_E} = \varphi(x) \tag{11.11}$$

    for all $\varphi \in \mathscr{H}_E$, and all $x \in V$, where o is a fixed base-point in the vertex set V, and we normalize in (11.11) such that $\varphi(o) = 0$.

Hint: For (1), use Corollary 2.2; and for (2), use Lemma 11.2.

Definition 11.5. Let $\mathscr{H}$ be a Hilbert space with inner product denoted $\langle \cdot, \cdot \rangle$, or $\langle \cdot, \cdot \rangle_{\mathscr{H}}$ when there is more than one possibility to consider. Let J be a countable index set, and let $\{w_j\}_{j \in J}$ be an indexed family of non-zero vectors in $\mathscr{H}$. We say that $\{w_j\}_{j \in J}$ is a *frame* for $\mathscr{H}$ iff (Def.) there are two finite positive constants b_1 and b_2 such that

$$b_1 \|u\|^2_{\mathscr{H}} \leq \sum_{j \in J} \left|\langle w_j, u \rangle_{\mathscr{H}}\right|^2 \leq b_2 \|u\|^2_{\mathscr{H}} \tag{11.12}$$

holds for all $u \in \mathscr{H}$. We say that it is a *Parseval frame* if $b_1 = b_2 = 1$.

For references to the theory and application of *frames*, see e.g., [HJL+13, KLZ09, CM13, SD13, KOPT13, EO13].

Lemma 11.3. *If $\{w_j\}_{j \in J}$ is a Parseval frame in $\mathscr{H}$, then the (analysis) operator $A = A_{\mathscr{H}} : \mathscr{H} \longrightarrow l^2(J)$,*

$$Au = \left(\langle w_j, u \rangle_{\mathscr{H}}\right)_{j \in J} \tag{11.13}$$

is well-defined and isometric. Its adjoint $A^* : l^2(J) \longrightarrow \mathscr{H}$ is given by

$$A^*\left((\gamma_j)_{j \in J}\right) := \sum_{j \in J} \gamma_j w_j \tag{11.14}$$

and the following hold:

1. *The sum on the RHS in (11.14) is norm-convergent;*

2. $A^* : l^2(J) \longrightarrow \mathscr{H}$ *is co-isometric; and for all* $u \in \mathscr{H}$*, we have*

$$u = A^* A u = \sum_{j \in J} \langle w_j, u \rangle w_j \tag{11.15}$$

where the RHS in (11.15) is norm-convergent.

Proof. The details are standard in the theory of frames; see the cited papers above. Note that (11.12) for $b_1 = b_2 = 1$ simply states that A in (11.13) is isometric, and so $A^* A = I_{\mathscr{H}}$ = the identity operator in $\mathscr{H}$, and AA^* = the projection onto the range of A. $\qquad\square$

Theorem 11.2. *Let* $G = (V, E, c)$ *be an infinite network. Choose an orientation on the edges, denoted by* $E^{(ori)}$*. Then the system of vectors*

$$\left\{ w_{xy} := \sqrt{c_{xy}} v_{xy}, \ (xy) \in E^{(ori)} \right\} \tag{11.16}$$

is a Parseval frame for the energy Hilbert space $\mathscr{H}_E$*. For all* $u \in \mathscr{H}_E$*, we have the following representation*

$$u = \sum_{(xy) \in E^{(ori)}} c_{xy} \langle v_{xy}, u \rangle v_{xy}, \ and \tag{11.17}$$

$$\|u\|_{\mathscr{H}_E}^2 = \sum_{(xy) \in E^{(ori)}} c_{xy} |\langle v_{xy}, u \rangle|^2 . \tag{11.18}$$

Proof. See [JT15a, CH08]. $\qquad\square$

Frames in $\mathscr{H}_E$ consisting of our system (11.16) are not ONBs when resisters are configured in non-linear systems of vertices, for example, resisters in parallel. See Figure 11.2, and 11.1.

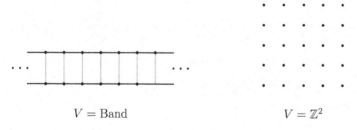

$V = \text{Band}$ $\qquad\qquad\qquad\qquad V = \mathbb{Z}^2$

Figure 11.2: non-linear system of vertices.

Example 11.1. Let c_{01}, c_{02}, c_{12} be positive constants, and assign conductances on the three edges (see Figure 11.3) in the triangle network.

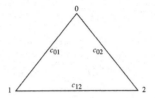

Figure 11.3: The set $\{v_{xy} : (xy) \in E\}$ is not orthogonal.

In this case, $w_{ij} = \sqrt{e_{ij}}v_{ij}$, $i < j$, in the cyclic order is a Parseval frame but not an ONB in $\mathscr{H}_E$ [JT15a].

Note the corresponding Laplacian $\Delta\,(= \Delta_c)$ has the following matrix representation

$$M := \begin{bmatrix} c(0) & -c_{01} & -c_{02} \\ -c_{01} & c(1) & -c_{12} \\ -c_{02} & -c_{12} & c(2) \end{bmatrix}. \tag{11.19}$$

The dipoles $\{v_{xy} : (xy) \in E^{(ori)}\}$ as 3-D vectors are the solutions to the equation

$$\Delta v_{xy} = \delta_x - \delta_y.$$

Hence,

$$Mv_{01} = \begin{bmatrix} 1 & -1 & 0 \end{bmatrix}^{tr}$$

$$Mv_{02} = \begin{bmatrix} 1 & 0 & -1 \end{bmatrix}^{tr}$$

$$Mv_{12} = \begin{bmatrix} 0 & 1 & -1 \end{bmatrix}^{tr}.$$

The Parseval frame from Lemma 11.3 is

$$w_{01} = \sqrt{c_{01}}v_{01} = \left[\frac{\sqrt{c_{01}}\,c_{12}}{c_{01}c_{02} + c_{01}c_{12} + c_{02}c_{12}}, -\frac{\sqrt{c_{01}}\,c_{02}}{c_{01}c_{02} + c_{01}c_{12} + c_{02}c_{12}}, 0 \right]^{tr}$$

$$w_{12} = \sqrt{c_{12}}v_{12} = \left[0, \frac{\sqrt{c_{12}}\,c_{02}}{c_{01}c_{02} + c_{01}c_{12} + c_{02}c_{12}}, -\frac{\sqrt{c_{12}}\,c_{01}}{c_{01}c_{02} + c_{01}c_{12} + c_{02}c_{12}} \right]^{tr}$$

$$w_{20} = \sqrt{c_{20}}v_{20} = \left[\frac{-\sqrt{c_{20}}\,c_{12}}{c_{01}c_{02} + c_{01}c_{12} + c_{02}c_{12}}, 0, \frac{\sqrt{c_{20}}\,c_{01}}{c_{01}c_{02} + c_{01}c_{12} + c_{02}c_{12}} \right]^{tr}.$$

Remark 11.2. The dipole v_{xy} is unique in $\mathscr{H}_E$ as an equivalence class, not a function on V. Note $\ker M$ = harmonic functions = constant (see (11.19)), and so $v_{xy} + \text{const} = v_{xy}$ in $\mathscr{H}_E$. Thus, the above frame vectors have non-unique representations as functions on V.

11.3   The Graph-Laplacian

Here we include some technical lemmas for *graph Laplacian* in the energy Hilbert space $\mathscr{H}_E$.

Let $G = (V, E, c)$ be as above; assume G is connected; i.e., there is a base point o in V such that every $x \in V$ is connected to o via a finite path of edges.

If $x \in V$, we set

$$\delta_x(y) = \begin{cases} 1 & \text{if } y = x \\ 0 & \text{if } y \neq x. \end{cases} \tag{11.20}$$

Definition 11.6. Let (V, E, c, o, Δ) be as above. Let $V' := V \backslash \{o\}$, and set

$$v_x := v_{x,o}, \ \forall x \in V'.$$

Further, let

$$\mathscr{D}_2 := span \left\{ \delta_x \mid x \in V \right\}, \text{ and} \tag{11.21}$$

$$\mathscr{D}_E := \left\{ \sum\nolimits_{x \in V'} \xi_x v_x \mid \text{finite support} \right\}; \tag{11.22}$$

where by "span" we mean of all *finite* linear combinations.

Lemma 11.4 below summarizes the key properties of Δ as an operator, both in $l^2(V)$ and in $\mathscr{H}_E$.

Lemma 11.4. *The following hold:*

1. $\langle \Delta u, v \rangle_{l^2} = \langle u, \Delta v \rangle_{l^2}, \ \forall u, v \in \mathscr{D}_2;$

2. $\langle \Delta u, v \rangle_{\mathscr{H}_E} = \langle u, \Delta v \rangle_{\mathscr{H}_E}, \ \forall u, v \in \mathscr{D}_E;$

3. $\langle u, \Delta u \rangle_{l^2} \geq 0, \ \forall u \in \mathscr{D}_2,$ *and*

4. $\langle u, \Delta u \rangle_{\mathscr{H}_E} \geq 0, \ \forall u \in \mathscr{D}_E.$

Moreover, we have

5. $\langle \delta_x, u \rangle_{\mathscr{H}_E} = (\Delta u)(x), \ \forall x \in V, \ \forall u \in \mathscr{H}_E.$

6. $\Delta v_{xy} = \delta_x - \delta_y, \ \forall v_{xy} \in \mathscr{H}_E.$ *In particular,* $\Delta v_x = \delta_x - \delta_o, \ x \in V' = V \backslash \{o\}.$

7.

$$\delta_x(\cdot) = c(x) v_x(\cdot) - \sum_{y \sim x} c_{xy} v_y(\cdot), \ \forall x \in V'.$$

8.

$$\langle \delta_x, \delta_y \rangle_{\mathscr{H}_E} = \begin{cases} c(x) = \sum_{t \sim x} c_{xt} & \text{if } y = x \\ -c_{xy} & \text{if } (xy) \in E \\ 0 & \text{if } (xy) \notin E, \ x \neq y. \end{cases}$$

Proof. See [JP10, JP11a, JT15a]. $\qquad\qquad\qquad\qquad\qquad\qquad\qquad\qquad\qquad\square$

11.4   The Friedrichs Extension of Δ, the Graph Laplacian

Fix a *conductance function* c. In this section we turn to some technical lemmas we will need for the Friedrichs extension of $\Delta \, (= \Delta_c)$.

It is known the graph-Laplacian Δ is automatically essentially selfadjoint as a densely defined operator in $l^2(V)$, but not as a $\mathscr{H}_E$ operator [Jor08, JP11b]. Since Δ defined on $\mathscr{D}_E$ is semibounded, it has the Friedrichs extension Δ_{Fri} (in $\mathscr{H}_E$).

Lemma 11.5. *Consider* Δ *with* $dom\,(\Delta) := span\,\{v_{xy} : x, y \in V\}$, *then*

$$\langle \varphi, \Delta \varphi \rangle_{\mathscr{H}_E} = \sum_{(xy) \in E} c_{xy}^2 \left| \langle v_{xy}, \varphi \rangle_{\mathscr{H}_E} \right|^2.$$

Proof. Suppose $\varphi = \sum \varphi_{xy} v_{xy} \in dom(\Delta)$. Note the edges are not oriented, and a direct computation shows that

$$\langle \varphi, \Delta \varphi \rangle_{\mathscr{H}_E} = 4 \sum_{x,y} |\varphi_{xy}|^2.$$

Using the Parseval frames in Theorem 11.2, we have the following representation

$$\varphi = \sum_{(xy) \in E} \underbrace{\frac{1}{2} c_{xy} \langle v_{xy}, \varphi \rangle_{\mathscr{H}_E} v_{xy}}_{=: \varphi_{xy}}.$$

Note $\varphi \in span\,\{v_{xy} : x, y \in V\}$, so the above equation contains a finite sum.

It follows that

$$\langle \varphi, \Delta \varphi \rangle_{\mathscr{H}_E} = 4 \sum_{(xy) \in E} |\varphi_{xy}|^2 = \sum_{(xy) \in E} c_{xy}^2 \left| \langle v_{xy}, \varphi \rangle_{\mathscr{H}_E} \right|^2$$

which is the assertion. $\square$

Theorem 11.3. *Let* $G = (V, E, c)$ *be an infinite network. If the deficiency indices of* $\Delta \, (= \Delta_c)$ *are* (k, k), $k > 0$, *where* $dom(\Delta) = span\,\{v_{xy}\}$, *then the Friedrichs extension* $\Delta_{Fri} \supset \Delta$ *is the restriction of* Δ^* *to*

$$dom(\Delta_{Fri}) := \left\{ u \in \mathscr{H}_E \mid \sum\nolimits_{(xy) \in E} c_{xy}^2 \left| \langle v_{xy}, \varphi \rangle_E \right|^2 < \infty \right\}. \qquad (11.23)$$

Proof. Follows from Lemma 11.5, and the characterization of Friedrichs extensions of semibounded Hermitian operators (Chapter 10); see, e.g., [DS88b, AG93, RS75]. $\square$

11.5   A 1D Example

Consider $G = (V, E, c)$, where $V = \{0\} \cup \mathbb{Z}_+$. Observation: Every sequence $a_1, a_2, \ldots$ in $\mathbb{R}_+$ defines a conductance $c_{n-1,n} := a_n$, $n \in \mathbb{Z}_+$, i.e.,

$$0 \xleftrightarrow[a_1]{} 1 \xleftrightarrow[a_2]{} 2 \xleftrightarrow[a_3]{} 3 \qquad \cdots \qquad n \xleftrightarrow[a_{n+1}]{} n+1 \cdots$$

The dipole vectors v_{xy} (for $x, y \in \mathbb{N}$) are given by

$$v_{xy}(z) = \begin{cases} 0 & \text{if } z \leq x \\ -\sum_{k=x+1}^{z} \frac{1}{a_k} & \text{if } x < z < y \\ -\sum_{k=x+1}^{y} \frac{1}{a_k} & \text{if } z \geq y. \end{cases}$$

See Figure 11.4.

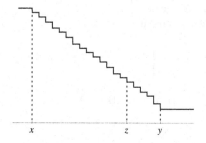

Figure 11.4: The dipole v_{xy}.

The corresponding graph Laplacian has the following matrix representation:

$$\begin{bmatrix} a_1 & -a_1 & & & & & & & \\ -a_1 & a_1 + a_2 & -a_2 & & & & & & \\ & -a_2 & a_2 + a_3 & -a_3 & & & & \mathbf{0} & \\ & & -a_3 & a_3 + a_4 & \ddots & & & & \\ & & & \ddots & \ddots & -a_n & & & \\ & & & & -a_n & a_n + a_{n+1} & -a_{n+1} & & \\ & \mathbf{0} & & & & -a_{n+1} & \ddots & \ddots & \\ & & & & & & \ddots & \ddots & \\ & & & & & & & \ddots & \ddots \end{bmatrix} \qquad (11.24)$$

That is,

$$\begin{cases} (\Delta u)_0 & = a_1 (u_0 - u_1) \\ (\Delta u)_n & = a_n (u_n - u_{n-1}) + a_{n+1} (u_n - u_{n+1}) \\ & = (a_n + a_{n+1}) u_n - a_n u_{n-1} - a_{n+1} u_{n+1}, \ \forall n \in \mathbb{Z}_+. \end{cases} \tag{11.25}$$

Lemma 11.6. *Let $G = (V, c, E)$ be as above, where $a_n := c_{n-1,n}$, $n \in \mathbb{Z}_+$. Then $u \in \mathscr{H}_E$ is the solution to $\Delta u = -u$ (i.e., u is a defect vector of Δ) if and only if u satisfies the following equation:*

$$\sum_{n=1}^{\infty} a_n \langle v_{n-1,n}, u \rangle_{\mathscr{H}_E} (\delta_{n-1}(s) - \delta_n(s) + v_{n-1,n}(s)) = 0, \ \forall s \in \mathbb{Z}_+; \tag{11.26}$$

where

$$\|u\|_{\mathscr{H}_E}^2 = \sum_{n=1}^{\infty} a_n \left| \langle v_{n-1,n}, u \rangle_{\mathscr{H}_E} \right|^2 < \infty. \tag{11.27}$$

Proof. By Theorem 11.2, the set $\left\{ \sqrt{a_n} v_{n-1,n} \right\}_{n=1}^{\infty}$ forms a Parseval frame in $\mathscr{H}_E$. In fact, the dipole vectors are

$$v_{n-1,n}(s) = \begin{cases} 0 & s \leq n-1 \\ -\frac{1}{a_n} & s \geq n \end{cases} ; n = 1, 2, \dots \tag{11.28}$$

and so $\left\{ \sqrt{a_n} v_{n-1,n} \right\}_{n=1}^{\infty}$ forms an ONB in $\mathscr{H}_E$; and $u \in \mathscr{H}_E$ has the representation

$$u = \sum_{n=1}^{\infty} a_n \langle v_{n-1,n}, u \rangle_{\mathscr{H}_E} v_{n-1,n}$$

see (11.15). Therefore, $\Delta u = -u$ if and only if

$$\sum_{n=1}^{\infty} a_n \langle v_{n-1,n}, u \rangle_{\mathscr{H}_E} (\delta_{n-1}(s) - \delta_n(s)) = - \sum_{n=1}^{\infty} a_n \langle v_{n-1,n}, u \rangle_{\mathscr{H}_E} v_{n-1,n}(s)$$

for all $s \in \mathbb{Z}_+$, which is the assertion. $\square$

Below we compute the deficiency space in an example with index values $(1, 1)$.

Lemma 11.7. *Let $(V, E, c = \{a_n\})$ be as above. Let $Q > 1$ and set $a_n := Q^n$, $n \in \mathbb{Z}_+$; then Δ has deficiency indices $(1, 1)$.*

Proof. Suppose $\Delta u = -u$, $u \in \mathscr{H}_E$. Then,

$$-u_1 = Q (u_1 - u_0) + Q^2 (u_1 - u_2) \Longleftrightarrow u_2 = \left(\frac{1}{Q^2} + \frac{1+Q}{Q} \right) u_1 - \frac{1}{Q} u_0$$

$$-u_2 = Q^2 (u_2 - u_1) + Q^3 (u_2 - u_3) \Longleftrightarrow u_3 = \left(\frac{1}{Q^3} + \frac{1+Q}{Q} \right) u_2 - \frac{1}{Q} u_1$$

and by induction,

$$u_{n+1} = \left(\frac{1}{Q^{n+1}} + \frac{1+Q}{Q} \right) u_n - \frac{1}{Q} u_{n-1}, \ n \in \mathbb{Z}_+$$

i.e., u is determined by the following matrix equation:

$$\begin{bmatrix} u_{n+1} \\ u_n \end{bmatrix} = \begin{bmatrix} \frac{1}{Q^{n+1}} + \frac{1+Q}{Q} & -\frac{1}{Q} \\ 1 & 0 \end{bmatrix} \begin{bmatrix} u_n \\ u_{n-1} \end{bmatrix}.$$

The eigenvalues of the coefficient matrix are

$$\lambda_{\pm} = \frac{1}{2} \left(\frac{1}{Q^{n+1}} + \frac{1+Q}{Q} \pm \sqrt{\left(\frac{1}{Q^{n+1}} + \frac{1+Q}{Q} \right)^2 - \frac{4}{Q}} \right)$$

$$\sim \frac{1}{2} \left(\frac{1+Q}{Q} \pm \left(\frac{Q-1}{Q} \right) \right) = \begin{cases} 1 \\ \frac{1}{Q} \end{cases} \quad \text{as } n \to \infty.$$

Equivalently, as $n \to \infty$, we have

$$u_{n+1} \sim \left(\frac{1+Q}{Q} \right) u_n - \frac{1}{Q} u_{n-1} = \left(1 + \frac{1}{Q} \right) u_n - \frac{1}{Q} u_{n-1}$$

and so

$$u_{n+1} - u_n \sim \frac{1}{Q} \left(u_n - u_{n-1} \right).$$

Therefore, for the tail-summation, we have:

$$\sum_n Q^n \left(u_{n+1} - u_n \right)^2 = \text{const} \sum_n \frac{(Q-1)^2}{Q^{n+2}} < \infty$$

which implies $\|u\|_{\mathscr{H}_E} < \infty$. $\qquad\square$

Next, we give a random walk interpretation of Lemma 11.7. See Remark 11.1, and Figure 11.1.

Remark 11.3 (Harmonic functions in $\mathscr{H}_E$). Note that in Section 11.5 (Lemma 11.7), the space of harmonic functions in $\mathscr{H}_E$ is one-dimensional; in fact if $Q > 1$ is fixed, then

$$\left\{ u \in \mathscr{H}_E \mid \Delta u = 0 \right\}$$

is spanned by $u = (u_n)_{n=0}^{\infty}$, $u_n = \frac{1}{Q^n}$, $n \in \mathbb{N}$; and of course $\|1/Q^n\|_{\mathscr{H}_E}^2 < \infty$.

Remark 11.4. For the domain of the Friedrichs extension Δ_{Fri}, we have:

$$dom(\Delta_{Fri}) = \left\{ f \in \mathscr{H}_E \mid (f(x) - f(x+1)) Q^x \in l^2(\mathbb{Z}_+) \right\} \qquad (11.29)$$

i.e.,

$$dom(\Delta_{Fri}) = \left\{ f \in \mathscr{H}_E \mid \sum_{x=0}^{\infty} |f(x) - f(x+1)|^2 Q^{2x} < \infty \right\}.$$

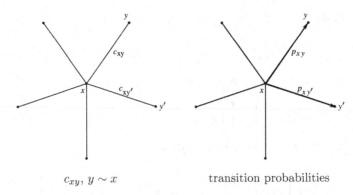

Figure 11.5: Neighbors of x.

Proof. By Theorem 11.2, we have the following representation, valid for all $f \in \mathscr{H}_E$:

$$f = \sum_x \langle f, Q^{\frac{x}{2}} v_{(x,x+1)} \rangle_{\mathscr{H}_E} Q^{\frac{x}{2}} v_{(x,x+1)}$$
$$= \sum_x (f(x) - f(x+1)) Q^x v_{(x,x+1)};$$

and

$$\langle f, \Delta f \rangle_{\mathscr{H}_E} = \sum_x |f(x) - f(x+1)|^2 Q^{2x}.$$

The desired conclusion (11.29) now follows from Theorem 11.3. Also see e.g. [DS88b, AG93].   □

Definition 11.7. Let $G = (V, E, c)$ be a connected graph. The set of transition probabilities (p_{xy}) is said to be reversible if there exists $c : V \to \mathbb{R}_+$ such that

$$c(x) p_{xy} = c(y) p_{yx}; \tag{11.30}$$

and then

$$c_{xy} := c(x) p_{xy} \tag{11.31}$$

is a system of conductance. Conversely, for a system of conductance (c_{xy}) we set

$$c(x) := \sum_{y \sim x} c_{xy}, \text{ and} \tag{11.32}$$

$$p_{xy} := \frac{c_{xy}}{c(x)} \tag{11.33}$$

and so (p_{xy}) is a set of transition probabilities. See Figure 11.5.

Lemma 11.8. *Let (V, E, c, p) be as in Definition 11.7. For functions f on V, set*

$$(Pf)(x) := \sum_{y \sim x} p_{xy} f(y), \quad x \in V. \tag{11.34}$$

Then P is bounded and selfadjoint in the weighted l^2-space

$$l^2(c) = \left\{ f : V \longrightarrow \mathbb{C} \mid \sum_{x \in V} |f(x)|^2 c(x) = \|f\|_c^2 < \infty \right\},$$

corresponding to the weight function c.

Proof. (Sketch) For $f, g \in l^2(c)$, we have

$$
\begin{aligned}
\langle f, Pg \rangle_c &= \sum_x \sum_y \overline{f(x)} g(y) p_{xy} c(x) \\
&\underset{\text{by (11.30)}}{=} \sum_x \sum_y \overline{f(x)} g(y) p_{yx} c(y) \\
&= \sum_y \sum_x \overline{p_{yx} f(x)} g(y) c(y) \\
&= \sum_y \overline{(Pf)(y)} g(y) c(y) = \langle Pf, g \rangle_c.
\end{aligned}
$$

(In the double summations above it is understood that we are summing only over those pairs x, y of vertices such that (x, y) constitute an edge in the given graph.)

Boundedness: From Lemma 11.4 (part 1), we directly get the following estimate, valid for all $f \in l^2(c)$:

$$- \|f\|_c^2 \le \langle f, Pf \rangle_c \le \|f\|_c^2. \tag{11.35}$$

$\square$

Example 11.2. We now return to the special case of (G, V, E) corresponding to the "linear" graph which was introduced at the outset of Section 11.5.

Recall the graph Laplacian in (11.25) can be written as

$$(\Delta u)_n = c(n)(u_n - p_-(n) u_{n-1} - p_+(n) u_{n+1}), \ \forall n \in \mathbb{Z}_+; \tag{11.36}$$

where

$$c(n) := a_n + a_{n+1} \tag{11.37}$$

and

$$p_-(n) := \frac{a_n}{c(n)}, \ p_+(n) := \frac{a_{n+1}}{c(n)} \tag{11.38}$$

are the left/right transition probabilities, as shown in Figure 11.6.

In the case $a_n = Q^n$, $Q > 1$, as in Lemma 11.7, we have

$$c(n) := Q^n + Q^{n+1}, \text{ and} \tag{11.39}$$

$$p_+ := p_+ (n) = \frac{Q^{n+1}}{Q^n + Q^{n+1}} = \frac{Q}{1 + Q} \tag{11.40}$$

$$p_- := p_- (n) = \frac{Q^n}{Q^n + Q^{n+1}} = \frac{1}{1 + Q}. \tag{11.41}$$

For all $n \in \mathbb{Z}_+ \cup \{0\}$, set

$$(Pu)_n := p_- u_{n-1} + p_+ u_{n+1}. \tag{11.42}$$

Note $(Pu)_0 = u_1$. By (11.36), we have

$$\Delta = c(1 - P). \tag{11.43}$$

In particular, $p_+ > \frac{1}{2}$, i.e., a random walker has probability $> \frac{1}{2}$ of moving to the right. It follows that

$$\underbrace{\text{travel time} (n, \infty)}_{= \text{ dist to } \infty} < \infty;$$

and so Δ is not essentially selfadjoint, i.e., indices $(1, 1)$.

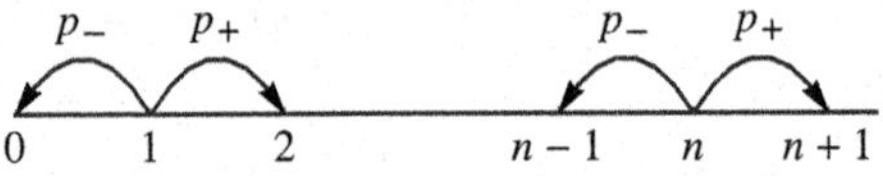

Figure 11.6: The transition probabilities p_+, p_-, in the case of constant transition probabilities, i.e., $p_+ (n) = p_+$, and $p_- (n) = p_-$ for all $n \in \mathbb{Z}_+$.

Lemma 11.9. *Let* $(V, E, \Delta (= \Delta_c))$ *be as above, where the conductance* c *is given by* $c_{n-1,n} = Q^n$, $n \in \mathbb{Z}_+$, $Q > 1$ *(see Lemma 11.7). For all* $\lambda > 0$, *there exists* $f_\lambda \in \mathscr{H}_E$ *satisfying* $\Delta f_\lambda = \lambda f_\lambda$.

Proof. By (11.43), we have $\Delta f_\lambda = \lambda f_\lambda \Longleftrightarrow P f_\lambda = \left(1 - \frac{\lambda}{c}\right) f_\lambda$, i.e.,

$$\frac{1}{1 + Q} f_\lambda (n - 1) + \frac{Q}{1 + Q} f_\lambda (n + 1) = \left(1 - \frac{\lambda}{Q^{n-1} (1 + Q)}\right) f_\lambda (n)$$

and so

$$f_\lambda (n + 1) = \left(\frac{1 + Q}{Q} - \frac{\lambda}{Q^n}\right) f_\lambda (n) - \frac{1}{Q} f_\lambda (n - 1). \tag{11.44}$$

This corresponds to the following matrix equation:

$$\begin{bmatrix} f (n + 1) \\ f (n) \end{bmatrix} = \begin{bmatrix} \frac{1+Q}{Q} - \frac{\lambda}{Q^n} & -\frac{1}{Q} \\ 1 & 0 \end{bmatrix} \begin{bmatrix} f (n) \\ f (n - 1) \end{bmatrix}$$

$$\sim \begin{bmatrix} \frac{1+Q}{Q} & -\frac{1}{Q} \\ 1 & 0 \end{bmatrix} \begin{bmatrix} f (n) \\ f (n - 1) \end{bmatrix}, \text{ as } n \to \infty.$$

The eigenvalues of the coefficient matrix are given by

$$\lambda_{\pm} \sim \frac{1}{2}\left(\frac{1+Q}{Q} \pm \left(\frac{Q-1}{Q}\right)\right) = \begin{cases} 1 \\ \frac{1}{Q} \end{cases} \quad \text{as } n \to \infty.$$

That is, as $n \to \infty$,

$$f_\lambda(n+1) \sim \left(\frac{1+Q}{Q}\right) f_\lambda(n) - \frac{1}{Q} f_\lambda(n-1);$$

i.e.,

$$f_\lambda(n+1) \sim \frac{1}{Q} f_\lambda(n); \tag{11.45}$$

and so the tail summation of $\|f_\lambda\|^2_{\mathscr{H}_E}$ is finite. (See the proof of Lemma 11.7.) We conclude that $f_\lambda \in \mathscr{H}_E$. □

Corollary 11.1. *Let* (V, E, Δ) *be as in the lemma. The Friedrichs extension* Δ_{Fri} *has continuous spectrum* $[0, \infty)$.

Proof. Fix $\lambda \geq 0$. We prove that if $\Delta f_\lambda = \lambda f_\lambda$, $f \in \mathscr{H}_E$, then $f_\lambda \notin dom(\Delta_{Fri})$.

Note for $\lambda = 0$, f_0 is harmonic, and so $f_0 = k\left(\frac{1}{Q^n}\right)_{n=0}^\infty$ for some constant $k \neq 0$. See Remark 11.3. It follows from (11.29) that $f_0 \notin dom(\Delta_{Fri})$.

The argument for $\lambda > 0$ is similar. Since as $n \to \infty$, $f_\lambda(n) \sim \frac{1}{Q^n}$ (eq. (11.45)), so by (11.29) again, $f_\lambda \notin dom(\Delta_{Fri})$.

However, if $\lambda_0 < \lambda_1$ in $[0, \infty)$ then

$$\int_{\lambda_0}^{\lambda_1} f_\lambda(\cdot)\, d\lambda \in dom(\Delta_{Fri}) \tag{11.46}$$

and so every f_λ, $\lambda \in [0, \infty)$, is a generalized eigenfunction, i.e., the spectrum of Δ_{Fri} is purely continuous with Lebesgue measure, and multiplicity one.

The verification of (11.46) follows from (11.44), i.e.,

$$f_\lambda(n+1) = \left(\frac{1+Q}{Q} - \frac{\lambda}{Q^n}\right) f_\lambda(n) - \frac{1}{Q} f_\lambda(n-1). \tag{11.47}$$

Set

$$F_{[\lambda_0, \lambda_1]} := \int_{\lambda_0}^{\lambda_1} f_\lambda(\cdot)\, d\lambda. \tag{11.48}$$

Then by (11.47) and (11.48),

$$F_{[\lambda_0, \lambda_1]}(n+1) = \frac{1+Q}{Q} F_{[\lambda_0, \lambda_1]}(n) - \frac{1}{Q^n}\int_{\lambda_0}^{\lambda_1} \lambda f_\lambda(n)\, d\lambda - \frac{1}{Q} F_{[\lambda_0, \lambda_1]}(n-1)$$

and $\int_{\lambda_0}^{\lambda_1} \lambda f_\lambda d\lambda$ is computed using integration by parts. □

A summary of relevant numbers from the Reference List

For readers wishing to follow up sources, or to go in more depth with topics above, we suggest: [Jor08, JP10, JP11a, JP11b, JP13a, JP13b, JT15a, RAKK05, Yos95, BKS13, Str12, JP14, LPW13, JP14, CZ07, AJSV13, SS09, JP12, JT15b].

Chapter 12

Reproducing Kernel Hilbert Space

"The simplicities of natural laws arise through the complexities of the language we use for their expression."
— Eugene Wigner

"... an apt comment on how science, and indeed the whole of civilization, is a series of incremental advances, each building on what went before."
— Stephen Hawking

"At a given moment there is only a fine layer between the 'trivial' and the impossible. Mathematical discoveries are made in this layer."
— Andrey Kolmogorov

"... the distinction between two sorts of truths, profound truths recognized by the fact that the opposite is also a profound truth; by contrast to trivialities, where opposites are obviously absurd."
— Niels Bohr

12.1 Fundamentals

A special family of Hilbert spaces $\mathscr{H}$ are reproducing kernel Hilbert spaces (RKHSs). We say that $\mathscr{H}$ is a RKHS if $\mathscr{H}$ is a Hilbert space of functions on some set S such that for every x in S, the linear mapping $f \longmapsto f(x)$ is continuous in the norm of $\mathscr{H}$.

We begin with a general

403

Definition 12.1. Let S be a set. We say that $\mathcal{H}$ is a S-reproducing kernel Hilbert space if:

1. $\mathcal{H}$ is a Hilbert space of functions on S; and

2. For all $s \in S$, the mapping $\mathcal{H} \longrightarrow \mathbb{C}$, by $E_s : h \longmapsto h(s)$ is continuous on $\mathcal{H}$, i.e., by Riesz, there is a $K_s \in \mathcal{H}$ such that $h(s) = \langle K_s, h \rangle$, $\forall h \in \mathcal{H}$.

Notation. The system of functions $\{K_s : s \in S\} \subset \mathcal{H}$ is called the associated reproducing kernel.

Theorem 12.1 (Aronszajn [Aro50]). *Let S be a set, and let $K : S \times S \longrightarrow \mathbb{C}$ be a positive definite function, i.e., for all finite subset $F \subset S$,*

$$\sum_{(x,y) \in F \times F} \sum \overline{c_x} c_y K(x,y) \geq 0 \tag{12.1}$$

holds for $\forall (c_x)_{x \in F} \in \mathbb{C}^{\#F}$.
  Then there is a Hilbert space $\mathcal{H} = \mathcal{H}(K)$ such that the functions

$$K_x(\cdot) = K(\cdot, x), \quad x \in S \tag{12.2}$$

span a dense subspace in $\mathcal{H}(K)$, called the reproducing kernel Hilbert space (RKHS), and for all $f \in \mathcal{H}(K)$, we have

$$f(x) = \langle K_x, f \rangle_{\mathcal{H}}, \quad x \in S. \tag{12.3}$$

Proof. See [Aro50] and Corollary 2.2; but also note that the present argument is almost identical to the proof of the GNS-construction, see Theorem 5.3 in Section 5.2.   □

Remark 12.1. If φ is a state on a $*$-algebra $\mathfrak{A}$, then

$$K_\varphi(A, B) = \varphi(A^* B) \text{ on } \mathfrak{A} \times \mathfrak{A} \tag{12.4}$$

is positive definite.

   Let $\mathcal{H}$ be a RKHS, see Definition 12.1, hence S is a set, and $\mathcal{H}$ is a Hilbert space of functions on S such that (2) holds. Fix $s \in S$, and note that $E_s : \mathcal{H} \longrightarrow \mathbb{C}$ is then a bounded linear operator, where $\mathbb{C}$ is a 1-dimensional Hilbert space. Hence its adjoint $E_s^* : \mathbb{C} \longrightarrow \mathcal{H}^* \simeq \mathcal{H}$ is well-defined.

Claim 12.1. The kernel K_s in Definition 12.1 is $E_s^*(1) = K_s$.

Proof. For $\lambda \in \mathbb{C}$ and $h \in \mathcal{H}$, we have

$$\langle E_s^*(\lambda), h \rangle_{\mathcal{H}} = \overline{\lambda} E_s(h) = \overline{\lambda} h(s) \underset{\text{(by (2))}}{=} \overline{\lambda} \langle K_s, h \rangle_{\mathcal{H}} = \langle \lambda K_s, h \rangle_{\mathcal{H}};$$

and therefore $E_s^*(\lambda) = \lambda K_s$ as desired.   □

Example 12.1. For $s, t \in [0, 1]$, set

$$K(s, t) = s \wedge t = \min(s, t), \text{ i.e.,} \tag{12.5}$$

the covariance kernel for Brownian motion on the interval $[0, 1]$, see Section 7.1.

Exercise 12.1 (The Fundamental Theorem of Calculus and a RKHS). Show that the RKHS for the kernel K in (12.5) is

$$\mathscr{H} = \left\{ f : \begin{array}{l} \text{locally integrable with distribution-} \\ \text{derivative } f' \in L^2(0, 1) \text{ and } f(0) = 0 \end{array} \right\}$$

with the inner product

$$\langle f, g \rangle_{\mathscr{H}} := \int_0^1 \overline{f'(x)} g'(x) \, dx.$$

Hint:

Step 1. Show that $K_t = K(t, \cdot)$ is in $\mathscr{H}$

Step 2. Show that for all $f \in \mathscr{H}$, we have

$$\langle K_t, f \rangle_{\mathscr{H}} = \int_0^t f'(x) \, dx = f(t).$$

Suppose $K : [0, 1] \times [0, 1] \longrightarrow \mathbb{C}$ is positive definite and *continuous*, i.e., for all finite sums:

$$\sum_j \sum_k \overline{c_j} c_k K(t_j, t_k) \geq 0, \tag{12.6}$$

$\{c_j\} \subset \mathbb{C}, \{t_j\} \subset [0, 1]$.

Exercise 12.2 (Mercer [GBD94, Wit74]). Show that the operator T_K,

$$(T_K \varphi)(t) = \int_0^1 K(t, s) \varphi(s) \, ds \tag{12.7}$$

in $L^2(0, 1)$ is trace-class, and

$$trace(T_K) = \int_0^1 K(t, t) \, dt. \tag{12.8}$$

Hint: Apply weak-compactness, and the Spectral Theorem for compact self-adjoint operators, Theorem 4.6. Combine this with a choice of an ONB in the RKHS defined from (12.6).

Exercise 12.3 (The Szegő-kernel). Let $\mathbb{H}_2$ be the Hardy space of the disk $\mathbb{D} = \{z \in \mathbb{C} : |z| < 1\}$, see Section 5.9. Show that $\mathbb{H}_2$ is a RKHS with reproducing kernel (the Szegő-kernel):

$$K(z, w) = \frac{1}{1 - \overline{z}w} \tag{12.9}$$

i.e., that we have:

$$\langle K(z, \cdot), f \rangle_{\mathbb{H}_2} = f(z), \ \forall f \in \mathbb{H}_2, \ \forall z \in \mathbb{D}. \tag{12.10}$$

Hint: Substitute (12.9) into the formula from $\langle \cdot, \cdot \rangle_{\mathbb{H}_2}$ inner product on the LHS in (12.10), and recall our convention: Inner products are linear in the second variable.

Definition 12.2. Let S be a set and $K : S \times S \to \mathbb{C}$ a fixed positive definite function; and let $\mathcal{H}_K$ be the corresponding RKHS; (see Definition 12.1). Let $\varphi : S \to \mathbb{C}$ be a function on S; then we say φ is a *multiplier*, written "$\varphi \in \text{Multp}(\mathcal{H}_K)$" if multiplication by φ defines a bounded linear operator in $\mathcal{H}_K$, so

$$\begin{aligned} M_\varphi : \mathcal{H}_K &\longrightarrow \mathcal{H}_K \\ (M_\varphi f)(x) = \varphi(x) f(x), &\ \forall f \in \mathcal{H}_K, \ \forall x \in S. \end{aligned} \tag{12.11}$$

Exercise 12.4 (Multp. $(\mathcal{H}_K)$).

1. Show the following equivalence:

$$\varphi \ \in \ \text{Multp}(\mathcal{H}_K) \tag{12.12}$$
$$\Updownarrow$$

$\exists$ constant $B < \infty$ such that we have the following estimate for all finite sums:

$$\sum_i \sum_j \overline{c_i} c_j \left(B - \overline{\varphi(x_i)}\varphi(x_j) \right) K(x_i, x_j) \geq 0; \tag{12.13}$$

i.e., computed for all systems $\{c_i\} \subset \mathbb{C}$, and $\{x_i\} \subset S$.

2. Let $\varphi \in \text{Multp}(\mathcal{H}_K)$, and let $\{K_x\}_{x \in S}$ be the kernel functions. Prove that

$$M_\varphi^*(K_x) = \overline{\varphi(x)} K_x, \ \forall x \in S, \tag{12.14}$$

where M_φ^* denotes the adjoint operator.

Hint: Verify the following identity:

$$\left\langle \overline{\varphi(x)} K_x, f \right\rangle_{\mathcal{H}_K} = \langle K_x, M_\varphi f \rangle_{\mathcal{H}_K}, \ \forall x \in S, \ f \in \mathcal{H}_K. \tag{12.15}$$

3. Show directly from (1) and (2) that $\text{Multp}(\mathcal{H}_K)$ is an algebra.

4. Apply (1) to the Hardy space $\mathbb{H}_2$ of the disk to conclude that

$$\text{Multp}(\mathbb{H}_2) = \mathbb{H}_\infty. \tag{12.16}$$

From the postulates of quantum physics, we know that measurements of observables are computed from associated selfadjoint operators–observables. From the corresponding spectral resolutions, we get probability measures, and of course uncertainty. There are many philosophical issues (which we bypass here), and we do not yet fully understand quantum reality. See for example, [Sla03, CJK+12].

The axioms are as follows: An *observable* is a Hermitian (selfadjoint) linear operator mapping a Hilbert space, the space of *states*, into itself. The values obtained in a *physical measurement* are, in general, described by a probability distribution; and the distribution represents a suitable "average" (or "*expectation*") in a measurement of values of some quantum observable in a state of some prepared system. The states are (up to phase) unit vectors in the Hilbert space, and a measurement corresponds to a *probability distribution* (derived from a projection-valued spectral measure). The spectral type may be *continuous* (such as position and momentum) or *discrete* (such as spin).

Information about the measures μ are computed with the use of *generating functions* (on $\mathbb{R}$), i.e., spectral (Bochner/Fourier) transforms of the corresponding measure. Generating functions are positive definite continuous functions $F (= F_\mu)$ on $\mathbb{R}$. One then tries to recover μ from information about F. In this chapter we explore the cases when information about $F(x)$ is only available for x in a bounded interval.

In probability theory, normalized continuous positive definite functions F, i.e., $F(0) = 1$, arise as *generating functions* for probability measures, and one passes from information about one to the other; — from generating function to probability measure is called "the inverse problem", see e.g., [DM72]. Hence the study of partially defined p.d. functions addresses the inverse question: ambiguity of measures when only partial information for a possible generating function is available.

In the details below, the following concepts and terms will play an important role; (for reference, see also Appendix F): Reproducing kernel Hilbert space, locally defined positive definite functions, positive definite extensions, deficiency spaces of the operator D_F.

12.2   Application to Optimization

One of the more recent applications of kernels and the associated reproducing kernel Hilbert spaces (RKHS) is to optimization, also called kernel-optimization. In the context of machine learning, it refers to training-data and feature spaces. In the context of numerical analysis, a popular version of the method is used to produce splines from sample points; and to create best spline-fits. In statistics, there are analogous optimization problems going by the names "least-square fitting," and "maximum-likelihood" estimation. In the latter instance, the object to be determined is a suitable probability distribution which makes "most likely" the occurrence of some data which arises from experiments, or from testing.

What these methods have in common is a minimization (or a max problem) involving a "quadratic" expression Q with two terms. The first in Q measures a suitable $L^2 (\mu)$-square applied to a difference of a measurement and a "best fit." The latter will then to be chosen from anyone of a number of suitable reproducing kernel Hilbert spaces (RKHS). The choice of kernel and RKHS will serve to select desirable features. So we will minimize a quantity Q which is the sum of two terms as follows: (i) a L^2-square applied to a difference, and (ii) a penalty term which is a RKHS norm-squared. (See eq. (12.18).) The term in (ii) is often called the penalty term. In the application to determination of splines, the penalty term may be a suitable Sobolev normed-square; i.e., L^2 norm-squared applied to a chosen number of derivatives. Hence non-differentiable choices will be "penalized."

In all of the cases, discussed above, there will be a good choice of (i) and (ii), and we show that there is then an explicit formula for the optimal solution; see eq (12.21) in Theorem 12.2 below.

Let S be a set, and let $K : S \times S \longrightarrow \mathbb{C}$ be a positive definite (p.d.) kernel. Let $\mathscr{H}_K$ be the corresponding reproducing kernel Hilbert space (RKHS). Let $\mathscr{F}$ be a sigma-algebra of subsets of S, and let μ be a positive measure on the corresponding measure space $(S, \mathscr{F})$. We assume that μ is sigma-finite. We shall further assume that the associated operator T given by

$$\mathscr{H}_K \ni f \xrightarrow{\ T\ } (f(s))_{s \in S} \in L^2(\mu) \tag{12.17}$$

is densely defined and closable.

Fix $\beta > 0$, and $\psi \in L^2(\mu)$, and set

$$Q_{\psi,\beta}(f) = \|\psi - Tf\|^2_{L^2(\mu)} + \beta \|f\|^2_{\mathscr{H}_K} \tag{12.18}$$

defined for $f \in \mathscr{H}_K$, or in the dense subspace $dom(T)$ where T is the operator in (12.17). Let

$$L^2(\mu) \xrightarrow{\ T^*\ } \mathscr{H}_K \tag{12.19}$$

be the corresponding adjoint operator, i.e.,

$$\langle F, T^*\psi \rangle_{\mathscr{H}_K} = \langle Tf, \psi \rangle_{L^2(\mu)} = \int_S \overline{f(s)}\psi(s)\, d\mu(s). \tag{12.20}$$

Theorem 12.2. *Let* K, μ, ψ, β *be as specified above; then the optimization problem* $\inf_{f \in \mathscr{H}_K} Q_{\psi,\beta}(f)$ *has a unique solution* F *in* $\mathscr{H}_K$, *it is*

$$F = (\beta I + T^*T)^{-1} T^*\psi \tag{12.21}$$

where the operator T *and* T^* *are as specified in (12.17)-(12.20).*

Proof. (Sketch) We fix F, and assign $f_\varepsilon := F + \varepsilon h$ where h varies in the dense domain $dom(T)$ from (12.17). For the derivative $\frac{d}{d\varepsilon}\big|_{\varepsilon=0}$ we then have:

$$\frac{d}{d\varepsilon}\big|_{\varepsilon=0} Q_{\psi,\beta}(f_\varepsilon) = 2\Re \langle h, (\beta I + T^*T) F - T^*\psi \rangle_{\mathscr{H}_K} = 0$$

for all h in a dense subspace in $\mathscr{H}_K$. The desired conclusion follows. □

12.2.1 Application: Least square-optimization

We now specialize the optimization formula from Theorem 12.2 to the problem of minimize a "quadratic" quantity Q. It is still the sum of two individual terms: (i) a L^2-square applied to a difference, and (ii) a penalty term which is the RKHS norm-squared. But the least-square term in (i) will simply be a sum of a finite number of squares of differences; hence "least-squares." As an application, we then get an easy formula (Theorem 12.3) for the optimal solution.

Let K be a positive definite kernel on $S \times S$ where S is an arbitrary set, and let $\mathscr{H}_K$ be the corresponding reproducing kernel Hilbert space (RKHS). Let $m \in \mathbb{N}$, and consider sample points:

$\{t_j\}_{j=1}^m$ as a finite subset in S, and

$\{y_i\}_{i=1}^m$ as a finite subset in $\mathbb{R}$, or equivalently, a point in $\mathbb{R}^m$.

Fix $\beta > 0$, and consider $Q = Q_{(\beta,t,y)}$, defined by

$$Q(f) = \underbrace{\sum_{i=1}^m |f(t_i) - y_i|^2}_{\text{least square}} + \beta \underbrace{\|f\|^2_{\mathscr{H}_K}}_{\text{penalty form}} , \quad f \in \mathscr{H}_K. \tag{12.22}$$

We introduce the associated dual pair of operators as follows:

$$\begin{aligned} T : \mathscr{H}_K &\longrightarrow \mathbb{R}^m \simeq l_m^2, \text{ and} \\ T^* : l_m^2 &\longrightarrow \mathscr{H}_K \end{aligned} \tag{12.23}$$

where

$$Tf = (f(t_i))_{i=1}^m, \quad f \in \mathscr{H}_K; \text{ and} \tag{12.24}$$

$$T^*y = \sum_{i=1}^m y_i K(\cdot, t_i) \in \mathscr{H}_K, \tag{12.25}$$

for all $\vec{y} = (y_i) \in \mathbb{R}^m$.

Note that the duality then takes the following form:

$$\langle T^*y, f \rangle_{\mathscr{H}_K} = \langle y, Tf \rangle_{l_m^2}, \quad \forall f \in \mathscr{H}_K, \forall y \in l_m^2; \tag{12.26}$$

consistent with (12.20).

Applying Theorem 12.2 to the counting measure

$$\mu = \sum_{i=1}^m \delta_{t_i} = \delta_{\{t_i\}}$$

for the set of sample points $\{t_i\}_{i=1}^m$, we get the two formulas:

$$T^*Tf = \sum_{i=1}^m f(t_i) K(\cdot, t_i) = \sum_{i=1}^m f(t_i) K_{t_i}, \text{ and} \tag{12.27}$$

$$TT^*y = K_m \vec{y} \tag{12.28}$$

where K_m denotes the $m \times m$ matrix

$$K_m = (K(t_i, t_j))_{i,j=1}^m = \begin{pmatrix} K(t_1, t_1) & \cdots & \cdots & K(t_1, t_m) \\ K(t_2, t_1) & \cdots & \cdots & K(t_2, t_m) \\ \vdots & & & \\ K(t_m, t_1) & & & K(t_m, t_m) \end{pmatrix}. \tag{12.29}$$

Theorem 12.3. *Let K, S, $\{t_i\}_{i=1}^m$, and $\{y_i\}_{i=1}^m$ be as above, and let K_m be the induced sample matrix (12.29).*

Fix $\beta > 0$; then the unique solution to the optimization problem with

$$Q_{\beta, \{t_i\}, \{y_i\}}(f) = \sum_{i=1}^m |y_i - f(t_i)|^2 + \beta \|f\|_{\mathscr{H}_K}^2, \quad f \in \mathscr{H}_K, \tag{12.30}$$

is

$$F(\cdot) = \sum_{i=1}^m (K_m + \beta I_m)_i^{-1} K(\cdot, t_i) \text{ on } S; \tag{12.31}$$

i.e., $F = \arg\min Q$ on $\mathscr{H}_K$.

Proof. From Theorem 12.2, we get that the unique solution $F \in \mathscr{H}$ is given by:

$$\beta F + T^* T F = T^* y,$$

and by (12.27)-(12.28), we further get

$$\beta F(\cdot) = \sum_{i=1}^m (y_i - F(t_i)) K(\cdot, t_i) \tag{12.32}$$

where the dot $\cdot$ refers to a free variable in S. An evaluation of (12.32) on the sample points yields:

$$\beta \vec{F} = K_m \left(\vec{y} - \vec{F} \right) \tag{12.33}$$

where $\vec{F} := (F(t_i))_{i=1}^m$, and $\vec{y} = (y_i)_{i=1}^m$. Hence

$$\vec{F} = (\beta I_m + K_m)^{-1} K_m \vec{y}. \tag{12.34}$$

Now substitute (12.34) into (12.33), and the desired conclusion in the theorem follows. We used the matrix identity

$$I_m - (\beta I_m + K_m)^{-1} K_m = \beta (\beta I_m + K_m)^{-1}.$$

$\square$

Exercise 12.5 (Kernels and differentiable functions). For the two kernels below, verify that the corresponding RKHSs are as states.

1. Spline: $S = [0, 1]$, $K : S \times S \longrightarrow \mathbb{R}$, $K(t_1, t_2) = t_1 \wedge t_2 \, (= \text{minimum})$,

$$\mathscr{H}_K = \left\{ f \text{ on } S \, ; \, f' \in L^2(0, 1), \, f(0) = 0, \, \|f\|_{\mathscr{H}_K}^2 = \int_0^1 |f'(t)|^2 \, dt < \infty \right\}.$$

2. Gaussian: $S = \mathbb{R}$, $K : S \times S \longrightarrow \mathbb{R}$, $v > 0$ fixed. If

$$K\left(t_1, t_2\right) = \exp\left(-\frac{\left|t_1 - t_2\right|^2}{2v}\right), \quad \forall t_1, t_2 \in \mathbb{R}, \text{ then}$$

$$\mathscr{H}_K = \left\{ f \in C^\infty\left(\mathbb{R}\right) \; ; \; f^{(n)} = \left(\frac{d}{dx}\right)^n f \in L^2\left(\mathbb{R}\right) \text{ for all } n \in \mathbb{N}_0, \right.$$

$$\left. \text{and } \|f\|_{\mathscr{H}_K}^2 = \sum_{n=0}^{\infty} \frac{v^n}{n! 2^n} \int_{\mathbb{R}} \left|f^{(n)}\left(x\right)\right|^2 dx < \infty \right\}.$$

Exercise 12.6 (New RKHS from "old" ones).

1. Restriction: If $K : S \times S \longrightarrow \mathbb{C}$ is a positive definite kernel on $S \times S$ where S is a set, and $S_0 \subset S$ is a subset, then show that the RKHS-norm $\|\cdot\|_{\mathscr{H}_{K_0}}$ of the restriction is as follows. The norm of the RKHS of the *restriction* $K_0 = K\big|_{S_0 \times S_0}$ satisfies:
For functions f_0 on S_0,

$$\|f_0\|_{\mathscr{H}_{K_0}}^2 = \inf\left\{ \|F\|_{\mathscr{H}_K}^2 \; ; \; F \in \mathscr{H}_K, \text{ and } F\big|_{S_0} = f_0 \right\},$$

and $f_0 \in \mathscr{H}_{K_0} \iff \|f_0\|_{\mathscr{H}_{K_0}}^2 < \infty$.

2. Sums and products: For $i = 1, 2$, and sets S_i, consider fixed p.d. kernels $K_i : S_i \times S_i \longrightarrow \mathbb{C}$. Show that then the RKHSs of the respective "new" kernels are as follows:

$$K_1\left(x_1, y_1\right) + K_2\left(x_2, y_2\right) \quad \rightsquigarrow \quad \mathscr{H}_{K_1} \oplus \mathscr{H}_{K_2}; \text{ and}$$
$$\underset{\text{(Hadamard product)}}{K_1\left(x_1, y_1\right) K_2\left(x_2, y_2\right)} \quad \rightsquigarrow \quad \mathscr{H}_{K_1} \otimes \mathscr{H}_{K_2};$$

i.e., for "+", the RKHS is the direct sum $\oplus$ of the corresponding RKHSs; and, for the *Hadamard product* (also known as the Schur product or the entry-wise product), it is the tensor product $\otimes$ of the factors, the two RKHSs $\mathscr{H}_{K_i}$, $i = 1, 2$.

Example 12.2 (Brownian motion). Here we shall discuss standard Brownian motion (B_t) indexed by time $t \in [0, \infty)$. See e.g., Chapter 7, and also [Hid80, Itô06].

Its properties may be described in terms of a system of joint distributions as follows. Reference is made to an underlying probability space $(M, \mathfrak{M}, \mathbb{P})$ where $\mathfrak{M}$ is a sigma-algebra of subsets of a given sample space M, and $\mathbb{P}$ is a probability measure defined on $\mathfrak{M}$. We set

$$\mathbb{E}\left(\cdots\right) = \int_M \cdots d\mathbb{P}$$

where "$\cdots$" refers to measurable functions on M, i.e., to random variables.

When $(M, \mathfrak{M}, \mathbb{P})$ is given and X is a random variable, there is a unique probability measure μ_X on $\mathbb{R}$ such that

$$\mathbb{E}(f \circ X) = \int_{\mathbb{R}} f \, d\mu_X \qquad (12.35)$$

holds for all continuous functions $f : \mathbb{R} \longrightarrow \mathbb{R}$. The measure μ_X is called the *distribution* of X.

The Gaussian (distribution) with mean m, and variance σ^2, i.e.,

$$\frac{1}{\sigma\sqrt{2\pi}} \exp\left(-\frac{(x-m)^2}{2\sigma^2}\right), \quad x \in \mathbb{R} \qquad (12.36)$$

is called $N\left(m, \sigma^2\right)$.

The standard Brownian motion $(B_t)_{t \in [0,\infty)}$ is characterized by the following axioms:

(i) $B_0 = 0$;

(ii) for all $s < t$, the *distribution* of $B_t - B_s$ is $N(0, t-s)$; and

(iii) if $u \le s < t$, then B_u is *independent* of $B_t - B_s$.

Lemma 12.1. *The covariance kernel K of standard Brownian motion is as follows: If $s, t \in [0, \infty)$, then*

$$K(s, t) = \mathbb{E}(B_s B_t)$$

$$= \int_M B_s B_t \, d\mathbb{P} = s \wedge t \, (= minimum \; of \; s \; and \; t). \qquad (12.37)$$

Proof. This follows from (i)-(iii) above and the easy formula:

$$s \wedge t = \frac{s + t - |s - t|}{2}. \qquad (12.38)$$

$\square$

Example 12.3 (Samples of standard Brownian motion). Let

$$0 < t_1 < t_2 < \cdots < t_m \qquad (12.39)$$

be fixed. Below we compute the sampled covariance matrix K_m for the case when K is the Brownian motion covariance function K on $[0, \infty) \times [0, \infty)$ given by (12.37); see (12.29). For K_m we have

$$K_m = \begin{pmatrix} t_1 & t_1 & t_1 & t_1 & \cdots & \cdots & t_1 \\ t_1 & t_2 & t_2 & t_2 & \cdots & \cdots & t_2 \\ t_1 & t_2 & t_3 & t_3 & \cdots & \cdots & t_3 \\ \vdots & \vdots & \vdots & \ddots & & & \vdots \\ \vdots & \vdots & \vdots & & \ddots & & \vdots \\ \vdots & \vdots & \vdots & & & t_{m-1} & t_{m-1} \\ t_1 & t_2 & t_3 & \cdots & \cdots & t_{m-1} & t_m \end{pmatrix} = L_m U_m \qquad (12.40)$$

i.e., the lower/upper factorization, so Lower × Upper matrix-product. Here,

$$
L_m = \begin{pmatrix}
\sqrt{t_1} & 0 & 0 & & & \\
\sqrt{t_1} & \sqrt{t_2 - t_1} & 0 & & \mathbf{0} & \\
\sqrt{t_1} & \sqrt{t_2 - t_1} & \sqrt{t_3 - t_2} & & & \\
\vdots & \vdots & \vdots & \ddots & & \\
\vdots & \vdots & \vdots & \ddots & \ddots & \\
\sqrt{t_1} & \sqrt{t_2 - t_1} & \sqrt{t_3 - t_2} & \cdots & \cdots & \sqrt{t_m - t_{m-1}}
\end{pmatrix}
$$

and

$$
U_m = \begin{pmatrix}
\sqrt{t_1} & \sqrt{t_1} & \sqrt{t_1} & \cdots & \cdots & \sqrt{t_1} \\
0 & \sqrt{t_2 - t_1} & \sqrt{t_2 - t_1} & \cdots & \cdots & \sqrt{t_2 - t_1} \\
0 & 0 & \sqrt{t_3 - t_2} & \cdots & \cdots & \sqrt{t_3 - t_2} \\
 & \mathbf{0} & & \ddots & \ddots & \vdots \\
 & & & \ddots & \ddots & \vdots \\
 & & & 0 & \sqrt{t_m - t_{m-1}}
\end{pmatrix}.
$$

Proof. Using (i)-(iii) for Brownian motion, we note that this is a sampled Brownian motion $X_{t_1}, X_{t_2}, \cdots, X_{t_m}$ corresponding to a fixed finite sample (12.39), where

$$
X_{t_j} = \sqrt{t_1} Z_1 + \sqrt{t_2 - t_1} Z_2 + \cdots + \sqrt{t_j - t_{j-1}} Z_j, \tag{12.41}
$$

and $Z_1, Z_2, Z_3, \cdots$, is a system of *independently identically distributed* (i.i.d) $N(0, 1)$. Clearly then

$$
\mathbb{E}\left(X_{t_i} X_{t_j} \right) = t_i \wedge t_j = K_m(t_i, t_j)
$$

$$
= \mathbb{E}\left(\sum_{k \leq i} \sum_{l \leq j} \sqrt{t_k - t_{k-1}} \sqrt{t_l - t_{l-1}} Z_k Z_l \right)
$$

$$
= \sum_{k \leq i} \sum_{l \leq j} \sqrt{t_k - t_{k-1}} \sqrt{t_l - t_{l-1}} \delta_{k,l}
$$

$$
= (LU)_{ij}
$$

where LU is the matrix-multiplication in (12.40). □

12.3 A Digression: Stochastic Processes

Below we continue the discussion of stochastic processes started in Section 2.4.

The interest in positive definite functions has at least three roots: (i) Fourier analysis, and harmonic analysis more generally, including the non-commutative variant where we study unitary representations of groups; (ii) optimization and approximation problems, involving for example spline approximations as envisioned by I. Schöenberg; and (iii) the study of stochastic (random) processes.

Definition 12.3. A *stochastic process* is an indexed family of random variables $\{X_s\}_{s \in S}$ based on a fixed probability space $(\Omega, \mathscr{F}, \mathbb{P})$.

In our present analysis, the processes will be indexed by some group G; for example $G = \mathbb{R}$, or $G = \mathbb{Z}$ correspond to processes indexed by real time, respectively discrete time. A main tool in the analysis of stochastic processes is an associated covariance function, see (12.42).

A process $\{X_g \mid g \in G\}$ is called Gaussian if each random variable X_g is Gaussian, i.e., its distribution is Gaussian. For Gaussian processes we only need two moments. So if we normalize, setting the mean equal to 0, then the process is determined by the covariance function. In general the covariance function is a function on $G \times G$, or on a subset, but if the process is stationary, the covariance function will in fact be a positive definite function defined on G, or a subset of G. We will be using three stochastic processes in this book, Brownian motion, Brownian Bridge, and the Ornstein-Uhlenbeck process, all Gaussian, or Itō integrals.

We outline a brief sketch of these facts below.

Let G be a locally compact group, and let $(\Omega, \mathscr{F}, \mathbb{P})$ be a probability space, $\mathscr{F}$ a sigma-algebra, and $\mathbb{P}$ a probability measure defined on $\mathscr{F}$. A stochastic L^2-process is a system of random variables $\{X_g\}_{g \in G}$, $X_g \in L^2(\Omega, \mathscr{F}, \mathbb{P})$. The covariance function c_X of the process is the function $G \times G \to \mathbb{C}$ given by

$$c_X(g_1, g_2) = \mathbb{E}\left(\overline{X}_{g_1} X_{g_2}\right), \ \forall (g_1, g_2) \in G \times G. \tag{12.42}$$

To simplify we will assume that the mean $\mathbb{E}(X_g) = \int_\Omega X_g d\mathbb{P}(\omega) = 0$ for all $g \in G$.

Definition 12.4. We say that (X_g) is stationary iff

$$c_X(hg_1, hg_2) = c_X(g_1, g_2), \ \forall h \in G. \tag{12.43}$$

In this case c_X is a function of $g_1^{-1} g_2$, i.e.,

$$\mathbb{E}(X_{g_1}, X_{g_2}) = c_X\left(g_1^{-1} g_2\right), \ \forall g_1, g_2 \in G. \tag{12.44}$$

(Just take $h = g_1^{-1}$ in (12.43).)

We now recall the following theorem of Kolmogorov (see [PS75]). One direction is easy, and the other is the deep part:

Definition 12.5. A function c defined on a subset of G is said to be *positive definite* iff

$$\sum_i \sum_j \overline{\lambda_i} \lambda_j c\left(g_i^{-1} g_j\right) \geq 0$$

for all finite summation, where $\lambda_i \in \mathbb{C}$ and $g_i^{-1} g_j$ in the domain of c.

Theorem 12.4 (Kolmogorov). *A function $c : G \to \mathbb{C}$ is positive definite if and only if there is a stationary Gaussian process $(\Omega, \mathscr{F}, \mathbb{P}, X)$ with mean zero, such that $c = c_X$.*

Proof. To stress the idea, we include the easy part of the theorem, and we refer to [PS75] for the non-trivial direction:

Let $\lambda_1, \lambda_2, \ldots, \lambda_n \in \mathbb{C}$, and $\{g_i\}_{i=1}^N \subset G$, then for all finite summations, we have:

$$\sum_i \sum_j \overline{\lambda_i} \lambda_j c_X \left(g_i^{-1} g_j \right) = \mathbb{E} \left(\left| \sum_{i=1}^N \lambda_i X_{g_i} \right|^2 \right) \geq 0.$$

□

12.4 Two Extension Problems

While each of the two extension problems has received a considerable amount of attention in the literature, our emphasis here will be the interplay between the two problems: Our aim is a duality theory; and, in the case $G = \mathbb{R}^n$, and $G = \mathbb{T}^n = \mathbb{R}^n / \mathbb{Z}^n$, we will state our theorems in the language of Fourier duality of Abelian groups: With the time frequency duality formulation of Fourier duality for $G = \mathbb{R}^n$ we have that both the time domain and the frequency domain constitute a copy of $\mathbb{R}^n$. We then arrive at a setup such that our extension questions (i) are in time domain, and extensions from (ii) are in frequency domain. Moreover we show that each of the extensions from (i) has a variant in (ii). Specializing to $n = 1$, we arrive of a spectral theoretic characterization of all skew-Hermitian operators with dense domain in a separable Hilbert space, having deficiency-indices $(1,1)$.

A systematic study of densely defined Hermitian operators with deficiency indices $(1,1)$, and later (d,d), was initiated by M. Krein [Kre46], and is also part of de Branges' model theory; see [dB68, dBR66]. The direct connection between this theme and the problem of extending continuous positive definite (p.d.) functions F when they are only defined on a fixed open subset to $\mathbb{R}^n$ was one of our motivations. One desires continuous p.d. extensions to $\mathbb{R}^n$.

If F is given, we denote the set of such extensions $Ext(F)$. If $n = 1$, $Ext(F)$ is always non-empty, but for $n = 2$, Rudin gave examples in [Rud70, Rud63] when $Ext(F)$ may be empty. Here we extend these results, and we also cover a number of classes of positive definite functions on locally compact groups in general; so cases when $\mathbb{R}^n$ is replaced with other groups, both Abelian and non-Abelian.

The results in the framework of locally compact Abelian groups are more complete than their counterparts for non-Abelian Lie groups, one reason is the availability of Bochner's duality theorem for locally compact Abelian groups; — not available for non-Abelian Lie groups.

Remark 12.2. Even in one dimension the extension problem for locally defined positive definite functions is interesting. One reason is that among the Fourier transforms (generating functions) for finite positive Borel measures P on $\mathbb{R}$,

$$g_P(u) = \int_{\mathbb{R}} e^{iux} dP(x), \ u \in \mathbb{R}; \tag{12.45}$$

one wishes to identify the *infinitely divisible distributions*. We have:

Theorem 12.5 (Lévy-Khinchin[1] [Rit88]). *Infinite divisibility holds if and only if g_P has the following representation: $g_P(u) = e^{\eta(u)}$, such that for some $a \in \mathbb{R}$, $\sigma \in \mathbb{R}_+$, and Borel measure L on $\mathbb{R}\setminus\{0\}$, we have:*

$$\eta(u) = i\,a\,u - \frac{\sigma^2}{2}u^2 + \int_{\mathbb{R}\setminus\{0\}} \left(e^{i\,u\,x} - 1 - \frac{i\,u\,x}{1+x^2} \right) L(dx) \qquad (12.46)$$

and the measure L satisfying

$$\int_{\mathbb{R}\setminus\{0\}} \left(1 \wedge x^2 \right) L(dx) < \infty. \qquad (12.47)$$

12.5   The Reproducing Kernel Hilbert Space $\mathscr{H}_F$

Reproducing kernel Hilbert spaces were pioneered by Aronszajn [Aro50], and subsequently they have been used in a host of applications; e.g., [Sza04, SZ09, SZ07]. The reproducing kernel property appeared for the first time in Zaremba's paper [Zar07].

As for positive definite functions, their use and applications are extensive and includes such areas as stochastic processes, see e.g., [JP13a, AJSV13, JP12, AJ12]; harmonic analysis (see [JÓ00]), and the references there); potential theory [Fug74, KL14b]; operators in Hilbert space [Alp92, AD86]; and spectral theory [AH13, Nus75, Dev72, Dev59]. We stress that the literature is vast, and the above list is only a small sample.

Associated to a pair (Ω, F), where F is a prescribed continuous positive definite function defined on Ω, we outline a reproducing kernel Hilbert space $\mathscr{H}_F$ which will serve as a key tool in our analysis. The particular RKHSs we need here will have additional properties (as compared to a general framework); which allow us to give explicit formulas for our solutions.

Contents of the section.

In this chapter, we study two extension problems, and their interconnections. The first class of extension problems concerns (i) positive definite (p.d.) continuous functions on Lie groups G, and the second deals with (ii) Lie algebras of unbounded skew-Hermitian operators in a certain family of reproducing kernel Hilbert spaces (RKHS). The analysis is non-trivial even if $G = \mathbb{R}^n$, and even if $n = 1$. If $G = \mathbb{R}^n$, we are concerned in (ii) with the study of systems of n skew-Hermitian operators $\{S_i\}$ on a common dense domain in Hilbert space, and in deciding whether it is possible to find a corresponding system of strongly commuting selfadjoint operators $\{T_i\}$ such that, for each value of i, the operator T_i extends S_i.

---

[1] Appendix C includes a short biographical sketch of P. Lévy.

Definition 12.6. Let G be a Lie group. Fix $\Omega \subset G$, non-empty, open and connected. A continuous function

$$F : \Omega^{-1} \cdot \Omega \to \mathbb{C} \tag{12.48}$$

is *positive definite* (p.d.) if

$$\sum_i \sum_j \overline{c_i} c_j F\left(x_i^{-1} x_j\right) \geq 0, \tag{12.49}$$

for all finite systems $\{c_i\} \subset \mathbb{C}$, and points $\{x_i\} \subset \Omega$.

Equivalently,

$$\int_\Omega \int_\Omega \overline{\varphi(x)} \varphi(y) F\left(x^{-1} y\right) dx dy \geq 0, \tag{12.50}$$

for all $\varphi \in C_c(\Omega)$; where dx denotes a choice of left-invariant Haar measure on G.

For simplicity we focus on the case $G = \mathbb{R}$, indicating the changes needed for general Lie groups.

Definition 12.7. Fix $0 < a < \infty$, set $\Omega := (0, a)$. Let $F : \Omega - \Omega \to \mathbb{C}$ be a continuous p.d. function. The *reproducing kernel Hilbert space (RKHS)*, $\mathscr{H}_F$, is the completion of the space of functions

$$\sum_{\text{finite}} c_j F\left(\cdot - x_j\right) : c_j \in \mathbb{C} \tag{12.51}$$

with respect to the inner product

$$\langle F\left(\cdot - x\right), F\left(\cdot - y\right) \rangle_{\mathscr{H}_F} = F\left(x - y\right), \; \forall x, y \in \Omega, \text{ and}$$

$$\left\langle \sum_i c_i F\left(\cdot - x_i\right), \sum_j c_j F\left(\cdot - x_j\right) \right\rangle_{\mathscr{H}_F} = \sum_i \sum_j \overline{c_i} c_j F\left(x_i - x_j\right), \tag{12.52}$$

Remark 12.3. Throughout, we use the convention that the inner product is conjugate linear in the first variable, and linear in the second variable. When more than one inner product is used, subscripts will make reference to the Hilbert space.

Notation. Inner product and norms will be denoted $\langle \cdot, \cdot \rangle$, and $\|\cdot\|$ respectively. Often more than one inner product is involved, and subscripts are used for identification.

Lemma 12.2. *The reproducing kernel Hilbert space (RKHS), $\mathscr{H}_F$, is the Hilbert completion of the space of functions*

$$F_\varphi(x) = \int_\Omega \varphi(y) F\left(x - y\right) dy, \; \forall \varphi \in C_c^\infty(\Omega), x \in \Omega \tag{12.53}$$

with respect to the inner product

$$\langle F_\varphi, F_\psi \rangle_{\mathscr{H}_F} = \int_\Omega \int_\Omega \overline{\varphi(x)} \psi(y) \, F(x-y) \, dx dy, \ \forall \varphi, \psi \in C_c^\infty(\Omega).$$ (12.54)

In particular,

$$\|F_\varphi\|_{\mathscr{H}_F}^2 = \int_\Omega \int_\Omega \overline{\varphi(x)} \varphi(y) \, F(x-y) \, dx dy, \ \forall \varphi \in C_c^\infty(\Omega)$$ (12.55)

and

$$\langle F_\varphi, F_\psi \rangle_{\mathscr{H}_F} = \int_\Omega \overline{\varphi(x)} F_\psi(x) \, dx, \ \forall \phi, \psi \in C_c^\infty(\Omega).$$ (12.56)

Proof. Apply standard approximation, see Lemma 12.3 below.   □

The remaining of this section is devoted to a number of technical lemmas which will be used throughout the chapter. Given a locally defined continuous positive definite function F, the issues addressed below are: approximation (Lemma 12.3), a reproducing kernel Hilbert space (RKHS) $\mathscr{H}_F$ built from F, an integral transform, and a certain derivative operator $D^{(F)}$, generally unbounded in the RKHS $\mathscr{H}_F$. We will be concerned with boundary value problems for $D^{(F)}$, and in order to produce suitable orthonormal bases in $\mathscr{H}_F$, we be concerned with an explicit family of skew-adjoint extensions of $D^{(F)}$, as well as the associated spectra, see Corollaries 12.3 and 12.4.

Lemma 12.3. *Let φ be a function such that*

1. $\operatorname{supp}(\varphi) \subset (0, a)$;

2. $\varphi \in C_c^\infty(0, a)$, $\varphi \geq 0$;

3. $\int_0^a \varphi(t) \, dt = 1$.

Fix $x \in (0, a)$, and set $\varphi_{n,x}(t) := n\varphi(n(t-x))$. Then $\lim_{n\to\infty} \varphi_{n,x} = \delta_x$, i.e., the Dirac measure at x; and

$$\left\| F_{\varphi_{n,x}} - F(\cdot - x) \right\|_{\mathscr{H}_F} \to 0, \ as \ n \to \infty.$$ (12.57)

Hence $\{F_\varphi : \varphi \in C_c^\infty(0, a)\}$ spans a dense subspace in $\mathscr{H}_F$. See Figure 12.1.

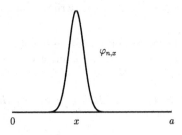

Figure 12.1: The approximate identity $\varphi_{n,x}$.

Recall, the following facts about $\mathscr{H}_F$, which follow from the general theory [Aro50] of RKHS:

- $F(0) > 0$, so we can always arrange $F(0) = 1$.

- $F(-x) = \overline{F(x)}$

- $\mathscr{H}_F$ consists of continuous functions $\xi : \Omega - \Omega \to \mathbb{C}$.

- The reproducing property:

$$\langle F(\cdot - x), \xi \rangle_{\mathscr{H}_F} = \xi(x), \; \forall \xi \in \mathscr{H}_F, \forall x \in \Omega,$$

is a direct consequence of (12.52).

Remark 12.4. It follows from the reproducing property that if $F_{\phi_n} \to \xi$ in $\mathscr{H}_F$, then F_{ϕ_n} converges uniformly to ξ in Ω. In fact

$$
\begin{aligned}
|F_{\phi_n}(x) - \xi(x)| &= \left| \langle F(\cdot - x), F_{\phi_n} - \xi \rangle_{\mathscr{H}_F} \right| \\
&\leq \|F(\cdot - x)\|_{\mathscr{H}_F} \|F_{\phi_n} - \xi\|_{\mathscr{H}_F} \\
&= F(0)^{1/2} \|F_{\phi_n} - \xi\|_{\mathscr{H}_F}.
\end{aligned}
$$

Lemma 12.4. *Let* $F : (-a, a) \to \mathbb{C}$ *be a continuous and p.d. function, and let* $\mathscr{H}_F$ *be the corresponding RKHS. Then:*

1. *the integral* $F_\varphi := \int_0^a \varphi(y) F(\cdot - y) \, dy$ *is convergent in* $\mathscr{H}_F$ *for all* $\varphi \in C_c(0, a)$; *and*

2. *for all* $\xi \in \mathscr{H}_F$, *we have:*

$$\langle F_\varphi, \xi \rangle_{\mathscr{H}_F} = \int_0^a \overline{\varphi(x)} \xi(x) \, dx. \tag{12.58}$$

Proof. For simplicity, we assume the following normalization $F(0) = 1$; then for all $y_1, y_2 \in (0, 1)$, we have

$$\|F(\cdot - y_1) - F(\cdot - y_2)\|_{\mathscr{H}_F}^2 = 2(1 - \Re\{F(y_1 - y_2)\}). \tag{12.59}$$

Now, view the integral in (1) as a $\mathscr{H}_F$-vector valued integral. If $\varphi \in C_c(0, a)$, this integral $\int_0^a \varphi(y) F(\cdot - y) \, dy$ is the $\mathscr{H}_F$-norm convergent. Since $\mathscr{H}_F$ is a RKHS, $\langle \cdot, \xi \rangle_{\mathscr{H}_F}$ is continuous on $\mathscr{H}_F$, and it passes under the integral in (1). Using

$$\langle F(y - \cdot), \xi \rangle_{\mathscr{H}_F} = \xi(y) \tag{12.60}$$

the desired conclusion (12.58) follows. $\qquad\square$

Corollary 12.1. *Let $F : (-a, a) \to \mathbb{C}$ be as above, and let $\mathscr{H}_F$ be the corresponding RKHS. For $\varphi \in C_c^1(0, a)$, set*

$$F_\varphi(x) = (T_F \varphi)(x) = \int_0^a \varphi(y) F(x - y) \, dy; \qquad (12.61)$$

then $F_\varphi \in C^1(0, a)$, and

$$\frac{d}{dx} F_\varphi(x) = (T_F(\varphi'))(x), \; \forall x \in (0, a). \qquad (12.62)$$

Proof. Since $F_\varphi(x) = \int_0^a \varphi(y) F(x - y) \, dy$, $x \in (0, a)$; the desired assertion (12.62) follows directly from the arguments in the proof of Lemma 12.4. $\square$

Theorem 12.6. *Fix $0 < a < \infty$. A continuous function $\xi : (0, a) \to \mathbb{C}$ is in $\mathscr{H}_F$ if and only if there exists a finite constant $A > 0$, such that*

$$\sum_i \sum_j \overline{c_i} c_j \overline{\xi(x_i)} \xi(x_j) \leq A \sum_i \sum_j \overline{c_i} c_j F(x_i - x_j) \qquad (12.63)$$

for all finite system $\{c_i\} \subset \mathbb{C}$ and $\{x_i\} \subset (0, a)$. Equivalently, for all $\varphi \in C_c^\infty(\Omega)$,

$$\left| \int_0^a \varphi(y) \xi(y) \, dy \right|^2 \leq A \int_0^a \int_0^a \overline{\varphi(x)} \varphi(y) F(x - y) \, dx dy. \qquad (12.64)$$

We will use these two conditions (12.63)($\Leftrightarrow$(12.64)) when considering for example the von Neumann deficiency-subspaces for skew Hermitian operators with dense domain in $\mathscr{H}_F$.

Proof of Theorem 12.6. Note, if $\xi \in \mathscr{H}_F$, then

$$\text{LHS}_{(12.64)} = \left| \langle F_\varphi, \xi \rangle_{\mathscr{H}_F} \right|^2,$$

and so (12.64) holds, since $\langle \cdot, \xi \rangle_{\mathscr{H}_F}$ is continuous on $\mathscr{H}_F$.

If ξ is continuous on $[0, a]$, and if (12.64) holds, then

$$\mathscr{H}_F \ni F_\varphi \longmapsto \int_0^a \varphi(y) \xi(y) \, dy$$

is well-defined, continuous, linear; and extends to $\mathscr{H}_F$ by density (see Lemma 12.3). Hence, by Riesz' theorem, $\exists! \; k_\xi \in \mathscr{H}_F$ such that

$$\int_0^a \varphi(y) \xi(y) \, dy = \langle F_\varphi, k_\xi \rangle_{\mathscr{H}_F}.$$

But using the reproducing property in $\mathscr{H}_F$, and $F_\varphi(x) = \int_0^a \varphi(x) F(x - y) \, dy$, we get

$$\int_0^a \overline{\varphi(x)} \xi(x) \, dx = \int_0^a \overline{\varphi(x)} k_\xi(x) \, dx, \; \forall \varphi \in C_c(0, a)$$

so

$$\int_0^a \varphi(x)\left(\xi(x) - k_\xi(x)\right)dx = 0, \ \forall \varphi \in C_c(0,a);$$

it follows that $\xi - k_\xi = 0$ on $(0,a) \implies \xi - k_\xi = 0$ on $[0,a]$. □

Definition 12.8 (The operator D_F). Let $D_F(F_\varphi) = F_{\varphi'}$, for all $\varphi \in C_c^\infty(0,a)$, where $\varphi' = \frac{d\varphi}{dt}$ and F_φ is as in (12.53).

Lemma 12.5. *The operator D_F defines a skew-Hermitian operator with dense domain in $\mathscr{H}_F$.*

Proof. By Lemma 12.3, $dom(D_F)$ is dense in $\mathscr{H}_F$. If $\psi \in C_c^\infty(0,a)$ and

$$|t| < \text{dist}\left(\text{supp}(\psi), \text{endpoints}\right),$$

then

$$\left\|F_{\psi(\cdot + t)}\right\|_{\mathscr{H}_F}^2 = \left\|F_\psi\right\|_{\mathscr{H}_F}^2 = \int_0^a \int_0^a \overline{\psi(x)}\psi(y) F(x-y)\,dx dy \qquad (12.65)$$

see (12.55), so

$$\frac{d}{dt}\left\|F_{\psi(\cdot+t)}\right\|_{\mathscr{H}_F}^2 = 0$$

which is equivalent to

$$\langle D_F F_\psi, F_\psi \rangle_{\mathscr{H}_F} + \langle F_\psi, D_F F_\psi \rangle_{\mathscr{H}_F} = 0. \qquad (12.66)$$

It follows that D_F is well-defined and skew-Hermitian in $\mathscr{H}_F$. □

Lemma 12.6. *Let F be a positive definite function on $(-a,a)$, $0 < a < \infty$ fixed. Let D_F be as in Definition 12.8, so that $D_F \subset D_F^*$ (Lemma 12.5), where D_F^* is the adjoint relative to the $\mathscr{H}_F$ inner product.*
Then $\xi \in \mathscr{H}_F$ (as a continuous function on $[0,a]$) is in $dom(D_F^*)$ iff

$$\xi' \in \mathscr{H}_F \text{ where } \xi' = \text{distribution derivative, and} \qquad (12.67)$$
$$D_F^*\xi = -\xi' \qquad (12.68)$$

Proof. By Theorem 12.6, a fixed $\xi \in \mathscr{H}_F$, i.e., $x \mapsto \xi(x)$ is a continuous function on $[0,a]$ such that $\exists C$, $\left|\int_0^a \varphi(x)\xi(x)\,dx\right|^2 \leq C \left\|F_\varphi\right\|_{\mathscr{H}_F}^2$.
ξ is in $dom(D_F^*) \iff \exists C = C_\xi < \infty$ such that

$$\left|\langle D_F(F_\varphi), \xi \rangle_{\mathscr{H}_F}\right|^2 \leq C\left\|F_\varphi\right\|_{\mathscr{H}_F}^2 = C\int_0^a \int_0^a \overline{\varphi(x)}\varphi(y) F(x-y)\,dx dy. \qquad (12.69)$$

But LHS of (12.69) under $|\langle \cdot, \cdot \rangle|^2$ is:

$$\left|\langle D_F(F_\varphi), \xi \rangle_{\mathscr{H}_F}\right|^2 = \langle F_{\varphi'}, \xi \rangle_{\mathscr{H}_F} \overset{(12.58)}{=} \int_0^a \overline{\varphi'(x)}\xi(x)\,dx, \ \forall \varphi \in C_c^\infty(0,a). \qquad (12.70)$$

So (12.69) holds $\Longleftrightarrow$

$$\left| \int_0^a \overline{\varphi'(x)} \xi(x)\, dx \right|^2 \le C\, \|F_\varphi\|^2_{\mathscr{H}_F},\ \forall \varphi \in C_c^\infty(0,a)$$

i.e.,

$$\left| \int_0^a \overline{\varphi(x)} \xi'(x)\, dx \right|^2 \le C\, \|F_\varphi\|^2_{\mathscr{H}_F},\ \forall \varphi \in C_c^\infty(0,a),\ \text{and}$$

ξ' as a distribution is in $\mathscr{H}_F$, and

$$\int_0^a \overline{\varphi(x)} \xi'(x)\, dx = \langle F_\varphi, \xi' \rangle_{\mathscr{H}_F}$$

where we use the characterization of $\mathscr{H}_F$ in (12.64), i.e., a function $\eta : [0,a] \to \mathbb{C}$ is in $\mathscr{H}_F \Longleftrightarrow \exists C < \infty, \left| \int_0^a \overline{\varphi(x)} \eta(x)\, dx \right| \le C\, \|F_\varphi\|_{\mathscr{H}_F}, \forall \varphi \in C_c^\infty(0,a)$, and then $\int_0^a \overline{\varphi(x)} \eta(x)\, dx = \langle F_\varphi, \eta \rangle_{\mathscr{H}_F}, \forall \varphi \in C_c^\infty(0,a)$. See Theorem 12.6. $\qquad\square$

Corollary 12.2. $h \in \mathscr{H}_F$ is in $dom\left(\left(D_F^2\right)^*\right)$ iff $h'' \in \mathscr{H}_F$ (h'' distribution derivative) and $\left(D_F^*\right)^2 h = \left(D_F^2\right)^* h = h''$.

Proof. Application of (12.70) to $D_F(F_\varphi) = F_{\varphi'}$, we have $D_F^2(F_\varphi) = F_{\varphi''} = \left(\frac{d}{dx}\right)^2 F_\varphi, \forall \varphi \in C_c^\infty(0,a)$, and

$$\left\langle D_F^2(F_\varphi), h \right\rangle_{\mathscr{H}_F} = \langle F_{\varphi''}, h \rangle_{\mathscr{H}_F} = \int_0^a \overline{\varphi''(x)} h(x)\, dx$$

$$= \int_0^a \overline{\varphi(x)} h''(x)\, dx = \left\langle F_\varphi, \left(D_F^2\right)^* h \right\rangle_{\mathscr{H}_F}.$$

$$\square$$

Definition 12.9. [DS88b]Let D_F^* be the adjoint of D_F relative to $\mathscr{H}_F$ inner product. The deficiency spaces $DEF^\pm$ consists of $\xi_\pm \in dom(D_F^*)$, such that $D_F^* \xi_\pm = \pm \xi_\pm$, i.e.,

$$DEF^\pm = \left\{ \xi_\pm \in \mathscr{H}_F : \langle F_{\psi'}, \xi_\pm \rangle_{\mathscr{H}_F} = \langle F_\psi, \pm \xi_\pm \rangle_{\mathscr{H}_F}, \forall \psi \in C_c^\infty(\Omega) \right\}.$$

Corollary 12.3. If $\xi \in DEF^\pm$ then $\xi(x) = \text{constant } e^{\mp x}$.

Proof. Immediate from Lemma 12.6. $\qquad\square$

The role of deficiency indices for the canonical skew-Hermitian operator D_F (Definition 12.8) in the RKHS $\mathscr{H}_F$ is as follows: using von Neumann's conjugation trick [DS88b], we see that the deficiency indices can be only $(0,0)$ or $(1,1)$.

We conclude that there exists proper skew-adjoint extensions $A \supset D_F$ in $\mathscr{H}_F$ (in case D_F has indices $(1,1)$). Then

$$D_F \subseteq A = -A^* \subseteq -D_F^* \tag{12.71}$$

(If the indices are $(0,0)$ then $\overline{D_F} = -D_F^*$; see [DS88b].)

Hence, set $U(t) = e^{tA} : \mathscr{H}_F \to \mathscr{H}_F$, and get the strongly continuous unitary one-parameter group

$$\{U(t) : t \in \mathbb{R}\}, \; U(s+t) = U(s)U(t), \; \forall s, t \in \mathbb{R};$$

and if

$$\xi \in dom(A) = \left\{ \xi \in \mathscr{H}_F : \text{s.t.} \lim_{t \to 0} \frac{U(t)\xi - \xi}{t} \text{ exists} \right\}$$

then

$$A\xi = \text{s.t.} \lim_{t \to 0} \frac{U(t)\xi - \xi}{t}. \tag{12.72}$$

Now use $F_x(\cdot) = F(x - \cdot)$ defined in $(0,a)$; and set

$$F_A(t) := \langle F_0, U(t)F_0 \rangle_{\mathscr{H}_F}, \; \forall t \in \mathbb{R} \tag{12.73}$$

then using (12.57), we see that F_A is a continuous positive definite extension of F on $(-a, a)$. This extension is in $Ext_1(F)$.

Corollary 12.4. *Assume* $\lambda \in \mathbb{R}$ *is in the point spectrum of* A*, i.e.,* $\exists \xi_\lambda \in dom(A)$*,* $\xi_\lambda \neq 0$*, such that* $A\xi_\lambda = i\lambda\xi_\lambda$ *holds in* $\mathscr{H}_F$*, then* $\xi_\lambda = const \cdot e_\lambda$*, i.e.,*

$$\xi_\lambda(x) = const \cdot e^{i\lambda x}, \; \forall x \in [0, a]. \tag{12.74}$$

Proof. Assume λ is in $spec_{pt}(A)$, and $\xi_\lambda \in dom(A)$ satisfying

$$(A\xi_\lambda)(x) = i\lambda\xi_\lambda(x) \text{ in } \mathscr{H}_F, \tag{12.75}$$

then since $A \subset -D_F^*$, we get $\xi \in dom(D_F^*)$ by Lemma 12.6 and (12.71), and $D_F^*\xi_\lambda = -\xi_\lambda'$ where ξ' is the distribution derivative (see (12.68)); and by (12.71)

$$(A\xi_\lambda)(x) = -(D_F^*\xi_\lambda)(x) = \xi_\lambda'(x) \overset{(12.75)}{=} i\lambda\xi_\lambda(x), \; \forall x \in (0, a) \tag{12.76}$$

so ξ_λ is the distribution derivative solution to

$$\xi_\lambda'(x) = i\lambda\xi_\lambda(x) \tag{12.77}$$

$$\Updownarrow$$

$$-\int_0^a \overline{\varphi'(x)}\xi_\lambda(x)\,dx = i\lambda \int_0^a \overline{\varphi(x)}\xi_\lambda(x)\,dx, \; \forall \varphi \in C_c^\infty(0, a)$$

$$\Updownarrow$$

$$-\langle D_F(F_\varphi), \xi_\lambda \rangle_{\mathscr{H}_F} = i\lambda \langle F_\varphi, \xi_\lambda \rangle_{\mathscr{H}_F}, \; \forall \varphi \in C_c^\infty(0, a).$$

But by Schwartz, the distribution solutions to (12.77) are $\xi_\lambda(x) = const \cdot e_\lambda(x) = const \cdot e^{i\lambda x}$. $\qquad \square$

In the considerations below, we shall be primarily concerned with the case when a fixed continuous p.d. function F is defined on a finite interval $(-a, a) \subset \mathbb{R}$. In this case, by a Mercer operator, we mean an operator T_F in $L^2(0, a)$ where $L^2(0, a)$ is defined from Lebesgue measure on $(0, a)$, given by

$$(T_F \varphi)(x) := \int_0^a \varphi(y) F(x - y) \, dy, \ \forall \varphi \in L^2(0, a), \forall x \in (0, a). \tag{12.78}$$

Lemma 12.7. *Under the assumptions stated above, the Mercer operator T_F is trace class in $L^2(0, a)$; and if $F(0) = 1$, then*

$$trace\,(T_F) = a. \tag{12.79}$$

Proof. This is an application of Mercer's theorem [LP89, FR42, FM13] to the integral operator T_F in (12.78). But we must check that F, on $(-a, a)$, extends uniquely by limit to a continuous p.d. function F_{ex} on $[-a, a]$, the closed interval. This is true, and easy to verify, see e.g. [JPT15]. $\qquad \square$

Corollary 12.5. *Let F and $(-a, a)$ be as in Lemma 12.7. Then there is a sequence $(\lambda_n)_{n \in \mathbb{N}}$, $\lambda_n > 0$, such that $\sum_{n \in \mathbb{N}} \lambda_n = a$, and a system of orthogonal functions $\{\xi_n\} \subset L^2(0, a) \cap \mathscr{H}_F$ such that*

$$F(x - y) = \sum_{n \in \mathbb{N}} \lambda_n \xi_n(x) \overline{\xi_n(y)}, \ and \tag{12.80}$$

$$\int_0^a \overline{\xi_n(x)} \xi_m(x) \, dx = \delta_{n,m}, \ n, m \in \mathbb{N}. \tag{12.81}$$

Proof. An application of Mercer's theorem [LP89, FR42, FM13]. See also Exercise 12.2. $\qquad \square$

Corollary 12.6. *For all $\psi, \varphi \in C_c^\infty(0, a)$, we have*

$$\langle F_\psi, F_\varphi \rangle_{\mathscr{H}_F} = \langle F_\psi, T_F^{-1} F_\varphi \rangle_2. \tag{12.82}$$

Consequently,

$$\|h\|_{\mathscr{H}_F} = \|T_F^{-1/2} h\|_2, \ \forall h \in \mathscr{H}_F. \tag{12.83}$$

Proof. Note

$$\langle F_\psi, T_F^{-1} F_\varphi \rangle_2 = \langle F_\psi, T_F^{-1} T_F \varphi \rangle_2 = \langle F_\psi, \varphi \rangle_2$$

$$= \int_0^a \overline{\left(\int_0^a \psi(x) F(y - x) \, dx \right)} \varphi(y) \, dy$$

$$= \int_0^a \int_0^a \overline{\psi(x)} \varphi(y) F(x - y) \, dx dy = \langle F_\psi, F_\varphi \rangle_{\mathscr{H}_F}.$$

$\qquad \square$

Corollary 12.7. *Let $\{\xi_n\}$ be the ONB in $L^2(0, a)$ as in Corollary 12.5; then $\{\sqrt{\lambda_n} \xi_n\}$ is an ONB in $\mathscr{H}_F$.*

Proof. The functions ξ_n are in $\mathscr{H}_F$ by Theorem 12.6. We check directly (Corollary 12.6) that

$$\left\langle \sqrt{\lambda_n}\xi_n, \sqrt{\lambda_m}\xi_m \right\rangle_{\mathscr{H}_F} = \sqrt{\lambda_n\lambda_m} \left\langle \xi_n, T^{-1}\xi_m \right\rangle_2$$

$$= \sqrt{\lambda_n\lambda_m}\lambda_m^{-1} \left\langle \xi_n, \xi_m \right\rangle_2 = \delta_{n,m}.$$

$\square$

12.6 Type I v.s. Type II Extensions

When a pair (Ω, F) is given, where F is a prescribed continuous positive definite function defined on Ω, we consider the possible continuous positive definite extensions to all of $\mathbb{R}^n$. The reproducing kernel Hilbert space $\mathscr{H}_F$ will play a key role in our analysis. In constructing various classes of continuous positive definite extensions to $\mathbb{R}^n$, we introduce operators in $\mathscr{H}_F$, and their *dilation* to operators, possibly acting in an enlargement Hilbert space [JPT15, KL14b]. Following techniques from dilation theory we note that every dilation contains a minimal one. If a continuous positive definite extensions to $\mathbb{R}^n$ has its minimal dilation Hilbert space equal to $\mathscr{H}_F$, we say it is type 1, otherwise we say it is type 2.

Definition 12.10. Let G be a locally compact group, and let Ω be an open connected subset of G. Let $F : \Omega^{-1} \cdot \Omega \to \mathbb{C}$ be a continuous positive definite function.

Definition 12.11. Consider a strongly continuous unitary representation U of G acting in some Hilbert space $\mathscr{K}$, containing the RKHS $\mathscr{H}_F$. We say that $(U, \mathscr{K}) \in Ext(F)$ iff there is a vector $k_0 \in \mathscr{K}$ such that

$$F(g) = \langle k_0, U(g)k_0 \rangle_{\mathscr{K}}, \ \forall g \in \Omega^{-1} \cdot \Omega. \tag{12.84}$$

1. The subset of $Ext(F)$ consisting of $(U, \mathscr{H}_F, k_0 = F_e)$ with

$$F(g) = \langle F_e, U(g)F_e \rangle_{\mathscr{H}_F}, \ \forall g \in \Omega^{-1} \cdot \Omega \tag{12.85}$$

is denoted $Ext_1(F)$; and we set

$$Ext_2(F) := Ext(F) \backslash Ext_1(F);$$

i.e., $Ext_2(F)$, consists of the solutions to problem (12.84) for which $\mathscr{K} \supsetneq \mathscr{H}_F$, i.e., unitary representations realized in an enlargement Hilbert space. (We write $F_e \in \mathscr{H}_F$ for the vector satisfying $\langle F_e, \xi \rangle_{\mathscr{H}_F} = \xi(e)$, $\forall \xi \in \mathscr{H}_F$, where e is the neutral (unit) element in G, i.e., $eg = g$, $\forall g \in G$.)

2. In the special case, where $G = \mathbb{R}^n$, and $\Omega \subset \mathbb{R}^n$ is open and connected, we consider

$$F : \Omega - \Omega \to \mathbb{C}$$

continuous and positive definite. In this case,

$$Ext(F) = \left\{ \mu \in \mathscr{M}_+(\mathbb{R}^n) \mid \widehat{\mu}(x) = \int_{\mathbb{R}^n} e^{i\lambda \cdot x} d\mu(\lambda) \right. \tag{12.86}$$

is a p.d. extensiont of $F \Big\}$.

Remark 12.5. Note that (12.86) is consistent with (12.84): For if $(U, \mathscr{K}, k_0)$ is a unitary representation of $G = \mathbb{R}^n$, such that (12.84) holds; then, by a theorem of Stone, there is a projection-valued measure (PVM) $P_U(\cdot)$, defined on the Borel subsets of $\mathbb{R}^n$ such that

$$U(x) = \int_{\mathbb{R}^n} e^{i\lambda \cdot x} P_U(d\lambda), \ x \in \mathbb{R}^n. \tag{12.87}$$

Setting

$$d\mu(\lambda) := \|P_U(d\lambda) k_0\|^2_{\mathscr{K}}, \tag{12.88}$$

it is then immediate that we have: $\mu \in \mathscr{M}_+(\mathbb{R}^n)$, and that the finite measure μ satisfies

$$\widehat{\mu}(x) = F(x), \ \forall x \in \Omega - \Omega. \tag{12.89}$$

Set $n = 1$: Start with a local p.d. continuous function F, and let $\mathscr{H}_F$ be the corresponding RKHS. Let $Ext(F)$ be the compact convex set of probability measures on $\mathbb{R}$ defining extensions of F.

We now divide $Ext(F)$ into two parts, say $Ext_1(F)$ and $Ext_2(F)$.

All continuous p.d. extensions of F come from strongly continuous unitary representations. So in the case of 1D, from unitary one-parameter groups of course, say $U(t)$.

Let $Ext_1(F)$ be the subset of $Ext(F)$ corresponding to extensions when the unitary representation $U(t)$ acts in $\mathscr{H}_F$ (internal extensions), and $Ext_2(F)$ denote the part of $Ext(F)$ associated to unitary representations $U(t)$ acting in a proper enlargement Hilbert space $\mathscr{K}$ (if any), i.e., acting in a Hilbert space $\mathscr{K}$ corresponding to a proper dilation of $\mathscr{H}_F$.

12.7 The Case of $e^{-|x|}$, $|x| < 1$

Our emphasis is on von Neumann indices, and explicit formulas for partially defined positive definite functions F, defined initially only on a symmetric interval $(-a, a)$. Among the cases of partially defined positive definite functions, the following example $F(x) = e^{-|x|}$, in the symmetric interval $(-1, 1)$, will play a special role. The present section is devoted to this example.

There are many reasons for this:

(i) It is of independent interest, and its type 1 extensions (see Section 12.6) can be written down explicitly.

(ii) Its applications include stochastic analysis [Itô06] as follows. Given a random variable X in a process; if μ is its distribution, then there are two measures of concentration for μ, one called "degree of concentration," and the other "dispersion," both computed directly from $F(x) = e^{-|x|}$ applied to μ.

(iii) In addition, there are analogous relative notions for comparing different samples in a fixed stochastic process. These notions are defined with the use of example $F(x) = e^{-|x|}$, and it will frequently be useful to localize the x-variable in a compact interval.

(iv) Additional reasons for special attention to example $F(x) = e^{-|x|}$, for $x \in (-1, 1)$ is its use in sampling theory, and analysis of de Branges spaces [DM72], as well as its role as a Green's function for an important boundary value problem.

(v) Related to this, the reproducing kernel Hilbert space $\mathscr{H}_F$ associated to this p.d. function F has a number of properties that also hold for wider families of locally defined positive definite function of a single variable. In particular, $\mathscr{H}_F$ has Fourier bases: The RKHS $\mathscr{H}_F$ has orthogonal bases of complex exponentials e_λ with aperiodic frequency distributions, i.e., frequency points $\{e_\lambda\}$ on the real line which do not lie on any arithmetic progression, see Figure 12.3. For details on this last point, see Corollaries 12.13, 12.14, 12.15, and 12.18.

12.7.1 The Selfadjoint Extensions $A_\theta \supset -iD_F$

The notation "$\supset$" above refers to containment of operators, or rather of the respective graphs of the two operators; see [DS88b].

Lemma 12.8. *Let* $F(x) = e^{-|x|}$, $|x| < 1$. *Set* $F_x(y) := F(x - y)$, $\forall x, y \in (0, 1)$; *and* $F_\varphi(x) = \int_0^1 \varphi(y) F(x - y)\, dy$, $\forall \varphi \in C_c^\infty(0, 1)$. *Define* $D_F(F_\varphi) = F_{\varphi'}$ *on the dense subset*

$$dom\,(D_F) = \{F_\varphi : \varphi \in C_c^\infty(0, 1)\} \subset \mathscr{H}_F. \tag{12.90}$$

Then the skew-Hermitian operator D_F *has deficiency indices* $(1, 1)$ *in* $\mathscr{H}_F$, *where the defect vectors are*

$$\xi_+(x) = F_0(x) = e^{-x} \tag{12.91}$$

$$\xi_-(x) = F_1(x) = e^{x-1}; \tag{12.92}$$

moreover,

$$\|\xi_+\|_{\mathscr{H}_F} = \|\xi_+\|_{\mathscr{H}_F} = 1. \tag{12.93}$$

Proof. (Note if Ω is any bounded, open and connected domain in $\mathbb{R}^n$, then a locally defined continuous p.d. function, $F : \Omega - \Omega :\to \mathbb{C}$, extends uniquely to the boundary $\partial\Omega := \overline{\Omega}\backslash\Omega$ by continuity [JPT15].)

In our current settings, $\Omega = (0,1)$, and $F_x(y) := F(x - y)$, $\forall x, y \in (0,1)$. Thus, $F_x(y)$ extends to all $x, y \in [0,1]$. In particular,

$$F_0(x) = e^{-x}, \ F_1(x) = e^{x-1}$$

are the two defect vectors, as shown in Corollary 12.3. Moreover, using the reproducing property, we have

$$\|F_0\|_{\mathscr{H}_F}^2 = \langle F_0, F_0 \rangle_{\mathscr{H}_F} = F_0(0) = F(0) = 1$$
$$\|F_1\|_{\mathscr{H}_F}^2 = \langle F_1, F_1 \rangle_{\mathscr{H}_F} = F_1(1) = F(0) = 1$$

and (12.93) follows. For more details, see [JPT15, lemma 2.10.14]. □

Lemma 12.9. *Let F be any continuous p.d. function on $(-1,1)$. Set*

$$h(x) = \int_0^1 \varphi(y) F(x - y)\, dy, \ \forall \varphi \in C_c^\infty(0,1);$$

then

$$h(0) = \int_0^1 \varphi(y) F(-y)\, dy, \qquad h(1) = \int_0^1 \varphi(y) F(1 - y)\, dy \qquad (12.94)$$

$$h'(0) = \int_0^1 \varphi(y) F'(-y)\, dy, \qquad h'(1) = \int_0^1 \varphi(y) F'(1 - y)\, dy; \qquad (12.95)$$

where the derivatives F' in (12.94)-(12.95) are in the sense of distribution.

Proof. Note that

$$h(x) = \int_0^x \varphi(y) F(x - y)\, dy + \int_x^1 \varphi(y) F(x - y)\, dy;$$
$$h'(x) = \int_0^x \varphi(y) F'(x - y)\, dy + \int_x^1 \varphi(y) F'(x - y)\, dy$$

and so (12.94)-(12.95) follow. □

We now specialize to the function $F(x) = e^{-|x|}$ defined in $(-1,1)$.

Corollary 12.8. *For $F(x) = e^{-|x|}$, $|x| < 1$, set $h = T_F\varphi$, i.e.,*

$$h := F_\varphi = \int_0^1 \varphi(y) F(\cdot - y)\, dy, \ \forall \varphi \in C_c^\infty(0,1);$$

then

$$h(0) = \int_0^1 \varphi(y) e^{-y}\, dy, \qquad h(1) = \int_0^1 \varphi(y) e^{y-1}\, dy \qquad (12.96)$$

$$h'(0) = \int_0^1 \varphi(y) e^{-y}\, dy, \qquad h'(1) = -\int_0^1 \varphi(y) e^{y-1}\, dy. \qquad (12.97)$$

In particular,

$$h(0) - h'(0) = 0 \tag{12.98}$$
$$h(1) + h'(1) = 0. \tag{12.99}$$

Proof. Immediately from Lemma 12.9. Specifically,

$$h(x) = e^{-x} \int_0^x \varphi(y) e^y dy + e^x \int_x^1 \varphi(y) e^{-y} dy$$

$$h'(x) = -e^{-x} \int_0^x \varphi(y) e^y dy + e^x \int_x^1 \varphi(y) e^{-y} dy.$$

Setting $x = 0$ and $x = 1$ gives the desired conclusions. $\qquad\square$

Remark 12.6. The space

$$\left\{ h \in \mathscr{H}_F \mid h(0) - h'(0) = 0,\ h(1) + h'(1) = 0 \right\}$$

is dense in $\mathscr{H}_F$. This is because it contains $\left\{ F_\varphi \mid \varphi \in C_c^\infty(0,1) \right\}$. Note

$$F_0 + F_0' = -\delta_0, \text{ and}$$
$$F_1 - F_1' = -\delta_1;$$

however, $\delta_0, \delta_1 \notin \mathscr{H}_F$.

By von Neumann's theory [DS88b] and Lemma 12.6, the family of selfadjoint extensions of the Hermitian operator $-iD_F$ is characterized by

$$A_\theta \left(h + c \left(e^{-x} + e^{i\theta} e^{x-1} \right) \right) = -i\, h' + i\, c \left(e^{-x} - e^{i\theta} e^{x-1} \right), \text{ where}$$
$$dom(A_\theta) := \left\{ h + c \left(e^{-x} + e^{i\theta} e^{x-1} \right) \mid h \in dom(D_F), c \in \mathbb{C} \right\}. \tag{12.100}$$

Remark 12.7. In (12.100), $h \in dom(D_F)$ (see (12.90)), and by Corollary 12.8, h satisfies the boundary conditions (12.98)-(12.99). Also, by Lemma 12.8, $\xi_+ = F_0 = e^{-x}$, $\xi_- = F_1 = e^{x-1}$, and $\|\xi_+\|_{\mathscr{H}_F} = \|\xi_-\|_{\mathscr{H}_F} = 1$.

Proposition 12.1. *Let A_θ be a selfadjoint extension of $-iD$ as in (12.100). Then,*

$$\psi(1) + \psi'(1) = e^{i\theta} \left(\psi(0) - \psi'(0) \right), \ \forall \psi \in dom(A_\theta). \tag{12.101}$$

Proof. Any $\psi \in dom(A_\theta)$ has the decomposition

$$\psi(x) = h(x) + c \left(e^{-x} + e^{i\theta} e^{x-1} \right)$$

where $h \in dom(D_F)$, and $c \in \mathbb{C}$. An application of Corollary 12.4 gives

$$\psi(1) + \psi'(1) = \underbrace{h(1) + h'(1)}_{=0 \text{ (by (12.99))}} + c \left(e^{-1} + e^{i\theta} \right) + c \left(-e^{-1} + e^{i\theta} \right) = 2c\, e^{i\theta}$$

$$\psi(0) - \psi'(0) = \underbrace{h(0) - h'(0)}_{=0 \text{ (by (12.98))}} + c \left(1 + e^{-1} e^{i\theta} \right) - c \left(-1 + e^{-1} e^{i\theta} \right) = 2c$$

which is the assertion in (12.101). $\qquad\square$

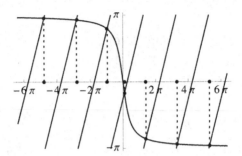

Figure 12.2: Fix $\theta = 0.8$, $\Lambda_\theta = \{\lambda_n(\theta)\}$ = intersections of two curves. (spectrum from curve intersections)

Corollary 12.9. *Let A_θ be a selfadjoint extension of $-iD_F$ as in (12.100). Fix $\lambda \in \mathbb{R}$, then $\lambda \in spec_{pt}(A_\theta) \iff e_\lambda(x) := e^{i\lambda x} \in dom(A_\theta)$, and λ is a solution to the following equation:*

$$\lambda = \theta + \tan^{-1}\left(\frac{2\lambda}{\lambda^2 - 1}\right) + 2n\pi, \ n \in \mathbb{Z}. \tag{12.102}$$

Proof. By assumption, $e^{i\lambda x} \in dom(A_\theta)$, so $\exists h_\lambda \in dom(D_F)$, and $\exists c_\lambda \in \mathbb{C}$ such that

$$e^{i\lambda x} = h_\lambda(x) + c_\lambda\left(e^x + e^{i\theta}e^{x-1}\right). \tag{12.103}$$

Applying the boundary condition in Proposition 12.1, we have

$$e^{i\lambda} + i\lambda e^{i\lambda} = e^{i\theta}(1 - i\lambda); \text{ i.e.,}$$

$$e^{i\lambda} = e^{i\theta}\frac{1 - i\lambda}{1 + i\lambda} = e^{i\theta}e^{i\arg\left(\frac{1-i\lambda}{1+i\lambda}\right)} \tag{12.104}$$

where

$$\arg\left(\frac{1 - i\lambda}{1 + i\lambda}\right) = \tan^{-1}\left(\frac{2\lambda}{\lambda^2 - 1}\right)$$

and (12.102) follows. For a discrete set of solutions, see Figure 12.2. $\square$

Corollary 12.10. *If $A_\theta \supset -iD_F$ is a selfadjoint extension in $\mathscr{H}_F$, then*

$$spect(A_\theta) = \left\{\lambda \in \mathbb{R} \mid e_\lambda \in \mathscr{H}_F \text{ satisfying } (12.101)\right\}$$
$$= \left\{\lambda \in \mathbb{R} \mid e_\lambda \in \mathscr{H}_F, \ e_\lambda = h_\lambda + c_\lambda\left(e^x + e^{i\theta}e^{x-1}\right),\right.$$
$$\left. h_\lambda \in dom(D_F), \ c_\lambda \in \mathbb{C}\right\}.$$

Remark 12.8. The corollary holds for *all* continuous p.d. functions $F : (-a, a) \to \mathbb{C}$.

Corollary 12.11. *All selfadjoint extensions $A_\theta \supset -iD_F$ have purely atomic spectrum; i.e.,*

$$\Lambda_\theta := spect\,(A_\theta) = discrete\ subset\ in\ \mathbb{R}. \tag{12.105}$$

And for all $\lambda \in \Lambda_\theta$,

$$ker\,(A_\theta - \lambda I_{\mathscr{H}_F}) = \mathbb{C}e_\lambda,\ where\ e_\lambda\,(x) = e^{i\lambda x} \tag{12.106}$$

i.e., all eigenvalues have multiplicity 1. (The set Λ_θ will be denoted $\{\lambda_n\,(\theta)\}_{n\in\mathbb{Z}}$ following Figure 12.2.)

Proof. This follows by solving eq. (12.102). □

Corollary 12.12. *Let A be a selfadjoint extension of $-iD_F$ as before. Suppose $\lambda_1, \lambda_2 \in spec\,(A)$, $\lambda_1 \neq \lambda_2$, then $e_{\lambda_i} \in \mathscr{H}_F$, $i = 1, 2$; and $\langle e_{\lambda_1}, e_{\lambda_2}\rangle_{\mathscr{H}_F} = 0$.*

Proof. Let λ_1, λ_2 be as in the statement, then

$$(\lambda_1 - \lambda_2)\,\langle e_{\lambda_1}, e_{\lambda_2}\rangle_{\mathscr{H}_F} = \langle Ae_{\lambda_1}, e_{\lambda_2}\rangle_{\mathscr{H}_F} - \langle e_{\lambda_1}, Ae_{\lambda_2}\rangle_{\mathscr{H}_F} = 0;$$

so since $\lambda_1 - \lambda_2 \neq 0$, we get $\langle e_{\lambda_1}, e_{\lambda_2}\rangle_{\mathscr{H}_F} = 0$. □

For explicit computations regarding these points, see also Corollaries 12.16, 12.17, and 12.18 below.

12.7.2 The Spectra of the s.a. Extensions $A_\theta \supset -iD_F$

Let $F\,(x) = e^{-|x|}$, $|x| < 1$. Define $D_F(F_\varphi) = F_{\varphi'}$ as before, where

$$F_\varphi\,(x) = \int_0^1 \varphi\,(y)\,F\,(x - y)\,dy$$

$$= \int_0^1 \varphi\,(y)\,e^{-|x-y|}dy,\ \forall\varphi \in C_c^\infty\,(0, 1).$$

And let $\mathscr{H}_F$ be the RKHS of F.

Lemma 12.10. *For all $\varphi \in C_c^\infty\,(0, 1)$, and all $h, h'' \in \mathscr{H}_F$, we have*

$$\langle F_\varphi, h\rangle_{\mathscr{H}_F} = \left\langle F_\varphi, \tfrac{1}{2}\,(h - h'')\right\rangle_2 - \tfrac{1}{2}\,[W]_0^1 \tag{12.107}$$

where

$$W = \det \begin{bmatrix} h & F_\varphi \\ h' & F_{\varphi'} \end{bmatrix}. \tag{12.108}$$

Setting $l := F_\varphi$, we have

$$[W]_0^1 = -\bar{l}\,(1)\,(h\,(1) + h'\,(1)) - \bar{l}\,(0)\,(h\,(0) - h'\,(0)). \tag{12.109}$$

Proof. Note

$$\langle F_\varphi, h \rangle_{\mathscr{H}_F} = \int_0^1 \varphi(x) h(x) \, dx \quad \text{(reproducing property)}$$
$$= \left\langle \tfrac{1}{2} \left(I - \left(\tfrac{d}{dx} \right)^2 \right) F_\varphi, h \right\rangle_2$$
$$= \left\langle F_\varphi, \tfrac{1}{2} (h - h'') \right\rangle_2 - \tfrac{1}{2} [W]_0^1.$$

Set $l := F_\varphi \in \mathscr{H}_F$, $\varphi \in C_c^\infty(0,1)$. Recall the boundary condition in Corollary 12.8:

$$l(0) - l'(0) = l(1) + l'(1) = 0.$$

Then

$$[W]_0^1 = \left(\overline{l'} h - \overline{l} h' \right)(1) - \left(\overline{l'} h - \overline{l} h' \right)(0)$$
$$= -\overline{l}(1) h(1) - \overline{l}(1) h'(1) - \overline{l}(0) h(0) + \overline{l}(0) h'(0)$$
$$= -\overline{l}(1) (h(1) + h'(1)) - \overline{l}(0) (h(0) - h'(0))$$

which is (12.109).   □

Corollary 12.13. $e_\lambda \in \mathscr{H}_F$, $\forall \lambda \in \mathbb{R}$.

Proof. By Theorem 12.6, we need the following estimate: $\exists C < \infty$ such that

$$\left| \int_0^1 \varphi(x) e_\lambda(x) \, dx \right|^2 \leq C \, \|F_\varphi\|_{\mathscr{H}_F}^2. \tag{12.110}$$

But

$$\int_0^1 \varphi(x) e_\lambda(x) \, dx$$
$$= \left\langle \tfrac{1}{2} \left(I - \left(\tfrac{d}{dx} \right)^2 \right) F_\varphi, e_\lambda \right\rangle_2$$
$$= \left\langle F_\varphi, \tfrac{1}{2} (e_\lambda - e_\lambda'') \right\rangle_2 - \tfrac{1}{2} [W]_0^1$$
$$= \tfrac{1}{2} (1 + \lambda^2) \langle F_\varphi, e_\lambda \rangle_2 - \tfrac{1}{2} \left(-l(1)(1 + i\lambda) e^{i\lambda} - l(0)(1 - i\lambda) \right);$$

see (12.107)-(12.109). Here, $l := F_\varphi$.

It suffices to show

(i) $\exists C_1 < \infty$ such that

$$|l(0)|^2 \text{ and } |l(1)|^2 \leq C_1 \|F_\varphi\|_{\mathscr{H}_F}^2.$$

(ii) $\exists C_2 < \infty$ such that

$$\left| \langle F_\varphi, e_\lambda \rangle_2 \right|^2 \leq C_2 \|F_\varphi\|_{\mathscr{H}_F}^2.$$

For (i), note that

$$|l(0)| = \left| \langle F_0, l \rangle_{\mathscr{H}_F} \right| \leq \|F_0\|_{\mathscr{H}_F} \|l\|_{\mathscr{H}_F} = \|F_0\|_{\mathscr{H}_F} \|F_\varphi\|_{\mathscr{H}_F}$$

$$|l(1)| = \left|\langle F_1, l\rangle_{\mathscr{H}_F}\right| \leq \|F_1\|_{\mathscr{H}_F} \|l\|_{\mathscr{H}_F} = \|F_1\|_{\mathscr{H}_F} \|F_\varphi\|_{\mathscr{H}_F}$$

and we have

$$\|F_0\|_{\mathscr{H}_F} = \|F_1\|_{\mathscr{H}_F} = 1$$
$$\|l\|_{\mathscr{H}_F}^2 = \|F_\varphi\|_{\mathscr{H}_F}^2 = \|T_F \varphi\|_2^2 \leq \lambda_1^2 \|\varphi\|_2^2 < \infty$$

where λ_1 is the top eigenvalue of the Mercer operator T_F (Lemma 12.7).

For (ii),

$$
\begin{aligned}
\left|\langle F_\varphi, e_\lambda\rangle_2\right|^2 &= \left|\langle T_F\varphi, e_\lambda\rangle_2\right|^2 \\
&= \left|\left\langle T_F^{1/2}\varphi, T_F^{1/2}e_\lambda\right\rangle_2\right|^2 \\
&\leq \left\|T_F^{1/2}\varphi\right\|_2^2 \left\|T_F^{1/2}e_\lambda\right\|_2^2 \quad \text{(by Cauchy-Schwarz)} \\
&= \langle\varphi, T_F\varphi\rangle_2 \left\|T_F^{1/2}e_\lambda\right\|_2^2 \\
&\leq \|F_\varphi\|_{\mathscr{H}_F}^2 \|e_\lambda\|_2^2 = \|F_\varphi\|_{\mathscr{H}_F}^2 ;
\end{aligned}
$$

where we used the fact that $\left\|T_F^{1/2}e_\lambda\right\|_2^2 \leq \lambda_1 \|e_\lambda\|_2^2 \leq 1$, since $\lambda_1 < 1 =$ the right endpoint of the interval $[0, 1]$ (see Lemma 12.7), and $\|e_\lambda\|_2 = 1$.

Therefore, the corollary follows. $\qquad\square$

Corollary 12.14. *For all $\lambda \in \mathbb{R}$, and all F_φ, $\varphi \in C_c^\infty(0, 1)$, we have*

$$\langle F_\varphi, e_\lambda\rangle_{\mathscr{H}_F} = \tfrac{1}{2}\left(1 + \lambda^2\right)\langle F_\varphi, e_\lambda\rangle_2 \qquad (12.111)$$
$$+ \tfrac{1}{2}\left(\bar{l}(1)(1 + i\lambda)e^{i\lambda} + \bar{l}(0)(1 - i\lambda)\right).$$

Proof. By Lemma 12.10,

$$\langle F_\varphi, e_\lambda\rangle_{\mathscr{H}_F} = \left\langle F_\varphi, \tfrac{1}{2}(e_\lambda - e_\lambda'')\right\rangle_2 - \tfrac{1}{2}[W]_0^1.$$

where

$$\tfrac{1}{2}(e_\lambda - e_\lambda'') = \tfrac{1}{2}\left(1 + \lambda^2\right)e_\lambda; \text{ and}$$

$$[W]_0^1 \overset{(12.109)}{=} -\bar{l}(1)(1 + i\lambda)e^{i\lambda} - \bar{l}(0)(1 - i\lambda), \; l := F_\varphi.$$

$\qquad\square$

Lemma 12.11. *For all F_φ, $\varphi \in C_c^\infty(0, 1)$, and all $\lambda \in \mathbb{R}$,*

$$\langle F_\varphi, e_\lambda\rangle_{\mathscr{H}_F} = \langle\varphi, e_\lambda\rangle_2. \qquad (12.112)$$

In particular, set $\lambda = 0$, we get

$$
\begin{aligned}
\langle F_\varphi, 1\rangle_{\mathscr{H}_F} &= \int_0^1 \varphi(x)\,dx = \frac{1}{2}\int_0^1 \left(F_\varphi - F_\varphi''\right)(x)\,dx \\
&= \frac{1}{2}\left(\langle F_\varphi, 1\rangle_2 - \langle F_\varphi'', 1\rangle_2\right) \\
&\leq C\|F_\varphi\|_{\mathscr{H}}.
\end{aligned}
$$

Proof. Eq. (12.112) follows from basic fact of the Mercer operator. See Lemma 12.7 and its corollaries. It suffices to note the following estimate:

$$\int_0^1 F_\varphi''(x)\,dx = F_\varphi'(1) - F_\varphi'(0)$$

$$= -e^{-1}\int_0^1 e^y \varphi(y)\,dy - \int_0^1 e^{-y}\varphi(y)\,dy$$

$$= -F_\varphi(1) - F_\varphi(0) \le 2\,\|F_\varphi\|_{\mathscr{H}}\,.$$

$\square$

Corollary 12.15. *For all* $\lambda \in \mathbb{R}$,

$$\langle e_\lambda, e_\lambda \rangle_{\mathscr{H}_F} = \frac{\lambda^2 + 3}{2}. \tag{12.113}$$

Proof. By Corollary 12.14, we see that

$$\langle F_\varphi, e_\lambda \rangle_{\mathscr{H}_F} = \frac{1}{2}\left(1 + \lambda^2\right)\langle F_\varphi, e_\lambda \rangle_2$$
$$+ \frac{1}{2}\left(\overline{l}(1)(1 + i\lambda)e^{i\lambda} + \overline{l}(0)(1 - i\lambda)\right); \ l := F_\varphi. \tag{12.114}$$

Since $\{F_\varphi : \varphi \in C_c^\infty(0,1)\}$ is dense in $\mathscr{H}_F$, $\exists F_{\varphi_n} \to e_\lambda$ in $\mathscr{H}_F$, so that

$$\langle F_{\varphi_n}, e_\lambda \rangle_{\mathscr{H}_F} \to \langle e_\lambda, e_\lambda \rangle_{\mathscr{H}_F}$$
$$= \frac{1}{2}\left(1 + \lambda^2\right) + \frac{1}{2}\left(e^{-i\lambda}(1 + i\lambda)e^{i\lambda} + (1 - i\lambda)\right)$$
$$= \frac{1}{2}\left(1 + \lambda^2\right) + 1 = \frac{\lambda^2 + 3}{2}.$$

The approximation is justified since all the terms in the RHS of (12.114) satisfy the estimate $|\cdots|^2 \le C\,\|F_\varphi\|_{\mathscr{H}_F}^2$. See the proof of Corollary 12.13 for details. $\square$

Note Lemma 12.10 is equivalent to the following:

Corollary 12.16. *For all* $h \in \mathscr{H}_F$, *and all* $k \in \mathrm{dom}\left(T_F^{-1}\right)$, *i.e.,* $k \in \{F_\varphi : \varphi \in C_c^\infty(0,1)\}$, *we have*

$$\langle h, k \rangle_{\mathscr{H}} = \frac{1}{2}\left(\langle h, k \rangle_0 + \langle h', k' \rangle_0\right) + \frac{1}{2}\left(\overline{h(0)}k(0) + \overline{h(1)}k(1)\right) \tag{12.115}$$

and eq. (12.115) extends to all $k \in \mathscr{H}_F$, *since* $\mathrm{dom}\left(T_F^{-1}\right)$ *is dense in* $\mathscr{H}_F$.

Example 12.4. Take $h = k = e_\lambda$, $\lambda \in \mathbb{R}$, then (12.115) gives

$$\langle e_\lambda, e_\lambda \rangle_{\mathscr{H}} = \frac{1}{2}\left(1 + \lambda^2\right) + \frac{1}{2}\left(1 + 1\right) = \frac{\lambda^2 + 3}{2}$$

as in (12.113).

Corollary 12.17. *Let $A_\theta \supset -iD$ be any selfadjoint extension in $\mathscr{H}_F$. If $\lambda, \mu \in$ spect (A_θ), such that $\lambda \neq \mu$, then $\langle e_\lambda, e_\mu \rangle_{\mathscr{H}_F} = 0$.*

Proof. It follows from (12.115) that

$$2 \langle e_\lambda, e_\mu \rangle_{\mathscr{H}} = \langle e_\lambda, e_\mu \rangle_0 + \lambda \mu \langle e_\lambda, e_\mu \rangle_0 + \left(1 + e^{i(\mu - \lambda)} \right)$$

$$= (1 + \lambda \mu) \langle e_\lambda, e_\mu \rangle_0 + \left(1 + e^{i(\mu - \lambda)} \right)$$

$$= (1 + \lambda \mu) \frac{e^{i(\mu - \lambda)} - 1}{i(\mu - \lambda)} + \left(1 + e^{i(\mu - \lambda)} \right). \tag{12.116}$$

By Corollary 12.9, eq. (12.104), we have

$$e^{i\lambda} = \frac{1 - i\lambda}{1 + i\lambda} e^{i\theta}, \quad e^{i\mu} = \frac{1 - i\mu}{1 + i\mu} e^{i\theta}$$

and so

$$e^{i(\mu - \lambda)} = \frac{(1 - i\mu)(1 + i\lambda)}{(1 + i\mu)(1 - i\lambda)}.$$

Substitute this into (12.116) yields

$$2 \langle e_\lambda, e_\mu \rangle_{\mathscr{H}} = \frac{-2(1 + \lambda \mu)}{(1 + i\mu)(1 - i\lambda)} + \frac{2(1 + \lambda \mu)}{(1 + i\mu)(1 - i\lambda)} = 0.$$

$\square$

Corollary 12.18. *Let $F(x) = e^{-|x|}$, $|x| < 1$. Let $D_F(F_\varphi) = F_{\varphi'}$, $\forall \varphi \in C_c^\infty(0, 1)$, and $A_\theta \supset -iD_F$ be a selfadjoint extension in $\mathscr{H}_F$. Set $e_\lambda(x) = e^{i\lambda x}$, and*

$$\Lambda_\theta := spect(A_\theta) (= discrete\ subset\ in\ \mathbb{R}\ by\ Cor.\ 12.11). \tag{12.117}$$

Then

$$\widetilde{F}_\theta(x) = \sum_{\lambda \in \Lambda_\theta} \frac{2}{\lambda^2 + 3} e_\lambda(x), \ \forall x \in \mathbb{R} \tag{12.118}$$

is a continuous p.d. extension of F to the real line. Note that both sides in eq. (12.118) depend on the choice of θ.

The type 1 extensions are indexed by $\theta \in [0, 2\pi)$ where Λ_θ is given in (12.117), see also (12.102) in Corollary 12.9.

Corollary 12.19 (Sampling property of the set Λ_θ). *Let $F(x) = e^{-|x|}$ in $|x| < 1$, $\mathscr{H}_F$, θ, and Λ_θ be as above. Let T_F be the corresponding Mercer operator. Then for all $\varphi \in L^2(0, 1)$, we have*

$$(T_F \varphi)(x) = 2 \sum_{\lambda \in \Lambda_\theta} \frac{\widehat{\varphi}(\lambda)}{\lambda^2 + 3} e^{i\lambda x}, \ for\ all\ x \in (0, 1).$$

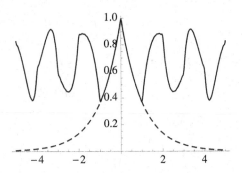

Figure 12.3: $\theta = 0$. A type 1 continuous p.d. extension of $F(x) = e^{-|x|}\big|_{[-1,1]}$ in $\mathscr{H}_F$.

Proof. This is immediate from Corollary 12.18.   □

Remark 12.9. Note that the system $\{e_\lambda \mid \lambda \in \Lambda_\theta\}$ is orthogonal in $\mathscr{H}_F$, but *not* in $L^2(0,1)$.

Proof. We saw that A_θ has pure atomic spectrum. By (12.15), the set

$$\left\{ \sqrt{\frac{2}{\lambda^2 + 3}} e_\lambda : \lambda \in \Lambda_\theta \right\}$$

is an ONB in $\mathscr{H}_F$. Hence, for $F = F_0 = e^{-|x|}$, we have the corresponding p.d. extension:

$$F_\theta(x) = \sum_{\lambda \in \Lambda_\theta} \frac{1}{\|e_\lambda\|_{\mathscr{H}_F}^2} \langle e_\lambda, F \rangle_{\mathscr{H}_F} e_\lambda(x)$$

$$= \sum_{\lambda \in \Lambda_\theta} \frac{2}{\lambda^2 + 3} e_\lambda(x), \ \forall x \in [0,1] \tag{12.119}$$

where $\langle e_\lambda, F \rangle_{\mathscr{H}_F} = \overline{e_\lambda(0)} = 1$ by the reproducing property. But the RHS of (12.119) extends to $\mathbb{R}$. See Figure 12.3.   □

Corollary 12.20. *Let $F(x) = e^{-|x|}$ in $(-1,1)$, and let $\mathscr{H}_F$ be the RKHS. Let $\theta \in [0, 2\pi)$, and let Λ_θ be as above; then $\left\{ \sqrt{\frac{2}{\lambda^2+3}} e_\lambda \mid \lambda \in \Lambda_\theta \right\}$ is an ONB in $\mathscr{H}_F$.*

A summary of relevant numbers from the Reference List

For readers wishing to follow up sources, or to go in more depth with topics above, we suggest:

The pioneering paper here is [Aro50] and the intervening decades have witnessed a host of applications. And by now there are books dealing with various aspects of reproducing kernel Hilbert spaces (RKHS). A more comprehensive citation list is: [AD86, JPT15, Nus75, Rud63, Alp01, CZ07, AJSV13, Aro50, Nel59b, Sch64a, SZ07, SZ09].

Part V

Appendix

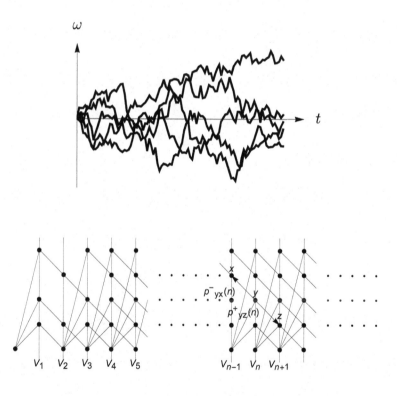

Appendix A

An Overview of Functional Analysis Books (Cast of Characters)

"If people do not believe that mathematics is simple, it is only because they do not realize how complicated life is."
— John von Neumann

Below we offer a list of related Functional Analysis books; they cover a host of diverse areas of functional analysis and applications, some different from what we select here: Our comments are telegraphic-review form (by P.J.):

Akhiezer and Glazman, *"Theory of linear operators in Hilbert space"* [AG93]
– a classic book set covering the detailed structure of unbounded operators and their applications; and now in a lovely Dover edition.

Arveson, *"An invitation to C^*-algebras"* [Arv76]
– an introduction to C^*-algebras and their representations on Hilbert spaces.
– covers the most basic ideas, as simply and concretely as we could. – Hilbert spaces are separable and C^*-algebras are GCR. Representations are given a concrete parametric description, including the irreducible representations of any C^*-algebra, even if not GCR. For someone interested in Borel structures, see Chapter 3. Chapter 1 is a bare-bones introduction to C^*-algebras.

Bachman and Narici, *"Functional analysis"* [BN00]
The book by Bachman and Narici's is a systematic introduction to the fundamentals of functional analysis. It is easier to follow than say Rudin's Functional Analysis book, but it doesn't go as far either. Rather it helps readers reinforcing

topics from real analysis and other masters level courses. It serves to bridge the
gap between more difficult treatments of functional analysis. (Dover reprints
classics in a cheap paper back format.)

Bratteli and Robinson, *"Operator algebras and quantum statistical mechanics"* [BR79, BR81b]
This is a widely cited two volume book-set, covering the theory of operator
algebras, and its applications to quantum statistical mechanics. It is one of the
more authoritative treatments of the many exciting applications of functional
analysis to physics. Both books are self-contained; with complete proofs; – a
useful text for students with a prior exposure to basic functional analysis. One
of the main themes in v1 is decomposition theory, and the use of Choquet sim-
plices. Example: the set of KMS-states for a C^*-algebraic dynamical system
typically forms a Choquet simplex. An introductory chapter covers algebraic
techniques and their use in statistical physics; this is followed up in v2. Indeed,
a host of applications are covered in v2. The new edition has a more comprehensive discussion of dissipative operators and analytic elements; – and it includes
a positive resolution of the question of whether maximal orthogonal probability
measure on the state space of algebra is automatically maximal among all the
probability measures on the space.

Conway, *"A course in functional analysis"* [Con90]
– a comprehensive introduction to functional analysis. The style is formal
and rigorous. – is designed to be used in grad courses. Through its eleven chapters, J. Conway masterfully wrote a beautiful exposition of this core subject.

Dunford and Schwartz, *"Linear operators"* [DS88a, DS88b, DS88c]
This classic three-volume book set, the first Functional analysis, and the
second the theory of linear operators. And for the theory of unbounded operators it is unsurpassed. – written by two notable mathematicians, it constitutes
a comprehensive survey of the general theory of linear operations, and their
diverse applications. Dunford and Schwartz are influenced by von Neumann,
and they emphasize the significance of the relationships between the abstract
theory and its applications. The two first volumes are for the students. – treatment is relatively self-contained. Now a paperback edition of the original work,
unabridged, in three volumes.

Kolmogorov and Fomin, *"Introductory real analysis"* [KF75]
This book is two books bound as one; and in the lovely format from Dover.
Part 1: metric spaces, and normed linear spaces. Part 2: Lebesgue integration
and basic functional analysis. Numerous examples are sprinkled through the
text. To get the most out of this book, it helps if you have already seen many
of the results presented elsewhere. History: The book came from original notes
from Andrei Kolmogorov's lectures given at Moscow's Lomonosov University in
the 1940's, and it still stands as timely introduction to real and functional analysis. Strengths: step by step presentation of all the key concepts needed in the

subject; proceeding all the way from set theory to Fredholm integral equations. Offers a wonderful and refreshing insight. Contents (sample): Elements of Set Theory; Metric and Topological Spaces; Normed and Topological Linear Spaces; Linear Functionals and Linear Operators; Elements of Differential Calculus in Linear Spaces; Measure, Measurable Functions, Integral; Indefinite Lebesgue Integral, Differentiation Theory; Spaces of Summable Functions; Trigonometric Series, Fourier Transformation; Linear Integral Equations.

Kadison and Ringrose, *"Fundamentals of the theory of operator algebras"* [KR97a, KR97b]
Here we cite the first two volumes in a 4-volume book-set. It begins with the fundamentals in functional analysis, and it aims at a systematic presentations of the main areas in the theory of operator algebras, both C^*-algebras, von Neumann algebras, and their applications, so including subfactors, Tomita-Takesaki theory, spectral theory, decomposition theory, and applications to ergodic theory, representations of groups, and to mathematical physics.

Lax, *"Functional analysis"* [Lax02]
The subject of functional analysis, while fundamental and central in the landscape of mathematics, really started with seminal theorems due to Banach, Hilbert, von Neumann, Herglotz, Hausdorff, Friedrichs, Steinhouse, ... and many other of, the perhaps less well known, founding fathers, in Central Europe (at the time), in the period between the two World Wars. It gained from there because of its many success stories, – in proving new theorems, in unifying old ones, in offering a framework for quantum theory, for dynamical systems, and for partial differential equations. The Journal of Functional Analysis, starting in the 1960ties, broadened the subject, reaching almost all branches of science, and finding functional analytic flavor in theories surprisingly far from the original roots of the subject. Peter Lax has himself, – alone and with others, shaped some of greatest successes of the period, right up to the present. That is in the book!! And it offers an upbeat outlook for the future. It has been tested in the class room, – it is really user-friendly. At the end of each chapter P. Lax offers personal recollections; – little known stories of how several of the pioneers in the subject have been victims, – in the 30ties and the 40ties, of Nazi atrocities. The writing is crisp and engaged.

MacCluer, *"Elementary functional analysis"* [Mac09]
I received extremely positive student-feedback on MacCluer's very nice book. It covers elementary functional analysis, is great for self-study, and easy to follow. It conveys the author's enthusiasm for her subject. It includes apposite quotes, anecdotes, and historical asides, all making for a wonderful personal touch and drawing the reader into dialogue in a palpable way. Contents: six chapters, each introduced by a well-chosen quote, often hinting in a very useful manner at the material that is to follow. I particularly like MacCluer's choice of Dunford and Schwartz to start off her third chapter: "In linear spaces with a suitable topology one encounters three far-reaching principles concerning contin-

uous linear transformations. . ." We find out quickly that these "Big Three" (as the chapter is titled) are uniform boundedness, the open mapping theorem, and Hahn-Banach. MacCluer quickly goes on to cover these three gems in a most effective and elegant manner, as well as a number of their corollaries or, in her words, "close cousins," such as the closed graph theorem and Banach-Steinhaus. The book takes the reader from Hilbert space preliminaries to Banach- and C^*-algebras and, to the spectral theorem.

Nelson, *"Topics in Dynamics I: Flows"* [Nel69]
This is a book in the Princeton Math Lecture Notes series, appearing first in 1972, but since Prof Nelson kindly made it available on his website. In our opinion, it is the best account of general multiplicity for normal operators, bounded and unbounded, and for Abelian $*$-algebras. In addition it contains a number of applications of functional analysis to geometry and to physics.

Riesz et al., *"Functional analysis"* [RSN90]
A pioneering book in F.A., first published in the early 50s, and now in a Dover edition, very readable. The book starts with an example of a continuous function which is not differentiable and then proves Lebesgue's theorem which tells you when a function does have a derivative. The 2nd part of the book is about integral equations which again starts with some examples of problems from the 19th century mathematicians. The presentation of Fredholm's method is a gem.

Rudin, *"Functional analysis"* [Rud91]
"Modern analysis" used to be a popular name for the subject of this lovely book. It is as important as ever, but perhaps less "modern". The subject of functional analysis, while fundamental and central in the landscape of mathematics, really started with seminal theorems due to Banach, Hilbert, von Neumann, Herglotz, Hausdorff, Friedrichs, Steinhouse, ... and many other of, the perhaps less well known, founding fathers, in Central Europe (at the time), in the period between the two World Wars. In the beginning it generated awe in its ability to provide elegant proofs of classical theorems that otherwise were thought to be both technical and difficult. The beautiful idea that makes it all clear as daylight: Wiener's theorem on absolutely convergent (AC) Fourier series of $1/f$ if you can divide, and if f has AC Fourier series, is a case in point. The new subject gained from there because of its many success stories, – in proving new theorems, in unifying old ones, in offering a framework for quantum theory, for dynamical systems, and for partial differential equations. And offering a language that facilitated interdisciplinary work in science! The topics in Rudin's book are inspired by harmonic analysis. The later part offers one of the most elegant compact treatment of the theory of operators in Hilbert space, I can think of. Its approach to unbounded operators is lovely.

Sakai, *"C^*-algebras and W^*-algebras"* [Sak98]
The presentation is succinct, theorem, proof, ... qed; but this lovely book

had a profound influence on the subject. It's scope cover nearly all major results in the subject up until that time. In order to accomplish this goal (without expanding into multiple volumes), the author omits examples, motivation,.... . It is for students who already have an interest in operator theory. As a student, myself (PJ), I learned a lot from this wonderful book.

Shilov, *"Elementary functional analysis"* [Shi96]
Elementary Functional Analysis by Georgi E. Shilov is suitable for a beginning course in functional analysis and some of its applications, e.g., to Fourier series, to harmonic analysis, to partial differential equations (PDEs), to Sobolev spaces, and it is a good supplement and complement to two other popular books in the subject, one by Rudin, and another by Edwards. Rudin's book is entitled "Functional Analysis" includes new material on unbounded operators in Hilbert space. Edwards' book "Functional Analysis: Theory and Applications;" is in the Dover series, and it is twice as thick as Shilov's book. Topics covered in Shilov: Function spaces, L^p-spaces, Hilbert spaces, and linear operators; the standard Banach, and Hahn-Banach theorems. It includes many exercises and examples. Well motivated with applications. Book Comparison: Shilov book is gentler on students, and it is probably easier to get started with: It stresses motivation a bit more, the exercises are easier, and finally Shilov includes a few applications; fashionable these days.

Stein et al., *"Functional analysis"* [SS11]
This book is the fourth book in a series: Elias Stein's and Rami Shakarchi's Princeton lectures in analysis. Elias Stein is a world authority on harmonic analysis. The book is of more recent vintage than the others from our present list. The book on functional analysis is actually quite different from other texts in functional analysis. For instance Rudin's textbook on functional analysis has quite a different emphasis from Stein's. Stein devotes a whole chapter to applications of the Baire category theory while Rudin devotes a page. Stein does this because it provides some insights into establishing the existence of a continuous but nowhere differentiable function as well as the existence of a continuous function with Fourier series diverging a point. A special touch in Stein: Inclusion of Brownian motion, and of process with independent increments, a la Doob's. Stein's approach to the construction of Brownian motion is different and closer to the approaches taken in books on financial math. Stein et al develop Brownian motion in the context of solving Dirichlet's problem.

Stone, *"Linear Transformations in Hilbert Space and Their Applications to Analysis"* [Sto90]
Stone's book is a classic, came out in 1932, and was the unique source on spectral multiplicity, and a host of applications of the theory of unbounded operators to analysis, to approximation theory, and to special functions. The last two chapters illustrate the theory with a systematic study of (infinite × infinite) Jacobi matrices; i.e., tri-diagonal infinite matrices; assumed formally selfadjoint (i.e., Hermitian). Sample results: A dichotomy: Their von Neumann indices

must be $(0, 0)$ or $(1, 1)$. Some of the first known criteria for when they are one or the other are given; plus a number of applications to classical analysis.

Takesaki, *"Theory of operator algebras"* [Tak79]
– written by one of the most prominent researchers of the area, provides an introduction to this rapidly developing theory. ... These books are recommended to every graduate student interested in this exciting branch of mathematics. Furthermore, they should be on the bookshelf of every researcher of the area.

Trèves, *"Topological vector spaces, distributions and kernels"* [Trè06b]
Covers topological vector spaces and their applications, and it is a pioneering book. It is antidote for those who mistakenly believe that functional analysis is about Banach and Hilbert spaces. It's also about Fréchet spaces, LF spaces, Schwartz distributions (generalized functions), nuclear spaces, tensor products, and the Schwartz Kernel Theorem (proved by Grothendieck). Trèves's book provides the perfect background for advanced work in linear differential, pseudodifferential, or Fourier integral operators.

Yosida *"Functional analysis"* [Yos95]
Yosida's book is based on lectures given decades ago at the University of Tokyo. It is intended as a textbook to be studied by students on their own or to be used in a course on Functional Analysis, i.e., the general theory of linear operators in function spaces together with salient features of its application to diverse fields of modern and classical analysis. Necessary prerequisites for the reading of this book are summarized, with or without proof, in Chapter 0 under titles: Set Theory, Topological Spaces, Measure Spaces and Linear Spaces. Then, starting with the chapter on Semi-norms, a general theory of Banach and Hilbert spaces is presented in connection with the theory of generalized functions of S.L. Sobolev and L. Schwartz. The reader may pass, e.g., from Chapter IX (Analytical Theory of Semi-groups) directly to Chapter XIII (Ergodic Theory and Diffusion Theory) and to Chapter XIV (Integration of the Equation of Evolution). Such materials as "Weak Topologies and Duality in Locally Convex Spaces" and "Nuclear Spaces" are presented in the form of the appendices to Chapter V and Chapter X, respectively.

Some relevant books: Classics, and in the Dover series:

Banach, *"Theory of Linear Operations"* [Ban93]
Georgi, *"Weak Interactions and Modern Particle Theory"* [Geo09]
Prenter, *"Splines and Variational Methods"* [Pre89]

In Chapters 5 and 9 above we have cited pioneers in quantum physics, the foundations of quantum mechanics. The most central here are Heisenberg (matrix mechanics), Schrödinger (wave mechanics, the Schrödinger equation), and Dirac (Dirac's equation is a relativistic wave equation, describes all spin-$\frac{1}{2}$ massive particles free form, as well as electromagnetic interactions). We further sketched von Neumann's discovery of the equivalence of the answers given by Heisenberg and Schrödinger, and the Stone-von Neumann uniqueness theorem. The relevant papers and books are as follows: [Hei69, Sch32, vN31, HN28, Dir35, Dir47].

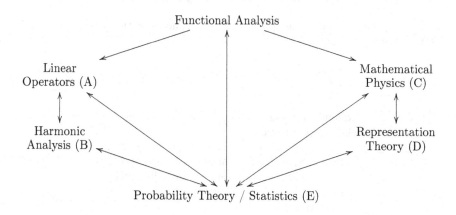

Table A.1: An overview of connections between Functional Analysis and some of its applications.

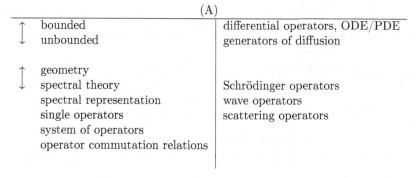

(A)	
↑ bounded ↓ unbounded	differential operators, ODE/PDE generators of diffusion
↑ geometry ↓ spectral theory spectral representation single operators system of operators operator commutation relations	Schrödinger operators wave operators scattering operators

Table A.2: Linear operators in Hilbert space and their applications.

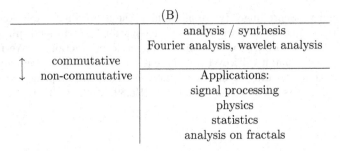

Table A.3: Harmonic analysis (commutative and non-commutative), and a selection of applications.

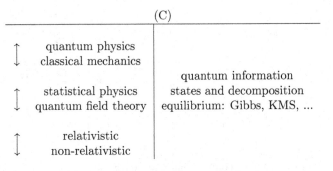

Table A.4: Topics from quantum physics.

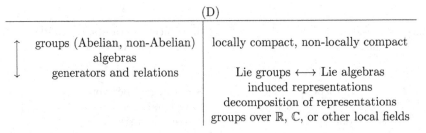

Table A.5: Representations of groups and algebras.

	(E)
stochastic processes	discrete continuous Gaussian Brownian motion non-Gaussian Lévy
	solutions of diffusion equations with the use of functional integrals (i.e., probability measure on infinite dimensional spaces such as $C(\mathbb{R})$ or Schwartz space $\mathcal{S}$)

Table A.6: Analysis of stochastic processes and some of their applications.

Appendix B

Terminology from Neighboring Areas

Classical Wiener measure/space Classical Wiener space (named Norbert Wiener) is the sample-space part of a probability space, a triple (sample-space, sigma-algebra, and probability measure). The sample space may be taken to be the collection of all continuous functions on a given domain (usually an interval for time). So sample paths are continuous functions. The sigma-algebra is generated by cylinder-sets, and the probability measure is called the Wiener measure; (its construction is subtle, see Chapter 7 above.) It has the property that the stochastic processes which samples the paths in the model is a Gaussian process with independent increments, the so called Brownian motion. It is also called the Wiener process. And it should perhaps be named after L. Bachelier, whose work predates that of Einstein.

It, and the related process "white noise", are important in pure and applied mathematics. It is a core ingredient in stochastic analysis: the study of stochastic calculus, diffusion processes, and potential theory. Applications include engineering and physics: models of noise in electronics engineering, in instrument errors in filtering theory, and in control theory. In atomic physics, it is used in the study of diffusion, the Fokker-Planck and Langevin equations; and in path-space integrals; the Feynman-Kac formula, in the solution of the Schrödinger equation. In finance, it is used in the solution to the Black-Scholes equation for option prices. (For details, see [AJ12, AJL13, ARR13, CW14, GJ87, Gro70, Hid80, HØUZ10, Itô06, Loè63, Jor06, Jør14, Nel64, Nel67, Sch32, Sch58, SS09].) Ref. Sections 2.4 (pg. 39), 7.2 (pg. 259), 7.3 (pg. 266), and 12.3 (pg. 413).

Hilbert's sixth problem This is not a "yes/no problem"; rather the 6th asks for a mathematical axiomatization of physics. In a common English translation, it reads: 6. Give a Mathematical Treatment of the Axioms of Physics. A parallel is drawn to the foundations of geometry: *To treat in the same manner, by means of axioms, those physical sciences where mathematics plays an* *important part; in the first rank are the theory of probabilities and mechanics.*

Hilbert: "As to the axioms of the theory of probabilities, it seems to me desirable that their logical investigation should be accompanied by a rigorous and satisfactory development of the method of mean values in mathematical physics, and in particular in the kinetic theory of gases. ... Boltzmann's work on the principles of mechanics suggests the problem of developing mathematically the limiting processes, there merely indicated, which lead from the atomistic view to the laws of motion of continua."

In the 1930s, probability theory was put on an axiomatic and sound foundation by Andrey Kolmogorov. In the 1960s, we have the work of A. Wightman[1], R. Haag, J. Glimm, and A. Jaffe on quantum field theory. This was followed by the Standard Model in particle physics and general relativity. Still unresolved is the theory of quantum gravity.

Ref. Sections 1.6 (pg. 17), 2.5.2 (pg. 70), 7.2 (pg. 259); see also [Wig76, MSS13].

Infinite-dimensional analysis Understood in a wide sense, this header includes almost all of analysis, but it has become customary to reserve the name for theories of calculus inspired in turn by stochastic processes, of one form or another. Understood this way, it encompasses Itō calculus, Malliavin calculus, and free probability theory.

Itō calculus has been especially successful in extending traditional methods of calculus to Gaussian stochastic processes, with Brownian motion (Wiener process) as its foundation. It has applications in mathematical finance and in stochastic differential equations (SDE). In mathematical finance, various Itō stochastic integrals represent the payoff of continuous-time trading strategies based on hedging in some risk-neutral model.

Parallel to this, but of a more recent vintage, Malliavin calculus (also called the stochastic calculus of variations) extends classical calculus of variations now to random variables and associated martingale-stochastic processes. In particular, it allows the computation of an infinite-dimensional variant of a derivative (for random variables), and, with an Ito integral, it includes an integration by parts formula; — keeping in mind that Malliavin calculus takes place in path space. Malliavin calculus was motivated initially by the need for a probabilistic proof of hypoellipticity from Hörmander's condition, and, with this, thereby to derive existence and smoothness of a density for the solution of a stochastic differential equation.

Free probability is an infinite-dimensional theory adapted to non-commutative random variables. The "freeness" here refers to free independence, a property which is analogues to "independence" in the commutative theory. It entails free products and free convolution. In the classical case, we know that the sum of several independent random variables has a distribution which is the convolution of the laws of the individual random variables in the sum. The beginning of the free theory therefore entails a notion of "free convolution." The theory was invented by Dan Voiculescu in the 1980ties; and was motivated by the study

---

[1]Appendix C includes a short biographical sketch of A. Wightman. See also [Wig76].

free group factors, and isomorphism questions. But it has since found a host of applications. (For details, see [AJS14, JPS01, Voi85, Voi91, Voi06].)
Ref. Sections 1.2.3 (pg. 8), 2.1 (pg. 25), 2.4 (pg. 39), and Chapter 7 (pg. 249).

Monte Carlo (MC) simulation (or Monte Carlo random number generators) refers to algorithms which generate repeated random numbers, or samples. Since the "randomness" is in fact computer generated, the term often used is "pseudo-randomness." Applications of MC-simulations include such areas as numerical analysis in very high dimensions, numerical integration, in Karhunen-Loève expansions based on the generation of i.i.d. sets of random variables, in estimation of parameters (for example in financial math), in optimization, and for generation of random paths; more generally, computer generated draws, or samples, from a prescribed probability distribution, uniform, Gaussian, Poisson etc. MC-simulation is useful for modeling phenomena subject to uncertainty or volatility. By the law of large numbers, integrals can be estimated as expected values of suitable random variables. When the size of a system under consideration tends to infinity, random empirical measurements converge to the distribution random states of Markov chains (some non-linear).

For example, the following Mathematica code yields a simulation of 5 sample paths of standard Brownian motion:

```
n = 10^3; dt = 1/n;
dB := Sqrt[dt] (2*RandomInteger[1, n] - 1);
ListPlot[Table[Accumulate[dB], {s, 1, 5}], ImageSize -> 350]
```

Ref. Sections 1.2.4 (pg. 9), 7.2 (pg. 259); and Chapters 7 (pg. 249), 8 (pg. 273).

Multiresolution analysis (MRA) refers to structures having an inherent structure of scales, and self-similarity. While it has received a large amount of attention in wavelet analysis, it has since found applications in a host of disparate areas of computational mathematics. In a wavelet context, it was introduced in the 1980ties by Stephane Mallat and Yves Meyer; but more generally it encompasses such numerical tools as pyramid schemes/algorithms, resolution scales in image processing, and renormalization in physics. They have in common the use of hierarchies, and masking points; for example, in computer graphics, in signal/image processing, and in image compression and processing. Many MRAs are based on a suitable variant of a discrete wavelet transform. (For details, see [Mal01, MM09, Mey06, BJ02, CM07, Jor06, JS09, Str12].)
Ref. Sections 2.4 (pg. 39), 6.6 (pg. 241); and Chapter 6 (pg. 223).

Quantum field theory (QFT) is a mathematical framework used in physics for constructing models of subatomic particles (quantum mechanical). It covers such areas as particle physics and condensed matter physics. A QFT treats particle-wave duality as excited states of an underlying physical field, called field quanta. Of interest are quantum mechanical interactions between particles and the corresponding underlying fields. (For details, see [GJ87].)

Ref. Section 1.6 (pg. 17).

Quantum information (QI) A key idea which makes QI different from classical information science is this: As opposed to the classical information theory bases on the binary states, the bit-computer. By contrast a quantum computer is based on quantum states, and manipulation of quantum systems, by the rules of quantum mechanics. And quantum programs involve instead manipulating quantum information processing. Quantum states allow superpositions (e.g., a continuum of convex combinations of two pure states), a feature excluded in the binary model. Superposition of quantum states (qubits) in turn entail the possibility of coherence of states; again a feature not feasible in the binary model. The starting point in quantum computation may still be binary quantum states, e.g., polarized spins (up/down), or photon states aligned by magnetic fields, e.g., horizontal vs vertical. Classical information can be processed with digital computers, transmitted from place to place, manipulated with algorithms, and analyzed with the mathematics of computer science. But only some of these concepts apply to quantum information; and with changes; superposition and coherence being two major ones. The notion of coherence from quantum theory in turn relies on the mathematics of tensor products of Hilbert space, and on an analysis of states and operators realized in these tensor product Hilbert spaces.

Some features of QI: A unit of quantum information is the qubit, a certain quantum state. Unlike classical digital states (which are discrete), a qubit is continuous-valued, in quantum theory taking the form of a direction on the Bloch sphere. A qubit is the smallest possible unit of quantum information. The reason for this indivisibility is the Heisenberg uncertainty principle: despite the qubit state being continuously-valued, measurement is counter intuitive. If a binary state in superposition is subjected to measurement, it will become one of the two binary pure states; the Schrödinger's cat problem.

Originally, Schrödinger intended his thought experiment as a discussion of the EPR article – named after its authors Einstein, Podolsky, and Rosen (1935.) The EPR article highlighted this strange nature of quantum superpositions: a quantum system (e.g., an atom or photon) can exist as a combination of multiple states corresponding to different possible outcomes. Thus the EPR experiment showed that a system with multiple particles separated by large distances could be in such a superposition.

A qubit cannot be (wholly) converted into classical bits; that is, it cannot be "read". This is the no-teleportation theorem. An arbitrary qubit cannot be copied. This is the content of the no cloning theorem. Qubits can be changed, by applying linear transformations, or quantum gates, to them. Two or more qubits can be arranged in such a way as to convey classical bits. The best understood such configuration is the Bell state, consisting of two qubits and four classical bits. Because of entanglement, quantum algorithms have a different computational complexity than classical algorithms. (For details, see [BJ02, Par09, Arv09a, Arv09b, Arv09b, Kho15, Pho15, NC00].)
Ref. Sections 2.5 (pg. 63), 4.4 (pg. 140), 5.4 (pg. 174), 8.2 (pg. 280); and Chapters 5 (pg. 155), 6 (pg. 223), 9 (pg. 333).

Quantum mechanics (QM);   (quantum physics, or quantum theory) is a branch of physics which describes physical phenomena at "small" scales, atomic and subatomic length scales. The action is on the order of the Planck constant. QM deals with observation of physical quantities that can only change and interact, by discrete amounts or "steps" (hence "quantum"), and behave probabilistically rather than deterministically. The "steps" are too tiny even for microscopes. Any description must be given in terms of a wave function, as opposed to particles. (For details, see [Dir47, BR81b].)
Ref. Sections 2.5.2 (pg. 70), 3.1 (pg. 90), 4.1 (pg. 120), and Chapter 9 (pg. 333).

Quantum probability (QP) was invented only recently. It is a noncommutative analog of probability and stochastic processes as per Kolmogorov and Itō. The non-commutativity derives from a statistical interpretation of quantum theory. One of the physics issues involves is that of finding suitable dynamical solutions to the quantum measurement problem. This in turn involves coming up with constructive models of quantum observation-processes. Mathematical tools include: quantum filtering, feedback control theory, and quantum stochastic calculus. In classical probability theory, random variables are commuting (measurable functions relative to an underlying probability space), as opposed to the non-commutativity in QP, directly dictated by laws of quantum theory. Quantum observables take the form of selfadjoint operators, typically non-commuting (for example, position and momentum, or quantum fields), and as a result we arrive at the study of algebras of operators in Hilbert space, von Neumann algebras, or C^*-algebras. In classical probability one relies on independence, while the corresponding notion for systems of non-commuting random variables is "freeness" in the sense of Voiculescu's free probability. In classical probability theory, the limit law is Gaussian (as per the Law of Large Numbers), while, in the free case, it is the semicircle law. (For details, see [AJ12, AJS14, ARR13, Arv09b, Fan10, Jør14, JPS01, Kho15, Maa10, Par09, Wig58, Voi06, Voi91, Voi85].)
Ref. Sections 1.2.4 (pg. 9), 2.2 (pg. 27), 2.5.2 (pg. 70), 4.1 (pg. 120), 12.3 (pg. 413); and Chapters 2 (pg. 23), 5 (pg. 155), 7 (pg. 249), 12 (pg. 403).

Signal processing (SP) is an engineering discipline dealing with transmission of signals (information, speech or images, over wires, or wireless. An important tool in the area involves subdivision of time signals into frequency bands, and it involves effective algorithms for implementations of processing or transferring information contained in a variety of different symbolic, or abstract formats broadly designated as signals. SP uses mathematical, statistical, computational tools. (For details, see [Wol51, BJ02].)
Ref. Section 6.4 (pg. 233); and Chapters 6 (pg. 223), 9 (pg. 333).

Stochastic processes (SP), or random process, are part of probability theory. They are used when deterministic quantities are not feasible: random

variables are measures in their respective probability distributions (also called "laws.") A SP is an indexed family of random variables (representing measurements, or samples), for example, if a SP is indexed by time, it represents the evolution or dynamics of some system. A SP is the probabilistic counterpart to a deterministic process (or a deterministic system). An example is Brownian motion (BM), the random motion of particles (e.g., pollen) suspended in a fluid (a liquid or a gas). BM results from their collision of the pollen with atoms or molecules making up the gas or liquid. BM also refer to the mathematical model used to describe such random movements. (For details, see [Gro64, Itô04, Itô06, Sch58, SS09].)
Ref. Sections 2.2.1 (pg. 33), 12.3 (pg. 413); and Chapters 7 (pg. 249), 8 (pg. 273).

Uncertainty quantification (UQ) is an interdisciplinary area, spanning risk-modeling, re-insurance, tail-ends of extreme events, failure-analysis (engineering), and more. Its aim is a systematic quantitative features of uncertainties, keeping in mind both the underlying mathematics and the practicalities of implementation. By its nature UQ deals with a "very large" number of variables, but statistical features are often captured best in continuous models, hence limits when the size of relevant data tends to infinity. And hence the relevance of our present chapters dealing with infinite-dimensional analysis. Naturally, UQ entails such tools from probability and stochastic analysis, inverse problems, as covered in the individual chapters inside the present book; but a glance of the literature also shows that the list of tools includes a host of functional analytic topics, and theorems from non-commutative analysis; referring to non-commuting linear operators in Hilbert space. One aims to determine how likely certain outcomes are subject to uncertainty; for example if some aspects of the system are not exactly known; or are governed by volatility. The nature of "uncertainty" is diverse, e.g., parametric, structural, algorithmic, experimental, sampling, or interpolation uncertainty (in data obtained from simulations, measurements or sampling; incompleteness, lack of data etc.) Readers interested in an accessible beginning book on UQ are referred to [Sul15] or the papers [OSS+13, OSS15]. The sources of error and uncertainty, include systematic and stochastic measurement errors; even ignorance; also limitations of theoretical models; limitations of numerical representations; limitations of the accuracy and reliability of computations, and human errors. (For details, see [Akh65, CW14, Gro64, Jor11, Itô06, AJ12, AJS14].)
Ref. Sections 1.2.3 (pg. 8), 1.2.4 (pg. 9), 2.1 (pg. 25), 2.4 (pg. 39), 6.6 (pg. 241), and Chapters 7 (pg. 249), 8 (pg. 273).

Unitary representations (UR) of a groups G are homomorphisms from the group G in question into the group of all unitary operators in some Hilbert space; the Hilbert space depending on the UR. The theory is best understood in in the case of strongly continuous URs of locally compact (topological) groups. Applications include quantum mechanics. Books by Hermann Weyl, Pontryagin, and George Mackey have influenced our presentation. The theory of UR is closely connected with harmonic analysis; – for non-commutative groups, non-Abelian harmonic analysis. Important groups in physics are non-commutative, so this case is extremely important, although it is also rather technical. There is a vast literature, though. Important papers are cited in the books by George Mackey. In fact, versions of the Plancherel theorem exist for some non-commutative Lie groups (of direct relevance to physics), but they are subtle, and the non-commutative analysis is carried out on a case-by-case basis. The best known special case it that of compact groups where we have the Peter-Weyl theorem. But the important symmetry groups in relativistic physics are non-compact, see [GJ87].

Ref. Sections 2.4 (pg. 39), 3.1 (pg. 90), and Chapters 5 (pg. 155), 6 (pg. 223), 8 (pg. 273).

Wavelets   are wave-like functions; they can typically be visualized as "brief oscillations" as one might see recorded in seismographs, or in heart monitors. What is special about wavelet functions is that they allow for effective algorithmic construction of bases in a variety of function spaces. The algorithms in turn are based on a notion of resolution and scale-similarity. The last two features make wavelet decompositions more powerful than comparable Fourier analyses. Wavelets can be localized, while Fourier bases cannot. Wavelets are designed to have specific properties that make them of practical use in signal processing. (For details, see [BJ02].)

Ref. Sections 2.4.2 (pg. 46), 5.8 (pg. 198), 6.4 (pg. 233), and Chapter 6 (pg. 223).

Appendix C

Often Cited

"**mathematical ideas originate in empirics.** But, once they are conceived, the subject begins to live a peculiar life of its own and is ... governed by almost entirely aesthetical motivations. In other words, at a great distance from its empirical source, or after much 'abstract' inbreeding, a mathematical subject is in danger of degeneration. Whenever this stage is reached the only remedy seems to me to be the rejuvenating return to the source: the reinjection of more or less directly empirical ideas."
— von Neumann

Inside the book, the following authors are cited frequently, W. Arveson, M. Atiyah, L. Bachelier, S. Banach, V. Bargmann, H. Bohr, M. Born, N. Bohr, G. Choquet, A. Connes, P. Dirac, W. Döblin, F. Dyson, W. Feller, K. Friedrichs, L. Gårding, I. Gelfand, Piet Hein, W. Heisenberg, D. Hilbert, E. Hille, K. Itō, R. Kadison, S. Kakutani, K. Karhunen, A. Kolmogorov, M. Krein, P. Lax, P. Lévy, M. Loève, G. Mackey, P. Malliavin, A. Markov, E. Nelson, R. Phillips, M. Plancherel, F. Riesz, M. Riesz, D. Robinson, I. Schoenberg, R. Schrader, E. Schrödinger, I. Schur, H.A. Schwarz, L. Schwartz, J. Schwartz, I. Segal, I. Singer, S. Smale, M. Stone, G. Szegö, J. von Neumann, N. Wiener, and A. Wightman. Below a short bio for each of them.

William Arveson (1934 – 2011) [Arv72, Arv98] Cited in connection with C^*-algebras and their states and representations.

W. Arveson, known for his work on completely positive maps, and their extensions; powerful generalizations of the ideas of Banach, Krein, and Stinespring. An early results in this area is an extension theorem for completely positive maps with values in the algebra of all bounded operators. This theorem led to injectivity of von Neumann algebras in general, and work by Alain Connes relating injectivity to hyperfiniteness. In a series of papers in the 60's and 70's, Arveson introduced non-commutative analogues of several concepts from classical harmonic analysis including the Shilov and. Choquet boundaries. Ref. Section 5.1 (pg. 158), and Chapter 6 (pg. 223).

Sir Michael Francis Atiyah (1929 –) Of the Atiyah-Singer Index Theorem. The Atiyah-Singer index of a partial differential operator (PDO) is related to the Fredholm index; – it equates an index, i.e., the difference of the number of independent solutions of two geometric, homogeneous PDEs (one for the operator and the other for its adjoint) to an associated list of invariants in differential geometry. It applies to many problems in mathematics after they are translated into the problem of finding the number of independent solutions of some PDE. The Atiyah-Singer index theorem gives a formula for the index of certain differential operators, in terms of geometric and topological invariants.

The Hirzebruch-Riemann-Roch theorem is a special cases of the Atiyah-Singer index theorem. In fact the index theorem gave a more powerful result, because its proof applied to all compact complex manifolds, while Hirzebruch's proof only worked for projective manifolds.

Related: In 1959 by Gelfand noticed homotopy invariance via an index, and he asked for more general formulas for topological invariants. For spin manifolds, Atiyah suggested that integrality could be explained as an index of a Dirac operator (Atiyah and Singer, 1961). Ref. Section 2.1 (pg. 25).

Louis Bachelier (1870 – 1946) A French probabilist, is credited with being the inventor of the stochastic process, now called Brownian motion; it was part of his PhD thesis, The Theory of Speculation, (1900). It discusses use of random walks, and Brownian motion, to evaluate stock options, and it is considered the first paper in mathematical finance. Even though Bachelier's work was more mathematical, and predates Einstein's Brownian motion paper by five years, it didn't receive much attention at the time, and it was only "discovered" much later by the MIT economists Paul Samuelson, in the 1960ties. Ref. Chapter 7 (pg. 249).

Stefan Banach (1892 – 1945) [Ban93] The Banach of "Banach space." Banach called them "B-spaces" in his book. They were also formalized by Norbert Wiener (who traveled in Europe in the 1920ties.) But the name "Banach space" stuck.

S. Banach, one of the founders of modern functional analysis and one of the original members of the Lwów School of Mathematics, in Poland between the two World Wars. His 1932 book, *Théorie des opérations linéaires* (Theory of Linear Operations), is the first monograph on the general theory of functional analysis. Ref. Sections 2.2 (pg. 27), 2.5.1 (pg. 67), 2.B (pg. 86), 5.3 (pg. 168), and Chapters 5 (pg. 155), 8 (pg. 273).

Valentine "Valya" Bargmann (1908 – 1989) [Bar47] German-American mathematical physicist. PhD from the University of Zürich, under Gregor Wentzel. Worked at the Institute for Advanced Study in Princeton (1937–46) as an assistant to Albert Einstein. Was Professor at Princeton University since 1946, in the Mathematics and Physics departments. He found the irreducible unitary representations of $SL_2(\mathbb{R})$ and of the Lorentz group (1947). And he found the corresponding holomorphic representation realized in Segal-Bargmann space (1961). The latter is a reproducing kernel Hilbert space (RKHS), with associated kernel, now called the Bargmann kernel, (— different from the Bergman kernel which is named after Stefan Bergman. It refers to a quite different RKHS, the Bergman space.) With Eugene Wigner, Valya Bargmann found what is now called the Bargmann-Wigner equations for particles of arbitrary spin; and he discovered the Bargmann-Michel-Telegdi equation, describing relativistic precession of the maximum number of bound states of quantum mechanical potentials (1952).

Ref. Section 8.9.1 (pg. 319).

Harald August Bohr (1887 – 1951) A Danish mathematician and soccer player. Best known for his theory of almost periodic functions. In modern language it became the Bohr-compactification. (Different from the alternative compactifications we discussed above.) He is the brother of the physicist Niels Bohr.

Ref. Section 5.3 (pg. 168).

Niels Henrik David Bohr (1885 – 1962) A Danish physicist who made foundational contributions to understanding atomic structure and quantum theory, the "Bohr-atom", justifying the Balmer series for the visible spectral lines of the hydrogen atom; received the Nobel Prize in Physics in 1922; — "for his services in the investigation of the structure of atoms, and of the radiation emanating from them". Based on his liquid drop model of the nucleus, Bohr concluded that it was the uranium-235 isotope, and not the more abundant uranium-238, that was primarily responsible for fission. In September 1941, at the start of WWII, Heisenberg, who had become head of the German nuclear energy project, visited Bohr in Copenhagen. During this meeting the two had discussions about possible plans by the two sides in the War, for a fission bomb, the content of the discussions have caused much speculation. Michael Frayn's 1998 play "Copenhagen" explores what might have happened at the 1941 meeting between Heisenberg and Bohr.

Ref. Section 2.1 (pg. 25), and Chapter 10 (pg. 349).

Max Born (1882 – 1970) [BP44] A German physicist and mathematician, a pioneer in the early development of quantum mechanics; also in solid-state physics, and optics. Won the 1954 Nobel Prize in Physics for his "fundamental research in Quantum Mechanics, especially in the statistical interpretation of the wave function." His assistants at Göttingen, between the two World Wars, included Enrico Fermi, Werner Heisenberg, and Eugene Wigner, among others. His early education was at Breslau, where his fellow students included Otto Toeplitz and Ernst Hellinger.

In 1926, he formulated the now-standard interpretation of the probability density function for states (represented as equivalence classes of solutions to the Schrödinger equation.) After the Nazi Party came to power in Germany in 1933, Born was suspended. Subsequently he held positions at Johns Hopkins University, at Princeton University, and he settled down at St John's College, Cambridge (UK). A quote: "I believe that ideas such as absolute certitude, absolute exactness, final truth, etc. are figments of the imagination which should not be admissible in any field of science. On the other hand, any assertion of probability is either right or wrong from the standpoint of the theory on which it is based." Max Born (1954.)

Ref. Sections 2.4 (pg. 39), and Chapter 9 (pg. 333).

Gustave Choquet (1915 – 2006) A French mathematician. Is perhaps best known for his result giving a boundary-integral representation for compact convex sets K in locally convex topological vector spaces (TVS); thereby giving a concrete representation for the set of extreme-points in an arbitrary compact convex set K. The result (Choquet integral) found applications in direct integral-decomposition theory, operator algebras; and to the mathematics of states in physics, where the pure states take the form of extreme points in compact convex sets of "mixed" states. His work left a mark in functional analysis, potential theory, and measure theory; as well as in their applications: The Choquet integral, and the theory of capacities.

Ref. Section 5.8 (pg. 198).

Alain Connes (1947 –) [Con07] A French mathematician, Professor at the Collège de France, IHÉS. Connes was awarded the Fields Medal in 1982, the Crafoord Prize in 2001, for his pioneering research in operator algebras, classification of von Neumann factors, non-commutative geometry, and in physics (Higgs boson, M Theory, the standard model).

Ref. Chapters 5 (pg. 155), 6 (pg. 223), 9 (pg. 333).

Paul Adrien Maurice Dirac (1902 – 1984) [Dir35, Dir47] Cited in connection with the "Dirac equation" and especially our notation for vectors and operators in Hilbert space, as well as the axioms of observables, states and measurements. P. Dirac, an English theoretical physicist; fundamental contributions to the early development of both quantum mechanics and quantum electrodynamics. He was the Lucasian Professor of Mathematics at the University of Cambridge. Notable discoveries, the Dirac equation, which describes the behavior of fermions and predicted the existence of antimatter. Dirac shared the Nobel Prize in Physics for 1933 with Erwin Schrödinger, "for the discovery of new productive forms of atomic theory." A rare interview with Dirac; see the link: http://www.math.rutgers.edu/~greenfie/mill_courses/math421/int.html

Ref. Sections 2.5 (pg. 63), 5.3 (pg. 168), and Chapters 4 (pg. 119), 9 (pg. 333).

Wolfgang Döblin (1915 – 40) [IR07] French-German mathematician, and probabilist. Studied probability theory in Paris, under Fréchet. Served in the French army in the Ardennes when World War II broke out in 1939. There, he wrote down his work on the Chapman-Kolmogorov equation. And he sent it in a sealed envelope to the French Academy of Sciences. In 1940, after burning his mathematical notes, he took his own life as the German troops came in sight. In 2000, the sealed envelope was opened, revealing that, at the time, Döblin had anticipated the theory of Markov processes, Itō's lemma (now the Itō-Döblin lemma), and parts of stochastic calculus.

Freeman John Dyson (1923 –) Theoretical physicist and mathematician; – known for his contributions to quantum electrodynamics, solid-state physics, astronomy, and to nuclear engineering. Within mathematics, he is known for his work on random matrices; his discovery of a perturbation expansion (the Dyson expansion.) He is a regular contributor to The New York Review of Books. Awards: the Lorentz Medal, the Max Planck Medal, and the Enrico Fermi Award.

William "Vilim" Feller (1906 – 1970) [Fel68, Fel71] American mathematician, and probabilist, Croatian born. Feller was a Professor at Princeton University till his death. Through his two-volume book set, "An Introduction to Probability Theory and its Applications," Feller was perhaps more than anyone else in the US responsible for turning probability theory into a main area of mathematical analysis. His two books were the Bible in probability theory for some time, but by now there are many current books covering a host of sub-disciplines; as well as cutting-edge advances. Feller's research pioneered the analysis of Markov chains and stochastic processes (especially diffusion processes) with the use of differential equations, and the theory of semigroups, via infinitesimal generators of associated one-parameter semigroups. His work on random walk, diffusion processes, and the law of the iterated logarithm was ground breaking.

Ref. Sections 2.4 (pg. 39), 7.2 (pg. 259), 7.3 (pg. 266), and 12.3 (pg. 413).

Kurt Otto Friedrichs (1901 – 1982) [Fri80, FL28] Is the Friedrichs of the Friedrichs extension; referring to the following Theorem: Every semibounded operator S with dense domain in Hilbert space has a selfadjoint extension having the same lower bound as S. There are other semibounded and selfadjoint extensions of S; – they were found later by M. Krein.

K. Friedrichs, a noted German-American mathematician; a co-founder of The Courant Institute at New York University and recipient of the National Medal of Science.

A story: Selfadjoint operators, and the gulf between the lingo and culture
of mathematics and of physics:
Peter Lax relates the following conversation in German between
K.O. Friedrichs and W. Heisenberg, to have been taken place in the late 1950ties,
in New York, when Heisenberg visited The Courant Institute at NYU. (The two
had been colleagues in Germany before the war.) As a gracious host, Friedrichs
praised Heisenberg for having created quantum mechanics. – After an awkward
silence, Friedrichs went on: "..and we owe to von Neumann our understanding
of the crucial difference between a selfadjoint operator and one that is merely
symmetric." Another silence, and then – Heisenberg: "What is the difference?"
Ref. Sections 10.4 (pg. 371), 10.5 (pg. 376), and Chapter 11 (pg. 385).

Lars Gårding (1919 – 2014)   The "G" in Gårding vectors (representations
of Lie groups), and in Gårding-Wightman quantum fields.
Ref. Section 3.1 (pg. 90), and Chapter 8 (pg. 273).

Israel Moiseevich Gelfand (1913 – 2009) [GJ60, GS60, GG59]   Is the
"G" in GNS (Gelfand-Naimark-Segal), the correspondence between states and
cyclic representations.
I. Gelfand, also written Israïl Moyseyovich Gel'fand, or Izrail M. Gelfand,
a Russian-American mathematician; major contributions to many branches of
mathematics: representation theory and functional analysis. The recipient of
numerous awards and honors, including the Order of Lenin and the Wolf Prize,
– a lifelong academic, serving decades as a professor at Moscow State University
and, after immigrating to the United States shortly before his 76th birthday, at
Rutgers University.
Ref. Section 2.2 (pg. 27), and Chapter 5 (pg. 155).

Piet Hein (1905 – 1996)   A Danish scientist, mathematician, inventor, de-
signer, author, and poet. Pseudonym "Kumbel" (Old Norse for "tombstone".
His poems, known as grooks (Danish: gruk), appeared first in the Danish news-
paper "Politiken" after the start of Nazi occupation of Denmark in April 1940.
In 1940, Piet Hein joined the Danish resistance movement during the five years
of occupation. Piet Hein, in his own words, "played mental ping-pong" with
Niels Bohr in the inter-War period.

Werner Karl Heisenberg (1901 – 1976) [Hei69]   Is the Heisenberg of
the Heisenberg uncertainty principle for the operators P (momentum) and Q
(position), and of matrix mechanics, as the first mathematical formulation of
quantum observables. In Heisenberg's picture, the dynamics, the observables are
studied as function of time; by contrast to Schrödinger's model which have the
states (wavefunctions) functions of time, and satisfying a PDE wave equation,
now called the Schrödinger equation. In the late 1920ties, the two pictures, that
of Heisenberg and of Schrödinger were thought to be irreconcilable. Work of
von Neumann in 1932 demonstrated that they in fact are equivalent.

W. Heisenberg; one of the key creators of quantum mechanics. A 1925 paper was a breakthrough. In the subsequent series of papers with Max Born and Pascual Jordan, this matrix formulation of quantum mechanics took a mathematical rigorous formulation. In 1927 he published his uncertainty principle. Heisenberg was awarded the Nobel Prize in Physics for 1932 "for the creation of quantum mechanics." He made important contributions to the theories of the hydrodynamics of turbulent flows, the atomic nucleus.

Ref. Sections 1.2.1 (pg. 5), 1.6 (pg. 17), 2.1 (pg. 25), 2.5.2 (pg. 70), and Chapters 8 (pg. 273), 10 (pg. 349).

David Hilbert (1862 – 1943) [Hil24, Hil22, Hil02] Cited in connection with the early formulations of the theory of operators in (what is now called) Hilbert space. The name Hilbert space was suggested by von Neumann who studied with Hilbert in the early 1930ties, before he moved to the USA. (The early papers by von Neumann are in German.)

D. Hilbert is recognized as one of the most influential and universal mathematicians of the 19th and early 20th centuries. Discovered and developed invariant theory and axiomatization of geometry. In his 1900 presentation of a collection of research problems, he set the course for much of the mathematical research of the 20th century.

Ref. Section 1.2.4 (pg. 9), and Appendix B (pg. 451).

Einar Hille (1894 – 1980) A Swedish-American mathematician. His work includes integral equations, differential equations, special functions, analytic functions, and Dirichlet series. He is the "H" in the Hille-Yosida theorem, and is the author of one of the first books on functional analysis; in its second edition, co-authored with R. S. Phillips [HP57]. Hille served as president of the American Mathematical Society (1937–38).

Ref. Sections 7.2 (pg. 264), 10.1 (pg. 350).

Kiyoshi Itō (1915 – 2008) [Itô07, Itô04] Cited in connection with Brownian motion, Itō-calculus, and stochastic processes. Making connection to functional analysis via the theory of semigroups of operators (Hille and Phillips.)

Ref. Section 2.4 (pg. 39), and Chapter 7 (pg. 249).

Richard V. Kadison [KS59, KR97a] (1925 –) ... known for his contributions to the study of operator algebras. Is the "K" in the Kadison-Singer problem (see [MSS15]); and the "K" in the Fuglede-Kadison determinant. He is a Gustave C. Kuemmerle Professor in the Department of Mathematics of the University of Pennsylvania; was awarded the Leroy P. Steele Prize for Lifetime Achievement, in 1999.

Ref. Section 5.1 (pg. 158) and Chapter 9 (pg. 333).

Shizuo Kakutani (1911 – 2004) [Kak48] Of his theorems in functional analysis, there is the Kakutani fixed-point theorem (a generalization of Brouwer's fixed-point theorem); – with such applications as to the Nash equilibrium in game theory. Also notable is his solution of the Poisson equation using the methods of stochastic analysis; as well as his pioneering advances in our understanding of two-dimensional Brownian motion; and its applications to PDE, and to potential theory.

Kari Karhunen (1915–1992)   A Finnish probabilist, University of Helsinki, and the K in the Karhunen-Loève theorem. His thesis advisor was Rolf Nevanlinna. He served from 1955 on the Finnish Committee for Mathematical Machines (computers.)
    Ref. Sections 7.3 (pg. 266), 7.4 (pg. 270).

Andrey Nikolaevich Kolmogorov (1903 – 1987),   Russian (Soviet at the time). Kolmogorov was a leading mathematician of the 20th Century, with pioneering contributions in analysis, probability theory, turbulence, information theory, and in computational complexity. A number of central notions in mathematics are named after him. The following list covers those of relevance to the topics in the present book, but it is only a fraction of the full list: Kolmogorov's axioms for probability theory, the Kolmogorov-Fokker-Planck equations, the Chapman-Kolmogorov equation, the Kolmogorov-Arnold-Moser (KAM) theorem, Kolmogorov's extension theorem, Kolmogorov complexity, and Kolmogorov's zero-one law. (Most of Kolmogorov's work was done during the Cold War, and the scientists on either side of the Iron Curtain at the time had very little access to parallel developments on the other side.)
    Ref. Sections 1.2 (pg. 5), 5.8 (pg. 198), 12.3 (pg. 413), and Chapter 7 (pg. 249).

Mark Grigorievich Krein [Kre46, Kre55] (1907 – 1989)   Is the Krein of Krein-Milman on convex weak ∗-compact sets. Soviet mathematician; known for pioneering works in operator theory, mathematical physics, the problem of moments, functional and classical analysis, and representation theory. Winner of the Wolf Prize, 1982. His list of former students includes David Milman, Mark Naimark, Izrail Glazman, Moshe Livshits.
    Ref. Sections 5.8 (pg. 5.8), 8.8 (pg. 315), and Chapters 5 (pg. 155), 9 (pg. 333).

Peter David Lax (1926 –)   The "L" in Lax-Phillips scattering theory, and the Lax-Milgram lemma. A pioneer in PDE, and in many areas of applied mathematics; – especially as they connect to functional analysis.
    Ref. Sections 2.1 (pg. 2.1), 9.3 (pg. 9.3).

Paul Pierre Lévy (1886 – 1971)   French mathematician, one of the founders of modern probability theory. Among his innovations are: martingales, Lévy processes, Lévy measures, Lévy's constant, Lévy's distribution, the Lévy area, the Lévy arcsine law, and Lévy flight. He was a Professor at the École Polytechnique. His former PhD students include Benoît Mandelbrot. He was a member of the French Academy of Sciences. Among his works, we highlight [Lév53, Lév51, Lév57, Lév63, Lév62] of special relevance for the present book. Levy is best known for his work on stochastic processes, but item [Lév51] is in fact a book in functional analysis, an early, very original, and influential book in modern analysis.

Ref. Chapter 7 (pg. 249).

Michel Loève (1907 – 1979) Born in Jaffa, now present day Israel, through his career he was a French-American probabilist. He is the L in the Karhunen-Loève theorem, and the author of a very influential book [Loè63]. Before the War, he studied mathematics at the Université de Paris, under Paul Lévy. And after he moved to the US, he became professor of Mathematics and Statistics at Berkeley, starting from 1955.

Ref. Sections 7.3 (pg. 266), 7.4 (pg. 270).

George Whitelaw Mackey (1916 – 2006) The first "M" in the "Mackey-machine," a systematic tool for constructing unitary representations of Lie groups, as induced representations. A pioneer in non-commutative harmonic analysis, and its applications to physics, to number theory, and to ergodic theory.

Ref. Section 8.4 (pg. 288).

Paul Malliavin (1925 – 2010) A French mathematician. His research covered harmonic analysis, functional analysis, and stochastic processes. Is perhaps best known for what is now called Malliavin calculus, an infinite-dimensional variational calculus for stochastic processes. Applications include a probabilistic proof of the existence and smoothness of densities for the solutions to certain stochastic differential equations. Malliavin's calculus applies to, for example, random variables arising in computation of financial derivatives (financial products created from underlying securities), to the Clark-Ocone formula for martingale representations, and to stochastic filtering.

Ref. Section 3.3.1 (pg. 109).

Andrey Andreyevich Markov (also spelled Markoff) (1856 – 1922) A Russian mathematician, known for his work on stochastic processes, and especially for what we now call Markov chains (discrete "time"), and Markov processes (in short, memory-free transition processes.) Both have applications to a host of statistical models. Loosely speaking, a Markov process is 'memoryless': the future of the process depends only on its present state, as opposed to the process's full history. So, conditioned on the present state, future and past are independent. With his younger brother Vladimir Andreevich Markov, he proved what is now called the Markov brothers' inequality. His son, Andrei Andreevich Markov (1903 – 1979), made important contributions to recursive function theory.

Ref. Section 2.4 (pg. 39), and Chapter 7 (pg. 249).

Edward (Ed) Nelson (1932 – 2014) [Nel69, Nel59a] Cited in connection with spectral representation, and Brownian motion.
Ref. Sections 4.2 (pg. 125), 5.11 (pg. 214), 8.10 (pg. 322), 10.1 (pg. 350).

Ralph Saul Phillips (1913 – 1998) [LP89, Phi94] Cited in connection with the foundations of functional analysis, especially the theory of semigroups of bounded operators acting on Banach space.
Ref. Sections 7.2 (pg. 264), 10.1 (pg. 350).

Michel Plancherel (1885 – 1967) A Swiss mathematician, and he is best known for the Plancherel theorem PT) in harmonic analysis, in signal processing, and in physics. In its current form, the Plancherel theorem applies in a general context of dual variables, for example, time-frequency, or some P-Q momentum-position pair. In the language of the associated L^2-spaces, it asserts that the appropriate transform (generalized Fourier transform) is a unitary isometric isomorphisms between corresponding two L^2-spaces, one for each of the two variables in the Fourier-dual pair at hand. There are versions both for Abelian and for non-Abelian locally compact groups G, but in the case of non-Abelian groups G, things are much more subtle as the dual to G (the set of equivalence classes of irreps) is then not a group. And moreover, for non-compact and non-Abelian groups G, the irreps are typically infinite-dimensional.
Ref. Chapter 8 (pg. 273).

Frigyes Riesz (1880 – 1956) Made fundamental contributions to functional analysis, and to the theory of operators in Hilbert space. We frequently use his Riesz representation theorem. He also did some of the fundamental work, developing functional analysis for applications, especially to spectral theory, and ergodic theory; both important in physics. And with his brother, Marcel Riesz, work in harmonic analysis.

Marcel Riesz (1886 – 1969) Born in Hungary, was the younger brother of the mathematician Frigyes Riesz (the two are known for the F. and M. Riesz theorem). Both are pioneers in Functional Analysis. M. Riesz moved to Sweden in 1911 where he taught at Stockholm University and at Lund University. His former students include Harald Cramér, Einar Hille (of Hille-Phillips), Otto Frostman (potential theory), Lars Hörmander (PDE), and Olaf Thorin (harmonic analysis).
Ref. Chapters 2 (pg. 23), 4 (pg. 119), 5 (pg. 155).

Derek William Robinson (1935 –) [Rob71, BR79, BR81b] A British-Australian mathematical physicist and mathematician. He was a professor in Switzerland, Germany, United States, France (Marseille), and in Australia, at University of New South Wales, and then at the Australian National University (current.) Was President of the Australian Mathematical Society. PhD from University of Oxford (in nuclear physics). His research includes quantum field theory, quantum statistical mechanics (phase-transition theorems), C^*-algebras, analysis on Lie groups, Gaussian estimates of heat kernels on Lie groups. His book "Thermodynamic pressure in Quantum Statistical Mechanics" (Springer 1971), and his two-volume book-set with Ola Bratteli "Operator algebras and quantum Statistical Mechanics" are widely cited.

Ref. Section 7.2 (pg. 264), and Chapter 5 (pg. 155).

Isaac Jacob Schoenberg (1903 – 1990) A Romanian-American mathematician, perhaps best known for his discovery of splines. Conditionally negative definite functions, and their use in metric geometry, are also due to him. Also known for his influential work on total positivity, and on variation-diminishing linear transformations. A totally positive matrix is a square matrix for which all square submatrices have non-negative determinant.

Robert Schrader (1939 – 2015) Was a mathematical physicist, Professor at Freien Universität Berlin, and he is the "S" in the Osterwalder-Schrader (OS) axioms, OS-positivity, and OS-Euclidean fields. This is part of the bigger picture of quantum field theory (QFT); and it is based on a certain analytic continuation (or reflection) of the Wightman distributions (from the Wightman axioms). In this analytic continuation, OS induces Euclidean random fields; and Euclidean covariance. For the unitary representations of the respective symmetry groups, we therefore change these groups as well: OS-reflection applied to the Poincaré group of relativistic fields yields the Euclidean group as its reflection. The starting point of the OS-approach to QFT is a certain positivity condition called "reflection positivity."

Ref. Sections 1.7 (pg. 18), 5.4 (pg. 174).

Erwin Rudolf Josef Alexander Schrödinger (1887 – 1961) [Sch99, Sch40, Sch32] Is the Schrödinger of the Schrödinger equation; the PDE which governs the dynamics of quantum states (as wavefunctions).

E. Schrödinger, a Nobel Prize in physics. – quantum theory forming the basis of wave mechanics: he formulated the wave equation (stationary and time-dependent Schrödinger equation) , and he Schrödinger proposed an original interpretation of the physical meaning of the wave function; formalized the notion of entanglement. He was critical of the conventional Copenhagen interpretation of quantum mechanics (using e.g. the paradox of Schrödinger's cat).

Ref. Sections 2.5.2 (pg. 70), 3.1 (pg. 90), and Chapter 8 (pg. 273).

Issai Schur (1875 – 1941) A German mathematician, and a pioneer in representation theory (The Schur decomposition, Schur duality, and Schur's lemma). He was a PhD student of Frobenius. And both worked on group representations. Schur's work is also important in the theory of integral equations, combinatorics, number theory, and in physics.

Ref. Sections 5.7 (pg. 189), 12.2 (pg. 407), and Chapters 5 (pg. 155), 8 (pg. 273).

Karl Hermann Amandus Schwarz (1843 – 1921) [Sch72]   Is the Schwarz of the Cauchy-Schwarz inequality. H.A. Schwarz is German and is a contemporary of K. Weierstrass. H.A. Schwarz, a German mathematician, known for his work in complex analysis. At Göttingen, he pioneered of function theory, differential geometry and the calculus of variations.
Ref. Chapters 2 (pg. 23), 4 (pg. 119).

Laurent-Moïse Schwartz (1915 – 2002) [Sch95, Sch58, Sch57]   Is the Schwartz (French) of the theory of distributions (dating the 1950ties), also now named "generalized functions" in the books by Gelfand et al. Parts of this theory were developed independently on the two sides of the Iron-Curtain; – in the time of the Cold War.
Ref. Sections 2.2 (pg. 27), 3.1 (pg. 90), 4.1 (pg. 120), and Chapter 8 (pg. 273).

Jacob Theodore "Jack" Schwartz (1930 – 2009) [DS88a, DS88b, DS88c]   Is the Schwartz of the book set "linear operators" by Dunford and Schwartz. Vol II [DS88b] is one of the best presentation of the theory of unbounded operators.

Irving Ezra Segal (1918 – 1998) [Seg50]   Cited in connection with the foundations of functional analysis, and pioneering research in mathematical physics. Is the "S" in GNS (Gelfand-Naimark-Segal). Segal proved the Plancherel theorem in a very general framework: locally compact unimodular groups. For any locally compact unimodular group, Segal established a Plancherel formula; see [Seg50]. Segal showed that there is a Plancherel formula, despite the fact that it may not be feasible, for all locally compact unimodular groups, to "write down" all the irreducible unitary representations.
Ref. Chapters 5 (pg. 155), 6 (pg. 223), and 8 (pg. 273).

Isadore Manuel Singer (1924 –)   Is an Institute Professor at the Massachusetts Institute of Technology; He is the "S" in the Atiyah–Singer index theorem (1962), Michael Atiyah is the "A." Also of note: The Atiyah–Hitchin–Singer theorem, and The Atiyah–Patodi–Singer eta-invariant.
Ref. Section 2.1 (pg. 25).

Stephen Smale (1930 –) [SZ07, SZ09]   An American mathematician, Fields Medal in 1966. Was for three decades on the mathematics faculty of the University of California, Berkeley. His research areas include topology, dynamical systems (e.g., the Smale horseshoe), integral operators, reproducing kernel Hilbert spaces (RKHS) and their use in optimization problems from machine learning, the study of geometry on probability spaces, and their applications.

In 1958, he gave a proof of sphere eversion, and in 1960, a proof of the Poincaré conjecture for all dimensions greater than or equal to 5; published in 1961. In the past two decades, he has turned his attention to a number of areas of applied mathematics, the later items in the list above.

Of relevance to our present book themes is his work on geometry and analysis of probability spaces. This work takes advantage of both classical analysis, and of newer graph theoretic ideas. The starting point for this work is a suitable space X equipped with a kernel (e.g., a Mercer kernel). The resulting integral operator in turn induces a metric on X. And it further leads to a construction of a certain Laplacian, to an associated heat equation, and to a diffusion distance. A main focus is then establishing useful and explicit *a priori* bounds on the error terms of various empirical approximation schemes for this Laplacian on X.

Ref. Chapter 12 (pg. 403).

Marshall Harvey Stone (1903 – 1989) [Sto51, Sto90] Is the "S" in the Stone-Weierstrass theorem; and in the Stone-von Neumann uniqueness theorem (see [vN32b, vN31]); the latter to the effect that any two representations of Heisenberg's commutation relations in the same (finite!) number of degrees of freedom are unitarily equivalent. Stone was the son of Harlan Fiske Stone, Chief Justice of the United States in 1941-1946. Marshall Stone completed a Harvard Ph.D. in 1926, with a thesis supervised by George David Birkhoff. He taught at Harvard, Yale, and Columbia University. And he was promoted to a full Professor at Harvard in 1937. In 1946, he became the chairman of the Mathematics Department at the University of Chicago. His 1932 monograph titled "Linear transformations in Hilbert space and their applications to analysis" develops the theory of selfadjoint operators, turning it into a form which is now a central part of functional analysis. Theorems that carry his name: The Banach-Stone theorem, The Glivenko-Stone theorem, Stone duality, The Stone-Weierstrass theorem, Stone's representation theorem for Boolean algebras, Stone's theorem for one-parameter unitary groups, Stone-Čech compactification, and The Stone-von Neumann uniqueness theorem. M. Stone and von Neumann are the two pioneers who worked at the same period. They were born at about the same time. Stone died at 1980's, and von Neumann died in the 1950's. For a bio and research overview, see [Kad90].

Ref. Sections 2.1 (pg. 25), 3.2 (pg. 97), 4.2 (pg. 125), and Chapters 6 (pg. 223), 8 (pg. 273), 9 (pg. 333).

Gábor Szegö (1895 – 1985) Was a Hungarian-American mathematician; made fundamental contributions to analysis (the Szegö kernel), and especially to the theory of Toeplitz matrices, and to orthogonal polynomials. His monograph Orthogonal polynomials, (1939) [Sze59] has had a profound influence on applied mathematics, including theoretical physics, stochastic processes, and numerical analysis. Szegö had a long time collaboration with George Pólya, e.g., [PS98]; and both were professors at Stanford University after the Second World War.

Their collaboration had started in Europe before that. Szegö was chairman of the mathematics department at Stanford from 1938 to 1966.
Ref. Section 5.9 (pg. 203), and Chapter 12 (pg. 403).

John von Neumann (1903 – 1957) [vN31, vN32a] Cited in connection with the Stone-von Neumann uniqueness theorem, the deficiency indices which determine parameters for possible selfadjoint extensions of given Hermitian (formally selfadjoint, symmetric) with dense domain in Hilbert space.
J. von Neumann, Hungarian-American; inventor and polymath. He made major contributions to: foundations of mathematics, functional analysis, ergodic theory, numerical analysis, physics (quantum mechanics, hydrodynamics, and economics (game theory), computing (von Neumann architecture, linear programming, self-replicating machines, stochastic computing (Monte-Carlo[1])), – was a pioneer of the application of operator theory to quantum mechanics, a principal member of the Manhattan Project and the Institute for Advanced Study in Princeton. – A key figure in the development of game theory, cellular automata, and the digital computer.
Ref. Sections 1.2 (pg. 5), 2.6 (pg. 77), and Chapters 2 (pg. 23), 3 (pg. 89), 4 (pg. 119), 8 (pg. 273), 10 (pg. 349).

Norbert Wiener (1894 – 1964) [Wie53, WS53] Cited in connection with Brownian motion, Wiener measure, and stochastic processes. And more directly, the "Wiener" of Paley-Wiener spaces; – at the crossroads of harmonic analysis and functional analysis. Also the Wiener of filters in signal processing; high-pass/low-pass etc.
Ref. Section 2.1 (pg. 25), and Chapter 7 (pg. 249).

Arthur Strong Wightman (1922 – 2013) [Wig76] the "W" in the Wightman axioms for quantum field theory [SW00] (also called the Gårding-Wightman axioms). Wightman was a leading mathematical physicist, a 1949 Princeton-PhD, and later a Professor at Princeton. He was one of the founders of quantum field theory. His former PhD-students include Arthur Jaffe, Robert T. Powers, and Alan Sokal. The "PCT" in the title of [SW00] refers to the combined symmetry of a quantum field theory under P for parity, C charge, and T time. "Spin and statistics" refers to a theorem to the effect that spin $1/2$ particles obey Fermi-Dirac statistics, whereas integer spin 0, 1, 2 particles obey Bose-Einstein statistics. The Wightman axioms provide a basis for a mathematically rigorous perturbative approach to quantum fields. One of the Millennium Problems [Fad05] asks for a realization of the axioms in the case of Yang-Mills fields. A fundamental idea behind the axioms is that there should be a Hilbert space upon which the Poincaré group (of relativistic space-time) acts as a unitary representation. With this we get energy, momentum, angular momentum and center of mass (corresponding to boosts) realized as selfadjoint operators (unbounded) in this Hilbert space. A stability assumption places a restriction on the spectrum

---

[1]"Monte-Carlo" means "simulation" with computer generated random number.

of the four-momentum, that it falls in the positive light cone (and its boundary). Then quantum fields entail realizations in the form of covariant representations of the Poincaré group. Further, the Wightman axioms entail operator valued distributions; in physics lingo, a smearing over Schwartz test functions. Here "operator-valued" refers to operators which are both non-commuting and unbounded. Hence the necessity for common dense domains; for example Gårding domains. The causal structure of the theory entails imposing either commutativity (CCR), or anticommutativity (CAR) rules for spacelike separated fields; and it further postulates the existence of a Poincaré-invariant and cyclic state, called the vacuum.

Ref. Section 1.2 (pg. 5).

Der skal et par dumheder
med i en bog ....
for at også de dumme
skal syns, den er klog.
— Piet Hein.

Translation:
Your book should include a few stupidities
mixing them in, – this is art.
so that also the stupid will think it is smart.
(Translated by P.J. from the Danish original.)

Appendix D

Prizes and Fame

Abel Prize:

- Sir Michael Atiyah, P. Lax, I.M. Singer.

AMS Steele Prize:

- I. Gelfand, R. Kadison, E. Nelson, R. Phillips, J. Schwartz, I.M. Singer.

Enrico Fermi Award:

- F. Dyson, J. von Neumann.

Fields Medal (Math):

- Sir Michael Atiyah, A. Connes, D. Mumford, L. Schwartz, S. Smale.

Gauss Prize:

- K. Itō.

Henri Poincaré Prize:

- F. Dyson, A. Wightman.

Kyoto Prize in Mathematical Sciences:

- I. Gelfand, K. Itō.

Lenin Prize:

- A.N. Kolmogorov.

Max Planck Medal:

- V. Bargmann, W. Heisenberg, E. Schrödinger.

National Medal of Science:

- F. Dyson, K. Friedrichs, P. Lax, I.M. Singer, S. Smale, M. Stone, N. Wiener, E. Wigner.

Nobel Prize in Physics:

- N. Bohr, M. Born, P. Dirac, A. Einstein, W. Heisenberg, E. Schrödinger, E. Wigner.

Order of Lenin:

- I. Gelfand (three times).

Presidential Medal of Freedom:

- J. von Neumann.

Wigner Medal:

- V. Bargmann, I. Gelfand, I.M. Singer.

Wolf Prize:

- F. Dyson, I. Gelfand, K. Itō, A.N. Kolmogorov, M.G. Krein, P. Lax, S. Smale.

Quotes: Index of Credits

Arveson, W., (1934-2011), v, 23, 223

Bachelier, L., (1870-1946), 250
Bohr, N., (1885-1962), 27, 349
Born, M., (1882-1970), 89

Connes, A., (1947-), 9, 223

Dirac, P.A.M., (1902-1984), 23, 89, 333

Einstein, A., (1879-1955), 119, 249

Gelfand, I.M., (1913-2009), 155

Hadamard, J., (1865-1963), 349
Hawking, S., (1942-), 403
Heaviside, O., (1850-1925), 385
Hein, P., (1905-1996), 23, 473
Heisenberg, W., (1901-1976), 155

Kolmogorov, A.N., (1903-1987), 9, 203, 263, 273, 414
Kraus, K., (1938-1988), 349

Lax, P.D., (1926-), 54, 85, 280, 343
Lie, S., (1842-1899), ix

Mackey, G.W., (1916-2006), 273
Mumford, D., (1937-), 249

Sakai, S., (1928-), 26, 193, 194, 444
Szegö, G., (1895-1985), 17, 209, 405

Toeplitz, O., (1881-1940), 89

von Neumann, J., (1903-1957), 119, 441, 459

Wiener, N., (1894-1964), 333
Wightman, A. S., (1922-2013), 452
Wigner, E.P., (1902-1995), 273, 385, 403

Appendix E

List of Exercises

We further emphasize that the Exercises are diverse, some easy, and some difficult. Some are meant to give the novice a familiarity with definitions, and other Exercises have the flavor of Theorems. But we chose to present them in the form of Exercises as opposed to Theorems. For those, the reader is then asked to supply proofs.

2.1	An ergodic theorem	32
2.2	weak-* neighborhoods	32
2.3	weak-* vs norm	32
2.4	Be careful with weak-* limits	32
2.5	Transformation of measures	34
2.6	Gelfand triple	37
2.7	Hilbert completion	41
2.8	L^2 of a measure-space	41
2.9	Product of two positive definite functions	44
2.10	An Itō-isometry	45
2.11	Fischer	48
2.12	ONBs and cardinality	49
2.13	$l^2(A)$	49
2.14	A functor from sets to Hilbert space	49
2.15	The $C^*property$	51
2.16	The group of all unitary operators	52
2.17	Riesz	54
2.18	Lax-Milgram	54
2.19	Generating functions	57
2.20	Recursive identities	59
2.21	Legendre, Chebyshev, Hermite, and Jacobi	60
2.22	The orthogonality rules	60
2.23	M_x in Jacobi form	60

2.24   The Haar wavelet, and multiplication by t   61
2.25   A duality   63
2.26   Finite rank reduction   64
2.27   Numerical Range and Toeplitz-Hausdorff   65
2.28   Matrix product of $\infty \times \infty$ banded matrices   66
2.29   A transform   67
2.30   The identity-operator in infinite dimension   68
2.31   Comparing norms   68
2.32   Matrix entries in infinite dimensions   69
2.33   The three steps   69
2.34   The canonical commutation relations   72
2.35   Raising and lowering operators   73
2.36   Infinite banded matrices   74
2.37   The Hilbert matrix   75
2.38   Sums of projections   81
2.39   Multiplication operators   83
2.40   Moment theory   83
3.1   The resolvent identity   97
3.2 $A^{**} = \overline{A}$ 101
3.3   Leibniz-Malliavin   110
4.1   Multiplication operators, continued   127
4.2   Continuous spectrum   127
4.3   Direct integral Hilbert space   132
4.4   Multiplicity free   135
4.5   The Cayley transform   136
4.6   A concrete PVM   136
4.7   An application to numerical range   143
4.8   Heisenberg's uncertainty   144
4.9   An eigenspace   145
4.10   Attaining the sup   145
4.11   A Green's function   146
4.12   Pinned Brownian motion from reflection   147
4.13   An application of Arzelà-Ascoli   150
4.14   Powers-Størmer   151
4.15   Parseval frames   151
4.16   A dilation space   151
5.1   Irreducible representations   160
5.2   Infinite-product measures and representations of $\mathcal{O}_N$ 160
5.3   A representation of $\mathcal{O}_N$ 161
5.4   The regular representation of G   162
5.5   Unitary representations   162
5.6   A proof detail   162
5.7   Reduced C^*-algebra   163
5.8   The trace state on $\mathbb{C}[G]$ 164
5.9   Cyclic subspaces   165
5.10   The GNS construction   167

5.11	Identification by isometry	168
5.12	A pre-dual	172
5.13	The Bohr compactification	174
5.14	Unitary operators on a direct Hilbert sum	175
5.15	The Julia operator	177
5.16	An induced operator	180
5.17	Time-reflection	181
5.18	Renormalization	182
5.19	A Hilbert space of distributions	183
5.20	Taylor for distributions	184
5.21	A complete metric space	185
5.22	A distance formula	185
5.23	Iterated function systems	185
5.24	The Middle-Third Cantor-measure	185
5.25	Straightening out the Devil's staircase	186
5.26	The homomorphism of l^1	189
5.27	Pure states on $\mathbb{C}^n$	195
5.28	Extreme measures	201
5.29	Irrational rotation	201
5.30	Wold's decomposition	204
5.31	Substitution by z^N	204
5.32	The backwards shift	205
5.33	Infinite multiplicity	205
5.34	A shift is really a shift	205
5.35	Multiplication by z is a shift in $\mathbb{H}_2$	205
5.36	The two shifts in $\mathbb{H}_2$	206
5.37	A numerical range	206
5.38	The finite shift	206
5.39	The Hardy space $\mathbb{H}_2$; a transform	207
5.40	Multiplicity-one and $\mathbb{H}_2$	208
5.41	The Toeplitz matrices	208
5.42	Homomorphism mod $\mathscr{K}$	209
5.43	An element in $Rep\,(\mathscr{O}_N, \mathbb{H}_2)$	210
5.44	The multivariable Toeplitz algebra	210
5.45	No bounded solutions to the canonical commutation relations	212
5.46	The CAR$(\mathscr{H})$-Fock-state	214
6.1	Tensor with M_n	236
6.2	Column operators	237
6.3	Row-isometry	237
6.4	Tensor products	238
6.5	Using tensor product in representations	239
6.6	"Transpose" is not completely positive	239
6.7	$Rep(\mathscr{O}_N, \mathscr{H})$	242
6.8	Simplest Low/High Pass filter bank	243
6.9	Endomorphism vs representation	247
6.10	Convex sets that are not numerical ranges	247

7.1   Independent random variables 252
7.2   Quadratic variation 254
7.3   Geometric Brownian motion 255
7.4   A unitary one-parameter group 256
7.5   Infinitesimal generator 256
7.6   An ergodic action 257
7.7   Reflection of Brownian motion 260
7.8   Iterated Itō-integrals 261
7.9   Symmetric Fock space as a random Gaussian space 262
7.10   The fair-coin measure 263
7.11   The Ornstein-Uhlenbeck semigroup 264
7.12   The Central Limit Theorem 268
7.13   Karhunen-Loève 271
8.1   Semidirect product $G \circledS V$ 277
8.2   Positive definite functions and states 279
8.3   Contractions 279
8.4   Central Extension 280
8.5   von Neumann's ergodic theorem 281
8.6 $SL_2(\mathbb{R})$ and Lemma 8.1 290
8.7   The Heisenberg group as a semidirect product 290
8.8   The invariant complex vector fields from the Heisenberg group . 290
8.9   The Campbell-Baker-Hausdorff formula 291
8.10   The Lie algebra of a central extension 291
8.11   Weyl's commutation relation 296
8.12   The Schrödinger representation 305
8.13   The Gårding space for the Schrödinger representation 315
8.14   The Lie bracket 315
8.15   Normal subgroups 318
8.16   Multiplicity 327
8.17   A Hilbert space of intertwiners 327
8.18   An ONB in $Int(\pi, \rho)$ 327
8.19   A Hilbert space of intertwining operators 328
8.20   Specify the operators in $Int(\pi, \rho)$ 328
8.21   A formula from Peter-Weyl 328
9.1   Non-normal pure states on $\mathscr{B}(l^2)$ 338
9.2   The Calkin algebra and Non-normal states on $\mathscr{B}(l^2)$ 339
9.3   The pure state $\varphi = s \circ \pi$ on $\mathscr{B}(l^2)$ 339
9.4   The Stone-Čech compactification 340
9.5   The Dixmier trace 341
9.6   Parseval frames 342
9.7   Frames from Lax-Milgram 343
9.8   Positive operator valued measures from frames 344
9.9   The operator B 344
10.1   Dissipative generators 360
10.2   The index values $(d_+, 0)$, and $(0, d_-)$ 360

10.3   Absence of analytic vectors 361
10.4   From selfadjoint extension to unitary one-parameter group . . . 369
10.5   Realize index-values 369
10.6   Realize the cases (d, d) for semibounded operators 369
10.7   The N-covering surface of $\mathbb{R}^2 \setminus \{(0,0)\}$ 369
10.8   The logarithmic Riemann surface 370
10.9   Formally commuting Hermitian symmetric operators on a common dense domain which are not commuting in the strong sense 370
11.1   Gaussian free field (GFF) 390
12.1   The Fundamental Theorem of Calculus and a RKHS 405
12.2   Mercer 405
12.3   Szegö-kernel 405
12.4   Multp. $(\mathscr{H}_K)$ 406
12.5   Kernels and differentiable functions 410
12.6   New RKHS from old ones 411

Appendix F

Definitions of Frequently Occurring Terms

Def 2.1 Linear functional 28

Def 2.2 The dual space E^* 28

Def 2.3 Banach space 28

Def 2.4 Weak-∗ topology 31

Def 2.5 Bounded linear operator 33

Def 2.6 The adjoint operator 33

Def 2.7 Probability space 34

Ex 2.6 Gelfand triple 37

Def 2.10 Norm on vector spaces over $\mathbb{C}$ 40

Def 2.11 Inner product 40

Def 2.12 Hilbert space 41

Def 2.14 Isometry between Hilbert spaces 44

Ex 2.10 Itō-isometry 45

Def 2.15 Orthonormal basis 46

Def 2.16 Separable Hilbert space 48

Def 2.17 $l^2(A)$, where A is a given index set 49

Def 2.18 Dyadic fraction (rational) 50

Def 2.19 Bounded linear operators between Hilbert spaces 51

Def 2.20 Normal, selfadjoint, unitary operators, and projections 51

Def 2.21 Generating function 57

Ex 2.21 Legendre, Chebyshev, Hermite, and Jacobi polynomials 60

Def 2.22 Wavelets in $L^2(\mathbb{R})$ 61

Ex 2.27 Numerical Range 65

Def 2.25 Uniform, trace, and Hilbert-Schmidt norms 67

Ex 2.34 The canonical commutation relations 72

Ex 2.35 Raising and lowering operators 73

Ex 2.36 Infinite banded matrices 74

Ex 2.37 The Hilbert matrix 75

Def 3.3   Schrödinger representation 91
Def 3.4   Selfadjoint, essentially selfadjoint, normal, and regular operators 97
Def 3.5   Resolvent of a linear operator 97
Def 3.6   Linear operators (possibly unbounded) between Hilbert spaces 102
Def 3.7   The adjoint operator of a densely defined operator 103
Def 3.8   The characteristic projection 106
Def 3.9   Malliavin derivative 109
Def 3.10   Projection valued measure (PVM), see also Def 4.5 115
Def 3.11   Unitary one-parameter groups 115
Def 3.12   Essentially selfadjoint operators in Hilbert spaces 117
Def 4.1   Unitary equivalence of operators 121
Def 4.2   Measure transformation 128
Def 4.3   Disintegration of measures 131
Def 5.1 C^*-algebras 158
Def 5.2   Pure states on ∗-algebras 159
Def 5.3   The group of all bounded operators 161
Def 5.4   Group representations on Hilbert spaces 161
Ex 5.4   The regular representation 162
Def 5.5   The reduced C^*-algebra of a group G 163
Def 5.6   Representations of ∗-algebras on Hilbert spaces 163
Def 5.7   States on ∗-algebras 163
Def 5.8   Cyclic representations 164
Def 5.9   Trace class operators 171
Def 5.10   The trace of an operator 171
Def 5.11   Hardy space on the disk 173
Def 5.12   The BMO space 174
Ex 5.13   The Bohr compactification 174
Ex 5.18   Renormalization 182
Ex 5.24   The Middle-Third Cantor-measure 185
Def 5.13   Banach algebras 187
Def 5.14   Gelfand transform 188
Def 5.15   Locally convex spaces 198
Def 5.16   Toeplitz-operator 208
Def 5.17   Fredholm operators and their indices 209
Ex 5.44   The multivariable Toeplitz algebra 210
Def 5.18   Multiplicity of a representation 216
Def 6.1   Completely positive mappings 224
Ex 6.4   Tensor products 238
Def 6.2   Representations of Cuntz algebras 241
Ex 6.8   Simplest Low/High filter bank 243
Def 7.1   Probability space 251
Def 7.2   Independent events 251
Def 7.3   Mutually independent random variables 251
Def 7.4   Convolution of measures 251
Def 7.5   Normal (Gaussian) distribution. 252
Def 7.6   Brownian motion 252

Def 7.7   Log-normal distribution (dist. of geometric Brownian motion)   253
Ex 7.2   Quadratic variation 254
Ex 7.3   Geometric Brownian motion 255
Ex 7.5   Infinitesimal generator 256
Def 7.8   Ergodic group actions 256
Ex 7.12   The Central Limit Theorem 268
Def 8.2   Positive definite functions on groups 279
Ex 8.4   Central Extension 280
Def 8.3   Left (resp. right) Haar measures 282
Def 8.4   Modular function 283
Ex 8.9   The Campbell-Baker-Hausdorff formula 291
Def 8.5   The exponential mapping of Lie theory 310
Def 8.6   Intertwining operators 326
Def 8.7   Multiplicity of representations 327
Def 8.8   Unitary equivalence of systems of operators 329
Def 8.9   The Heisenberg group 330
Ex 9.2   The Calkin algebra 339
Ex 9.4   The Stone-Čech compactification 340
Def 9.1   Dixmier trace 341
Def 9.2   Frame 342
Def 10.1   Deficiency spaces, and deficiency indices 355
Def 10.2   Conjugation operator 361
Def 10.3   Boundary triple 365
Def 11.1   Conductance function 387
Def 11.2   Graph Laplacian 387
Def 11.5   Parseval frame 390
Def 11.6   Two dense domains for a fixed Graph Laplacian 393
Def 11.7   Reversible transition probabilities 398
Def 12.1 S-reproducing kernel Hilbert space 403
Ex 12.3   Szegö-kernel 405
Def 12.2   Multiplier of a positive definite function 406
Def 12.5   Positive definite functions on groups 414
Def 12.6   Locally defined positive definite functions on Lie groups . . . 417
Def 12.7   The RKHS of a locally defined positive definite function . . . 417
Def 12.8   The operator D_F 421
Def 12.9   Deficiency spaces of D_F 422
Def 12.10   Locally defined positive definite functions on l.c. groups . . . 425

Appendix G

List of Figures

1   Examples of infinite weighted network; see Chapter 11.   xiv

1.1   Two random variables X, Y, and their distributions.   10

1.2   The lines illustrate connections between the six themes, A through F; different chapters making separate connections between these related and intertwined themes. See also Table 1.1, and the discussion before Table 1.1.   16

2.1   A few spaces in analysis, and their inter-relations; ranging from general spaces to classes of linear spaces.   28

2.2   Dual exponents for the L^p spaces, $\frac{1}{p} + \frac{1}{q} = 1$.   31

2.3   A measurement X; $X^{-1}(B) = \{\omega \in \Omega : X(\omega) \in B\}$. A random variable and its distribution.   35

2.4 $\Psi^{-1}(E) = \{\omega \in \Omega_2 : \Psi(\omega) \in E\}$, pull-back.   35

2.5   Contra-variance (from point transformations, to transformation of functions, to transformation of measures.) See Exercise 2.5. .   35

2.6   Finite Tree (natural order on the set of vertices). Examples of maximal elements: $m_1, m_2, \ldots$ 37

2.7   Infinite Tree (no maximal element!) All finite words in the alphabet $\{0, 1\}$ continued indefinitely.   38

2.8   Covariance function of Brownian motion.   44

2.9   A set of Brownian sample-paths generated by a Monte-Carlo computer simulation.   44

2.10   The parallelogram law.   54

2.11 $\|x - Px\| = \inf\{\|x - w\| : w \in \mathscr{W}\}$. Projection from optimization in Hilbert space.   55

2.12   The first two steps in G-S.   56

2.13   Samples of the Legendre, Chebyshev, and Hermite polynomials.   57

2.14   An $\infty \times \infty$ matrix representation of M_x in Exercise 2.23.   60

2.15   Haar wavelet. Scaling properties, multiresolutions.   62

2.16 $\infty \times \infty$ *banded matrix*. Supported on a band around the diagonal.   67

2.17   The raising and lowering operators $A_\pm$. The lowering operator
        A_- kills e_0.   74
2.18 $A \subset A^*$ (a Hermitian Jacobi matrix).   76
2.19   Transition of quantum-states.   76
2.20   Gram-Schmidt: $V_n \longrightarrow V_{n+1}$ 81
2.21   Separation of K and y by the hyperplane H_c.   88

4.1   Illustration of forces: attractive vs repulsive. The case of "only
        bound states" (a), vs continuous Lebesgue spectrum (b).   125
4.2   The energy surface S_E for the quantum mechanical $H_4 = P^2 -$
        Q^4, with $P \rightsquigarrow x'(t)$.   126
4.3   Fix $E_1 < E_2 < E_3$, then $0 < t_{E_3}(\infty) < t_{E_2}(\infty) < t_{E_1}(\infty)$. . . .   126
4.4   Application of the transform (4.28).   133
4.5   Pinned Brownian motion (with 5 sample paths).   146
4.6   Time-reflection of Brownian motion: $X_t = tB\left(\frac{1}{t}\right)$, $t \in \mathbb{R}_+$,
        reflection $B \longrightarrow X$.   147
4.7   A simulation of pinned Brownian motion as $Y(t) = \frac{1}{\sqrt{2}}(B(t) - X(t))$
        (black).   147

5.1   Circuit of a controlled U-gate, $U \in U(2)$.   178
5.2   A mixed state. (With apologies to Schrödinger). The two Dirac
        kets represent the two states, the "alive" state and the "dead"
        state, the second. By QM, the cat would be both dead and alive
        at the same time, but of course, the cat is not a quantum system.
        In QM, the presence of an observer entails the "collapse" of the
        wavefunction $|\psi\rangle$, assuming the cat could be in a superposition,
        as opposed to in one of the two states (decoherence).   179
5.4   Transformation of skew-adjoint A into selfadjoint semibounded
        L.   181
5.3   Reflection positivity. A unitary operator $\mathcal{U}$ transforms into a
        selfadjoint contraction $\widetilde{\mathcal{U}}$.   181
5.5   The Middle-Third Cantor set as a limit.   186
5.6   The cumulative distribution of the middle-third Cantor measure.   187
5.7   Bloch sphere   196
5.8   A simplex. The four extreme points are marked.   200
5.9   Irrational rotation.   202
5.10   The numerical range of T_3 vs T_3'.   207
5.11   The numerical range (NR) of the truncated finite matrices: Ex-
        panding truncations of the infinite matrix T corresponding to the
        backward shift, and letting the size $\longrightarrow \infty$: $T_3, T_4, \cdots, T_n, T_{n+1}, \cdots$;
        the limit-NR fills the open disk of radius 1.   207

6.1   Orthogonal bands in filter bank, "in" = "out". An application of
        representations of the Cuntz relations (6.27).   241
6.2   Low / high pass filters for the Haar wavelet (frequency mod 2π).   243
6.3   The Cuntz operators S_0 and S_1 in the Haar wavelet.   244

6.4   The two-channel analysis filter bank. 244
6.5   The two-channel synthesis filter bank. 245
6.6   The outputs of $(S_0^*)^n$, $n = 3, 4, 5$. 245
6.7   The outputs of high-pass filters. 246
6.8   A coarser resolution in three directions in the plane, filtering in
         directions, x, y, and diagonal; — corresponding dyadic scaling
         in each coordinate direction. (Image cited from M.-S. Song,
         "*Wavelet Image Compression*" in [HJL06].) 246
6.9   Illustration of eq. (6.34), with $y_i := \frac{S_i^* x}{\|S_i^* x\|}$, $i = 1, 2, \ldots, N$. . . . 247

7.1   The joint distribution of (B_s, B_t) where B is standard Brownian
         motion and $0 < s < t$. 253
7.2   The joint distribution of $(B_s, B_t) : \Omega \longrightarrow \mathbb{R}^2$, $0 < s < t$. 253
7.3   Geometric Brownian motion: 5 sample paths, with $\mu = 1$, $\sigma = $
         0.02, and $X_0 = 1$. 255
7.4   The process $\frac{1}{2} \left(B_T^2 - T \right)$ in (7.14): 5 sample paths, with $T = 1$.   256
7.5   Some commonly occurring probability distributions. See Ta-
         ble 7.1 for explicit formulas for the respective distributions, and
         their parameters. 258
7.6   Stochastic processes indexed by time: A cylinder set C is a spe-
         cial subset of the space of all paths, i.e., functions of a time
         variable. A fixed cylinder set C is specified by a finite set of
         sample point on the time-axis (horizontal), and a corresponding
         set of "windows" (intervals on the vertical axis). When sam-
         ple points and intervals are given, we define the corresponding
         cylinder set C to be the set of all paths that pass through the
         respective windows at the sampled times. In the figure we illus-
         trate sample points (say future relative to $t = 0$). Imagine the
         set C of all outcomes with specification at the points $t_1, t_2, \ldots$ etc.261
7.7   The cylinder sets C, C', and C''. 261
7.8   Simulation of the solution X_t to the SDE (7.31), i.e., the Ornstein-
         Uhlenbeck process. It governs a massive Brownian particle under
         the influence of friction. It is stationary, and Markovian. Over
         time, the process tends to drift towards its long-term mean: such
         a process is called mean-reverting. (Monte-Carlo simulation of
         OU-process with 5 sample paths. $\beta = \gamma = 1$, $v_0 = 1$, $t \in [0, 10]$)   266
7.9   Monte-Carlo simulation of the standard Brownian motion pro-
         cess $\{X_t : 0 \leq t \leq 1\}$, where $\mathbb{E}(X_t) = 0$, and $\mathbb{E}(X_s X_t) = s \wedge t = $
         $\min(s, t)$.
         For $n = 10, 50, 100, 500$, set $X_0 = 0$ and $X_{j/n}^{(n)} = n^{-1/2} \sum_{k=1}^{j} W_k$.
         Applying linear interpolation between sample points $\{j/n : j = 0, \ldots, n\}$
         yields the n-point approximation $X_t^{(n)}$, which converges in mea-
         sure to X_t (standard BM restricted to the unit interval), as
         $n \to \infty$. 269

8.1 G and $\mathfrak{g}$ (Lie algebra, Lie group, and exponential mapping). . . 292
8.2   Approximation of identity. 313
8.3 W intertwines S_1 and S_2. 330

9.1   Observable, state, measurement. Left column: An idealized
       physics experiment. Right: the mathematical counterpart, a
       selfadjoint operator H, its associated projection-valued measure
       P_H, and a norm-one vector v in Hilbert space. 339
9.2   The pure state $\varphi = s \circ \pi$ on $\mathscr{B}\left(l^2\right)$ 340

10.1 $\mathscr{D}_\pm = \mathrm{Ker}\,(A^* \mp i)$, $\mathscr{D} = dom\,(A)$ 355
10.2   A semigroup of isometries in $L^2\,(0, \infty)$. 361
10.3   Translation of a locally defined φ on M to different sheets. . . . 371
10.4   The operator J and its adjoint. 372

11.1   Transition probabilities p_{xy} at a vertex x (in V). 388
11.2   non-linear system of vertices. 391
11.3   The set $\{v_{xy} : (xy) \in E\}$ is not orthogonal. 392
11.4   The dipole v_{xy}. 395
11.5   Neighbors of x. 398
11.6   The transition probabilities p_+, p_-, in the case of constant tran-
       sition probabilities, i.e., $p_+\,(n) = p_+$, and $p_-\,(n) = p_-$ for all
       $n \in \mathbb{Z}_+$. 400

12.1   The approximate identity $\varphi_{n,x}$. 418
12.2   Fix $\theta = 0.8$, $\Lambda_\theta = \{\lambda_n\,(\theta)\} = $ intersections of two curves. (spec-
       trum from curve intersections) 430
12.3 $\theta = 0$. A type 1 continuous p.d. extension of $F\,(x) = e^{-|x|}\big|_{[-1,1]}$
       in $\mathscr{H}_F$. 436

List of Tables

1.1   Examples of Linear Spaces: Banach spaces, Banach algebras, Hilbert spaces $\mathscr{H}$, linear operators act in $\mathscr{H}$. 15

2.1   Examples of dual spaces. 30
2.2   ORTHOGONAL POLYNOMIALS. Legendre, Chebyshev, Hermite. 58
2.4   Lattice of projections in Hilbert space, Abelian vs non-Abelian. . 79

5.1   Examples of pure states. 160

7.1   Some commonly occurring probability measures on $\mathbb{R}$, or on $\mathbb{R}_+$, and their corresponding *generating functions*. P.S.: Note that the Lévy distribution has both its mean and its variance $= \infty$. See Figure 7.5. 257

8.2   Overview of five duality operations. 329

A.1   An overview of connections between Functional Analysis and some of its applications. 447
A.2   Linear operators in Hilbert space and their applications. 447
A.3   Harmonic analysis (commutative and non-commutative), and a selection of applications. 448
A.4   Topics from quantum physics. 448
A.5   Representations of groups and algebras. 448
A.6   Analysis of stochastic processes and some of their applications. . 449

Bibliography

[AA91]   Charles A. Akemann and Joel Anderson, *Lyapunov theorems for operator algebras*, Mem. Amer. Math. Soc. **94** (1991), no. 458, iv+88. MR 1086563 (92e:46113)

[AA02]   Y. A. Abramovich and C. D. Aliprantis, *An invitation to operator theory*, Graduate Studies in Mathematics, vol. 50, American Mathematical Society, Providence, RI, 2002. MR 1921782 (2003h:47072)

[AAR99]   George E. Andrews, Richard Askey, and Ranjan Roy, *Special functions*, Encyclopedia of Mathematics and its Applications, vol. 71, Cambridge University Press, Cambridge, 1999. MR 1688958 (2000g:33001)

[AAT14]   Charles A. Akemann, Joel Anderson, and Betül Tanbay, *Weak paveability and the Kadison-Singer problem*, J. Operator Theory **71** (2014), no. 1, 295–300. MR 3173062

[AD86]   Daniel Alpay and Harry Dym, *On applications of reproducing kernel spaces to the Schur algorithm and rational J unitary factorization*, I. Schur methods in operator theory and signal processing, Oper. Theory Adv. Appl., vol. 18, Birkhäuser, Basel, 1986, pp. 89–159. MR 902603 (89g:46051)

[AD03]   D. Alpay and C. Dubi, *Backward shift operator and finite dimensional de Branges Rovnyak spaces in the ball*, Linear Algebra Appl. **371** (2003), 277–285. MR 1997376 (2004e:46035)

[AG93]   N. I. Akhiezer and I. M. Glazman, *Theory of linear operators in Hilbert space*, Dover Publications Inc., New York, 1993, Translated from the Russian and with a preface by Merlynd Nestell, Reprint of the 1961 and 1963 translations, Two volumes bound as one. MR 1255973 (94i:47001)

[AH09]   Kendall Atkinson and Weimin Han, *Theoretical numerical analysis*, third ed., Texts in Applied Mathematics, vol. 39, Springer, Dordrecht, 2009, A functional analysis framework. MR 2511061 (2010b:65001)

[AH13]   Giles Auchmuty and Qi Han, *Spectral representations of solutions of linear elliptic equations on exterior regions*, J. Math. Anal. Appl. **398** (2013), no. 1, 1–10. MR 2984310

[AJ12]   Daniel Alpay and Palle E. T. Jorgensen, *Stochastic processes induced by singular operators*, Numer. Funct. Anal. Optim. **33** (2012), no. 7-9, 708–735. MR 2966130

[AJL13]   Daniel Alpay, Palle Jorgensen, and Izchak Lewkowicz, *Extending wavelet filters: infinite dimensions, the nonrational case, and indefinite inner product spaces*, Excursions in harmonic analysis. Volume 2, Appl. Numer. Harmon. Anal., Birkhäuser/Springer, New York, 2013, pp. 69–111. MR 3050315

[AJLM13]   Daniel Alpay, Palle Jorgensen, Izchak Lewkowicz, and Itzik Marziano, *Representation formulas for Hardy space functions through the Cuntz relations and new interpolation problems*, Multiscale signal analysis and modeling, Springer, New York, 2013, pp. 161–182. MR 3024468

[AJS14]   Daniel Alpay, Palle Jorgensen, and Guy Salomon, *On free stochastic processes and their derivatives*, Stochastic Process. Appl. **124** (2014), no. 10, 3392–3411. MR 3231624

[AJSV13]   Daniel Alpay, Palle Jorgensen, Ron Seager, and Dan Volok, *On discrete analytic functions: products, rational functions and reproducing kernels*, J. Appl. Math. Comput. **41** (2013), no. 1-2, 393–426. MR 3017129

[Akh65]   N. I. Akhiezer, *The classical moment problem and some related questions in analysis*, Translated by N. Kemmer, Hafner Publishing Co., New York, 1965. MR 0184042 (32 #1518)

[Alp92]   Daniel Alpay, *On linear combinations of positive functions, associated reproducing kernel spaces and a non-Hermitian Schur algorithm*, Arch. Math. (Basel) **58** (1992), no. 2, 174–182. MR 1143167 (92m:46039)

[Alp01]   ———, *The Schur algorithm, reproducing kernel spaces and system theory*, SMF/AMS Texts and Monographs, vol. 5, American Mathematical Society, Providence, RI; Société Mathématique de France, Paris, 2001, Translated from the 1998 French original by Stephen S. Wilson. MR 1839648 (2002b:47144)

[And74]   Joel Anderson, *On compact perturbations of operators*, Canad. J. Math. **26** (1974), 247–250. MR 0333774 (48 #12098)

[And79a]   Joel Anderson, *Extreme points in sets of positive linear maps on $B(H)$*, Journal of Functional Analysis **31** (1979), no. 2, 195 – 217.

[And79b] Joel Anderson, *The Haar functions almost diagonalize multiplication by x*, Proc. Amer. Math. Soc. **73** (1979), no. 3, 361–362. MR 518520 (80f:47022)

[AØ15] Nacira Agram and Bernt Øksendal, *Malliavin Calculus and Optimal Control of Stochastic Volterra Equations*, J. Optim. Theory Appl. **167** (2015), no. 3, 1070–1094. MR 3424704

[Aro50] N. Aronszajn, *Theory of reproducing kernels*, Trans. Amer. Math. Soc. **68** (1950), 337–404. MR 0051437 (14,479c)

[ARR13] Luigi Accardi, Habib Rebei, and Anis Riahi, *The quantum decomposition of random variables without moments*, Infin. Dimens. Anal. Quantum Probab. Relat. Top. **16** (2013), no. 2, 1350012, 28. MR 3078823

[Arv72] William Arveson, *Lattices of invariant subspaces*, Bull. Amer. Math. Soc. **78** (1972), 515–519. MR 0298451 (45 #7503)

[Arv76] _____, *An invitation to C*-algebras*, Springer-Verlag, New York-Heidelberg, 1976, Graduate Texts in Mathematics, No. 39. MR 0512360 (58 #23621)

[Arv98] _____, *Subalgebras of C*-algebras. III. Multivariable operator theory*, Acta Math. **181** (1998), no. 2, 159–228. MR 1668582 (2000e:47013)

[Arv09a] _____, *Maximal vectors in Hilbert space and quantum entanglement*, J. Funct. Anal. **256** (2009), no. 5, 1476–1510. MR 2490227 (2010b:46047)

[Arv09b] _____, *The probability of entanglement*, Comm. Math. Phys. **286** (2009), no. 1, 283–312. MR 2470932 (2010d:81049)

[Arv09c] _____, *Quantum channels that preserve entanglement*, Math. Ann. **343** (2009), no. 4, 757–771. MR 2471599 (2010b:47234)

[AW14] Charles Akemann and Nik Weaver, *A Lyapunov-type theorem from Kadison-Singer*, Bull. Lond. Math. Soc. **46** (2014), no. 3, 517–524. MR 3210706

[AZ07] Sheldon Axler and Dechao Zheng, *Toeplitz algebras on the disk*, J. Funct. Anal. **243** (2007), no. 1, 67–86. MR 2291432 (2008d:30080)

[Ban93] Stefan Banach, *Théorie des opérations linéaires*, Éditions Jacques Gabay, Sceaux, 1993, Reprint of the 1932 original. MR 1357166 (97d:01035)

[Bar47] V. Bargmann, *Irreducible unitary representations of the Lorentz group*, Ann. of Math. (2) **48** (1947), 568–640. MR 0021942 (9,133a)

[BC05]   R. Burioni and D. Cassi, *Random walks on graphs: ideas, techniques and results*, J. Phys. A **38** (2005), no. 8, R45–R78. MR 2119174 (2006b:82059)

[BCD06]   Lali Barrière, Francesc Comellas, and Cristina Dalfó, *Fractality and the small-world effect in Sierpinski graphs*, J. Phys. A **39** (2006), no. 38, 11739–11753. MR 2275879 (2008i:28008)

[BD91]   Paul Baum and Ronald G. Douglas, *Relative K homology and C* algebras*, K-Theory **5** (1991), no. 1, 1–46. MR 1141333 (92m:19008)

[BDM⁺97]   Detlev Buchholz, Sergio Doplicher, Gianni Morchio, John E. Roberts, and Franco Strocchi, *A model for charges of electromagnetic type*, Operator algebras and quantum field theory (Rome, 1996), Int. Press, Cambridge, MA, 1997, pp. 647–660. MR 1491148

[BGV12]   Philippe Biane, Alice Guionnet, and Dan-Virgil Voiculescu, *Noncommutative probability and random matrices at Saint-Flour*, Probability at Saint-Flour, Springer, Heidelberg, 2012. MR 2905931

[BHS05]   Michael Barnsley, John Hutchinson, and Örjan Stenflo, *A fractal valued random iteration algorithm and fractal hierarchy*, Fractals **13** (2005), no. 2, 111–146. MR 2151094 (2006b:28014)

[BJ97a]   O. Bratteli and P. E. T. Jorgensen, *Endomorphisms of $\mathcal{B}(\mathcal{H})$. II. Finitely correlated states on $\mathcal{O}_n$*, J. Funct. Anal. **145** (1997), no. 2, 323–373. MR 1444086 (98c:46128)

[BJ97b]   Ola Bratteli and Palle E. T. Jorgensen, *Isometries, shifts, Cuntz algebras and multiresolution wavelet analysis of scale N*, Integral Equations Operator Theory **28** (1997), no. 4, 382–443. MR 1465320 (99k:46094b)

[BJ02]   Ola Bratteli and Palle Jorgensen, *Wavelets through a looking glass*, Applied and Numerical Harmonic Analysis, Birkhäuser Boston Inc., Boston, MA, 2002, The world of the spectrum. MR 1913212 (2003i:42001)

[BJKR84]   O. Bratteli, P. E. T. Jorgensen, A. Kishimoto, and D. W. Robinson, *A C*-algebraic Schoenberg theorem*, Ann. Inst. Fourier (Grenoble) **34** (1984), no. 3, 155–187. MR 762697 (86b:46105)

[BJO04]   Ola Bratteli, Palle E. T. Jorgensen, and Vasyl' Ostrovs'kyĭ, *Representation theory and numerical AF-invariants*, Mem. Amer. Math. Soc. **168** (2004), no. 797, xviii+178. MR 2030387 (2005i:46069)

[BKS13]   Matthew Begué, Tristan Kalloniatis, and Robert S. Strichartz, *Harmonic functions and the spectrum of the Laplacian on the Sierpinski carpet*, Fractals **21** (2013), no. 1, 1350002, 32. MR 3042410

[BM13]   John P. Boyd and Philip W. McCauley, *Quartic Gaussian and inverse-quartic Gaussian radial basis functions: the importance of a nonnegative Fourier transform*, Comput. Math. Appl. **65** (2013), no. 1, 75–88. MR 3003386

[BN00]   George Bachman and Lawrence Narici, *Functional analysis*, Dover Publications, Inc., Mineola, NY, 2000, Reprint of the 1966 original. MR 1819613 (2001k:46001)

[Boc08]   Florin P. Boca, *Rotation algebras and continued fractions*, Operator algebras, operator theory and applications, Oper. Theory Adv. Appl., vol. 181, Birkhäuser Verlag, Basel, 2008, pp. 121–142. MR 2681883 (2012f:46094)

[BØSW04]   Francesca Biagini, Bernt Øksendal, Agnès Sulem, and Naomi Wallner, *An introduction to white-noise theory and Malliavin calculus for fractional Brownian motion*, Proc. R. Soc. Lond. Ser. A Math. Phys. Eng. Sci. **460** (2004), no. 2041, 347–372, Stochastic analysis with applications to mathematical finance. MR 2052267 (2005a:60107)

[BP44]   Max Born and H. W. Peng, *Quantum mechanics of fields. III. Electromagnetic field and electron field in interaction*, Proc. Roy. Soc. Edinburgh. Sect. A. **62** (1944), 127–137. MR 0011453 (6,167c)

[BR60]   Garrett Birkhoff and Gian-Carlo Rota, *On the completeness of Sturm-Liouville expansions*, Amer. Math. Monthly **67** (1960), 835–841. MR 0125274 (23 #A2577)

[BR79]   Ola Bratteli and Derek W. Robinson, *Operator algebras and quantum statistical mechanics. Vol. 1*, Springer-Verlag, New York-Heidelberg, 1979, C^*- and W^*-algebras, algebras, symmetry groups, decomposition of states, Texts and Monographs in Physics. MR 545651 (81a:46070)

[BR81a]   ——— , *Equilibrium states of a Bose gas with repulsive interactions*, J. Austral. Math. Soc. Ser. B **22** (1981), no. 2, 129–147. MR 593999 (82b:82022)

[BR81b]   ——— , *Operator algebras and quantum-statistical mechanics. II*, Springer-Verlag, New York-Berlin, 1981, Equilibrium states. Models in quantum-statistical mechanics, Texts and Monographs in Physics. MR 611508 (82k:82013)

[BS94]   Albrecht Böttcher and Bernd Silbermann, *Operator-valued Szegő-Widom limit theorems*, Toeplitz operators and related topics (Santa Cruz, CA, 1992), Oper. Theory Adv. Appl., vol. 71, Birkhäuser, Basel, 1994, pp. 33–53. MR 1300213 (95j:47030)

[BT91]   J. Bourgain and L. Tzafriri, *On a problem of Kadison and Singer*, J. Reine Angew. Math. **420** (1991), 1–43. MR 1124564 (92j:46104)

[Cas13]   Peter G. Casazza, *The Kadison-Singer and Paulsen problems in finite frame theory*, Finite frames, Appl. Numer. Harmon. Anal., Birkhäuser/Springer, New York, 2013, pp. 381–413. MR 2964016

[Cas14]   Peter G. Casazza, *Consequences of the Marcus/Spielman/Stivastava solution to the Kadison-Singer Problem*, arXiv:1407.4768 (2014)

[CCG⁺04]   A. Connes, J. Cuntz, E. Guentner, N. Higson, J. Kaminker, and J. E. Roberts, *Noncommutative geometry*, Lecture Notes in Mathematics, vol. 1831, Springer-Verlag, Berlin; Centro Internazionale Matematico Estivo (C.I.M.E.), Florence, 2004, Lectures given at the C.I.M.E. Summer School held in Martina Franca, September 3–9, 2000, Edited by S. Doplicher and R. Longo, Fondazione C.I.M.E.. [C.I.M.E. Foundation]. MR 2067646

[CD78]   M. J. Cowen and R. G. Douglas, *Complex geometry and operator theory*, Acta Math. **141** (1978), no. 3-4, 187–261. MR 501368 (80f:47012)

[CDR01]   Roberto Conti, Sergio Doplicher, and John E. Roberts, *Superselection theory for subsystems*, Comm. Math. Phys. **218** (2001), no. 2, 263–281. MR 1828981

[CE07]   Peter G. Casazza and Dan Edidin, *Equivalents of the Kadison-Singer problem*, Function spaces, Contemp. Math., vol. 435, Amer. Math. Soc., Providence, RI, 2007, pp. 123–142. MR 2359423 (2008k:46181)

[CFMT11]   Peter G. Casazza, Matthew Fickus, Dustin G. Mixon, and Janet C. Tremain, *The Bourgain-Tzafriri conjecture and concrete constructions of non-pavable projections*, Oper. Matrices **5** (2011), no. 2, 351–363. MR 2830604 (2012g:42056)

[CH08]   D. A. Croydon and B. M. Hambly, *Local limit theorems for sequences of simple random walks on graphs*, Potential Anal. **29** (2008), no. 4, 351–389. MR 2453564 (2010c:60218)

[Cho75]   Man Duen Choi, *Completely positive linear maps on complex matrices*, Linear Algebra and Appl. **10** (1975), 285–290. MR 0376726 (51 #12901)

[Chr96]   Ole Christensen, *Frames containing a Riesz basis and approximation of the frame coefficients using finite-dimensional methods*, J. Math. Anal. Appl. **199** (1996), no. 1, 256–270. MR 1381391 (97b:46020)

[CJK⁺12]   Jianxin Chen, Zhengfeng Ji, David Kribs, Zhaohui Wei, and Bei Zeng, *Ground-state spaces of frustration-free Hamiltonians*, J. Math. Phys. **53** (2012), no. 10, 102201, 15. MR 3050570

[CM06]   Alain Connes and Matilde Marcolli, *From physics to number theory via noncommutative geometry*, Frontiers in number theory, physics, and geometry. I, Springer, Berlin, 2006, pp. 269–347. MR 2261099 (2007k:58010)

[CM07]   _____, *Renormalization, the Riemann-Hilbert correspondence, and motivic Galois theory*, Frontiers in number theory, physics, and geometry. II, Springer, Berlin, 2007, pp. 617–713. MR 2290770 (2008g:81156)

[CM13]   B. Currey and A. Mayeli, *The Orthonormal Dilation Property for Abstract Parseval Wavelet Frames*, Canad. Math. Bull. **56** (2013), no. 4, 729–736. MR 3121682

[Cob67]   L. A. Coburn, *The C^*-algebra generated by an isometry*, Bull. Amer. Math. Soc. **73** (1967), 722–726. MR 0213906 (35 #4760)

[Con90]   John B. Conway, *A course in functional analysis*, second ed., Graduate Texts in Mathematics, vol. 96, Springer-Verlag, New York, 1990. MR 1070713 (91e:46001)

[Con07]   Alain Connes, *Non-commutative geometry and the spectral model of space-time*, Quantum spaces, Prog. Math. Phys., vol. 53, Birkhäuser, Basel, 2007, pp. 203–227. MR 2382238 (2009c:81048)

[CP82]   F. Coester and W. N. Polyzou, *Relativistic quantum mechanics of particles with direct interactions*, Phys. Rev. D (3) **26** (1982), no. 6, 1348–1367. MR 675039 (84d:81024)

[CRKS79] P. Cotta-Ramusino, W. Krüger, and R. Schrader, *Quantum scattering by external metrics and Yang-Mills potentials*, Ann. Inst. H. Poincaré Sect. A (N.S.) **31** (1979), no. 1, 43–71. MR 557051 (81h:81129)

[Cun77]   Joachim Cuntz, *Simple C^*-algebras generated by isometries*, Comm. Math. Phys. **57** (1977), no. 2, 173–185. MR 0467330 (57 #7189)

[CW14]   K. L. Chung and R. J. Williams, *Introduction to stochastic integration*, second ed., Modern Birkhäuser Classics, Birkhäuser/Springer, New York, 2014. MR 3136102

[CZ07]   Felipe Cucker and Ding-Xuan Zhou, *Learning theory: an approximation theory viewpoint*, Cambridge Monographs on Applied and Computational Mathematics, vol. 24, Cambridge University Press, Cambridge, 2007, With a foreword by Stephen Smale. MR 2354721 (2009a:41001)

[dB68]   Louis de Branges, *Hilbert spaces of entire functions*, Prentice-Hall Inc., Englewood Cliffs, N.J., 1968. MR 0229011 (37 #4590)

[dBR66]   Louis de Branges and James Rovnyak, *Canonical models in quan-
          tum scattering theory*, Perturbation Theory and its Applications in
          Quantum Mechanics (Proc. Adv. Sem. Math. Res. Center, U.S.
          Army, Theoret. Chem. Inst., Univ. of Wisconsin, Madison, Wis.,
          1965), Wiley, New York, 1966, pp. 295–392. MR 0244795 (39 #6109)

[Dev59]   Allen Devinatz, *On the extensions of positive definite functions*,
          Acta Math. **102** (1959), 109–134. MR 0109992 (22 #875)

[Dev72]   ———, *The deficiency index of a certain class of ordinary self-
          adjoint differential operators*, Advances in Math. **8** (1972), 434–473.
          MR 0298102 (45 #7154)

[DHL09]   Dorin Dutkay, Deguang Han, and David Larson, *A duality princi-
          ple for groups*, J. Funct. Anal. **257** (2009), no. 4, 1133–1143. MR
          2535465 (2010i:22006)

[Die75]   Joseph Diestel, *Geometry of Banach spaces—selected topics*, Lecture
          Notes in Mathematics, Vol. 485, Springer-Verlag, Berlin-New York,
          1975. MR 0461094 (57 #1079)

[Dir35]   P. A. M. Dirac, *The electron wave equation in de-Sitter space*, Ann.
          of Math. (2) **36** (1935), no. 3, 657–669. MR 1503243

[Dir47]   ———, *The Principles of Quantum Mechanics*, Oxford, at the
          Clarendon Press, 1947, 3d ed. MR 0023198 (9,319d)

[Dix81]   Jacques Dixmier, *von Neumann algebras*, North-Holland Mathemat-
          ical Library, vol. 27, North-Holland Publishing Co., Amsterdam-
          New York, 1981, With a preface by E. C. Lance, Translated from
          the second French edition by F. Jellett. MR 641217 (83a:46004)

[DJ08]   Dorin Ervin Dutkay and Palle E. T. Jorgensen, *A duality approach to
          representations of Baumslag-Solitar groups*, Group representations,
          ergodic theory, and mathematical physics: a tribute to George W.
          Mackey, Contemp. Math., vol. 449, Amer. Math. Soc., Providence,
          RI, 2008, pp. 99–127. MR 2391800 (2009j:22007)

[DM72]   H. Dym and H. P. McKean, *Fourier series and integrals*, Aca-
          demic Press, New York-London, 1972, Probability and Mathemati-
          cal Statistics, No. 14. MR 0442564 (56 #945)

[DM91]   V. A. Derkach and M. M. Malamud, *Generalized resolvents and
          the boundary value problems for Hermitian operators with gaps*, J.
          Funct. Anal. **95** (1991), no. 1, 1–95. MR 1087947 (93d:47046)

[dO09]   César R. de Oliveira, *Intermediate spectral theory and quantum dy-
          namics*, Progress in Mathematical Physics, vol. 54, Birkhäuser Ver-
          lag, Basel, 2009. MR 2723496

[Dop95]   S. Doplicher, *Quantum physics, classical gravity and noncommutative spacetime*, XIth International Congress of Mathematical Physics (Paris, 1994), Int. Press, Cambridge, MA, 1995, pp. 324–329. MR 1370688

[Dou80]   Ronald G. Douglas, C^*-*algebra extensions and K-homology*, Annals of Mathematics Studies, vol. 95, Princeton University Press, Princeton, N.J.; University of Tokyo Press, Tokyo, 1980. MR 571362 (82c:46082)

[DS88a]   Nelson Dunford and Jacob T. Schwartz, *Linear operators. Part I*, Wiley Classics Library, John Wiley & Sons Inc., New York, 1988, General theory, With the assistance of William G. Bade and Robert G. Bartle, Reprint of the 1958 original, A Wiley-Interscience Publication. MR 1009162 (90g:47001a)

[DS88b]   ———, *Linear operators. Part II*, Wiley Classics Library, John Wiley & Sons Inc., New York, 1988, Spectral theory. Selfadjoint operators in Hilbert space, With the assistance of William G. Bade and Robert G. Bartle, Reprint of the 1963 original, A Wiley-Interscience Publication. MR 1009163 (90g:47001b)

[DS88c]   ———, *Linear operators. Part III*, Wiley Classics Library, John Wiley & Sons, Inc., New York, 1988, Spectral operators, With the assistance of William G. Bade and Robert G. Bartle, Reprint of the 1971 original, A Wiley-Interscience Publication. MR 1009164 (90g:47001c)

[Dud14]   F. A. Dudkin, *On the embedding of Baumslag-Solitar groups into the generalized Baumslag-Solitar groups*, Sibirsk. Mat. Zh. **55** (2014), no. 1, 90–96. MR 3220588

[Emc84]   Gérard G. Emch, *Mathematical and conceptual foundations of 20th-century physics*, North-Holland Mathematics Studies, vol. 100, North-Holland Publishing Co., Amsterdam, 1984, Notas de Matemática [Mathematical Notes], 100. MR 777146 (86i:00032)

[EN12]   George A. Elliott and Zhuang Niu, *Extended rotation algebras: adjoining spectral projections to rotation algebras*, J. Reine Angew. Math. **665** (2012), 1–71. MR 2908740

[Enf73]   Per Enflo, *A counterexample to the approximation problem in Banach spaces*, Acta Math. **130** (1973), 309–317. MR 0402468 (53 #6288)

[EO13]   Martin Ehler and Kasso A. Okoudjou, *Probabilistic frames: an overview*, Finite frames, Appl. Numer. Harmon. Anal., Birkhäuser/Springer, New York, 2013, pp. 415–436. MR 2964017

[Fad05]   L. D. Faddeev, *Mass in quantum Yang-Mills theory (comment on a Clay Millennium Problem)*, Perspectives in analysis, Math. Phys. Stud., vol. 27, Springer, Berlin, 2005, pp. 63–72. MR 2206769 (2007c:81121)

[Fan10]   Mark Fannes, *An introduction to quantum probability*, Theoretical foundations of quantum information processing and communication, Lecture Notes in Phys., vol. 787, Springer, Berlin, 2010, pp. 1–38. MR 2762151 (2012d:81174)

[Fef71]   Charles Fefferman, *Characterizations of bounded mean oscillation*, Bull. Amer. Math. Soc. **77** (1971), 587–588. MR 0280994 (43 #6713)

[Fel68]   William Feller, *An introduction to probability theory and its applications. Vol. I*, Third edition, John Wiley & Sons, Inc., New York-London-Sydney, 1968. MR 0228020 (37 #3604)

[Fel71]   _____, *An introduction to probability theory and its applications. Vol. II.*, Second edition, John Wiley & Sons, Inc., New York-London-Sydney, 1971. MR 0270403 (42 #5292)

[FL28]   Kurt Friedrichs and Hans Lewy, *Über die Eindeutigkeit und das Abhängigkeitsgebiet der Lösungen beim Anfangswertproblem linearer hyperbolischer Differentialgleichungen*, Math. Ann. **98** (1928), no. 1, 192–204. MR 1512400

[FM13]   J. C. Ferreira and V. A. Menegatto, *Positive definiteness, reproducing kernel Hilbert spaces and beyond*, Ann. Funct. Anal. **4** (2013), no. 1, 64–88. MR 3004212

[FR42]   W. H. J. Fuchs and W. W. Rogosinski, *A note on Mercer's theorem*, J. London Math. Soc. **17** (1942), 204–210. MR 0008270 (4,272g)

[Fri80]   Kurt Otto Friedrichs, *Spectral theory of operators in Hilbert space*, Applied Mathematical Sciences, vol. 9, Springer-Verlag, New York-Berlin, 1980, Corrected reprint. MR 635783 (82j:47001)

[Fug74]   Bent Fuglede, *Boundary minimum principles in potential theory*, Math. Ann. **210** (1974), 213–226. MR 0357827 (50 #10293b)

[Fug82]   _____, *Conditions for two selfadjoint operators to commute or to satisfy the Weyl relation*, Math. Scand. **51** (1982), no. 1, 163–178. MR 681266 (84a:81013)

[GBD94]   José González-Barrios and R. M. Dudley, *On extensions of Mercer's theorem*, Third Symposium on Probability Theory and Stochastic Processes (Spanish) (Hermosillo, 1994), Aportaciones Mat. Notas Investigación, vol. 11, Soc. Mat. Mexicana, México, 1994, pp. 91–97. MR 1356475 (96i:47090)

[GDV06] Michael Gastpar, Pier Luigi Dragotti, and Martin Vetterli, *The distributed Karhunen-Loève transform*, IEEE Trans. Inform. Theory **52** (2006), no. 12, 5177–5196. MR 2300686 (2007m:94045)

[Geo09] H. Georgi, *Weak interactions and modern particle theory*, Dover Books on Physics Series, Dover Publications, Incorporated, 2009

[GG59] I. M. Gel'fand and M. I. Graev, *Geometry of homogeneous spaces, representations of groups in homogeneous spaces and related questions of integral geometry. I*, Trudy Moskov. Mat. Obšč. **8** (1959), 321–390; addendum **9** (1959), 562. MR 0126719 (23 #A4013)

[GG91] V. I. Gorbachuk and M. L. Gorbachuk, *Boundary value problems for operator differential equations*, Mathematics and its Applications (Soviet Series), vol. 48, Kluwer Academic Publishers Group, Dordrecht, 1991, Translated and revised from the 1984 Russian original. MR 1154792 (92m:34133)

[GG02] M. Gadella and F. Gómez, *A unified mathematical formalism for the Dirac formulation of quantum mechanics*, Found. Phys. **32** (2002), no. 6, 815–869. MR 1917012 (2003j:81070)

[GIS90] James Glimm, John Impagliazzo, and Isadore Singer (eds.), *The legacy of John von Neumann*, Proceedings of Symposia in Pure Mathematics, vol. 50, American Mathematical Society, Providence, RI, 1990. MR 1067743 (91e:00024)

[GJ60] I. M. Gel'fand and A. M. Jaglom, *Integration in functional spaces and its applications in quantum physics*, J. Mathematical Phys. **1** (1960), 48–69. MR 0112604 (22 #3455)

[GJ87] James Glimm and Arthur Jaffe, *Quantum physics*, second ed., Springer-Verlag, New York, 1987, A functional integral point of view. MR 887102 (89k:81001)

[Gli60] James G. Glimm, *On a certain class of operator algebras*, Trans. Amer. Math. Soc. **95** (1960), 318–340. MR 0112057 (22 #2915)

[Gli61] James Glimm, *Type I C^*-algebras*, Ann. of Math. (2) **73** (1961), 572–612. MR 0124756 (23 #A2066)

[GLS12] Martin J. Gander, Sébastien Loisel, and Daniel B. Szyld, *An optimal block iterative method and preconditioner for banded matrices with applications to PDEs on irregular domains*, SIAM J. Matrix Anal. Appl. **33** (2012), no. 2, 653–680. MR 2970224

[GN94] I. Gel'fand and M. Neumark, *On the imbedding of normed rings into the ring of operators in Hilbert space*, C^*-algebras: 1943–1993 (San Antonio, TX, 1993), Contemp. Math., vol. 167, Amer. Math. Soc., Providence, RI, 1994, Corrected reprint of the 1943 original [MR **5**, 147], pp. 2–19. MR 1292007

[Gro64]   Leonard Gross, *Classical analysis on a Hilbert space*, Proc. Conf. on Theory and Appl. of Analysis in Function Space (Dedham, M ass., 1963), The M.I.T. Press, Cambridge, Mass., 1964, pp. 51–68. MR 0184066 (32 #1542)

[Gro67]   _____, *Abstract Wiener spaces*, Proc. Fifth Berkeley Sympos. Math. Statist. and Probability (Berkeley, Calif., 1965/66), Vol. II: Contributions to Probability Theory, Part 1, Univ. California Press, Berkeley, Calif., 1967, pp. 31–42. MR 0212152 (35 #3027)

[Gro70]   _____, *Abstract Wiener measure and infinite dimensional potential theory*, Lectures in Modern Analysis and Applications, II, Lecture Notes in Mathematics, Vol. 140. Springer, Berlin, 1970, pp. 84–116. MR 0265548 (42 #457)

[GS60]   I. M. Gel'fand and Do-šin Sya, *On positive definite distributions*, Uspehi Mat. Nauk **15** (1960), no. 1 (91), 185–190. MR 0111991 (22 #2849)

[GS77]   I. M. Gel'fand and G. E. Shilov, *Generalized functions. Vol. 2*, Academic Press [Harcourt Brace Jovanovich, Publishers], New York-London, 1968 [1977], Spaces of fundamental and generalized functions, Translated from the Russian by Morris D. Friedman, Amiel Feinstein and Christian P. Peltzer. MR 0435832 (55 #8786b)

[GvN51]   Herman H. Goldstine and John von Neumann, *Numerical inverting of matrices of high order. II*, Proc. Amer. Math. Soc. **2** (1951), 188–202. MR 0041539 (12,861b)

[Hal64]   P. R. Halmos, *Numerical ranges and normal dilations*, Acta Sci. Math. (Szeged) **25** (1964), 1–5. MR 0171168 (30 #1399)

[Hal82]   Paul Richard Halmos, *A Hilbert space problem book*, second ed., Graduate Texts in Mathematics, vol. 19, Springer-Verlag, New York-Berlin, 1982, Encyclopedia of Mathematics and its Applications, 17. MR 675952 (84e:47001)

[Hal13]   Brian C. Hall, *Quantum theory for mathematicians*, Graduate Texts in Mathematics, vol. 267, Springer, New York, 2013. MR 3112817

[Hal15]   Brian Hall, *Lie groups, Lie algebras, and representations*, second ed., Graduate Texts in Mathematics, vol. 222, Springer, Cham, 2015, An elementary introduction. MR 3331229

[HC57]   Harish-Chandra, *A formula for semisimple Lie groups*, Amer. J. Math. **79** (1957), 733–760. MR 0096138 (20 #2633)

[HdSS12] Seppo Hassi, Hendrik S. V. de Snoo, and Franciszek Hugon Szafraniec (eds.), *Operator methods for boundary value problems*, London Mathematical Society Lecture Note Series, vol. 404, Cambridge University Press, Cambridge, 2012. MR 3075434

[Hei69] W Heisenberg, *Über quantentheoretische umdeutung kinematischer und mechanischer beziehungen.(1925) in:* G, Ludwig, Wellen-mechanik, Einführung und Originaltexte, Akademie-Verlag, Berlin (1969), 195.

[Hel13] Bernard Helffer, *Spectral theory and its applications*, Cambridge Studies in Advanced Mathematics, vol. 139, Cambridge University Press, Cambridge, 2013. MR 3027462

[Hid80] Takeyuki Hida, *Brownian motion*, Applications of Mathematics, vol. 11, Springer-Verlag, New York, 1980, Translated from the Japanese by the author and T. P. Speed. MR 562914 (81a:60089)

[Hil02] David Hilbert, *Mathematical problems*, Bull. Amer. Math. Soc. 8 (1902), no. 10, 437–479. MR 1557926

[Hil22] _____, *Die logischen Grundlagen der Mathematik*, Math. Ann. 88 (1922), no. 1-2, 151–165. MR 1512123

[Hil24] _____, *Die Grundlagen der Physik*, Math. Ann. 92 (1924), no. 1-2, 1–32. MR 1512197

[HJL06] Deguang Han, Palle E. T. Jorgensen, and David Royal Larson (eds.), *Operator theory, operator algebras, and applications*, Contemporary Mathematics, vol. 414, American Mathematical Society, Providence, RI, 2006. MR 2277225 (2007f:46001)

[HJL+13] Deguang Han, Wu Jing, David Larson, Pengtong Li, and Ram N. Mohapatra, *Dilation of dual frame pairs in Hilbert C*-modules*, Results Math. 63 (2013), no. 1-2, 241–250. MR 3009685

[HKLW07] Deguang Han, Keri Kornelson, David Larson, and Eric Weber, *Frames for undergraduates*, Student Mathematical Library, vol. 40, American Mathematical Society, Providence, RI, 2007. MR 2367342 (2010e:42044)

[HMU80] H. Hudzik, J. Musielak, and R. Urbański, *Riesz-Thorin theorem in generalized Orlicz spaces of nonsymmetric type*, Univ. Beograd. Publ. Elektrotehn. Fak. Ser. Mat. Fiz. (1980), no. 678-715, 145–158 (1981). MR 623243 (83j:46044)

[HN28] D. Hilbert and J. v. Neumann, *Über die Grundlagen der Quanten-mechanik*, Math. Ann. 98 (1928), no. 1, 1–30. MR 1512390

[HØUZ10] Helge Holden, Bernt Øksendal, Jan Ubøe, and Tusheng Zhang, *Stochastic partial differential equations*, second ed., Universitext, Springer, New York, 2010, A modeling, white noise functional approach. MR 2571742 (2011c:60201)

[HP57] Einar Hille and Ralph S. Phillips, *Functional analysis and semi-groups*, American Mathematical Society Colloquium Publications, vol. 31, American Mathematical Society, Providence, R. I., 1957, rev. ed. MR 0089373 (19,664d)

[HS68] Melvin Hausner and Jacob T. Schwartz, *Lie groups; Lie algebras*, Gordon and Breach Science Publishers, New York-London-Paris, 1968. MR 0235065 (38 #3377)

[Hut81] John E. Hutchinson, *Fractals and self-similarity*, Indiana Univ. Math. J. **30** (1981), no. 5, 713–747. MR 625600 (82h:49026)

[IR07] Peter Imkeller and Sylvie Roelly, *Die Wiederentdeckung eines Mathematikers: Wolfgang Döblin*, Mitt. Dtsch. Math.-Ver. **15** (2007), no. 3, 154–159. MR 2385126

[Itô04] Kiyosi Itô, *Stochastic processes*, Springer-Verlag, Berlin, 2004, Lectures given at Aarhus University, Reprint of the 1969 original, Edited and with a foreword by Ole E. Barndorff-Nielsen and Ken-iti Sato. MR 2053326 (2005e:60002)

[Itô06] Kiyoshi Itô, *Essentials of stochastic processes*, Translations of Mathematical Monographs, vol. 231, American Mathematical Society, Providence, RI, 2006, Translated from the 1957 Japanese original by Yuji Ito. MR 2239081 (2007i:60001)

[Itô07] Kiyosi Itô, *Memoirs of my research on stochastic analysis*, Stochastic analysis and applications, Abel Symp., vol. 2, Springer, Berlin, 2007, pp. 1–5. MR 2397781

[Jac79] Nathan Jacobson, *Lie algebras*, Dover Publications, Inc., New York, 1979, Republication of the 1962 original. MR 559927 (80k:17001)

[JL01] William B. Johnson and Joram Lindenstrauss, *Basic concepts in the geometry of Banach spaces*, Handbook of the geometry of Banach spaces, Vol. I, North-Holland, Amsterdam, 2001, pp. 1–84. MR 1863689 (2003f:46013)

[JLS96] W. B. Johnson, J. Lindenstrauss, and G. Schechtman, *Banach spaces determined by their uniform structures*, Geom. Funct. Anal. **6** (1996), no. 3, 430–470. MR 1392325 (97b:46016)

[JM80] Palle T. Jørgensen and Paul S. Muhly, *Selfadjoint extensions satisfying the Weyl operator commutation relations*, J. Analyse Math. **37** (1980), 46–99. MR 583632 (82k:47058)

[JM84] Palle E. T. Jorgensen and Robert T. Moore, *Operator commutation relations*, Mathematics and its Applications, D. Reidel Publishing Co., Dordrecht, 1984, Commutation relations for operators, semigroups, and resolvents with applications to mathematical physics and representations of Lie groups. MR 746138 (86i:22006)

[JÓ00]   Palle E. T. Jorgensen and Gestur Ólafsson, *Unitary representa-
            tions and Osterwalder-Schrader duality*, The mathematical legacy of
            Harish-Chandra (Baltimore, MD, 1998), Proc. Sympos. Pure Math.,
            vol. 68, Amer. Math. Soc., Providence, RI, 2000, pp. 333–401. MR
            1767902 (2001f:22036)

[Joh88]   William B. Johnson, *Homogeneous Banach spaces*, Geometric as-
            pects of functional analysis (1986/87), Lecture Notes in Math., vol.
            1317, Springer, Berlin, 1988, pp. 201–203. MR 950981 (89g:46032)

[Jør80]   Palle E. T. Jørgensen, *Unbounded operators: perturbations and com-
            mutativity problems*, J. Funct. Anal. **39** (1980), no. 3, 281–307. MR
            600620 (82e:47003)

[Jor88]   Palle E. T. Jorgensen, *Operators and representation theory*, North-
            Holland Mathematics Studies, vol. 147, North-Holland Publishing
            Co., Amsterdam, 1988, Canonical models for algebras of operators
            arising in quantum mechanics, Notas de Matemática [Mathematical
            Notes], 120. MR 919948 (89e:47001)

[Jor94]   _____, *Quantization and deformation of Lie algebras*, Lie algebras,
            cohomology, and new applications to quantum mechanics (Spring-
            field, MO, 1992), Contemp. Math., vol. 160, Amer. Math. Soc.,
            Providence, RI, 1994, pp. 141–149. MR 1277380 (95d:46074)

[Jor02]   _____, *Diagonalizing operators with reflection symmetry*, J. Funct.
            Anal. **190** (2002), no. 1, 93–132, Special issue dedicated to the mem-
            ory of I. E. Segal. MR 1895530 (2003e:47069)

[Jor06]   _____, *Analysis and probability: wavelets, signals, fractals*, Gradu-
            ate Texts in Mathematics, vol. 234, Springer, New York, 2006. MR
            2254502 (2008a:42030)

[Jor08]   _____, *Essential self-adjointness of the graph-Laplacian*, J. Math.
            Phys. **49** (2008), no. 7, 073510, 33. MR 2432048 (2009k:47099)

[Jor11]   _____, *Representations of Lie algebras built over Hilbert space*, In-
            fin. Dimens. Anal. Quantum Probab. Relat. Top. **14** (2011), no. 3,
            419–442. MR 2847247

[Jør14]   Palle E. T. Jørgensen, *A universal envelope for Gaussian processes
            and their kernels*, J. Appl. Math. Comput. **44** (2014), no. 1-2, 1–38.
            MR 3147727

[JP10]   Palle E. T. Jorgensen and Erin Peter James Pearse, *A Hilbert space
            approach to effective resistance metric*, Complex Anal. Oper. Theory
            4 (2010), no. 4, 975–1013. MR 2735315 (2011j:05338)

[JP11a]   Palle E. T. Jorgensen and Erin P. J. Pearse, *Resistance bound-aries of infinite networks*, Random walks, boundaries and spectra, Progr. Probab., vol. 64, Birkhäuser/Springer Basel AG, Basel, 2011, pp. 111–142. MR 3051696

[JP11b]   _____, *Spectral reciprocity and matrix representations of un-bounded operators*, J. Funct. Anal. **261** (2011), no. 3, 749–776. MR 2799579

[JP12]   P. E. T. Jorgensen and A. M. Paolucci, *q-frames and Bessel func-tions*, Numer. Funct. Anal. Optim. **33** (2012), no. 7-9, 1063–1069. MR 2966144

[JP13a]   Palle E. T. Jorgensen and Erin P. J. Pearse, *A discrete Gauss-Green identity for unbounded Laplace operators, and the transience of ran-dom walks*, Israel J. Math. **196** (2013), no. 1, 113–160. MR 3096586

[JP13b]   _____, *Multiplication operators on the energy space*, J. Operator Theory **69** (2013), no. 1, 135–159. MR 3029492

[JP14]   _____, *Spectral comparisons between networks with different con-ductance functions*, J. Operator Theory **72** (2014), no. 1, 71–86. MR 3246982

[JPS01]   Palle Jorgensen, Daniil Proskurin, and Yurii Samoilenko, *A family of ∗-algebras allowing Wick ordering: Fock representations and uni-versal enveloping C^*-algebras*, Noncommutative structures in math-ematics and physics (Kiev, 2000), NATO Sci. Ser. II Math. Phys. Chem., vol. 22, Kluwer Acad. Publ., Dordrecht, 2001, pp. 321–329. MR 1893475 (2003c:46080)

[JPS05]   Palle E. T. Jørgensen, Daniil P. Proskurin, and Yuriĭ S. Samoĭlenko, *On C^*-algebras generated by pairs of q-commuting isometries*, J. Phys. A **38** (2005), no. 12, 2669–2680. MR 2132080 (2005m:46093)

[JPT12]   Palle Jorgensen, Steen Pedersen, and Feng Tian, *Translation repre-sentations and scattering by two intervals*, J. Math. Phys. **53** (2012), no. 5, 053505, 49. MR 2964262

[JPT13]   Palle E. T. Jorgensen, Steen Pedersen, and Feng Tian, *Momentum operators in two intervals: spectra and phase transition*, Complex Anal. Oper. Theory **7** (2013), no. 6, 1735–1773. MR 3129890

[JPT14]   Palle Jorgensen, Steen Pedersen, and Feng Tian, *Restrictions and extensions of semibounded operators*, Complex Anal. Oper. Theory **8** (2014), no. 3, 591–663. MR 3167762

[JPT15]   _____, *Harmonic analysis of a class of reproducing kernel hilbert spaces arising from groups*, Contemp. Math., vol. 650, Amer. Math. Soc., Providence, RI, 2015, pp. 157–198.

[JS07]   Palle E. T. Jorgensen and Myung-Sin Song, *Entropy encoding,
           Hilbert space, and Karhunen-Loève transforms*, J. Math. Phys. **48**
           (2007), no. 10, 103503, 22. MR 2362796 (2008k:94016)

[JS09]   _____, *Analysis of fractals, image compression, entropy encoding,
           Karhunen-Loève transforms*, Acta Appl. Math. **108** (2009), no. 3,
           489–508. MR 2563494 (2011d:42071)

[JT15a]   Palle Jorgensen and Feng Tian, *Frames and factorization of graph
           Laplacians*, Opuscula Math. **35** (2015), no. 3, 293–332. MR 3296811

[JT15b]   _____, *Infinite networks and variation of conductance functions in
           discrete Laplacians*, J. Math. Phys. **56** (2015), no. 4, 043506, 27.
           MR 3390972

[Kad90]   Richard V. Kadison, *Operator algebras—an overview*, The legacy of
           John von Neumann (Hempstead, NY, 1988), Proc. Sympos. Pure
           Math., vol. 50, Amer. Math. Soc., Providence, RI, 1990, pp. 61–89.
           MR 1067753 (92d:46141)

[Kak48]   Shizuo Kakutani, *On equivalence of infinite product measures*, Ann.
           of Math. (2) **49** (1948), 214–224. MR 0023331 (9,340e)

[Kan58]   L. Kantorovitch, *On the translocation of masses*, Management Sci.
           5 (1958), 1–4. MR 0096552 (20 #3035)

[Kat95]   Tosio Kato, *Perturbation theory for linear operators*, Classics in
           Mathematics, Springer-Verlag, Berlin, 1995, Reprint of the 1980 edi-
           tion. MR 1335452 (96a:47025)

[KF75]   A. N. Kolmogorov and S. V. Fomīn, *Introductory real analysis*, Dover
           Publications, Inc., New York, 1975, Translated from the second Rus-
           sian edition and edited by Richard A. Silverman, Corrected reprint-
           ing. MR 0377445 (51 #13617)

[Kho15]   A. S. Kholevo, *Gaussian optimizers and the additivity problem
           in quantum information theory*, Uspekhi Mat. Nauk **70** (2015),
           no. 2(422), 141–180. MR 3353129

[Kir88]   A. A. Kirillov, *Introduction to representation theory and noncommu-
           tative harmonic analysis*, Current problems in mathematics. Funda-
           mental directions, Vol. 22 (Russian), Itogi Nauki i Tekhniki, Akad.
           Nauk SSSR, Vsesoyuz. Inst. Nauchn. i Tekhn. Inform., Moscow,
           1988, pp. 5–162. MR 942947 (90a:22005)

[KL14a]   Richard V. Kadison and Zhe Liu, *The Heisenberg relation—
           mathematical formulations*, SIGMA Symmetry Integrability Geom.
           Methods Appl. **10** (2014), Paper 009, 40. MR 3210626

[KL14b]   Mark G. Krein and Heinz Langer, *Continuation of Hermitian Positive Definite Functions and Related Questions*, Integral Equations Operator Theory **78** (2014), no. 1, 1–69. MR 3147401

[Kla15]   E. Klarreich, *'Outsiders' Crack 50-Year-Old Math Problem*, Quanta Magazine (2015)

[KLR09]   V. G. Kravchenko, A. B. Lebre, and J. S. Rodríguez, *Factorization of singular integral operators with a Carleman backward shift: the case of bounded measurable coefficients*, J. Anal. Math. **107** (2009), 1–37. MR 2496397 (2010b:47126)

[KLZ09]   Victor Kaftal, David R. Larson, and Shuang Zhang, *Operator-valued frames*, Trans. Amer. Math. Soc. **361** (2009), no. 12, 6349–6385. MR 2538596 (2010h:42060)

[KM04]   Debasis Kundu and Anubhav Manglick, *Discriminating between the Weibull and log-normal distributions*, Naval Res. Logist. **51** (2004), no. 6, 893–905. MR 2079449 (2005b:62033)

[KMRS05]   Norio Konno, Naoki Masuda, Rahul Roy, and Anish Sarkar, *Rigorous results on the threshold network model*, J. Phys. A **38** (2005), no. 28, 6277–6291. MR 2166622 (2006j:82054)

[KOPT13]   Gitta Kutyniok, Kasso A. Okoudjou, Friedrich Philipp, and Elizabeth K. Tuley, *Scalable frames*, Linear Algebra Appl. **438** (2013), no. 5, 2225–2238. MR 3005286

[KR57]   L. V. Kantorovič and G. Š. Rubinšteĭn, *On a functional space and certain extremum problems*, Dokl. Akad. Nauk SSSR (N.S.) **115** (1957), 1058–1061. MR 0094707 (20 #1219)

[KR97a]   Richard V. Kadison and John R. Ringrose, *Fundamentals of the theory of operator algebras. Vol. I*, Graduate Studies in Mathematics, vol. 15, American Mathematical Society, Providence, RI, 1997, Elementary theory, Reprint of the 1983 original. MR 1468229 (98f:46001a)

[KR97b]   _____, *Fundamentals of the theory of operator algebras. Vol. II*, Graduate Studies in Mathematics, vol. 16, American Mathematical Society, Providence, RI, 1997, Advanced theory, Corrected reprint of the 1986 original. MR 1468230 (98f:46001b)

[Kra83]   Karl Kraus, *States, effects, and operations*, Lecture Notes in Physics, vol. 190, Springer-Verlag, Berlin, 1983, Fundamental notions of quantum theory, Lecture notes edited by A. Böhm, J. D. Dollard and W. H. Wootters. MR 725167 (86j:81008)

[Kre46]   M. Krein, *Concerning the resolvents of an Hermitian operator with the deficiency-index (m, m)*, C. R. (Doklady) Acad. Sci. URSS (N.S.) **52** (1946), 651–654. MR 0018341 (8,277a)

[Kre55]   M. G. Krein, *On some cases of the effective determination of the density of a nonuniform string by its spectral function*, 2 Pine St., West Concord, Mass., 1955, Translated by Morris D. Friedman. MR 0075403 (17,740f)

[Kru07]   Natan Kruglyak, *An elementary proof of the real version of the Riesz-Thorin theorem*, Interpolation theory and applications, Contemp. Math., vol. 445, Amer. Math. Soc., Providence, RI, 2007, pp. 179–182. MR 2381892 (2009j:46056)

[KS59]   Richard V. Kadison and I. M. Singer, *Extensions of pure states*, Amer. J. Math. **81** (1959), 383–400. MR 0123922 (23 #A1243)

[KS02]   Vadim Kostrykin and Robert Schrader, *Statistical ensembles and density of states*, Mathematical results in quantum mechanics (Taxco, 2001), Contemp. Math., vol. 307, Amer. Math. Soc., Providence, RI, 2002, pp. 177–208. MR 1946030 (2004g:82007)

[KS03]   Hui-Hsiung Kuo and Ambar N. Sengupta (eds.), *Finite and infinite dimensional analysis in honor of Leonard Gross*, Contemporary Mathematics, vol. 317, American Mathematical Society, Providence, RI, 2003. MR 1966883 (2003j:00025)

[KW12]   Greg Kuperberg and Nik Weaver, *A von Neumann algebra approach to quantum metrics*, Mem. Amer. Math. Soc. **215** (2012), no. 1010, v, 1–80. MR 2908248

[Lax02]   Peter D. Lax, *Functional analysis*, Pure and Applied Mathematics (New York), Wiley-Interscience [John Wiley & Sons], New York, 2002. MR 1892228 (2003a:47001)

[Leb05]   H. Lebesgue, *Sur le problème des aires*, Bull. Soc. Math. France **33** (1905), 273–274. MR 1504529

[Lév51]   Paul Lévy, *Problèmes concrets d'analyse fonctionnelle. Avec un complément sur les fonctionnelles analytiques par F. Pellegrino*, Gauthier-Villars, Paris, 1951, 2d ed. MR 0041346 (12,834a)

[Lév53]   _____, *Random functions: General theory with special reference to Laplacian random functions*, Univ. California Publ. Statist. **1** (1953), 331–390. MR 0055607 (14,1099f)

[Lév57]   _____, *Remarques sur le processus de W. Feller et H. P. MacKean*, C. R. Acad. Sci. Paris **245** (1957), 1772–1774. MR 0093058 (19,1202h)

[Lév62]   _____, *Le déterminisme de la fonction brownienne dans l'espace de Hilbert*, Ann. Sci. École Norm. Sup. (3) **79** (1962), 377–398. MR 0150844 (27 #830)

[Lév63]   _____, *Le mouvement brownien fonction d'un ou de plusieurs paramètres*, Rend. Mat. e Appl. (5) **22** (1963), 24–101. MR 0163367 (29 #670)

[LF01]   M. Levanda and V. Fleurov, *A Wigner quasi-distribution function for charged particles in classical electromagnetic fields*, Ann. Physics **292** (2001), no. 2, 199–231. MR 1855029

[LMCL09]   F. E. A. Leite, Raúl Montagne, G. Corso, and L. S. Lucena, *Karhunen-Loève spectral analysis in multiresolution decomposition*, Comput. Geosci. **13** (2009), no. 2, 165–170. MR 2505483 (2010e:86011)

[Loè63]   Michel Loève, *Probability theory*, Third edition, D. Van Nostrand Co., Inc., Princeton, N.J.-Toronto, Ont.-London, 1963. MR 0203748 (34 #3596)

[LP89]   Peter D. Lax and Ralph S. Phillips, *Scattering theory*, second ed., Pure and Applied Mathematics, vol. 26, Academic Press Inc., Boston, MA, 1989, With appendices by Cathleen S. Morawetz and Georg Schmidt. MR 1037774 (90k:35005)

[LPS88]   A. Lubotzky, R. Phillips, and P. Sarnak, *Ramanujan graphs*, Combinatorica **8** (1988), no. 3, 261–277. MR 963118 (89m:05099)

[LPW13]   Michel L. Lapidus, Erin P. J. Pearse, and Steffen Winter, *Minkowski measurability results for self-similar tilings and fractals with monophase generators*, Fractal geometry and dynamical systems in pure and applied mathematics. I. Fractals in pure mathematics, Contemp. Math., vol. 600, Amer. Math. Soc., Providence, RI, 2013, pp. 185–203. MR 3203403

[Maa10]   H. Maassen, *Quantum probability and quantum information theory*, Quantum information, computation and cryptography, Lecture Notes in Phys., vol. 808, Springer, Berlin, 2010, pp. 65–108. MR 2768446

[Mac52]   George W. Mackey, *Induced representations of locally compact groups. I*, Ann. of Math. (2) **55** (1952), 101–139. MR 0044536 (13,434a)

[Mac85]   _____, *Quantum mechanics from the point of view of the theory of group representations*, Applications of group theory in physics and mathematical physics (Chicago, 1982), Lectures in Appl. Math., vol. 21, Amer. Math. Soc., Providence, RI, 1985, pp. 219–253. MR 789292 (86j:81051)

[Mac88]   _____, *Induced representations and the applications of harmonic analysis*, Harmonic analysis (Luxembourg, 1987), Lecture Notes in Math., vol. 1359, Springer, Berlin, 1988, pp. 16–51. MR 974302 (90c:22021)

[Mac92]   _____, *The scope and history of commutative and noncommutative harmonic analysis*, History of Mathematics, vol. 5, American Mathematical Society, Providence, RI; London Mathematical Society, London, 1992. MR 1171011 (93g:22006)

[Mac09]   Barbara D. MacCluer, *Elementary functional analysis*, Graduate Texts in Mathematics, vol. 253, Springer, New York, 2009. MR 2462971 (2010b:46001)

[Mal01]   Stéphane Mallat, *Applied mathematics meets signal processing*, Challenges for the 21st century (Singapore, 2000), World Sci. Publ., River Edge, NJ, 2001, pp. 138–161. MR 1875017

[Mey06]   Yves Meyer, *From wavelets to atoms*, 150 years of mathematics at Washington University in St. Louis, Contemp. Math., vol. 395, Amer. Math. Soc., Providence, RI, 2006, pp. 105–117. MR 2206895

[MJD+15]   Daniel Markiewicz, Palle E. T. Jorgensen, Kenneth R. Davidson, Ronald G. Douglas, Edward G. Effros, Richard V. Kadison, Marcelo Laca, Paul S. Muhly, David R. Pitts, Robert T. Powers, Geoffrey L. Price, Donald E. Sarason, Erling Stormer, and Lee Ann Kaskutas, *William B. Arveson: A Tribute*, Notices Amer. Math. Soc. **62** (2015), no. 07, 1

[MM09]   Basarab Matei and Yves Meyer, *A variant of compressed sensing*, Rev. Mat. Iberoam. **25** (2009), no. 2, 669–692. MR 2569549 (2010j:42023)

[MQ14]   Wen Mi and Tao Qian, *On backward shift algorithm for estimating poles of systems*, Automatica J. IFAC **50** (2014), no. 6, 1603–1610. MR 3214905

[MSS13]   Adam Marcus, Daniel A. Spielman, and Nikhil Srivastava, *Interlacing families I: Bipartite Ramanujan graphs of all degrees*, 2013 IEEE 54th Annual Symposium on Foundations of Computer Science— FOCS 2013, IEEE Computer Soc., Los Alamitos, CA, 2013, pp. 529– 537. MR 3246256

[MSS15]   Adam W. Marcus, Daniel A. Spielman, and Nikhil Srivastava, *Interlacing families II: Mixed characteristic polynomials and the Kadison-Singer problem*, Ann. of Math. (2) **182** (2015), no. 1, 327– 350. MR 3374963

[MT13]   Jan Mycielski and Grzegorz Tomkowicz, *The Banach-Tarski para-dox for the hyperbolic plane (II)*, Fund. Math. **222** (2013), no. 3, 289–290. MR 3104075

[MvN43]   F. J. Murray and J. von Neumann, *On rings of operators. IV*, Ann. of Math. (2) **44** (1943), 716–808. MR 0009096

[NC00]   Michael A. Nielsen and Isaac L. Chuang, *Quantum computation and quantum information*, Cambridge University Press, Cambridge, 2000. MR 1796805 (2003j:81038)

[Nel59a]   Edward Nelson, *Analytic vectors*, Ann. of Math. (2) **70** (1959), 572–615. MR 0107176 (21 #5901)

[Nel59b]   _____, *Regular probability measures on function space*, Ann. of Math. (2) **69** (1959), 630–643. MR 0105743 (21 #4479)

[Nel64]   _____, *Feynman integrals and the Schrödinger equation*, J. Mathematical Phys. **5** (1964), 332–343. MR 0161189 (28 #4397)

[Nel67]   _____, *Dynamical theories of Brownian motion*, Princeton University Press, Princeton, N.J., 1967. MR 0214150 (35 #5001)

[Nel69]   _____, *Topics in dynamics. I: Flows*, Mathematical Notes, Princeton University Press, Princeton, N.J., 1969. MR 0282379 (43 #8091)

[Nus75]   A. Edward Nussbaum, *Extension of positive definite functions and representation of functions in terms of spherical functions in symmetric spaces of noncompact type of rank 1*, Math. Ann. **215** (1975), 97–116. MR 0385473 (52 #6334)

[OH13]   Anatol Odzijewicz and Maciej Horowski, *Positive kernels and quantization*, J. Geom. Phys. **63** (2013), 80–98. MR 2996399

[OPS88]   B. Osgood, R. Phillips, and P. Sarnak, *Extremals of determinants of Laplacians*, J. Funct. Anal. **80** (1988), no. 1, 148–211. MR 960228 (90d:58159)

[OR07]   V. S. Olkhovsky and E. Recami, *Time as a quantum observable*, Internat. J. Modern Phys. A **22** (2007), no. 28, 5063–5087. MR 2371443 (2009b:81008)

[Ørs79]   Bent Ørsted, *Induced representations and a new proof of the imprimitivity theorem*, J. Funct. Anal. **31** (1979), no. 3, 355–359. MR 531137 (80d:22007)

[OS75]   Konrad Osterwalder and Robert Schrader, *Axioms for Euclidean Green's functions. II*, Comm. Math. Phys. **42** (1975), 281–305, With an appendix by Stephen Summers. MR 0376002 (51 #12189)

[OSS+13] H. Owhadi, C. Scovel, T. J. Sullivan, M. McKerns, and M. Ortiz, *Optimal uncertainty quantification*, SIAM Rev. **55** (2013), no. 2, 271–345. MR 3049922

[OSS15] Houman Owhadi, Clint Scovel, and Tim Sullivan, *On the brittleness of Bayesian inference*, SIAM Rev. **57** (2015), no. 4, 566–582. MR 3419871

[Par05] K. R. Parthasarathy, *Probability measures on metric spaces*, AMS Chelsea Publishing, Providence, RI, 2005, Reprint of the 1967 original. MR 2169627 (2006d:60004)

[Par09] ———, *An invitation to quantum information theory*, Perspectives in mathematical sciences. I, Stat. Sci. Interdiscip. Res., vol. 7, World Sci. Publ., Hackensack, NJ, 2009, pp. 225–245. MR 2581746 (2011d:81061)

[Phe01] Robert R. Phelps, *Lectures on Choquet's theorem*, second ed., Lecture Notes in Mathematics, vol. 1757, Springer-Verlag, Berlin, 2001. MR 1835574 (2002k:46001)

[Phi94] Ralph Phillips, *Reminiscences about the 1930s*, Math. Intelligencer **16** (1994), no. 3, 6–8. MR 1281747 (95e:01021)

[Pho15] Simon J. D. Phoenix, *Quantum information as a measure of multipartite correlation*, Quantum Inf. Process. **14** (2015), no. 10, 3723–3738. MR 3401051

[PK88] Wayne N. Polyzou and W. H. Klink, *The structure of Poincaré covariant tensor operators in quantum mechanical models*, Ann. Physics **185** (1988), no. 2, 369–400. MR 965583 (90c:81060)

[Pol02] W. N. Polyzou, *Cluster properties in relativistic quantum mechanics of N-particle systems*, J. Math. Phys. **43** (2002), no. 12, 6024–6063. MR 1939631 (2003j:81242)

[Pou72] Neils Skovhus Poulsen, *On C^∞-vectors and intertwining bilinear forms for representations of Lie groups*, J. Functional Analysis **9** (1972), 87–120. MR 0310137 (46 #9239)

[Pou73] Niels Skovhus Poulsen, *On the canonical commutation relations*, Math. Scand. **32** (1973), 112–122. MR 0327214 (48 #5556)

[Pow75] Robert T. Powers, *Simplicity of the C^*-algebra associated with the free group on two generators*, Duke Math. J. **42** (1975), 151–156. MR 0374334 (51 #10534)

[Pre89] P. M. Prenter, *Splines and variational methods*, Wiley Classics Library, John Wiley & Sons, Inc., New York, 1989, Reprint of the 1975 original, A Wiley-Interscience Publication. MR 1013116 (90j:65001)

[PS70]   Robert T. Powers and Erling Størmer, *Free states of the canonical anticommutation relations*, Comm. Math. Phys. **16** (1970), 1–33. MR 0269230 (42 #4126)

[PS75]   K. R. Parthasarathy and K. Schmidt, *Stable positive definite functions*, Trans. Amer. Math. Soc. **203** (1975), 161–174. MR 0370681 (51 #6907)

[PS98]   George Pólya and Gabor Szegő, *Problems and theorems in analysis. II*, Classics in Mathematics, Springer-Verlag, Berlin, 1998, Theory of functions, zeros, polynomials, determinants, number theory, geometry, Translated from the German by C. E. Billigheimer, Reprint of the 1976 English translation. MR 1492448

[RAKK05] G. J. Rodgers, K. Austin, B. Kahng, and D. Kim, *Eigenvalue spectra of complex networks*, J. Phys. A **38** (2005), no. 43, 9431–9437. MR 2187996 (2006j:05186)

[Rit88]   Gunter Ritter, *A note on the Lévy-Khinchin representation of negative definite functions on Hilbert spaces*, J. Austral. Math. Soc. Ser. A **45** (1988), no. 1, 104–116. MR 940528 (89d:60007)

[Rob71]   Derek W. Robinson, *The thermodynamic pressure in quantum statistical mechanics*, Springer-Verlag, Berlin-New York, 1971, Lecture Notes in Physics, Vol. 9. MR 0432122 (55 #5113)

[RS75]   Michael Reed and Barry Simon, *Methods of modern mathematical physics. II. Fourier analysis, self-adjointness*, Academic Press [Harcourt Brace Jovanovich, Publishers], New York-London, 1975. MR 0493420 (58 #12429b)

[RSN90]   Frigyes Riesz and Béla Sz.-Nagy, *Functional analysis*, Dover Books on Advanced Mathematics, Dover Publications, Inc., New York, 1990, Translated from the second French edition by Leo F. Boron, Reprint of the 1955 original. MR 1068530 (91g:00002)

[Rud63]   Walter Rudin, *The extension problem for positive-definite functions*, Illinois J. Math. **7** (1963), 532–539. MR 0151796 (27 #1779)

[Rud70]   _____, *An extension theorem for positive-definite functions*, Duke Math. J. **37** (1970), 49–53. MR 0254514 (40 #7722)

[Rud87]   _____, *Real and complex analysis*, third ed., McGraw-Hill Book Co., New York, 1987. MR 924157 (88k:00002)

[Rud90]   _____, *Fourier analysis on groups*, Wiley Classics Library, John Wiley & Sons Inc., New York, 1990, Reprint of the 1962 original, A Wiley-Interscience Publication. MR 1038803 (91b:43002)

[Rud91]   _____, *Functional analysis*, second ed., International Series in Pure and Applied Mathematics, McGraw-Hill, Inc., New York, 1991. MR 1157815 (92k:46001)

[Rüs07]   Ludger Rüschendorf, *Monge-Kantorovich transportation problem and optimal couplings*, Jahresber. Deutsch. Math.-Verein. **109** (2007), no. 3, 113–137. MR 2356041 (2009e:60013)

[Sak98]   Shôichirô Sakai, *C*-algebras and W*-algebras*, Classics in Mathematics, Springer-Verlag, Berlin, 1998, Reprint of the 1971 edition. MR 1490835 (98k:46085)

[Sch32]   E. Schrödinger, *Sur la théorie relativiste de l'électron et l'interprétation de la mécanique quantique*, Ann. Inst. H. Poincaré **2** (1932), no. 4, 269–310. MR 1508000

[Sch40]   _____, *A method of determining quantum-mechanical eigenvalues and eigenfunctions*, Proc. Roy. Irish Acad. Sect. A. **46** (1940), 9–16. MR 0001666 (1,277d)

[Sch55]   J. J. Schäffer, *On unitary dilations of contractions*, Proc. Amer. Math. Soc. **6** (1955), 322. MR 0068740 (16,934c)

[Sch57]   Laurent Schwartz, *Théorie des distributions à valeurs vectorielles. I*, Ann. Inst. Fourier, Grenoble **7** (1957), 1–141. MR 0107812 (21 #6534)

[Sch58]   _____, *La fonction aléatoire du mouvement brownien*, Séminaire Bourbaki; 10e année: 1957/1958. Textes des conférences; Exposés 152 à 168; 2e éd. corrigée, Exposé 161, vol. 23, Secrétariat mathématique, Paris, 1958. MR 0107312 (21 #6037)

[Sch64a]   I. J. Schoenberg, *Spline interpolation and the higher derivatives*, Proc. Nat. Acad. Sci. U.S.A. **51** (1964), 24–28. MR 0160064 (28 #3278)

[Sch64b]   Laurent Schwartz, *Sous-espaces hilbertiens d'espaces vectoriels topologiques et noyaux associés (noyaux reproduisants)*, J. Analyse Math. **13** (1964), 115–256. MR 0179587 (31 #3835)

[Sch72]   H. A. Schwarz, *Gesammelte mathematische Abhandlungen. Band I, II*, Chelsea Publishing Co., Bronx, N.Y., 1972, Nachdruck in einem Band der Auflage von 1890. MR 0392470 (52 #13287)

[Sch95]   Laurent Schwartz, *Les travaux de L. Gårding sur les équations aux dérivées partielles elliptiques*, Séminaire Bourbaki, Vol. 2, Soc. Math. France, Paris, 1995, pp. Exp. No. 67, 175–182. MR 1609224

[Sch97]   L. Schwartz, *Calcul infinitésimal stochastique* [*MR0956973 (89g:60180)]*, Fields Medallists' lectures, World Sci. Ser. 20th Century Math., vol. 5, World Sci. Publ., River Edge, NJ, 1997, pp. 33–51. MR 1622895

[Sch99]   E. Schrödinger, *About Heisenberg uncertainty relation (original annotation by A. Angelow and M.-C. Batoni)*, Bulgar. J. Phys. **26** (1999), no. 5-6, 193–203 (2000), Translation of Proc. Prussian Acad. Sci. Phys. Math. Sect. 19 (1930), 296–303. MR 1782215 (2001i:81002)

[Sch12]   Konrad Schmüdgen, *Unbounded self-adjoint operators on Hilbert space*, Graduate Texts in Mathematics, vol. 265, Springer, Dordrecht, 2012. MR 2953553

[SD13]   F. A. Shah and Lokenath Debnath, *Tight wavelet frames on local fields*, Analysis (Berlin) **33** (2013), no. 3, 293–307. MR 3118429

[Seg50]   I. E. Segal, *An extension of Plancherel's formula to separable unimodular groups*, Ann. of Math. (2) **52** (1950), 272–292. MR 0036765 (12,157f)

[Seg53]   _____, *Correction to "A non-commutative extension of abstract integration"*, Ann. of Math. (2) **58** (1953), 595–596. MR 0057306 (15,204h)

[Seg54]   _____, *Abstract probability spaces and a theorem of Kolmogoroff*, Amer. J. Math. **76** (1954), 721–732. MR 0063602 (16,149d)

[Seg58]   _____, *Distributions in Hilbert space and canonical systems of operators*, Trans. Amer. Math. Soc. **88** (1958), 12–41. MR 0102759 (21 #1545)

[Shi96]   Georgi E. Shilov, *Elementary functional analysis*, Dover Publications, Inc., New York, 1996, Revised English edition translated from the Russian and edited by Richard A. Silverman, Corrected reprint of the 1974 English translation. MR 1375236 (98a:46002)

[Sho36]   J. Shohat, *The Relation of the Classical Orthogonal Polynomials to the Polynomials of Appell*, Amer. J. Math. **58** (1936), no. 3, 453–464. MR 1507168

[Sla03]   D. A. Slavnov, *Quantum measurements and Kolmogorov's probability theory*, Teoret. Mat. Fiz. **136** (2003), no. 3, 436–443. MR 2025366 (2004m:81046)

[SN53]   Béla Sz.-Nagy, *Approximation properties of orthogonal expansions*, Acta Sci. Math. Szeged **15** (1953), 31–37. MR 0056732 (15,119d)

[Spe11]   Roland Speicher, *Free probability theory*, The Oxford handbook of random matrix theory, Oxford Univ. Press, Oxford, 2011, pp. 452–470. MR 2932642

[Sri13]   N. Srivastava, *Discrepancy, graphs, and the Kadison-Singer problem*, Asia Pac. Math. Newsl. **3** (2013), no. 4, 15–20. MR 3156130

[SS98]   Ya. G. Sinaĭ and A. B. Soshnikov, *A refinement of Wigner's semicircle law in a neighborhood of the spectrum edge for random symmetric matrices*, Funktsional. Anal. i Prilozhen. **32** (1998), no. 2, 56–79, 96. MR 1647832 (2000c:82041)

[SS09]   Oded Schramm and Scott Sheffield, *Contour lines of the two-dimensional discrete Gaussian free field*, Acta Math. **202** (2009), no. 1, 21–137. MR 2486487 (2010f:60238)

[SS11]   Elias M. Stein and Rami Shakarchi, *Functional analysis*, Princeton Lectures in Analysis, vol. 4, Princeton University Press, Princeton, NJ, 2011, Introduction to further topics in analysis. MR 2827930 (2012g:46001)

[Sti55]   W. Forrest Stinespring, *Positive functions on C^*-algebras*, Proc. Amer. Math. Soc. **6** (1955), 211–216. MR 0069403 (16,1033b)

[Sti59]   ———, *Integrability of Fourier transforms for unimodular Lie groups*, Duke Math. J. **26** (1959), 123–131. MR 0104161 (21 #2921)

[Sto51]   M. H. Stone, *On unbounded operators in Hilbert space*, J. Indian Math. Soc. (N.S.) **15** (1951), 155–192 (1952). MR 0052042 (14,565d)

[Sto90]   Marshall Harvey Stone, *Linear transformations in Hilbert space*, American Mathematical Society Colloquium Publications, vol. 15, American Mathematical Society, Providence, RI, 1990, Reprint of the 1932 original. MR 1451877 (99k:47001)

[Str12]   Robert S. Strichartz, *Spectral asymptotics revisited*, J. Fourier Anal. Appl. **18** (2012), no. 3, 626–659. MR 2921088

[Sul15]   T. J. Sullivan, *Introduction to uncertainty quantification*, Texts in Applied Mathematics, vol. 63, Springer, Cham, 2015. MR 3364576

[SW00]   R. F. Streater and A. S. Wightman, *PCT, spin and statistics, and all that*, Princeton Landmarks in Physics, Princeton University Press, Princeton, NJ, 2000, Corrected third printing of the 1978 edition. MR 1884336 (2003f:81154)

[SWZ11]   Dieter Schmidt, Gary Weiss, and Vrej Zarikian, *Paving small matrices and the Kadison-Singer extension problem II—computational results*, Sci. China Math. **54** (2011), no. 11, 2463–2472. MR 2859705 (2012j:46004)

[SZ07] Steve Smale and Ding-Xuan Zhou, *Learning theory estimates via integral operators and their approximations*, Constr. Approx. **26** (2007), no. 2, 153–172. MR 2327597 (2009b:68184)

[SZ09] _____, *Geometry on probability spaces*, Constr. Approx. **30** (2009), no. 3, 311–323. MR 2558684 (2011c:60006)

[Sza04] F.H. Szafraniec, *Przestrzenie hilberta z jądrem reprodukującym*, Matematyka - Uniwersytet Jagielloński, Wydaw. Uniwersytetu Jagiellońskiego, 2004

[Sza15] F. H. Szafraniec, *Operators of the quantum harmonic oscillator and its relatives*, pp. 59–120, John Wiley & Sons Inc., 2015

[Sze59] Gabor Szegö, *Orthogonal polynomials*, American Mathematical Society Colloquium Publications, Vol. 23. Revised ed, American Mathematical Society, Providence, R.I., 1959. MR 0106295 (21 #5029)

[Tak79] Masamichi Takesaki, *Theory of operator algebras. I*, Springer-Verlag, New York-Heidelberg, 1979. MR 548728 (81e:46038)

[Tay86] Michael E. Taylor, *Noncommutative harmonic analysis*, Mathematical Surveys and Monographs, vol. 22, American Mathematical Society, Providence, RI, 1986. MR 852988 (88a:22021)

[TD03] Christophe Texier and Pascal Degiovanni, *Charge and current distribution in graphs*, J. Phys. A **36** (2003), no. 50, 12425–12452. MR 2025876 (2004i:81264)

[Trè06a] François Trèves, *Basic linear partial differential equations*, Dover Publications Inc., Mineola, NY, 2006, Reprint of the 1975 original. MR 2301309 (2007k:35004)

[Trè06b] _____, *Topological vector spaces, distributions and kernels*, Dover Publications, Inc., Mineola, NY, 2006, Unabridged republication of the 1967 original. MR 2296978 (2007k:46002)

[Vár06] Joseph C. Várilly, *An introduction to noncommutative geometry*, EMS Series of Lectures in Mathematics, European Mathematical Society (EMS), Zürich, 2006. MR 2239597 (2007e:58011)

[vN31] J. v. Neumann, *Die Eindeutigkeit der Schrödingerschen Operatoren*, Math. Ann. **104** (1931), no. 1, 570–578. MR 1512685

[vN32a] J. von Neumann, *Über adjungierte Funktionaloperatoren*, Ann. of Math. (2) **33** (1932), no. 2, 294–310. MR 1503053

[vN32b] _____, *Über einen Satz von Herrn M. H. Stone*, Ann. of Math. (2) **33** (1932), no. 3, 567–573. MR 1503076 [vN32c] _____, *Zur Operatorenmethode in der klassischen Mechanik*, Ann. of Math. (2) **33** (1932), no. 3, 587–642. MR 1503078

[vN35] _____, *Charakterisierung des spektrums eines integraloperators*, Actualités Scientifique Industrielles. Exposés Mathématiques, Hermann, 1935

[Voi85] Dan Voiculescu, *Symmetries of some reduced free product C^*-algebras*, Operator algebras and their connections with topology and ergodic theory (Buşteni, 1983), Lecture Notes in Math., vol. 1132, Springer, Berlin, 1985, pp. 556–588. MR 799593 (87d:46075)

[Voi91] _____, *Free noncommutative random variables, random matrices and the II_1 factors of free groups*, Quantum probability & related topics, QP-PQ, VI, World Sci. Publ., River Edge, NJ, 1991, pp. 473–487. MR 1149846 (93c:46119)

[Voi06] _____, *Symmetries arising from free probability theory*, Frontiers in number theory, physics, and geometry. I, Springer, Berlin, 2006, pp. 231–243. MR 2261097 (2008e:46082)

[vS15] Walter D. van Suijlekom, *Noncommutative geometry and particle physics*, Mathematical Physics Studies, Springer, Dordrecht, 2015. MR 3237670

[VZ92] H. Vogt and A. Zippelius, *Invariant recognition in Potts glass neural networks*, J. Phys. A **25** (1992), no. 8, 2209–2226. MR 1162879 (93c:82044)

[Wea03] Nik Weaver, *A counterexample to a conjecture of Akemann and Anderson*, Bull. London Math. Soc. **35** (2003), no. 1, 65–71. MR 1934433 (2003i:46062)

[Wea04] _____, *The Kadison-Singer problem in discrepancy theory*, Discrete Math. **278** (2004), no. 1-3, 227–239. MR 2035401 (2004k:46093)

[Wei03] Joachim Weidmann, *Lineare Operatoren in Hilberträumen. Teil II*, Mathematische Leitfäden. [Mathematical Textbooks], B. G. Teubner, Stuttgart, 2003, Anwendungen. [Applications]. MR 2382320 (2008k:47002)

[Wie53] Norbert Wiener, *Ex-prodigy. My childhood and youth*, Simon and Schuster, New York, 1953. MR 0057817 (15,277f)

[Wig58] Eugene P. Wigner, *On the distribution of the roots of certain symmetric matrices*, Ann. of Math. (2) **67** (1958), 325–327. MR 0095527 (20 #2029)

[Wig76]   A. S. Wightman, *Hilbert's sixth problem: mathematical treatment of the axioms of physics*, Mathematical developments arising from Hilbert problems (Proc. Sympos. Pure Math., Northern Illinois Univ., De Kalb, Ill., 1974), Amer. Math. Soc., Providence, R. I., 1976, pp. 147–240. MR 0436800 (55 #9739)

[Wit74]   C. S. Withers, *Mercer's theorem and Fredholm revolvents*, Bull. Austral. Math. Soc. **11** (1974), 373–380. MR 0380303 (52 #1203)

[Woj08]   Radoslaw Krzysztof Wojciechowski, *Stochastic completeness of graphs*, ProQuest LLC, Ann Arbor, MI, 2008, Thesis (Ph.D.)–City University of New York. MR 2711706

[Woj09]   Radosław K. Wojciechowski, *Heat kernel and essential spectrum of infinite graphs*, Indiana Univ. Math. J. **58** (2009), no. 3, 1419–1441. MR 2542093 (2010k:35208)

[Wol51]   Herman O. A. Wold, *Stationary time series*, Trabajos Estadística **2** (1951), 3–74. MR 0042667 (13,144i)

[WS53]   Norbert Wiener and Armand Siegel, *A new form for the statistical postulate of quantum mechanics*, Physical Rev. (2) **91** (1953), 1551–1560. MR 0057763 (15,273f)

[Yos95]   Kōsaku Yosida, *Functional analysis*, Classics in Mathematics, Springer-Verlag, Berlin, 1995, Reprint of the sixth (1980) edition. MR 1336382 (96a:46001)

[Zar07]   S Zaremba, *L'équation biharmonique et une classe remarquable de fonctions fondamentales harmoniques*, Bulletin international de l'Académie des Sciences de Cracovie **3** (1907), 147–196

[Zim90]   Robert J. Zimmer, *Essential results of functional analysis*, Chicago Lectures in Mathematics, University of Chicago Press, Chicago, IL, 1990. MR 1045444 (91h:46002)

Index

C^*-algebra, 8, 26, 87, 133, 156, 158, 163, 167, 187, 190, 199, 202, 203, 205, 209–212

adjoint operator, xiv, 18, 117, 135, 160, 351, 408
affiliated with, 18, 102
algebra
    Calkin-, 339
    quotient-, 208, 339
    Toeplitz-, 208
algebras
    C^*-algebra, 26, 158, 167, 187, 337
    Banach algebra, 15, 187
    Cuntz algebra, 26, 158, 199, 214
    group algebra, 162, 281, 322
    Lie algebra, 293, 310, 311
    von Neumann algebra (W^*-algebra), 26, 102, 191, 212
analytic vector, 361, 369, 370, 444
Anderson paving, 336
anti-commutator, 213
approximate identity, 419
Aronszajn, N., 5, 211, 403, 407, 416, 417, 425, 437
Arzelà-Ascoli, 150
axioms, 18, 24, 26, 27, 31, 37, 40, 46, 70, 136, 158, 188, 337, 349, 452

backward shift, 206, 207
Banach space, 15, 28, 169
    double-dual, 168
    dual-, 26
    pre-dual, 26
    reflexive, 168

Banach-Alaoglu Theorem, 169, 200
Banach-Tarski paradox, 38
band-limited, 123, 241, 242, 245, 455
banded matrix, 67, 73, 75
Bloch sphere, 178, 195, 454
Bohr-compactification, 174
Borel measure, 30, 120, 136, 200, 381
boundary condition, 267, 321, 364, 368, 369, 429
boundary triple, 364–366
Brownian motion, 5, 10, 43, 252, 254, 259, 260, 262, 264, 267, 268, 405, 411, 412
    geometric-, 255
    pinned-, 146
Brownian sample-paths, 44

Calkin algebra, 339
Campbell-Baker-Hausdorff formula, 291
canonical commutation relations, 60, 212
Cantor
    -measure, 105, 185
    -middle third, 105, 185
    -set, 15
CAR, 213
cardinality, 48, 338
Cartesian product, 38, 39, 251, 263
Cauchy-Schwarz' inequality, 40, 49, 51, 144, 149, 167, 231, 470
CCR, 213
Central Limit Theorem, 268
characteristic matrix, 97, 101
characteristic projection, 106, 107, 109

Choquet, G., 198, 199
circuit, 178
closable operator, 91, 95, 351
closed operator, 91, 96, 101, 107, 108
closed range theorem, 4
co-adjoint orbit, 306
commutant, 101, 191, 192
commutation relations, 16, 26, 71, 72, 122, 212
commutator, 144, 212, 291
compact, 12, 24, 32, 33, 36, 39, 62, 68, 146, 150, 158, 159, 167, 171, 197, 198, 263, 267, 274, 282, 316, 340, 415
completely positive map, 224, 228, 297
completion
    C^*-, 158, 212
    Hilbert-, 41, 42, 159, 180, 183, 184, 327, 389, 417
    Hilbert-Schmidt, 68, 176
    norm-, 67, 94, 95
    trace-, 68
conditional expectation, 268, 297, 303
conductance, 385, 387, 391, 395, 398
Connes, A., 9, 223
continuous linear functional, 28, 168, 267
controlled U-gate, 177
convex, 39, 159, 167, 198
convex hull, 143, 198
convolution, 123, 189, 279, 313, 324
Coulomb's Law, 350
covariance, 43, 266, 302, 412, 414
Cuntz relations, 8, 241, 243
Cuntz-algebra, 26, 158, 210
cyclic
    -representation, 132, 135, 164, 165, 224
    -space, 138, 166
    -subspace, 133, 135, 165, 224
    -vector, 84, 132, 133, 164
cylinder-set, 43, 260

decomposition of Brownian motion, 267

decomposition of state, 198, 199
deficiency indices, 415, 422
deficiency space, 355, 396, 422
densely defined, 6, 90, 95, 97, 103, 109, 353, 355, 359, 363, 365, 372, 394, 415
diagonalization, 121, 129, 132, 187
dilation (See also unitary dilation.)
    minimal-, 425
    Stinespring, 5, 224, 228, 233, 236
    unitary-, 77, 115, 151, 177, 229, 280, 425
dipole, 390, 392, 395, 396
Dirac formalism, 5, 19, 48, 63, 67, 136, 172, 197, 225
Dirac, P.A.M
    bra-ket, 63
    ket-bra, 63
direct integral, 12, 130, 131, 156, 219, 325, 462
discrete Laplacian (See also graph Laplacian.), xiv, 5, 385–387, 392, 395
distribution, 29, 34, 35, 76, 183, 186, 197, 252, 272, 350, 407, 412, 414, 422, 423, 427, 428
    Gaussian-, 8, 9, 252, 414
    log-normal-, 254, 255
    probability-, 76, 197, 350, 427
    Schwartz-, 29, 30, 182, 184, 422
Dixmier trace, 340
double-commutant, 216
down-sampling, 243, 245, 268
dual, 168, 193, 308, 316, 338
dyadic rationals, 50
dyadic scaling, 246

eigenfunction, 92, 401
eigenvalue, 24, 92, 121, 127, 134, 151, 171, 197, 217, 272, 341, 352, 361, 401, 431, 433
electrical networks, 20, 386
endomorphism, 241
Energy Hilbert Space, xiv, 386, 387, 389, 392

ergodic, ix, 201, 257
ergodic theorem, 32, 109
essentially selfadjoint operator, 26,
    97, 275, 389, 394
essentially skew-adjoint, 93
exponential mapping, 291, 292, 310
extension
    -of functional, 39, 94
    -of operator, 5, 91, 349
    -of positive functional, 133
    -of state, 224, 338
    Friedrichs'-, 372, 385, 394
    selfadjoint-, 5, 119, 349, 363,
        364, 418
extensions of symmetric operators,
    18
extreme-point, 190, 199

filter bank, 18, 241, 242
formally selfadjoint, 70, 83, 119, 349,
    446, 472
Four Big Theorems in Functional Anal-
    ysis, x, 4
Fourier basis, 61
Fourier transform, 16, 60, 122, 123,
    129, 244, 323, 325
Fréchet topology, 36
frame, 337, 341, 390–392, 396
free probability, 8, 452
frequency band, 241, 245, 455
Friedrichs extension, 372, 382, 386,
    394, 397
functional, 28, 51, 133, 163, 168,
    192, 194, 234, 300, 374
functional calculus, 120, 254

Gårding
    -space, 93, 310, 313, 315
    -vector, 92, 93
Gaussian distribution, 8, 9, 252, 414
Gaussian free field (GFF), 390
Gaussian process, 43, 252, 266, 390
Gelfand space, 189, 191
Gelfand transform, 188, 211
Gelfand-Naimark-Segal, 156, 158, 224
Gelfand-triple, 37, 262

generalized eigenfunction, 401
generating function, 57, 60, 256, 407,
    416
Geometric Brownian motion, 255
GNS construction, 18, 42, 156, 163,
    164, 167, 190, 194, 200, 224,
    234
Gram-Schmidt orthogonalization, 55,
    124
graph
    - of operator, 91, 95, 98, 351,
        381
    network-, xiv, 20, 385, 386, 388,
        395
graph Laplacian, 385, 387, 392
Green's function, 146, 427
group algebra, 13, 162, 281, 286, 314,
    322, 323, 328
groups
    $ax + b$, 50, 286, 292, 294, 305,
        311, 316, 317
    Abelian, 26, 316, 415
    Baumslag-Solitar, 50
    compact, 282
    free, 199
    Heisenberg, 295, 302
    Lie, ix, 307, 311, 417
    locally compact, 279, 297, 302,
        317, 415, 425
    non-Abelian, 26
    Poincaré, 308

Haar wavelet, 61, 63, 244, 245
Hadamard product, 411
Hahn-Banach theorem, 4, 168, 338
Hahn-Hellinger theory, 127
Hardy space, 29, 39, 173, 204, 207,
    268, 406
harmonic, ix, 5, 26, 124, 203, 275,
    324, 326, 392, 397, 401, 413,
    444
harmonic analysis, xiii, 16, 70, 274,
    344
harmonic functions, 392, 397
harmonic oscillator, 124, 125, 326
Heisenberg's picture, 351, 464

Heisenberg, W.K.
  commutation relation, 26, 71,
    122, 212
  uncertainty principle, 72, 337
Hermite function, 124
Hermite polynomials, 124
Hilbert matrix, 75
Hilbert space
  L^2, 168, 200
  l^2, 50, 168
  direct sum, 175
  energy, 389
  reproducing kernel, 416
  rigged, 376
  tensor product, 176
Hilbert's sixth problem, 27, 452
homomorphism, 130, 134, 158, 161,
    188, 284
HYP-theorem, 264

ideal, 65, 167, 188, 210, 214, 339
IFS-measures, 105
independent
  -events, 251
  -increments, 251
  -random variables, 263, 271
index
  deficiency, 350, 394, 415
  Fredholm-, 209, 460
  Powers-, 241
  Szegö-, 209
  von Neumann-, 350, 355, 415
inequality
  Cauchy-Schwarz, 40, 49, 51, 144,
    149
  Hölder, 30
  Schwarz, 167, 231
infinite electrical network, 386
infinite graphs, 20, 119, 386
infinite network, xiii, xiv, 20, 385,
    386, 391, 394
integral
  Borel-, 120
  direct-, 131, 203
  Itō-, 45, 109, 414
  Riesz-, 133, 202

Stieltjes-, 30
Stochastic-, 45, 109, 263
irrational rotation, 201
isometry, 138, 203–205, 228, 229, 238
  Cayley-, 364
  Cuntz-, 203, 205, 227
  Itō-, 45, 109
  partial-, 112, 113, 170, 358
isomorphism, 12, 43, 48, 49, 130,
    134, 137, 177, 188, 211, 225,
    263, 315, 329
Itō, K.
  integral, 414
  Itō's lemma, 255, 463
Itō-calculus, 255, 265
iterated function systems, 105, 185
Iterated Itō-integrals, 261

Jacobi matrix, 60, 75, 76, 84
jointly Gaussian, 252

Kadison-Singer problem, 19, 334, 335
Kantorovich-Rubinstein
    Theorem, 184
kernel
  -of operator, 188, 218
  integral-, 267
  reproducing-, 211, 403, 405, 416,
    417, 425, 426
Kolmogorov, A.N., 9, 203, 263, 273,
    389, 414
Kolmogorov-
    consistency, 203, 263, 389,
    414
Krein-Milman, 39, 198, 338

Lévy-Khinchin, 416
Laplacian
  graph-, xiv, 5, 385–387, 392, 395
large networks, xiii, 386
lattice, 78, 82
laws of Kirchhoff and Ohm, 385
Lax-Milgram, 54, 343
Lebesgue dominated convergence
    theorem, 123, 142
Leibniz derivation, 110

Lie
  algebra, ix, 26, 275, 294, 306, 310, 311
  group, ix, 26, 93, 289, 306, 311, 319, 415
limit distributions, 271
locally convex topological space, 198, 199
log-normal distribution, 254, 255

machine learning, 6, 20, 407
Mackey machine, 297, 306, 320
Malliavin
  -calculus, ix, 25, 452, 467
  -derivative, 109–111
Marcus-Spielman-Srivastava solution, 335
Markov
  -chains, 453, 463
  -operator, xiv, 386
  -processes, 264, 388, 463
  -semigroup, 265
Markov-Kakutani fixed-point theorem, 282
matrix
  banded-, 73, 74, 83, 388
  block-, 192, 233
  characteristic, 97
  diagonal-, 73, 121
  Hilbert-, 75
  $\infty \times \infty$, 71, 72, 388
  Jacobi, 83, 446
  Pauli-, 177, 195
  raising/lowering-, 73
  random-, 271
  tri-diagonal-, 72, 396
  triangular-, 275, 330
  unitary-, 195
matrix representation, xiv, 18, 20, 60, 62, 73, 83, 392, 395
measure
  Borel, 163, 300
  cantor, 186
  Dirac, 200, 338
  ergodic, 201
  Gaussian, 84
  Haar, 279, 282, 319, 417
  independent, 251
  Lebesgue, 202, 218, 401
  Plancherel, 316
  probability, 164, 199, 263, 414
  projection-valued, 77, 136, 280, 339, 426
  quasi-invariant, 297
  Wiener, 266
measurement, 5, 27, 71, 337, 338, 349, 407
Mercer's Theorem, 405
metric, 40, 41, 184, 185, 369, 443
modular function, 283, 297
moment problem, 83
Monte-Carlo simulation, 269
multiplicity, 135, 187, 215, 235, 322
  -of an isometry, 204, 205
  -of spectrum, 62, 187, 192
multiplicity free, 135, 215
multiplier, 406
multivariable spectral theory, 7

Nelson, E., 132, 135, 236, 322
non-normal states, 339
noncommutative
  -Radon-Nikodym derivative, 202
  -analysis, 8, 12, 274
  -geometry, 9
  -probability, 8, 453, 455
  -spectral theory, 7
Norm
  C^*-, 158, 187, 190
  L^2-, 6, 34, 41, 129, 132, 143, 270, 301, 304
  L^p-, 29, 31, 168
  l^2-, 7, 30
  Banach space-, 28, 40, 168
  dual-, 28, 30, 34, 69, 168, 170, 174
  Hilbert space-, 41, 93, 103, 169, 301, 411
  Sobolev-, 15, 408
  trace-, 68, 69
  uniform-, 56, 67, 69

Norm-Completions, 41, 42, 67–69, 94, 95, 158, 166, 176, 180, 183, 235, 371, 389
normal distribution, 8, 9, 252, 414
normal operator, 18, 24, 90, 122, 125, 322
normal state, 172, 196, 338, 339
numerical range, 65, 241, 247

observable, 5, 71, 90, 120, 121, 156, 197, 223, 321, 325, 337, 406
open mapping theorem, 4
operators
    adjoint-, 26, 52, 72, 75, 81, 95, 96, 98, 160, 179, 193, 237
    bounded, 141
    closable-, 91, 351
    closed-, 102, 351, 353, 354, 358
    compact-, 62, 65, 68, 144, 148, 150, 170, 171, 196, 197, 208, 209, 267, 270, 339, 341, 424
    contractive, 176
    convolution-, 123, 324
    domain of, 91
    essentially selfadjoint-, 26, 74, 97, 275, 355, 400
    extension of, 350
    finite rank-, 64, 176
    formally selfadjoint, 350
    graph of, 91
    Hermitian, 121, 350
    Hilbert-Schmidt, 176
    integral-, 146, 150, 267, 270, 424, 446
    isometric, 231
    Laplace-, 326, 350, 385–387, 392, 399
    momentum-, 18, 72, 92, 122, 350, 407
    multiplication, 127
    normal, 111
    position-, 18, 72, 144
    regular-, 97, 112, 113
    selfadjoint, 52, 71, 125, 197, 349
    semibounded-, 181, 386, 394

symmetric-, 74, 83, 271, 351, 352, 359
trace class, 171
unbounded, 349
unitary, 71, 161
optimal solution, 408, 409
optimization, 6, 40, 407, 409, 414
Ornstein-Uhlenbeck
    -process, 266, 414
    -semigroup, 264
orthogonal
    -decomposition, 351, 356
    -subspaces, 165
    -vectors, 46, 55, 64, 263, 418
orthogonal polynomials
    Chebyshev, 57
    Hermite, 57, 124
    Legendre, 57
orthogonal projection, 24, 63, 70, 71, 77, 80, 114, 136, 281, 331, 343
orthogonality relations, 61
orthonormal basis (ONB), 39, 263, 337
oscillator Hamiltonian, 60, 124
Osterwalder-Schrader, 19

Parseval frame, 390–392, 396
Parseval identity, 48, 121, 171, 197
partial differential equations (PDE), 5, 39, 70, 90
partial isometry, 112, 113, 170, 359, 364, 365
partially ordered set, 37, 38, 46, 77, 81
path space, 11, 16, 17, 25, 43, 203, 251, 259, 266, 389, 451
paving conjecture, 336
Peano's axiom, 38
perturbation, 62, 351
Peter-Weyl theorem, 328
Poincaré group, 19, 275, 289, 321
polar decomposition, 113, 170, 171
polarization identity, 48, 148, 149
positive definite

-function, 42, 43, 233, 235, 279, 407, 416, 417, 425
-kernel, 267, 405, 406, 418
-matrix, 41
-operator, 75, 167
Powers-index, 241
pre-dual, 26, 168, 173
product
  infinite, 39
  semi-direct, 322
  semidirect, 50, 322
  tensor-, 176, 219, 224, 231, 238
product topology, 39, 169, 200, 263
projection, 63, 77
projection valued measure (PVM), 24, 70, 71, 77, 114, 136, 281, 331, 343
pure state, 159, 160, 190, 194, 197, 199, 201
  non-normal, 338
  normal, 339

quadratic form, 194, 235, 307, 311
quadratic variation, 254
quantum field theory, ix, 9, 452
quantum gate, 454
quantum mechanics
  composite system, 176
  energy operator, 124
  measurement, 71, 338
  momentum operator, 72, 92, 122
  observable, 71, 120, 321, 338
  position operator, 72, 92, 122
  uncertainty principle, 144, 337
quantum states, 18, 76, 143
qubit, 177, 178, 454

Radon-Nikodym derivative, 193, 202
random matrix, 8
random symmetric matrix, 271
random variable, 252, 263, 271, 386, 414
  distribution of-, 10, 34, 35, 252, 254, 263
  Gaussian-, 35, 43, 45, 252, 254, 269

independent-, 251, 263, 271
random walk, 9, 386, 388, 397, 400
recursive identities, 59
reflection of Brownian motion, 260
reflection positivity, 180
renormalization, 18, 19, 179, 182
representation
  - of the Cuntz algebra, 158, 199, 205, 210, 214, 242, 243
  adjoint-, 308
  co-adjoint, 306, 308
  cyclic, 132, 165
  Fock-, 210, 213
  GNS, 158, 190, 226
  Heisenberg, 302
  induced, 289
  irreducible, 190
  Lie algebra, 294
  multiplicity free, 215
  of $ax + b$, 286
  of algebra, 163
  of group, 278
  of Lie group, 278
  Schrödinger-, 92, 275, 303, 305, 312, 315, 317, 325, 330
  spectral, 120, 130
  strongly continuous, 91, 93, 162, 256, 310, 425, 426
  unitary, 13, 19, 24, 91, 158, 162, 275, 281, 286, 291, 293, 299, 302, 303, 310, 328, 425, 426
reproducing kernel Hilbert space (RKHS). See also kernel., 211, 404, 416, 426
resistance distance, 385
resolution
  multi-, 62, 246
resolvent identity, 97
resolvent operator, 97, 264
reversible, 388, 398
Riesz' theorem, 51, 54, 133, 149, 169, 202, 234, 266, 300, 376, 390, 420
rigged Hilbert space, 376
row-isometry, 232, 237

sample-path. See also Brownian sample-paths., 43, 259, 451
sampling
 down-sampling, 243, 268
 up-sampling, 243
Schrödinger equation, 90, 351, 446
Schrödinger representation, 92, 302, 303, 305, 312, 315, 317, 325
Schrödinger's picture, 351
Schrödinger, E. R., 71, 92, 306
Schur-Sakai-Nikodym, 193
Schwartz space, 92, 122, 125, 315, 325, 449
selfadjoint extensions, 355, 362, 366, 368, 427
selfadjoint operator, 18, 26, 62, 71, 75, 81, 84, 90, 99, 120, 125, 132, 163, 193, 275, 338, 349, 375, 382, 405, 407
semicircle law, 272
semidirect product, 50, 277, 290, 322
semigroup, 181, 182, 278
separable, 48, 67, 131, 143, 174, 184, 247, 344
sesquilinear form, 41, 42, 147, 148, 159, 180, 183, 231, 234, 352
shift, 204–206, 235
short exact sequence, 208
sigma-algebra, 43, 128, 234, 263, 414
signal, ix, 18, 19, 241, 337
signal processing, ix, 18, 40, 199, 242, 243, 337, 455, 472
simplex, 199, 442
Sobolev space, 15
space
 L^p-, 30, 168
 Banach-, 15, 26, 28, 31, 168, 169, 200, 338
 dual-, 26, 29, 158, 168, 172, 173, 200, 310, 338
 Fock-, 210
 Hardy-, 29, 173, 207, 211, 268, 405, 416
 Hilbert-, 5, 17, 19, 24, 29, 42, 48, 50, 63, 75, 90, 97, 121, 132, 145, 163, 168, 176, 182–

184, 211, 224, 228, 236, 279, 319, 328, 349, 356, 376, 389
 metric-, 184, 185
 Sobolev-, 15
 state-, 121, 159, 164, 338
spectral multiplicity, 127
Spectral Representation Theorem, 127
Spectral Theorem, 71, 82, 119, 121, 127, 136, 140, 144, 170, 197, 322
spectral theory, 7, 102, 127, 322, 446
spectral types, xiv, 127, 350, 387, 407
spectrum
 continuous, 125, 198, 401, 407
 continuous-, 62, 92, 127
 discrete, 407, 430
 singular-, 191
Standard Model, 9, 452
state
 Fock-, 213
 non-normal, 339
 normal-, 196, 339
 pure-, 19, 65, 159, 160, 167, 189, 190, 194, 197, 199, 201, 203, 337, 338
 vacuum-, 213
Stinespring, W. F., 224, 228
stochastic integrals, 45, 46, 109, 263, 452
stochastic process, 46, 266, 413
Stone's theorem, 70, 90, 114, 181, 251, 274, 281, 312, 331, 369, 370, 423, 426, 471
Stone, M. H., 57, 71, 97, 189, 471
Stone-von Neumann Uniqueness Theorem, 74
Stone-Čech
 compactification, 189, 338, 340
Stone-Weierstrass' theorem, 56, 133, 139
strong operator topology, 79, 137, 142, 191

strongly continuous, 90, 91, 93, 161, 256, 280, 310, 423, 426
strongly continuous representation, 93, 310
strongly continuous semigroup, 25, 264
Szegö's Index Theorem, 209
Szegö-kernel, 17, 406

tensor product. See also Hilbert space
    −, 176, 219, 224, 231, 238, 411, 446, 454
Theorem
    Banach's fixed-point-, 185
    Banach-Alaoglu-, 169, 200
    Central Limit-, 268
    Choquet-, 156, 199
    Ergodic-, 281
    Hahn-Hellinger-, 127
    HYP-, 264
    imprimitivity-, 302
    Karhunen-Loève-, 466
    Krein-Milman's-, 38, 156, 168, 198, 338
    Lax-Milgram-, 54, 343
    Markov-Kakutani fixed-point-, 282
    Mercer's-, 405, 424, 436
    Pontryagin duality-, 316
    Radon-Nikodym-, 193, 202
    Riesz-, 51, 54, 135, 149, 169, 202, 234, 266, 300, 374, 376, 390, 420
    Spectral Representation-, 127
    Spectral-, 5, 24, 71, 112, 119, 121, 124, 130, 137, 143, 151, 190, 268, 338, 406
    Stinespring's-, 224, 226, 228, 236
    Stone's-, 70, 90, 97, 114, 181, 251, 274, 281, 312, 331, 369, 370, 423, 426, 471
    Stone-von Neumann uniqueness-, 74, 275, 329
    Tychonoff-, 39, 389
    Wold decomposition-, 204

Toeplitz
    -algebra, 208
    -matrix, 208
    -operator, 208
    multivariable-algebra, 210
Toeplitz-Hausdorff Theorem, 65
trace, 67, 72, 169, 171, 295, 307, 340, 424
transform
    Cayley, 362
    Fourier, 122, 129, 189, 322
transition probabilities, 71, 76, 388, 398, 399
Tychonoff (compactness), 39

ultra-filters, 189, 340
uncertainty, 72, 144, 337
uncertainty quantification, 17, 456
uniform boundedness principle, 4
unimodular, 282, 297, 299, 316, 317
unitary dilation, 77, 115, 151, 177, 229, 280, 425
unitary equivalence, 121, 165, 224, 299
unitary one-parameter group, 281
up-sampling, 243, 245

voltage distribution, 385
von Neumann algebra, 9, 18, 26, 102, 112, 191, 212, 216
von Neumann's ergodic theorem, 281
von Neumann, J., 26, 40, 71, 111, 120

wavelet, 26, 61
Wavelet Image Compression, 246
wavelet-bases, 7, 50, 61–63
weak*-compact, 159, 169, 200
weak*-topology, 167, 188, 338
Weyl commutation relation, 331
Wiener, N., 43, 259
Wightman axioms, 472
Wold decomposition, 204

Zorn's lemma, 37, 39, 133, 165, 188